# ASTROPHYSICS AND
# SPACE SCIENCE LIBRARY

A SERIES OF BOOKS ON THE RECENT DEVELOPMENTS
OF SPACE SCIENCE AND OF GENERAL GEOPHYSICS AND ASTROPHYSICS
PUBLISHED IN CONNECTION WITH THE JOURNAL
SPACE SCIENCE REVIEWS

VOLUME 98
PROCEEDINGS

# BINARY AND MULTIPLE STARS

# AS TRACERS

# OF STELLAR EVOLUTION

PROCEEDINGS OF THE 69th COLLOQUIUM OF THE
INTERNATIONAL ASTRONOMICAL UNION, HELD IN BAMBERG, F.R.G.,
AUGUST 31 – SEPTEMBER 3, 1981

Edited by

ZDENĚK KOPAL

*Department of Astronomy, The University, Manchester, U.K.*

and

JÜRGEN RAHE

*Dr.-Remeis-Sternwarte Bamberg,*
*Astronomisches Institut der Universität Erlangen-Nürnberg, F.R.G.*

D. REIDEL PUBLISHING COMPANY

DORDRECHT : HOLLAND / BOSTON : U.S.A.

LONDON : ENGLAND

Library of Congress Cataloging in Publication Data

International Astronomical Union. Colloquium (69th . 1981 : Bamberg,
    Germany)
    Binary and multiple stars as tracers of stellar evolution.

    (Astrophysics and space science library. Proceedings ; v. 98)
    Includes bibliographical references.
    1.   Stars–Evolution–Congresses.   2.   Stars, Double–
Congresses.   3.   Stars, Variable–Congresses.   I.   Kopal, Zdeněk,
1914–       II.   Rahe, Jürgen.   III.   Title.   IV.   Series.
QB806.I57  1981            523.8                    82–9036
                                                    AACR2
ISBN-13: 978-94-009-7863-8          e-ISBN-13: 978-94-009-7861-4
DOI: 10.1007/978-94-009-7861-4

Published by D. Reidel Publishing Company,
P.O. Box 17, 3300 AA Dordrecht, Holland.

Sold and distributed in the U.S.A. and Canada
by Kluwer Boston Inc.,
190 Old Derby Street, Hingham, MA 02043, U.S.A.

In all other countries, sold and distributed
by Kluwer Academic Publishers Group,
P.O. Box 322, 3300 AH Dordrecht, Holland.

D. Reidel Publishing Company is a member of the Kluwer Group.

# TABLE OF CONTENTS

LIST OF PARTICIPANTS                                      xiii

WELCOME
R. Grafberger                                              xix
A. Bollée                                                  xxi
N. Fiebiger                                               xxiii
J. Rahe                                                    xxv

INTRODUCTION
Z. Kopal                                                  xxvii

## PART I
## EVOLUTION OF STARS IN THE POST-MAIN SEQUENCE STAGE

EVOLUTION OF STARS IN THE POST-MAIN SEQUENCE STAGE
R. Kippenhahn                                                3

DISK INSTABILITIES IN HUBBLE-SANDAGE VARIABLES?
F. Meyer, E. Meyer-Hofmeister                               19

THE FREQUENCY OF CEPHEID BINARIES
G. Russo                                                    23

SEARCH FOR PULSATING STARS IN MULTIPLE STELLAR SYSTEMS
E. Antonello, M. Fracassini, L.E. Pasinetti, L. Pastori     27

AN AD HOC HYPOTHESIS ON THE PULSATION OF DELTA SCUTI STARS
E. Antonello                                                33

ON THE INTERNAL STRUCTURE OF MAIN-SEQUENCE STARS
A. Giménez, J.M. García-Pelayo                              37

A STUDY OF THE ORBITAL PERIODS OF 22 ECLIPSING BINARIES
T. Panchatsaram, K.D. Abhyankar                             47

ON THE GROUP FORMATION OF STARS IN STELLAR ASSOCIATIONS
L.V. Mirzoyan                                               61

THE AGE OF THE TRIPLE SYSTEM HD 165590
G. A. Bakos, J. Tremko                                      67

PROPERTIES OF RELATIVISTIC LINEAR STELLAR MODELS
V. Ureche                                                   73

PART II
EVOLUTIONARY TRENDS IN WIDE BINARY SYSTEMS


EVOLUTIONARY TRENDS IN WIDE BINARIES
P. van de Kamp                                                           81


A NOTE ABOUT MULTIPLE SYSTEMS OF DENSE MOLECULAR CLOUDS
V. Vanýsek                                                              105


TRAPEZIUM TYPE MULTIPLE SYSTEMS AND FORMATION OF STARS
G. N. Salukvadze                                                       109


A HYPOTHESIS ON THE RICHNESS OF A GALAXY IN BINARY STELLAR
SYSTEMS
H. Zinnecker                                                           115


A MASS-ECCENTRICITY CORRELATION IN SPECTROSCOPIC AND
VISUAL BINARY ORBITS
J. Dommanget                                                           119


PHOTOMETRIC INVESTIGATION OF VISUAL BINARIES WITH A
COMPONENT ABOVE THE MAIN SEQUENCE
O.G. Franz                                                             123


ON THE DISTRIBUTION OF YOUNG SPECTROSCOPIC BINARY STARS
OVER THE MAJOR SEMIAXES OF THEIR ORBITS
E.I. Popova, A.V. Tutukov, B.M. Shustov, L.R. Yungelson                 129


OBSERVATIONAL SELECTION IN SPECTROSCOPIC BINARY
ECCENTRICITIES
M.S. Staniucha                                                         133


DOUBLE STARS: THE CLOSEST AND THE MOST DISTANT SYSTEMS
T.J. Herczeg                                                           145


THE EVOLUTIONARY STATE OF ZETA AURIGAE
R.D. Chapman                                                           153


PART III
EVOLUTIONARY PROCESSES IN CLOSE BINARY SYSTEMS


EVOLUTION OF CLOSE BINARY STARS: OBSERVATIONAL ASPECTS
M.J. Plavec                                                            159


STATISTICAL PROPERTIES OF ALGOL-TYPE SYSTEMS
G. Giuricin, F. Mardirossian, M. Mezzetti                              183

MASS LOSS IN ALGOL-TYPE STARS: IMPLICATION ON THEIR
EVOLUTIONARY STAGE
F. Mardirossian, G. Giuricin                                          187

ULTRAVIOLET LIGHT CURVES OF U GEMINORUM AND VW HYDRI
C.-C. Wu, R.J. Panek, A.V. Holm, F.H. Schiffer III                   191

STATISTICAL MODELS FOR CLOSE BINARIES
J.L. Halbwachs                                                       199

STRUCTURE AND EVOLUTION OF SV CENTAURI
H. Drechsel, J. Rahe, W. Wargau, B. Wolf                             205

AGE DETERMINATION IN CW ERI = BV 1000
H. Mauder                                                            217

NUMERICAL CALCULATIONS FOR ACCRETION DISKS IN CLOSE
BINARY SYSTEMS
G. Hensler                                                           219

QUASI-SIMULTANEOUS PHOTOMETRIC AND POLARIMETRIC
OBSERVATIONS OF T TAU AND RY TAU
N.P. Red'kina, G.P. Chernova                                         231

S Cnc AS A CASE STUDY OF CLASSICAL ALGOL EVOLUTION
N.S. Awadalla, E. Budding                                            239

STRUCTURAL MODELS FOR BETA LYRAE-TYPE DISKS
R.E. Wilson                                                          261

EVOLUTIONARY EFFECTS IN CONTACT BINARIES
H. Mauder                                                            275

EVOLUTIONARY POSSIBILITIES AND IMPOSSIBILITIES FOR
SOLAR TYPE CONTACT BINARIES IN NGC 188
F. van't Veer                                                        279

ORIGIN AND EVOLUTION OF CONTACT BINARIES OF W UMa TYPE
T. Rahunen, O. Vilhu                                                 289

SOME ASPECTS OF THE EVOLUTIONARY STATUS OF W URSAE MAJORIS
BINARIES DEDUCED FROM OBSERVATIONAL DATA
W. van Hamme                                                         301

A POSSIBLE EXPLANATION OF DISTORTIONS IN THE LIGHT CURVES
OF VW CEP BY STARSPOTS
A. Yamasaki                                                          305

THE IUE OBSERVATIONS OF W UMA
S.M. Rucinski, J.E. Pringle, J.A.J. Whelan                           309

MASS FLOW DUE TO HEATING IN A BINARY SYSTEM:
APPLICATION TO U CEPHEI
Y. Kondo, J.L. Modisette                                                              317

THE VERY UNUSUAL ULTRAVIOLET SPECTRA OF R ARAE
Y. Kondo, G.E. McCluskey                                                             321

ON A POSSIBLE LINKAGE BETWEEN W-TYPE W UMa SYSTEMS AND
THE SHORT PERIOD RSCVn-LIKE BINARIES
L. Milano, G. Russo, S. Mancuso                                                      327

RECENT OBSERVATIONAL EVIDENCE ON W URSAE MAJORIS STARS
H. Rovithis-Livaniou, P. Rovithis                                                    337

THE TIME CHANGE OF THE AW UMa SYSTEM
M. Kurpinska, K. Otmianowska                                                         343

THE SHORT-PERIOD NON-CONTACT BINARY SYSTEMS UU LYN AND GR
TAU
A. Yamasaki, A. Okazaki, M. Kitamura                                                 345

ON THE SPATIAL DENSITY OF W UMa TYPE STARS
E. Budding                                                                           351

PART IV
CATACLYSMIC BINARIES AND THEIR ROLE IN STELLAR EVOLUTION

BINARY STARS AND SS433
R. Ruffini                                                                           373

OBSERVATIONS OF SECULAR CHANGES IN THE KINEMATIC MODEL
OF SS433
G.W. Collins II, G.H. Newsom                                                         389

TIME RESOLVED CIRCULAR POLARIMETRY OF WHITE DWARF
PULSARS
W. Krzemiński, J.D. Landstreet, I. Thompson                                          399

THE MASS DISTRIBUTION OF WHITE DWARFS AND CENTRAL
STARS OF PLANETARY NEBULAE
V. Weidemann                                                                         403

ABUNDANCE PECULIARITIES IN WHITE DWARF ATMOSPHERES AS
A RESULT OF BINARY EVOLUTION?
I. Bues                                                                              409

THE MINIMUM ORBITAL PERIOD OF HYDROGEN-RICH CATACLYSMIC
BINARIES
R. Sienkiewicz, B. Paczyński                                                         413

REMARKS ON THE EVOLUTIONARY STATUS OF CATACLYSMIC
VARIABLES
N. Vogt                                                              415

SUPERNOVAE IN BINARY SYSTEMS: PRODUCTION OF RUN-
AWAY STARS AND PULSARS
J.-P. de Cuyper                                                      417

AN OPTICAL BURST IN SCO X-1
H. Mauder                                                            445

THE FORMATION OF MASSIVE WHITE DWARFS IN CATACLYSMIC
BINARIES
W.-Y. Law, H. Ritter                                                 447

UV SPECTROSCOPY OF THE NOVALIKE VARIABLE TT ARIETIS
W. Wargau, H. Drechsel, J. Rahe, G. Klare, B. Wolf,
J. Krautter, N. Vogt                                                 453

THE SUBDWARF ECLIPSING BINARY LB3459 (AA Dor)
R.W. Hilditch, G. Hill, D. Kilkenny                                  455

VARIABILITY OF SOFT X-RAY EMISSION OF EX HYDRAE
OBSERVED WITH EINSTEIN OBSERVATORY
A. Kruszewski, R. Mewe, J. Heise, T. Chlebowski,
W. van Dijk, R. Bakker                                               457

ON THE POSSIBLE SHORT-PERIOD IRREGULAR LIGHT
FLUCTUATIONS OF V1357 CYG = CYG X-1
M.I. Kumsiashvili, Z. Kraicheva                                      467

PU VULPECULAE (OBJECT HONDA-KUWANO 1979) - A POSSIBLE
SHORT-PERIOD RELATIVE OF THE SYMBIOTIC STARS
D. Chochol, J. Grygar                                                473

DID SU UMa UNDERGO A CLASSICAL NOVA OUTBURST?
M.F. Bode, A. Evans, A. Bruch                                        475

DOES THE CATACLYSMIC BINARY Z CHA CONTAIN A BLACK
DWARF SECONDARY?
J. Faulkner, H. Ritter                                               483


CONCLUDING REMARKS
Z. Kopal                                                             489

INDEX                                                                499

All's well that ends well ...

Group photograph of participants of the Colloquium.

# List of Participants

| | |
|---|---|
| Abhyankar, K.D. | Osmania University, Hyderabad, India |
| Adam, J. | Universität Heidelberg, Heidelberg, F.R.G. |
| Andersen, J. | Copenhagen University, Tølløse, Denmark |
| Antonello, E. | Osservatorio Astronomico di Brera, Merate, Italy |
| Awadalla, N. | University of Manchester, U.K.; Helwan Obs.,Egypt |
| Bakos, G.A. | University of Waterloo, Waterloo, Canada |
| Barbaro, G. | Osservatorio Astronomico, Padova, Italy |
| Beuermann, K.P. | Technical University, Berlin, F.R.G. |
| Bode, M.F. | University of Keele, Staffordshire, U.K. |
| Brancewicz, H. | University of Mining and Metallurgy, Krakow, Poland |
| Broglia, P. | Osservatorio Astronomico, Merate, Italy |
| Bruch, A. | University Münster, Münster, F.R.G. |
| Budding, E. | University of Manchester, Manchester, U.K. |
| Bues, I. | Universität Erlangen,Remeis-Sternwarte, Bamberg, F.R.G. |
| Chapman, R.D. | NASA-Goddard Space Flight Center,Greenbelt, U.S.A. |
| Chochol, D. | Astron.Institute, Slovak Acad.of Sc., Tatranska Lomnica, Czechoslovakia |
| Clausen, J. | Copenhagen University Observatory, Tølløse, Denmark |
| Collins II, C.W. | Ohio State University, Columbus, U.S.A. |
| Costa Boronat, V. da | Instituto de Astrofisica de Andalucia, Granada, Spain |
| Dallaporta, N. | Instituto Astronomia University, Padova, Italy |
| De Cuyper, J.P. | Vrije Universiteit Brussel, Brussels, Belgium |
| De Grève, J.P. | Vrije Universiteit Brussel, Brussels, Belgium |

De Kort, J.J.                  Nijmegen, The Netherlands

De Landtsheer, A.C.            Sterrewacht Sonnenborgh,Utrecht,The Netherlands

Delgado Sanchez, A.J.,Instituto de Astrofisica de Andalucia, Granada,
                               Spain

Dommanget, J.                  Observatoire Royal, Brussels, Belgium

Drechsel, H.                   Universität Erlangen, Remeis-Sternwarte, Bamberg,
                               F.R.G.

Franz, O.G.                    Lowell Observatory, Flagstaff, U.S.A.

Giménez, A.                    Instituto de Astrofisica de Andalucia, Granada,
                               Spain

Giuricin, G.                   Osservatorio Astronomico, Trieste, Italy

Glass, I.S.                    European Southern Observatory, Garching, F.R.G.;
                               South African Astron. Obs., South Africa

Grygar, J.                     Institute of Physics, Cz. Acad. of Sciences, Rez,
                               Czechoslovakia

Guarnieri, A.                  Istituto di Astronomia, Universita di Bologna,
                               Bologna, Italy

Halbwachs, J.L.                Observatoire de Strasbourg, Strasbourg, France

Hensler, G.                    Universitätssternwarte, Göttingen, F.R.G.

Hilditch, R.W.                 University Observatory, St.Andrews, Scotland,U.K.

Hopp, U.                       Technical University, Berlin, F.R.G.

Isserstedt, J.                 Universität Würzburg, Würzburg, F.R.G.

Jeffery, C.S.                  University Observatory, St. Andrews,Scotland,U.K.

Jørgensen, H.E.                University Observatory, Tølløse, Denmark

Kippenhahn, R.                 MPI für Astrophysik, Garching, F.R.G.

Knigge, R.                     Universität Erlangen, Remeis-Sternwarte, Bamberg,
                               F.R.G.

Kohoutek, L.                   Hamburger Sternwarte, Hamburg, F.R.G.

Kondo, Y.                      NASA-Goddard Space Flight Center, Greenbelt,U.S.A.

| | |
|---|---|
| Kopal, Z. | University of Manchester, U.K. |
| Kreiner, J.M. | Silesian University, Katowice, Poland |
| Kruszewski, A. | ESO, Garching, F.R.G.; Warsaw Univ.Obs.,Poland |
| Krzeminski, W. | Copernicus Astronomical Center, Warsaw, Poland |
| Kumsiashvili, M.I. | Astrophysical Observatory, Abastumany,U.S.S.R. |
| Kurpinska, M. | Uniwersytet Jagiellonski, Krakow, Poland |
| LaDous, C. | Universitäts-Sternwarte, München, F.R.G. |
| Law, W.Y. | MPI für Astrophysik, Garching, F.R.G. |
| Lukas, R. | W. Foerster-Sternwarte, Berlin, F.R.G. |
| Martin, R. | Universität Erlangen, Remeis-Sternwarte,Bamberg, F.R.G. |
| Mauder, H. | Universität Tübingen, Tübingen, F.R.G. |
| McLean, B. | University Observatory, St.Andrews,Scotland,U.K. |
| Meyer-Hofmeister, E. | MPI für Astrophysik, Garching, F.R.G. |
| Milano, L. | Capodimonte Astronomical Observatory, Napoli, Italy |
| Mirzoyan, L.V. | Astrophysical Observatory, Byurakan, U.S.S.R. |
| Nordström, B. | Copenhagen University Observatory, Tølløse, Denmark |
| Otmianowska, K. | Uniwersytet Jagiellonski, Krakow, Poland |
| Packet, W. | Vrije Universiteit Brussel, Brussels, Belgium |
| Pavlovski, K. | Hvar Observatory, Zagreb, Yugoslavia |
| Plavec, M. | University of California, Los Angeles, U.S.A. |
| Rahe, J. | Universität Erlangen, Remeis-Sternwarte,Bamberg, F.R.G. |
| Rahunen, T. | University of Helsinki, Helsinki, Finland |
| Ritter, H. | MPI für Astrophysik, Garching, F.R.G. |
| Ruffini, R. | Instituto di Fisica G.Marconi, Roma, Italy |

Rucinski, R.          MPI für Astrophysik, Garching, F.R.G.; Warsaw
                      Univ. Obs., Poland

Rupprecht, G.         Universität Erlangen, Remeis-Sternwarte,Bamberg,
                      F.R.G.

Russo, G.             Osservatorio Astronomico Capodimonte, Napoli,
                      Italy

Sadik, A.R.           Astron. and Space Research Center,Baghdad, Iraq

Salukvadze, G.N.      Astrophysical Observatory, Abastumani, U.S.S.R.

Schiener, B.          Universität Erlangen, Remeis-Sternwarte,Bamberg,
                      F.R.G.

Schnur, G.            European Southern Observatory, Santiago, Chile

Seggewiß, W.          Observatorium Hoher List, Daun, F.R.G.

Shustov, B.M.         Astronomical Council, Ac.of Sciences, Moscow,
                      U.S.S.R.

Sienkiewicz, R.       Copernicus Astronomical Center,Warszawa,Poland

Smith, R.C.           University of Sussex, Falmer Brighton, U.K.

Sreenivasan, S.R.     University of Calgary, Calgary, Canada

Staniucha, M.         Warsaw University Observatory, Warszawa, Poland

Strohmeier, W.        Universität Erlangen, Remeis-Sternwarte,Bamberg,
                      F.R.G.

Szeidl, B.            Konkoly Observatory, Budapest, Hungary

Tremko, J.            Skalnaté Pleso Observatory, Tatranská Lomnica,
                      Czechoslovakia

Vanbeveren, D.        Vrije Universiteit Brussel, Brussels, Belgium

Van de Kamp, P.       Sproul Observatory, Swarthmore, PA,U.S.A.

Van Hamme, W.         Sterrenkundig Obs., R.U.G., Gent, Belgium

Van't Veer, F.        Institut d'Astrophysique, Paris, France

Vanýsek, V.           Charles University, Praha, Czechoslovakia

Vogt, N.              European Southern Observatory, Santiago, Chile

| | |
|---|---|
| Walter, K. | Universität Tübingen, Tübingen, F.R.G. |
| Wargau, W. | Universität Erlangen, Remeis-Sternwarte,Bamberg, F.R.G. |
| Weidemann, V. | Universität Kiel, Kiel, F.R.G. |
| West, R. | European Southern Observatory, Garching, F.R.G. |
| Whitelock, P.A. | South African Astronomical Observatory, South Africa |
| Wilson, R.E. | Dept. of Astronomy, University of Florida, Gainesville, U.S.A. |
| Wood, F.B. | University of Florida, Gainesville, U.S.A. |
| Worg, R. | Universität Erlangen, Remeis-Sternwarte,Bamberg, F.R.G. |
| Wu, Ch.-Ch. | Computer Sciences Corporation, Silver Spring, U.S.A. |
| Yamasaki, A. | University of Tokyo, Tokyo, Japan |
| Zhou, Y. | MPI für Astrophysik, Garching, F.R.G.; Dept. of Astrophysics, Hoffei, China |
| Zinnecker, H. | MPI für Extraterrestr. Physik, Garching, F.R.G. |

WELCOME


R. Grafberger
Mayor of Bamberg


Sehr geehrter Herr Professor Dr. Kopal !
Sehr geehrter Herr Professor Dr. Rahe !
Meine sehr verehrten Damen und Herren der Internationalen Astronomischen
Union !

Als "Nabel der Welt" begriff sich Bamberg als es gegründet wurde.
Es war die Zeit des Ptolemäischen Weltbildes als die Erde noch Mittelpunkt des Universums war. Der Sternenmantel, den Kaiser Heinrich zum
Osterfest 1020 geschenkt bekam, zeugt heute noch als größte Kostbarkeit
im Diözesan-Museum von diesem Weltbild. Abt Gerhard von Seeon schrieb
damals über Bamberg "Haec caput est orbis, hic gloria conditur omnis"
(Hier ist die Hauptstadt der Welt, die Wiege jeglichen Ruhmes). Aus
diesem Denken der damaligen Welt ist diese Stadt weitgehend geprägt.

Heute ist Bamberg Mittelpunkt einer ganz anderen astronomischen
Welt. Es ist mir eine Ehre, Sie, die Teilnehmer des 69. Colloquiums
der Internationalen Astronomischen Union hier in Bamberg in Vertretung
des Oberbürgermeisters begrüßen zu dürfen. Die Bürger und der Rat der
Stadt Bamberg heißen Sie, die Gäste aus aller Welt, herzlich willkommen.
Wir sind stolz darauf, daß Ihre Organisation dieses Colloquium nach
Bamberg vergeben hat. Daß dies so ist, verdanken wir den guten Ver-
bindungen von Professor Dr. Rahe. Die Stadt Bamberg bedankt sich
bei Ihnen recht herzlich und spricht Ihnen und Ihrem Institut ihre
Anerkennung aus.

Heute geht es in der Astronomie um Doppelsterne, Mehrfachsysteme,
Quasare, Pulsare, Neutronensterne, Schwarze Löcher und andere wunder-
same Gebilde. Heute wird auf der Bamberger Sternwarte nicht mehr
beobachtet, sondern an dem modernen astronomischen Weltbild mitge-
arbeitet. Zwischen dem alten Weltbild des Sternenmantels und dem
faszinierenden modernen astronomischen Weltbild standen Kopernikus,
Keppler und Galilei. An diesem Umbruch wirkte auch der Bamberger
Mathematiker, Astronom und Computist Christoph Clavius Bambergensis
mit. Er war maßgeblich an der Kalenderreform von Pabst Gregor
beteiligt. Diese Männer begründeten die moderne Naturwissenschaft,

*Z. Kopal and J. Rahe (eds.), Binary and Multiple Stars as Tracers of Stellar Evolution, xix–xx.*
*Copyright © 1982 by D. Reidel Publishing Company.*

in dem sie die Frage nach dem "Warum" durch die Frage "Wie ist" er-
setzten.

Eine moderne theologische Genesis könnte wie folgt beginnen:
"Am Anfang schuf Gott die Einheit.  Diese Einheit war von unendlicher
Dichte und alles, was war und jemals sein wird steckte komprimiert in
ihr.  Und Finsternis lag über dem Universum.  Und Gott sprach:
"Es geschehe ein Urknall".  Und es gab einen Urknall.  Und so ent-
standen Materie und Strahlung.  Und Gott besah sich den Urknall, der
ein wahrhaft höllisches Ausmaß hatte..."
Eine solche theologische Genesis ist nicht ihre Aufgabe und doch sind
Ihre Forschungen Hilfen für die Menschen, sich in dieser faszinierenden
Welt zurechtzufinden.

Unsere Stadt gibt den Rahmen für solche wissenschaftlichen Ge-
spräche ab.  Es ist eine Stadt, die von einem anderen Weltbild geprägt
ist und doch von den Menschen unserer Zeit liebend angenommen wird.
Es ist eine Stadt, in der Sie Menschen finden, für die Gott, Bamberg
und der Mensch selbst immer noch die Mitte - der Nabel der Welt - sind.

In dieser Stadt wünschen wir Ihrer wissenschaftlichen Tagung
einen erfolgreichen Verlauf.  Ihnen, den Teilnehmern, wünschen wir
menschliche Begegnung untereinander und erlebnisreiche Tage in Bamberg.

WELCOME


       Prof. Dr. Annegret Bollée
       Vice-President, University of Bamberg


    Ladies and gentlemen, dear colleagues -

    It is an honor for the University of Bamberg and a great pleasure
for me to welcome you here today on behalf of President Oppolzer who
is at present absent from Bamberg. He has asked me to extend his
greetings to you and his best wishes for a successful Colloquium and
a pleasant stay in Bamberg.

    The International Astronomical Union has chosen the town of
Bamberg for its 69th Colloquium, and among the many old houses of this
town one that stands out as having a particularly colorful history: one
so full of memorable events that a little book could be written on the
subject; indeed one has been written by my colleague and former Rector
of the University, Elisabeth Roth. I refer you to her book for more
detailed information.

    The present building - in fact, two houses united into one - dates
back to the Middle Ages. Before becoming Hochzeitshaus - I shall explain this name in due course - it was the Gasthaus zum Wilden Mann, the
Wild-Man-Inn, the wild man being a kind of hairy giant, a figure which
is often found in medieval and Renaissance coats-of-arms and also as a
house sign. The inn Zum Wilden Mann was first mentioned in a document
of 1484, and in the 16th Century it must have been an inn of great re-
nown, otherwise the Prince Bishop would not have chosen this accomo-
dation for his famous guest Albrecht Dürer in 1520. On the photo in
your programme you can see the memorial tablet in memory of Dürer's
various sojourns in Bamberg in that year.

    In the times of the Renaissance, times of sumptuous festivities
and opulent meals, German towns used to provide a public building for
their citizens where they could celebrate private events, especially
weddings. In 1605, therefore, the Mayor and Council of Bamberg decided
to buy the Wild-Man-Inn and to transform it into a Hochzeitshaus for
the townspeople of Bamberg. Hochzeit, meaning 'wedding' in modern
German, did not have this restricted meaning in Old and Middle High

*Z. Kopal and J. Rahe (eds.), Binary and Multiple Stars as Tracers of Stellar Evolution, xxi–xxii.*
*Copyright © 1982 by D. Reidel Publishing Company.*

German.  It meant 'festivity' or 'feast' in general.  However, the
Hochzeitshaus zum Wilden Mann was in fact mainly used for big weddings.  This destination is documented by the painting of the Wedding
at Cana.  This painting by an unknown artist, dating from the early
17th century, was rediscovered in the course of the renovation of the
building in 1973/74 and thoroughly restored in 1975.

The Hochzeitshaus was among the very few houses of Bamberg to be
very badly damaged in the war, in 1945.  Rebuilt in 1950-52, it was
restored to its present shape in 1973-75.  The Hochzeitshaus is part
of our programme "Universität in der Altstadt", a programme which aims
at integrating a new university into an old town by restoring old
buildings and filling them with new life.  I said a new university because it was in fact founded in 1972.  But rather than a new foundation,
it is a re-foundation; it takes up a tradition dating back to the 17th
century.  In 1648 the Prince Bishop Melchior Otto Voit von Salzburg
raised a former seminary for priests to the rank of an Academy.  After
his patron saint Otto he named it Academia Ottoniana.  At the beginning
there were only two faculties, Theology and Philosophy, all the
professors and the rector belonging to the order of the Jesuits.  Prince
Bishops succeeding Melchior Otto gradually enlarged the Academy in the
course of the 18th century by adding faculties of Law and Medicine.  In
1773 the academic constitution was changed in accordance with those of
other German universities, and the name was changed to University of
Bamberg.  In the early 19th century, however, this promising development was interrupted.  In 1802 the reign of the Prince Bishops of
Bamberg came to an end and Bamberg became part of the state of Bavaria.
The university was dissolved in 1803, but theological and philosophical
studies have survived up to the present day.  I shall not bore you with
all the details of a rather complicated history.  Let me only mention
that the institutions succeeding the old university were transformed
into a Philosophisch-Theologische Hochschule in 1923.  Together with
a Teacher Training College founded in 1958, the Philosophisch-Theologische Hochschule became one of the corner-stones of the refounded
University in 1972.  The Universitas Bambergensis rediviva at first
offered courses only for future priests, teachers and social workers,
but in 1977 more faculties were added, and we hope that the period of
expansion has not yet come to an end.  We have now 2.800 students - we
expect to have over 3.000 in the winter term -, a staff of 520 and
almost 90 professors.  There are 6 faculties:  Theology, Philosphy-
Psychology-Education, Languages and Literature, History and Geography,
Social Sciences and Economics, and a department for the training of
social workers.

Ladies and gentlemen, may I wish you a successful meeting, and I
hope that you will enjoy your stay in Bamberg.  We have just said good-
bye to the students who have attended the International Summer Course
of German Language and Literature.  They seemed to like Bamberg very
much, so I hope that you will like it as well.

# WELCOME

Prof. Dr. Nikolaus Fiebiger
President, University Erlangen-Nürnberg

Mr. Mayor, Dear Colleague, Mr. Chairman, Ladies and Gentlemen -

On behalf of the Friedrich-Alexander-Universität Erlangen-Nürnberg, I welcome you here in Bamberg. Very special greetings go to our guests from abroad.

The fact that presidents of two universities welcome you, illustrates the unusual situation in Bamberg. The Remeis-Observatory is part of the Friedrich-Alexander-Universität and belonged to this University already before the new University in Bamberg was founded; and this has remained so until today.

The University Erlangen-Nürnberg is a classical German university with a Faculty of Theology, Law, Medicine, Humanities, Natural Science, and Economy. Fifteen years ago - and this is something new in Germany - a Technical Faculty was incorporated, which now has more than 3.000 students. This has been an effort to fill a gap that existed in a few areas between universities and technical colleges in the courses offered. We can now say that this experiment has been successful.

We know, of course, that the Remeis-Observatory with its experimental facilities, is today of no great significance in astronomical research. Astronomy needs international cooperation and a great amount of sophisticated, and consequently, expensive instruments which cannot be provided by a single university. But we can and want to be a place where astrophysics is practiced and where young people can get an education in this subject. Although the appropriate experimental observations are carried out at the large national and international centers, the preparation and evaluation of these experiments still have to take place at the university.

*Z. Kopal and J. Rahe (eds.), Binary and Multiple Stars as Tracers of Stellar Evolution, xxiii–xxiv.*
*Copyright © 1982 by D. Reidel Publishing Company.*

Frankonia boasts a great son in astronomy: a few years ago, we celebrated the 500th anniversary of the birth of Regiomontanus who was born in Königsberg, Haßberge, and whose computational tables were used not only by Copernicus in his work but also by Columbus in his voyage of discovery to America.


I wish the Bamberg conference a good start and productive discussions. In my opinion, no other city could offer a more suitable frame than Bamberg and I hope that in the course of your meeting you will also have the chance to see a few of the historical places.

WELCOME


Jürgen Rahe
Dr.-Remeis Sternwarte Bamberg
Astronomisches Institut
Universität Erlangen-Nürnberg


Mr. Mayor, Mrs. Vice-President, Mr. President, Ladies and Gentlemen, Dear Colleagues –

On behalf of the Dr.-Remeis Observatory Bamberg, Astronomical
Institute of the University Erlangen-Nürnberg, I would like to welcome
you to IAU-Colloquium No. 69 on "Binary and Multiple Stars as Tracers
of Stellar Evolution". It is the 7th International Astronomical
Conference organized by our Institute, and the fourth sponsored by the
International Astronomical Union. The first conference was organized
by Professor Strohmeier in 1959 and dealt with Variable Stars. The
last one was held in 1979 and dealt with space experiments.

The first IAU-Colloquium took place August 11-14, 1965, and dealt
with "The Position of Variable Stars in the Hertzsprung-Russell
Diagram". During the second, IAU-Colloquium No. 15, August 31 to
September 3, 1971, "New Directions and New Frontiers in Variable Star
Research" were discussed. The third, IAU Colloquium No. 42, was held
September 6-9, 1977, and dealt with "The Interaction of Variable Stars
with their Environment".

The idea for the present Colloquium arose actually last year
(1980) when Professor Kopal and I met on a beautiful Spring day in
Washington. When presenting this plan to several colleagues, we
received only favorable responses and we decided to pursue the idea.
A scientific organizing committee was soon established, consisting of
Drs. Z. Kopal (U.K.), Chairman; M.K.V. Bappu (India); J.D. Fernie
(Canada); R. Kippenhahn (F.R.G.); M. Kitamura (Japan); L. Milano (Italy);
R.E. Mustel (USSR); S.L. Piotrowski (Poland); M. Plavec (U.S.A.);
A. Rigutti (Italy); V. Vanýsek (Czechoslovakia); B. Warner (South
Africa).

Over the years our Institute has received the continuous support
of the University of Erlangen-Nürnberg and the city of Bamberg. This,
and the assistance rendered by the University of Bamberg and the IAU,
has enabled us to organize such meetings.

*Z. Kopal and J. Rahe (eds.), Binary and Multiple Stars as Tracers of Stellar Evolution, xxv–xxvi.*
*Copyright © 1982 by D. Reidel Publishing Company.*

I am pleased and honored to welcome here today, representatives of the city of Bamberg, the University of Bamberg, and the University of Erlangen-Nürnberg.

Since 1959, every meeting organized by our Institute has actively been supported by the city of Bamberg. You will especially enjoy the already traditional "Frankonian evening" in the historical surroundings of the city, offering a glimpse of the hospitality of Bamberg and its citizens. I am very pleased to welcome here today Mr. Rudolf Grafberger, Mayor of the City of Bamberg.

During the last conferences and this meeting, the University of Bamberg has allowed us to use their facilities, and I take this opportunity to express our sincere thanks. It is my pleasure to welcome here today the Vice-President of the University of Bamberg, Mrs. Professor Annegret Bollée.

I welcome and introduce to you, Professor Nikolaus Fiebiger, President of the University of Erlangen-Nürnberg, an internationally well-known experimental physicist. I am proud to say that he has for years supported the activities of our Institute. Some months ago Professor Fiebiger was re-elected as President of the University and I heartily congratulate him - and us - on this occasion.

Professor Zdenek Kopal as you know, is Chairman of the Scientific Organizing Committee for this meeting. He has been a very special friend of our Institute and it is an honor to prepare this meeting with him.

If one would write a Michelin guide - not to restaurants - but to astronomical meetings, I am sure one would easily agree on the following scheme: many astronomical meetings deserve at least one star - that is, according to the Michelin guide, they are worth a stop. Many other meetings deserve two stars - that is, they are worth a de-tour. But meetings organized by Professor Kopal, always deserve three stars - that is, these meetings are worth a whole trip. If I look at the audience, and the program, I am convinced you will all agree with me.

Finally, I hope we will all profit from these days of interesting scientific discussions among colleagues, who over the years have become friends, and add new ones when meetings such as this take place. I personally thank you for coming and with all members of our Institute, wish you a very pleasant and rewarding sojourn in this historical city.

The following institutions generously supported the meeting: Deutsche Forschungsgemeinschaft, Bonn. International Astronomical Union, Paris. Stadt Bamberg. Universität Bamberg. Universität Erlangen-Nürnberg.

# INTRODUCTION

Zdeněk Kopal
Department of Astronomy
University of Manchester

Your Magnificences, my Lord Mayor, ladies and gentlemen! It is a great pleasure for me to respond, on behalf of your foreign guests, to your gracious words of welcome; and to thank you for the wonderful reception which you have extended to us. The city of Bamberg and its Remeis Sternwarte has indeed been renowned all over the world for a great many years – as the place where your Observatory's first director, Professor Ernst Hartwig (1851-1923) – in addition to his other titles to fame – collaborated (with Gustav Müller of Potsdam) on the construction of the monumental Geschichte und Literatur des Lichtwechsels der Veränderlichen Sterne, which since 1918 has (together with its subsequent continuation) been a veritable vade-mecum of all students of variable stars; where the second director, Professor Ernst Zinner (1886-1970) prepared his valuable Katalog der Verdächtigen Veränderlichen Sterne (1926) which safeguarded many an astronomer (including the present speaker in the days of his innocence) from premature discovery claims; and whose third director, Professor Wolfgang Strohmeier, initiated in 1959 the tradition of the international colloquia of which ours is the latest successor. It is indeed a great pleasure to welcome Professor Strohmeier – now Emeritus – among us; and to congratulate him on the grace with which he is carrying his years. And – last but not least – it is a very pleasant duty to express our sincere thanks to his successor, Professor Jürgen Rahe, and to the Bamberg city authorities, for their invitation to hold our meetings in their lovely town, which for many years has been the place of periodic pilgrimage to so many students of variable stars – and we sincerely hope it remains so also in the future.

The presence of so many of you here this morning – with participants from 20 countries ranging geographically from California to Japan, and from Finland to South Africa, in the days when travel is no longer as easy to wage as it may have been before, bespeaks the attraction and affection which Bamberg holds for the international astronomical community. It demonstrates the truth which the great German poet Friedrich Schiller (whose last resting place is not too far north of us today ) expressed in his ode An die Freude – later rendered so

*Z. Kopal and J. Rahe (eds.), Binary and Multiple Stars as Tracers of Stellar Evolution, xxvii–xxx.*
*Copyright © 1982 by D. Reidel Publishing Company.*

strikingly into music by Beethoven - which could equally have been
addressed to our muse Urania... "Alle Menschen werden Brüder wo dein
sanfter Flügel weilt". This is certainly true today in this room;
and may it remain so anywhere in the world also in the future where-
ever students of science congregate to discuss their heavenly problems
- whether or not these are concerned with double stars!

The subject of our colloquium "Binary and Multiple Stars as
Tracers of Stellar Evolution", calls perhaps for a few words of introduction. As is well known, binary stars in the sky represent limiting
cases of stellar associations, consisting of gravitationally-bound
systems whose components originated virtually simultaneously and
from material of very much the same initial composition; although their
components could at the time of their origin have possessed arbitrarily
different masses. Ever since early 1950's - when the nature of
stellar evolution was placed on a (plausible) nuclear basis - we have
known that the progress of nuclear evolution depends essentially on
the star's mass (and, therefore, internal temperature); its rate being
the faster, the larger the mass of the respective configuration. The
occurrence, in the sky, of large number of the pairs of stars whose
components are of equal age and initial composition, but possibly of
different mass, offers a veritable royal road for tracing the effects
of differential evolution caused by a difference in mass alone. In
particular, on the commonly accepted basis of nuclear evolution we
should expect the evolution of more massive components to proceed more
rapidly than that of their less massive mates.

The observed properties of double stars seem, however, to bear out
a rather different story; for while theoretical expectations are
broadly in harmony with observations of binary systems whose both
components belong to the Main Sequence, for systems in which at least
one component evolved away from this sequence, the opposite of what
theory alone would lead us to expect turned out to be true: namely,
the component whose observed attributes identify it as a more evolved
object invariably happens to be the less massive of the two.

Since the mid 1950's, when this "evolutionary paradox" was first
pointed out (cf. Kopal, 1955; or Crawford, 1955) and made the subject
of specific discussions, the problem of accounting for this peculiar
behaviour has been exercising our minds; and many conferences were
devoted in the past in an effort to establish its satisfactory solution.
Our present colloquium will, however, differ from its predecessors in
one important aspect: namely, while all past conferences concerned
with evolutionary aspects of double (or multiple) star systems focussed attention almost exclusively on "close" binaries (mainly of those
which manifest their nature by mutual eclipses of their components),
we shall endeavour to bring into a common focus the evolutionary behaviour of all binary stars from which the relevant data are available
- whether these binaries are "close" or "wide" (i.e., "visual") in
the accepted sense of the word. For all such binaries - wide as well
as close - consist of components which are (most probably) of the same

age and initial composition; and whose mean lifetimes are longer than
the age of the Galaxy - regardless of whether their components are
separated by $10^{-2}$ or $10^4$ astronomical units (cf. Chandrasekhar, 1944;
or Kopal, 1978) - so that the effects of their differential evolution
going back to an initial difference in mass, can be traced for time
intervals up to $10^8 - 10^9$ years.

Do all such binaries - close as well as wide - exhibit indications
of the same "evolutionary paradox"? The answer - which did not seem to
have attracted sufficient attention - appears to be in the affirmative:
at least the extent of the evolutionary paradox in (say) Algol or
U Sagittae is no more pronounced than it is in wide ("visual")
binaries like Sirius or Procyon - to name only a few among the nearest
stars - whose (at present) less massive components far out-distanced
their more massive mates in their evolutionary courses; and while the
latter still linger on the Main Sequence today, their less massive
components have already become white dwarfs. Moreover, perhaps the
most flagrant paradox of this kind is encountered among X-ray
binaries, several of which were discovered in the sky in the last
decade; and about which more will be said in later parts of this
colloquium.

In order to bring about so profound a metamorphosis of the role
of the components, it is inevitable for the more evolved star to
have once been the more massive of the two; and since it is no longer
one today, it should have lost a large part of it (up to 90% for
X-ray sources) some time before. There is no room for doubt that this
must have been the case if we are to reconcile the presently observed
properties of binary systems with the basic properties of the current
stellar models, requiring that the more massive body is bound to
evolve at a faster rate.

The actual mechanisms by which a star in post-Main Sequence stage
can divest itself of a large fraction of its mass are still largely
uncertain; and so is the mode of disposal of excess mass – whether
it escapes from the system at high speed, or whether (in the case of
low-velocity escape) a part of it can be captured by its mate. Such
questions will no doubt come under discussion during subequent sessions of our colloquium. But whatever may be the more detailed outcome of such discussions, one more general conclusion may be of
overriding importance; namely, if the above-described "evolutionary
paradox" is characteristic of all binaries alike - be these close or
wide, and if (as seems logical) its origins are to be sought in terms
of the same physical process operative in all binaries alike - be these
close or wide - the proximity of the components cannot be the main
cause of the observed phenomena; or could, at best, play only a
subordinate role.

It is mainly to throw more light on these questions that our
colloquium has been convoked, and its programme organized. In the first
session which is to follow these introductory remarks, Professor

Rudolf Kippenhahn and the speakers following him will discuss the
evolutionary trends of single stars in their post-Main Sequence stage,
on the background of which we can attempt to trace also the evolution
of components in the binary systems.  The second session, to be
introduced by Professor Peter van de Kamp, will be concerned with a
survey of observed characteristics of wide binaries; while session 3,
which will be introduced by Professor Miroslav Plavec, will be similarly concerned with photometry and spectroscopy of close binaries; with
special attention devoted to "contact binaries" of W UMa-type.  Ultimately, the fourth (and last) session, to be introduced by Professor
Remo Ruffini, will be devoted to binary systems at the extreme end
of stellar evolution - and, in particular, to the enigmatic object
SS 433 - which provided in recent years more excitement in double-star
astronomy than any other type of systems - an excitement which will
no doubt be shared by all participants of this colloquium.  The
satisfactory solution of all problems arising in this connection will
no doubt have to await the acquisition of additional observed data,
and further theoretical developments will be necessary before their
solution can be placed on a truly satisfactory basis; to contribute
towards this goal should be the principal aim of our colloquium.

REFERENCES

Chandrasekhar, S.:  1944, Astrophys. J., 99, 54.
Crawford, J.A.:  1955, Astrophys. J., 121, 71.
Kopal, Z.:  1955, Mem. Roy. Soc. Sci. Liège, (4) 15, 684; Annales
      d'Astrophys., 18, 379.
Kopal, Z.:  1978, Dynamics of Close Binary Systems, D. Reidel Publ.
      Co., Dordrecht; pp. 10-11.

# Part I

# Evolution of Stars in the Post-Main Sequence Stage

# EVOLUTION OF STARS IN THE POST-MAIN SEQUENCE STAGE

*(Dedicated to Arnulf Schlüter on the occasion of his 60th birthday)*

R. Kippenhahn
Max-Planck-Institut für Physik und Astrophysik
Institut für Astrophysik, 8046 Garching b. München

I would like to start with a short historical introduction. In 1938 thermonuclear reaction rates for hydrogen burning became available. This made it possible to fit a convective core into the point source model of stellar structure integrated by Cowling three years earlier. The free parameter in the fitting process could be fixed with the thermonuclear reaction rates, the first realistic stellar model for a massive main sequence star was constructed! After the war electronic computers became available, and one was able to do more complicated models like those on the lower main sequence, like realistic models for the sun with its helium enriched interior, and one tried already to follow in time the exhaustion of hydrogen in the central regions of stars numerically. There was not too much progress for stars at the upper end of the main sequence. As soon as the stellar model tried to leave the main sequence and to march towards the region of the red supergiants the methods known at that time failed to produce models. For less massive stars the exhaustion of hydrogen could be followed up more easily and, in 1955, the great paper by Hoyle and Schwarzschild came out, which showed how these stars from the main sequence move into the red giant branch and move up parallel to what we now call the Hayashi line (which was not yet known at that time). But when helium started to burn the methods also failed.

The field stagnated for several years until, in 1961, the Henyey method became available. Henyey's famous talk at the IAU in Berkeley on that topic had its 20th anniversary about two weeks ago. We owe almost everything we have learned about stellar models on computers to this method; furthermore, if you look in recent issues of Ap. J. for stellar evolution calculations you will find that it has not yet been replaced by anything substantially better. With this method it was possible to follow the evolution of stars of intermediate mass and also of massive stars through several types of nuclear burning. It was possible to follow these stars through their different stages of Cepheid pulsation and into stages of very complicated evolution, with features about which I will report later. The method also made it possible to follow low mass stars through the helium flash. If, today, we get stuck with a computer simulation of stellar evolution it is normally not the numerical method,

3

*Z. Kopal and J. Rahe (eds.), Binary and Multiple Stars as Tracers of Stellar Evolution, 3–18.*
*Copyright © 1982 by D. Reidel Publishing Company.*

but rather our lack of understanding the physics which is responsible
for the failure. We do not know enough hydrodynamics to deal with convective overshooting, with semiconvection and with effects of stellar
rotation. We do not know all the mechanisms which bring material which
has just undergone a nuclear process to the surface, although the observations suggest it, and we do not know the mechanisms which produce
strange chemical abundances, even for stars on the main sequence. We do
not know the mechanisms which blow material from the stellar surface into space, sometimes causing mass loss which cannot be neglected in stellar evolution calculations.

In my report I will confine myself to the evolution of stars of
intermediate mass (1.4-10 $M_\odot$) and of low mass ($M < 1.4\ M_\odot$). A good review appeared in vol. 12 of Annual Review of A & A by Icko Iben seven
years ago (Iben, 1974).

## I. THE EVOLUTION OF STARS OF INTERMEDIATE MASS

It is well known from observation and from model calculations that stars,
when exhausting their hydrogen in the central regions, become red giants
or supergiants. Some evolutionary tracks are given in Figure 1 computed
by our group almost 20 years ago. Although opacities and nuclear reaction rates have improved in the meantime, nothing has changed qualitatively. The loops in the red giant region are extended more or less, de-

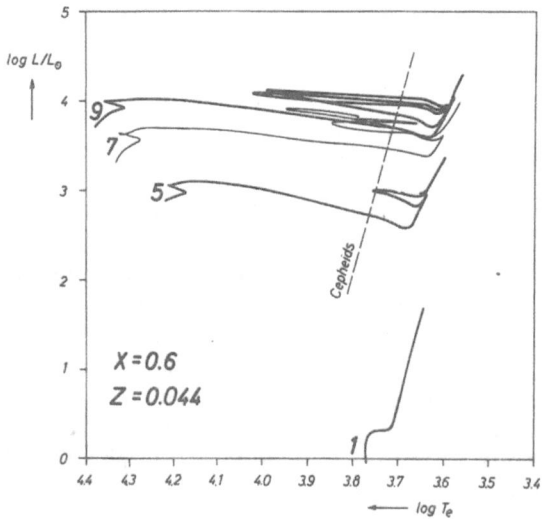

*Figure 1.* Evolutionary tracks for 1, 5, 7, 9 $M_\odot$ for an
extreme population 1 (according to v. Sengbusch, 1967;
Hofmeister et al., 1964b,c and Hofmeister, 1967).

pending on different parameters. They are also sensitive to inaccurately computed helium abundances in the central region. Anyone who is careless with his models on the main sequence as regards the central enrichment of helium, may later have to pay for this by receiving completely wrong loops in the red giant region (Lauterborn et al., 1971).

For a long time the boundaries between convective and radiative regions were determined by the Schwarzschild criterion. Also the computations on which the figures 1 and 2 are based use the Schwarzschild criterion and ignore the effect of overshooting. But, following the recent work of Shaviv and Salpeter (1973) one learned that convective overshooting at the surface of the convective core can have dramatic effects. This has been shown first by Maeder (1975) for a star of 2 $M_\odot$. If the mixing length is taken equal to the scale height ($\alpha$ = 1), overshooting increases the mass of the central core over which mixing occurs from 13 % to 17 % of the total mass. The results depend on the assumption of different parameters which appear in the application of the mixing length theory to overshooting. The result is more drastic for more massive stars. In his thesis at Hamburg University, Wassermann (1978) showed that for 5 $M_\odot$ the Shaviv-Salpeter type of overshooting increases the central region which after the exhaustion of central hydrogen is polluted by newly formed helium by 18 % to 26 % of the total mass of the star (again for $\alpha$ = 1). It is obvious that this effect increases the main sequence lifetime. But the more striking effect, already predicted by Weigert (1975), is that the chemical profile in the central region will have an effect on the occurence or disappearance of loops in the evolutionary stages after the onset of helium burning. This has the consequence that the overshooting has an influence on the mass range of those stars which during their life have the chance to stay long enough in the Cepheid strip. The latest result of the Hamburg school (Matraka et al. 1981) is rather disturbing. For stellar masses below 6 $M_\odot$ as soon as overshooting with $\alpha$ = 1 is taken into account the loops avoid the Cepheid strip. In order to make these stars to become Cepheids again, the ratio of mixing length to pressure scale height in the convective envelope has to be reduced below the canonical value of 1.5 down at least to 1. It seems difficult to obtain Cepheids with the *same* value of $\alpha$ for the convective envelope and the convective core. Unfortunately, in the treatment of overshooting in the very interior of stars the uncertainties in the mixing length theory become important, whereas in the old treatment, the mixing length uncertainties were important only in the outer convective zones.

The motion of a star in the HR diagram to the right and then back and forth is steered by events in the very interior, events which can be seen from Fig. 2 for a star of 5 $M_\odot$. There the evolution is described from the zero age main sequence (left) to the formation of a carbon oxygen core in a double shell model where, at the end (right), the hydrogen burning shell is exhausted but, sometime later, when the helium burning shell has eaten further outwards, hydrogen will reignite and both shells will move outwards parallel in the mass scale, as we will see later. Actually the carbon oxygen core is due to ignite carbon, but this process is delayed because of cooling by neutrinos.

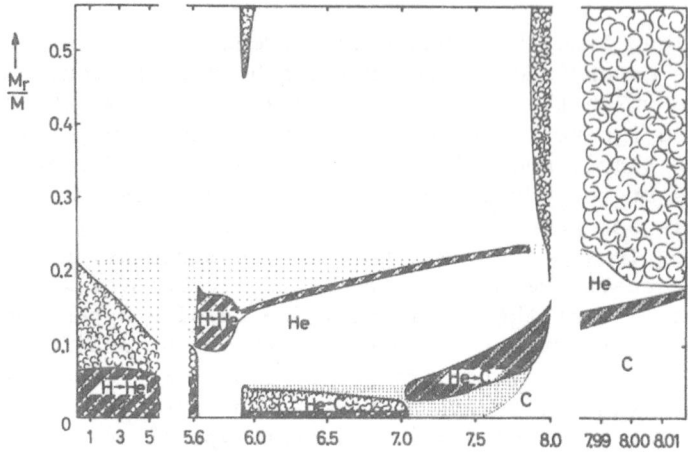

*Figure 2.* Variation in time of the interior of a star of 5 M☉ (Kippenhahn et al. 1965). The abscissae give the time after the zero age main sequence stage, the ordinate the mass $M_r$ in units of the total mass. "Cloudy" regions correspond to convective zones. Regions in which the nuclear energy generation rate exceeds $10^3$ c.g.s. units are hatched. Dotted regions indicate areas in which the hydrogen decreases inwards or where the helium abundance decreases inwards. Two dredge-up phases are seen: The first when the outerconvective zone reaches down to the $M_r/M \approx 0.47$, the second when it comes down below 0.2 and reaches the helium shell.

There are not too many ways of checking whether the events described in Fig. 2 will take place not only in computers but also in nature. Certainly, the evolutionary tracks, the isochrones which can be derived from them and the comparison with observed HR-diagrams of clusters are a help. Studying the pulsational properties of the models when they cross the instability strip can also indicate whether the computations have given the right picture, but these tests are not very compelling. They were not even able to tell us whether mass loss in the red supergiant region will reduce the mass in the hydrogen rich envelope appreciably during that phase of evolution or not.

Several authors have discussed the effect in which the outer convective zone (which becomes rather thick not only in radius but also in mass when the star comes close to the Hayashi line) mixes material which has been recently processed by nuclear reactions to the surface. This effect is called *dredge-up*. In Fig. 2 one can see two dredge-up phases for the star of 5 $M_\odot$. Although the first dredge-up process seems not to go very far into the interior, since it always stays away from the hydrogen burning shell, it is still very effective. The CNO cycle and its subreactions occur not only in the hatched regions in Fig. 2 but also in their neighbourhood. During the first dredge-up the outer convective zone reaches down to $M_r/M = 0.47$. At this layer transformation of $^{12}C$ into $^{14}N$ has already taken place and half of the carbon is already transformed into nitrogen. Therefore, the ratio of the two elements in the outer layers will be changed (Iben, 1964). Simultaneously, $^7Li$ as well as B and Be deplete, since they are destroyed at the hot bottom of the convective envelope (Boesgaard, 1977).

The second dredge-up phase occurs after the exhaustion of helium in the central region. Then, the outer convective zone eats even into the shell which contains pure helium (despite a small amount of heavier elements being already present in the interstellar material out of which the star has formed). Consequently the $^{14}N$ enrichment becomes even more pronounced and in addition helium will be enriched in the envelope although this is not too drastic an effect.

It was a great surprise when Schwarzschild and Härm (1965), Weigert (1966) and Rose (1966) found that there occur thermal runaways in shells of nuclear burning under nondegenerate conditions. After the phenomenon was first encountered it was easy to understand it. It is explained in the first of the three papers mentioned with an analytical approximation. The two other authors following up the thermal runaway showed that the phenomenon repeats cyclically. In intervals of some thousand years the luminosity of the helium shell temporarily increases by several powers of 10. The phenomenon is called *thermal pulses*, sometimes also *shell flashes*. The phenomenon is shown in Figures 3 and 4 which come from Weigert's classical paper. In Fig. 3 the convective shell which during each pulse develops above the helium burning core is just indicated by spikes, the first and the sixth pulse are plotted with a better time resolution in Fig. 4. From the first to the sixth pulse the phenomenon gets more and more violent. After the discovery of the pulses there was hope that after a few more pulses the convective shell and the convective envelope would merge bringing photons into a region where helium burning takes place and causing nuclear reactions which provide the local stellar material with neutrons and enable a build up of the s-process elements. Now, however, we know that this unification of the two convective regions does not take place. Instead a third dredge-up phase occurs.

With more computing time and with refined numerical methods it was possible to show that after the growing of the subsequent pulses a "steady state" occurs in which the pulse amplitude remains practically constant from pulse to pulse and the differences in subsequent pulses are only due to the fact that between two pulses the two shells have moved outwards together in mass a little bit. The fully developed pulse is described in Fig. 5 which is taken from Paczyński (1977). There

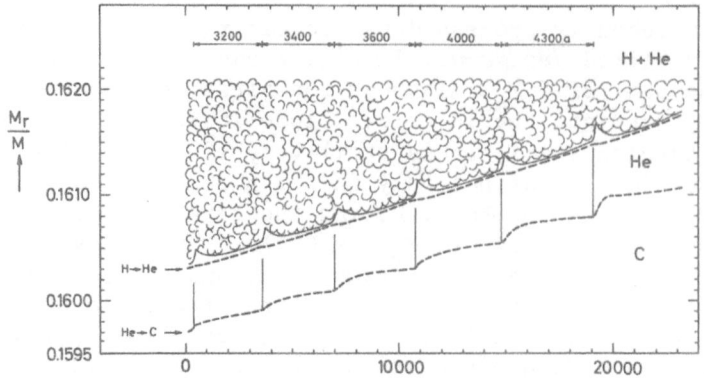

*Figure 3.* Weigert's 6 thermal pulses for a star of 5 M$_\odot$.
This evolutionary phase follows after that depicted on the
right-hand side of Figure 2. The outgoing helium burning
shell has reignited the hydrogen burning shell which was
extinguished for some time and both shells now move outwards parallel in mass scale. The thermal pulses are
sharp increases in the energy production of the helium
burning shell for a short time during which convective
shells develop which, due to the insufficient resolution
in time in the graph, are vertical spikes. The abscissa
is given in years, starting at an arbitrary zero point.
The first and the last thermal pulse with better time
resolution are given in Figure 4.

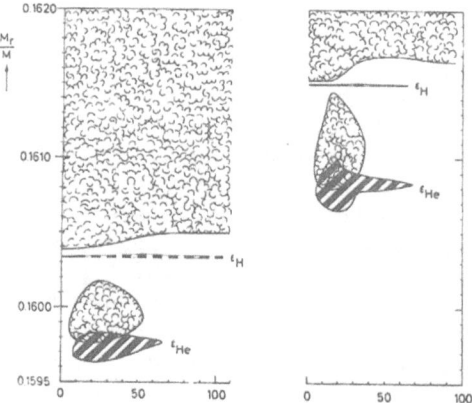

*Figure 4.* The first and the sixth pulse of Figure 3
with better time resolution. The abscissae give time in
years starting at arbitrary zero points. The convective
envelope and the convective shells are indicated by cloudy
regions. Hatched regions correspond to a nuclear energy
generation greater than 10 c.g.s. units.

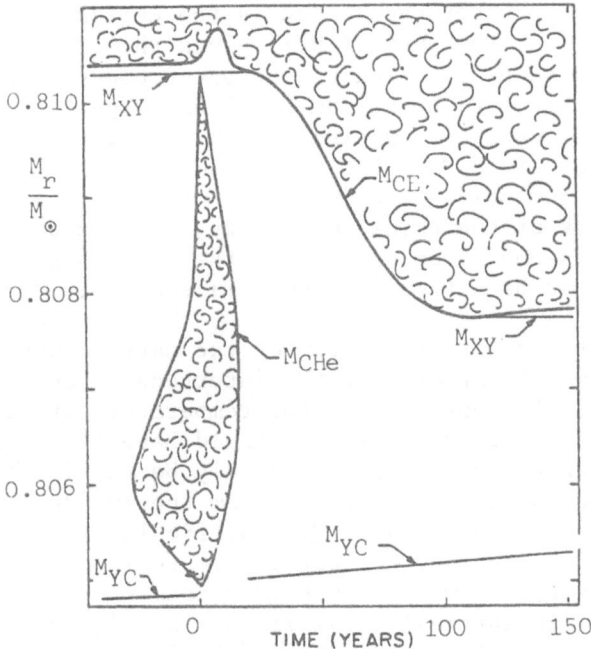

*Figure 5.* Paczyński's fully developed thermal pulse.
Cloudy regions correspond to the outer convective zone
and to the convective shell. $M_{CE}$ gives the base of the
convective envelope. $M_{XY}$ the transition between hydrogen
rich outer part and helium shell. $M_{YC}$ indicates the transition between helium envelope and carbon core while $M_{CHe}$
indicates the boundaries of the helium convective shell.
The total mass of the model is 8 $M_\odot$. The dredge-up phase
after the pulse is clearly indicated. The bottom of the
outer convective zone after the disappearance of the convective shell reaches down into a layer in which carbon
formed during the pulse has been brought up by the convective shell.

is one quantitative difference between Weigert's pulses and Paczyński's,
and if one looks carefully into Fig. 3 one can see that there is also a
tendency in this direction in Weigert's pulses. During the peak of the
pulse, a convective shell appears while the convective envelope is re-
treating in the mass scale. But after the peak of the pulse and after
the convective shell has disappeared the outer convective region pene-
trates inward into a region where the convective shell had released
some of the ashes of helium burning. This material is now dredged-up
by the outer convective zone. Here only half of the expected mixing
takes place: no protons are brought into the helium burning region but
matter from the helium burning region is brought to the surface. This
dredge-up process had first been seen by Gingold (1975) for low mass
stars which also show the phenomenon as we shall see later. The main

effect of the dredge-up after each pulse is the enrichment of the surface composition by $^{12}$C. Paczyński (1977) gives the following explanation for the post-pulse penetration of the outer convective zone. During
the pulse the luminosity of the helium shell has a peak. Consequently, a
heat wave propagates outwards, causing a peak in the total luminosity of
the star. If now (for given $M_{core}$) M is sufficiently high, then the
depth of the outer convective zone increases with luminosity, and therefore, during the peak of the total luminosity, the convective zone eats
inwards. The effect seems to occur only if the core mass is sufficiently
high, no dredge-up has been found for core masses below 0.6 $M_\odot$.

But if one wants to fix the bottom of the outer convetive zone one
encounters a problem. The opacity at the interface is determined by
electron scattering, and consequently the temperature gradient will be
discontinuous at the interface where it should have its adiabatic value.
Therefore the Schwarzschild criterion can only be fulfilled either coming from the outside to the interface *or* from the inside, but not, as
should be the case, from both sides. This difficulty is well-known from
the problems of semiconvection. A intermediate region will build up in
which the Ledoux criterion (which takes into account the gradient in
molecular rate) will be marginally fulfilled. This is not enough. While
such a region is being built, helium has to be lifted into outer layers
and up to the surface, whereas hydrogen is pushed deeper inwards. This
needs energy, and so during the process the luminosity, which is an
important parameter in the stability criteria, and which determines the
depth of the outer convective zone will be diminished, since part of the
energy is used for mixing.Although Paczyński guesses that these effects
reduce the dredge-up considerably, nevertheless a certain amount of
dredge-up is to be expected.

At the moment, it seems that the thermal pulses give the only
reasonable explanation for the formation of s-process elements. Fujimoto et al. (1976); Sugimoto and Nomoto (1975), and Iben (1975a,b) suggested that the process $^{22}$Ne$(\alpha,n)^{25}$Mg can take place at the peak temperature of the flash, which is about $3.1 \times 10^8$ $^\circ$K. At this temperature the
lifetime of $^{22}$Ne is of the order of 10 years and this process provides
the neutrons necessary to build up the series of s-process elements by
neutron capture of $^{56}$Fe and the subsequent nuclei. The dredge-up process described above will bring the newly formed elements to the surface.

As mentioned already, the post-pulse dredge-up phases will bring
carbon to the surface. This might be an explanation for the occurrence of
carbon stars, but it seems that quantitatively there are difficulties,
as has recently been pointed out by Iben (1981). Since the post-pulse
dredge-up occurs only for a stars with core masses sufficiently large,
calculations indicate that there should be no enrichment in carbon
for stars below the bolometic magnitude $-5^m$. But fainter carbon stars
are observed in the Magellanic clouds. The carbon enrichment seems not
to be present in stars brighter than $-6^m$, which cannot be explained by
theory unless one assumes that the brighter stars with carbon enriched
outer layers hide behind an opaque envelope.

Recently the question came up of whether the pulse phenomenon does really occur in stars. Prialnik et al. (1981) suggested that diffusion might become important in the helium burning shell of stars on the asymptotic branch. Diffusion is normally neglected in stellar evolution theory (although it might be of some importance in regions near the surface for stars which have no convection and no meridional circulation there). But in the stage during which the thermal pulses occur the transition zone between the helium shell and the hydrogen rich envelope contains very little mass and simple estimates indicate that diffusion will become important there, preventing the transition zone to become too narrow. It is known already from the paper by Schwarzschild and Härm (1965) that a necessary condition for the occurrence of the pulses is that the shell be geometrically thin. Prialnik et al. (1981) give estimates which indicate that, because of diffusion, the shell might never become thin enough to undergo a thermal runaway. I am sure that people soon will repeat their thermal pulse calculations including diffusion and learn whether or not the pulses disappear. It would be a pity if they vanish, a nice theory for the formation of s-process elements would disappear with them and the occurrence of carbon in the surface layers of carbon stars would need a new explanation.

I will stop here the section dealing with stars of intermediate mass.


## II. THE EVOLUTION OF LOW MASS STARS

Low mass stars, which, on the main sequence, have no appreciable central convective cores, develop degenerate helium cores and move slowly to the red giant branch, as was first pointed out by Hoyle and Schwarzschild (1955). More recent calculations are given in Fig. 6. The events in the interior are shown in Fig. 7. One can clearly see the first dredge-up phase which corresponds to the first dredge-up for stars of intermediate mass. Helium is mixed to the surface but there is no change in the $^{12}C$ and $^{14}N$ abundances as for more massive stars. In the HR-diagram the star moves up the red giant branch (see Fig. 9) until finally helium is ignited. Because of neutrino cooling the maximum temperature is not in the centre but in a sphere which in Thomas' model was at $M_r/M$ = 0.3. However, Thomas at that time used neutrino rates which have since been improved. (Another computation he carried out later with better neutrino rates showed that a star of 1.3 $M_\odot$ ignites helium at $M_r/M$ = 0.1, (Thomas 1970, see also Demarque, Mengel, 1971)). Figure 8 describes the events at and after the onset of helium burning.

After the ignition of helium in the degenerate region a thermal runaway starts and a convective shell is formed. Due to degeneracy the energy released in the helium burning shell is used to a large extent to increase the temperature and makes the phenomenon even more violent – until the temperature has increased so much that the degeneracy is reduced and the relevant shell can expand to enable nuclear burning to

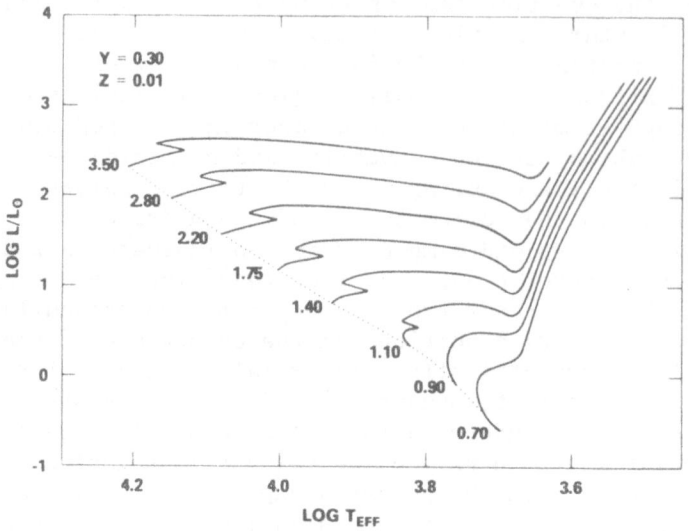

*Figure 6.* Evolutionary tracks for low mass stars according to Sweigart, 1978.

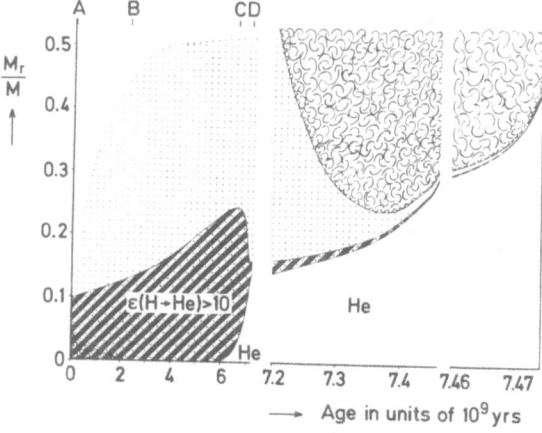

*Figure 7.* The changes in the interior of a star of 1.3 M$_\odot$ after exhaustion of hydrogen in the centre (after Thomas, 1967). The symbols have the same meaning as in Fig. 2. But hatched regions indicate the areas in which the nuclear energy generation exceeds 10 c.g.s. units. The first dredge-up occurs when the convective envelope reaches down below $M_r/M \approx 0.25$.

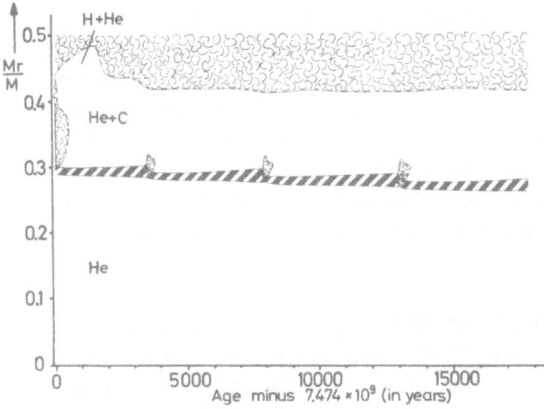

*Figure 8.* The events in the interior after Thomas'
off-centre helium flash (Thomas, 1967). The hatched
regions indicate areas in which the nuclear energy
generation exceeds $10^3$ c.g.s units.

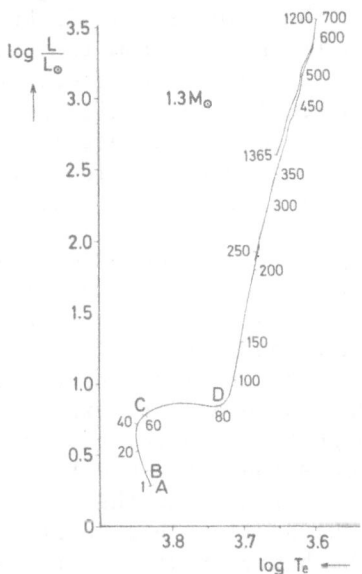

*Figure 9.* Thomas' evolutionary track (Thomas, 1967).
The letters A - D correspond to the phases indicated
in Fig. 7. The numbers correspond to model numbers.
The short phase during which the luminosity decreases
around model number 250 corresponds to the first dredge-
up. After the flash the luminosity decreases (number
1200 - 1365). There is no indication that the star tries
to settle down on the horizontal branch.

take place in a secularly stable way. The convective shell does not merge with the outer hydrogen-rich envelope, but a post-flash dredge-up phase occurs in which carbon is mixed to the surface. Thomas expects an enrichment of $X_C$ = 0.008 whereas, more recently, Pacyński and Tremaine (1977), by varying the depth of the layer in which the flash occurs, came to the conclusion that $X_C$ could reach 0.012. This could explain the carbon enrichment in stars in globular clusters. The luminosities of these stars reach down to the horizontal branch luminosity and they therefore cannot be explained by dredge-up caused by thermal pulses on the asymptotic branch.

After the flash nuclear burning takes place in a non-degenerate region more or less quietly interrupted by some smaller thermal pulses, as can be seen in Fig. 8. In the star therefore, a carbon-rich shell is formed, surrounding a helium core. There is material of higher mean molecular weight above a region of lower molecular weight and, consequently, a secular instability is expected, causing thermohaline mixing. It has been estimated by Kippenhahn et al. (1979) that this mixing is ineffective and will not change the chemical composition in a time shorter than the nuclear timescale.

In the HR-diagram the star moves down towards the horizontal branch luminosity but it always stays close to the ascending branch and no horizontal branch can be formed with these models. It is known that only if an appreciable amount of mass is taken away from the envelope, can agreement with the observed horizontal branch be obtained. This can be seen in Fig. 10. The models on the dashed line have all the same core mass, but they have undergone a different amount of mass loss. In a cluster with an age of $12.5 \times 10^9$ years, but without mass loss, there would be no stars to the left of the point indicated by an arrow. The more mass loss the star has undergone, the more its position is shifted towards the blue. The horizontal branch is direct evidence of mass loss! The branch crosses the instability strip where one has the RR Lyrae stars. In the post-horizontal branch evolution the star moves to the right and to the left and later again to the right. The instability strip is crossed in several directions. Van Albada and Baker (1972) have shown that these different directions of crossing can be used to explain the Oosterhoff groups of RR Lyrae stars. If in post-horizontal branch evolution the strip is crossed again, one expects there the W Vir stars of shorter periods (Kraft, 1972). When the stars move to their asymptotic branch nuclear burning takes place in two shells. This constitutes an ideal situation for thermal pulses! The behavior of the stars in the HR-diagram at this stage of evolution is not very clear. It depends on several different parameters. The stars during the thermal pulses may leave their ascending branch and move to the left for a short time (Schwarzschild, Härm, 1970; Schwarzschild, 1970). If they then on these loops cross the instability strip they start to pulsate and this might be the state in which they are seen as the W Vir stars of longer period (Kraft, 1972).

During the thermal pulses dredge-up occurs and one could ask whether observable changes in chemical composition occur. Several attempts have

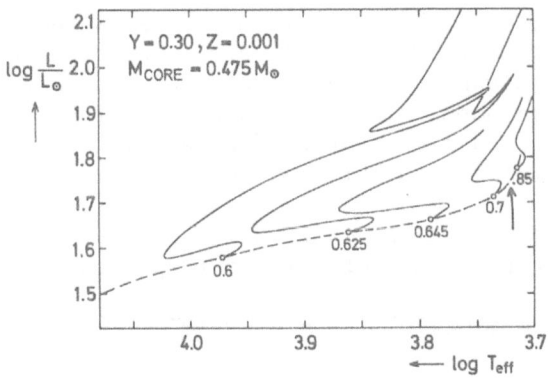

*Figure 10.* The zero age horizontal branch (broken line) and the post-horizontal branch evolution for different total masses (solid lines). The numbers at the horizontal branch give the stellar mass in solar units. The core mass is the same for all models. The models on the horizontal branch therefore can be considered as having their origin in stars of the same original mass but having undergone mass loss of different strength in their evolved stage. Without mass loss stars of globular clusters in the observed age-range would populate only the very right-hand side of the horizontal branch (after Iben, 1972).

been made but recently Kraft (1981) together with his co-workers have put forward some rather new ides. They investigated CH and NH bands in 139 giants in globular clusters and in field stars. It turned out that the C and N abundances are not in agreement with theory. Stars in two of the clusters have different abundances than those in the other two clusters and in the field stars. The authors come to the conclusion that in globular clusters carbon enrichment might be caused by accretion of material which has been expelled in the early phase by the supernova explosions of massive stars. Carbon enriched material blown off the exploding stars normally escapes, but in denser clusters gravity might hold back some of it which could then pollute the stars in the cluster.

Wood and Carro (1981) made attempts to relate period changes of Mira stars like in A Aql, R Hya, W Dra to changes in the stellar interior during thermal pulses, but the situation seems not to be so clear.

But is our picture of the post helium flash evolution correct? Did we compute the helium flash properly? There is a Ph.D. thesis at Oxford University by Wickett (1977). Unfortunately only the results are published. The author claims that, during the flash, convection is insufficient to carry away the energy and that the star explodes. As far as I know, nobody in stellar evolution theory has tried to clarify this point until now. Recently in another Ph.D. thesis by P.W. Cole (under

the supervision of R.G. Deupree (Boston)) the problem was taken up
again. (Cole, Deupree, 1980, 1981). It is the first attempt to deal with
convection in stars with a twodimensional code. Unfortunately the code
they used had only two meshpoints in $\hat{\lambda}$-direction. The convection they
can compute is something more akin to circulation. With the onset of
the flash the energy was carried away by "nonradial" convection and
inertia forces became important. This has to be clarified.

I have tried to make some estimates using Thomas' flash calculation (Thomas, 1970). At the shell of maximum temperature he finds

$$L_r = 4 \times 10^{44}, \ P = 4.02 \times 10^{21}, \ \varrho = 2.19 \times 10^5, \ r = 6 \times 10^8;$$

(all quantities in c.g.s. units). We now use two equations of mixing
length theory:

$$F_k = c_p \varrho \, Tv (\nabla - \nabla') \, \frac{\ell}{2H_p} \quad , \quad v^2 = \frac{g\ell^2}{8H_p} (\nabla - \nabla')$$

(see for instance Hofmeister et al., 1964a) where $F_k$ is the convective
energy flux, $\ell$ the mixing length which we put equal to $H_p$, $H_p = P/g\varrho$
is the pressure scale height, g the gravitational acceleration, $\nabla =$
$= dln \, T/dln \, P$, $\nabla' \approx \nabla_{ad}$ describes the temperature variation in the
rising or falling elements and $\delta = (\, ln\varrho/ \, lnT)_P$. From the two equations
we obtain

$$\nabla - \nabla' = \frac{32}{3} \frac{\psi}{\alpha^2} \frac{v^2}{P}$$

where we have made use of the approximate relation $\delta \approx 3/4\psi$ where $\psi$ is
the degeneracy parameter which has the value $\approx 3$ in the region of interest. With the numerical values given above and with $\alpha = 1.5$ we obtain
$\nabla - \nabla_{ad} \approx \nabla - \nabla' = 0.0116$. This means that the temperature gradient is only
slightly superadiabatic. It is not clear from this estimate whether this
effect is important or not. A slightly superadiabatic gradient means a
higher temperature which would increase the strength of the flash,
raising $\nabla$ even higher. This, in principle, could enhance the flash in
such a way that the inertia terms indeed become important. Up to now
one generally computed the helium flash by assuming that whenever convection occurs, the temperature gradient will be the adiabatic one,
since convection can easily carry away the energy generated there. This
is not necessarily always true.

In the last decade with the improvement of computational facilities
and with the development of better Henyey codes it has become rather
easy to compute stellar evolution by varying uncertain parameters and
trying to investigate their influence. I am not sure whether this was
always very effective. The new picture of overshooting and the consequences as they were investigated in Wasserman's thesis, the disturbing
results on the effects of diffusion in thin shells as obtained by the
Israeli group and the results of the two Ph.D. thesis' in Oxford and
Boston questioning the effectivity of convection in the helium flash

might cause *qualitative* changes in our picture. A large amount of stellar evolution models published and used to explain observations might turn out to need revision. Maybe we stellar evolution people should look into these problems ourself and not leave all the difficult problems to our graduate students.

I thank Dr. H.-C. Thomas for helpful discussions.

## REFERENCES

van Albada T.S., Baker N., 1972, The Evolution of Pop. II Stars, Philip A.G. Davis (ed.), p. 193.

Becker S.A., Iben I., 1980, Astrophys. J. 237, 111.

Boesgaard A., 1977, Publ. Astron. Soc. Pacific 88, 353.

Cole P.W., Deupree R.G., 1980, Astrophys. J. 239, 284.

Cole P.W., Deupree R.G., 1981, Astrophys. J. 247, 607.

Demarque P., Mengel J.G., 1971, Astrophys. J. 164, 317.

Fujimoto M.Y., Nomoto K., Sugimoto D., 1976, Publ. Astr. Soc. Japan 28, 89.

Gingold R.A., 1974, Astrophys. J. 193, 177.

Hofmeister, E., 1967, Z. f. Astrophys. 65, 194.

Hofmeister E., Kippenhahn R., Weigert A., 1964a, Z.f.Astrophys. 59, 215.
                                          1964b, Z.f.Astrophys. 59, 242.
                                          1964c, Z.f.Astrophys. 60, 57.

Hoyle F., Schwarzschild M., 1955, Astrophys. J. Suppl. 2, 1.

Iben I., 1964, Astrophys. J. 140, 1631.

Iben I. Jr., 1967a, Ann. Rev. Astr. Astrophys. 5, 571.

Iben I., 1974, Ann. Rev. Astr. Astrophys. 12, 215.

Iben I., 1975a, Astrophys. J. 196, 525.

Iben I., 1975b, Astrophys. J. 196, 549.

Iben I., 1976, Astrophys. J. 208, 165.

Iben I., 1972, The Evolution of Pop. II Stars, Philip A.G. Davis (ed.).

Iben I., 1981, Astrophys. J. 246, 278.

Kippenhahn R., Ruschenplatt G., Thomas H.-C., 1980, Astron. Astrophys. 91, 175.

Kippenhahn R., Thomas H.-C., Weigert A., 1965, Z.f.Astrophys. 61, 241.

Kraft R.P., 1972, Dudley Obs. Reports No. 4, Philip A.G. Davis (ed.).

Kraft R.P., 1981, personal communication.

Lauterborn D., Refsdal S., Weigert A., 1971, Astron. Astrophys. 10, 97.

Maeder A., 1975, Astron. Astrophys. 40, 303.

Matraka B., Wassermann C., Weigert A., 1981, to appear in Astronomy & Astrophysics.

Paczyński B.,1977, Astroph. J. 214, 812.

Paczyński B.,Tremaine S.D., 1977, Astrophys. J. 216, 57.

Prialnik D., Shaviv G., Koretz A., 1981, Astrophys. J. 247, 225.

Rose W.K., 1966, Astrophys. J. 146, 838.

Schwarzschild M., Härm R., 1962, Astrophys. J. 136, 158.

Schwarzschild M., Härm R., 1965, Astrophys. J. 142, 855.

Schwarzschild M., 1970, Quart. Journ. Roy. Astron. Soc. 11, 12.

Schwarzschild M., Härm R., 1970, Astrophys. J. 160, 341.

Sengbusch K.v., 1967, Ph.D. thesis, Göttingen University.

Sugimoto D., Nomoto K., 1975, Publ. Astr. Soc. Japan 27, 197.

Shaviv G., Salpeter E., 1973, Astrophys. J. 184, 191.

Sweigert A.V., 1978, The HR-Diagramm, IAU Symp. No. 80 (1977),
    A.G.Davis Philip and D.S. Hayes (eds.), Reidel, p. 333.

Thomas H.-C., 1967, Z. f. Astrophys. 67, 420.

Thomas H.-C., 1970, Astrophys. Sp. Science 6, 400.

Tomkin J., Lambert D.L., Luck R.E., 1975, Astrophys. J. 199, 436.

Wassermann C., 1979, Ph. D. Thesis, Hamburg University.

Weigert A., 1975, Proc. 19th. Intern. Coll. Liège (1974), 355.

Wickett A.J., 1977, Problems of Stellar Convection, ed. Spiegel E.A.
    and Zahn J.P., in: Lecture Notes in Physics 71.

Wood P.R., Zarro D.M., 1981, Astrophys. J. 247, 247.

# DISK INSTABILITIES IN HUBBLE-SANDAGE VARIABLES?

F. Meyer and E. Meyer-Hofmeister
Max-Planck-Institut für Physik und Astrophysik
Institut für Astrophysik
Karl-Schwarzschildstr. 1,
8046 Garching bei München

The brightest individual objects in extragalactic nebulae are the Hubble-Sandage variables. They were first investigated by Hubble and Sandage (1953). Observations by Tamman and Sandage (1968) and Rosino and Bianchini (1973) followed. The main characteristics are high luminosity $(L/L_\odot \approx 10^5)$, blue color indices, F type spectra and irregular variability.

Bath (1979) has suggested that the Hubble-Sandage variables contain an accreting main-sequence star with a Roche Lobe filling companion in a wide binary system. Based on this model we derive theoretical color indices for disks and determine the mass in the disk for different mass accretion rates. Further we discuss an instability of the disk which could explain the change in the color index observed for Var A in M33.

a) <u>COLOR INDICES</u>

The color index of a disk is determined by radiation from regions of different effective temperature. We derived the (absolute) visual luminosity for each ring of thickness $\Delta s$ in the disk at a distance $s$ from the centre

$$L_V(s) = 4\pi s \ F(s)\Delta s \ 10^{0.4 \ B.C.}$$

where $F(s)$ is the energy flux density

$$F(s) = \frac{3}{8\pi} \ \frac{GM_{m.s.}\dot{M}}{s^3} \ ,$$

$M_{m.s.}$ mass of main-sequence star, $\dot{M}$ mass accretion rate, B.C. bolometric corrections (Allen, 1973 p. 205). From the corresponding visual magnitude $M_V$ and the color index (Allen, 1973, p.206; because of the low surface gravity the data for supergiants were used) the blue magnitude $M_B$ can be derived for each ring.

$$M_B = M_V + (B - V)$$

and an energy flux $L_B$ in the blue spectral region can be defined

*Z. Kopal and J. Rahe (eds.), Binary and Multiple Stars as Tracers of Stellar Evolution, 19–22.*
*Copyright © 1982 by D. Reidel Publishing Company.*

$$\log L_B = \log L_V - 0.4(B - V).$$

The difference of the energy fluxes gives the theoretical quantity for a comparison with the color index of the whole disk

$$(B - V)_d = 2.5(\log \sum_s L_V(s) - \log \sum_s L_B(s))$$

As an example we give in Fig. 1 the bolometric and the visual luminosity from disk rings of thickness $\Delta s$ ($\Delta \log s = 0.25$, $M_{m.s.} = 15\ M_\odot$). To study the effect we have chosen a lower and a high mass accretion rate (already above the critical rate for the innermost disk region).

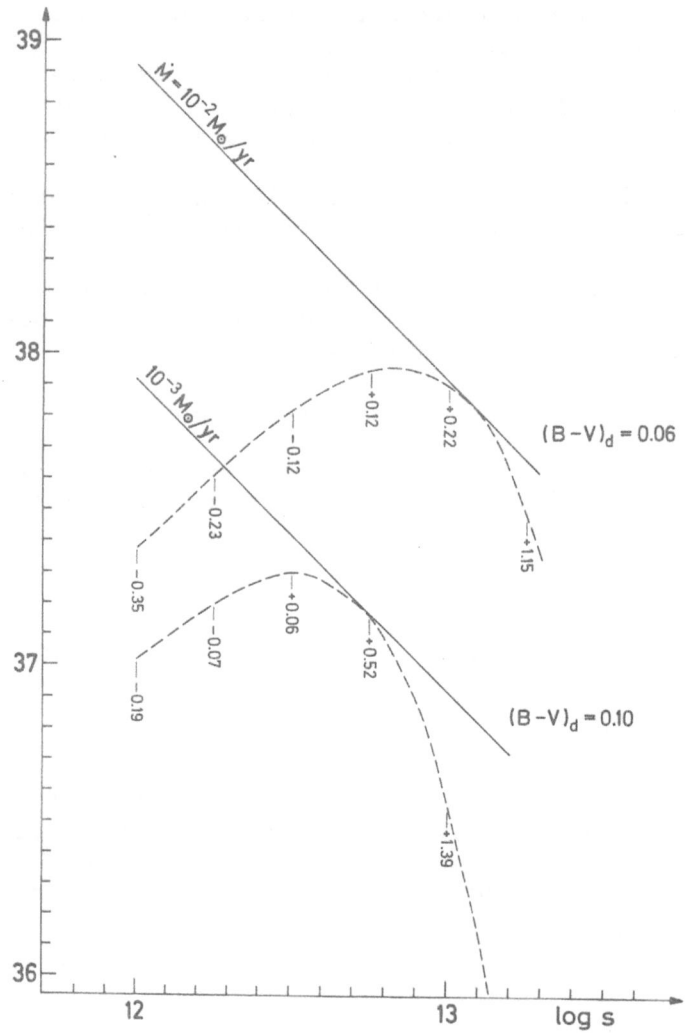

Fig. 1: Luminosity from disk regions of thickness $\Delta s$,
———— bolometric luminosity $\log L(s) = \log 4\pi sF(s)\Delta s$,
- - - - - visual luminosity $\log L_V(s)$ (see text). Color indices are added.

The total visual luminosity from the whole disk is $10^{38.5}$ ($\dot{M} = 10^{-2} M_\odot/yr$) and $10^{37.8}$ ($\dot{M} = 10^{-3} M_\odot/yr$). The color indices $(B-V)_d$ for the whole disk show very little dependence on M. We therefore conclude that the variation of color indices is related to the size of the disk; the smaller the disk the bluer the color index.

The color index 0.3 of Var A in M33 cannot be explained from our results. It could possibly come about by a contribution of a bright red giant companion (for example an 18 - 20 $M_\odot$ star) or by reddening from a dust cloud around the system.

b) <u>MASS CONTAINED IN THE DISK</u>

The vertical structure of an accretion disk around a 15 $M_\odot$ main-sequence star was computed with a code ($\alpha$-model, non-self-gravitating) which includes radiative and convective energy transport (Meyer and Meyer-Hofmeister, 1981). The computations show, as given in Fig. 2, that the mass in the disk rings, and also the total mass in the disk, do not vary essentially with $\dot{M}$, but depend very much on the size of the disk. Then non-monotonic increase of mass with the increase of M is due to a changeover from convective to radiative structure.

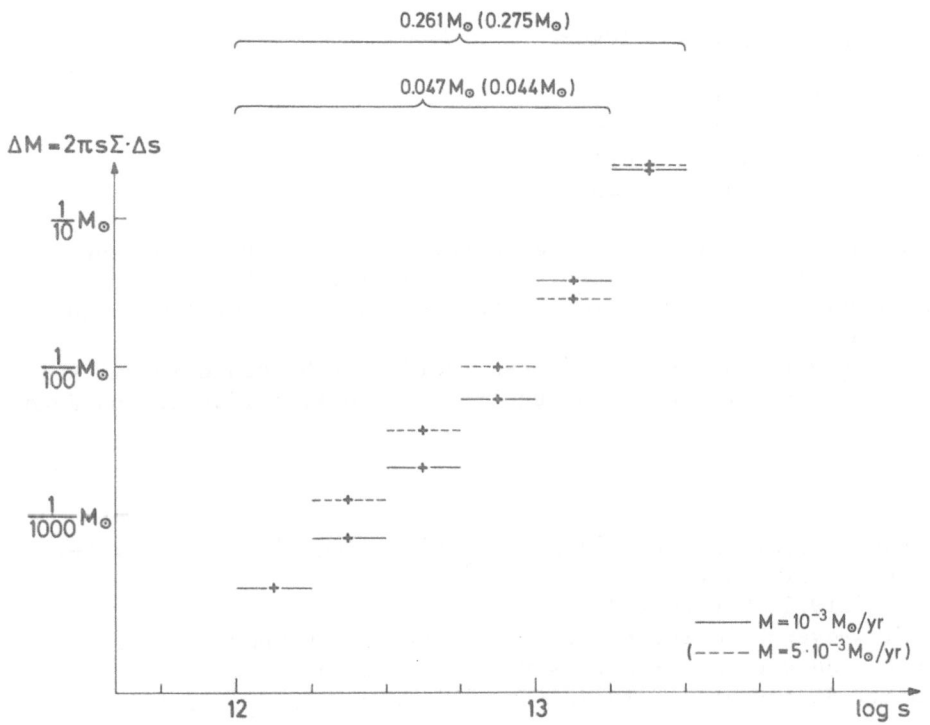

Fig. 2: Mass in regions $\Delta s$ and total mass in the disk for disk radius log r = 13.25 and 13.50 (values in paranthesis correspond to higher rate M). $\Sigma$ surface density.

## c) INSTABILITY OF THE DISK

In the innermost disk region the radiation pressure becomes important and it is possible that (maybe due to a fluctuation in $\dot{M}$) the mass, originally in the disk, forms a sphere around the main sequence star. From the disk structure computations we know the distribution of mass and entropy in the disk. We find that mass with lower entropy is then above mass with higher entropy, and there should be convective mixing in this innermost sphere. We assume that the restructuring into a sphere and the convective mixing are fast processes and that the total energy is approximately conserved. In this simplified model we will have an isentropic gas sphere around the main sequence star. If we now determine the radius of this sphere we find it to be larger than that of the disk region from which it formed. Thus this process can spread to the next disk region until all mass originally contained in the disk is transformed into an extended envelope. For a given disk radius $s_d$ we give, in the table, the radius $R_{is}$ of the corresponding isentropic gas sphere, and the settling time $\tau (= E/L_{Edd})$ after which the energy E (mostly gravitational) is used up by the radiative losses with about Eddington luminosity $L_{Edd}$. The values given are determined for $M_{m.s.} = 15 M_\odot$ and $\dot{M} = 5 \times 10^{-3} M_\odot/yr$.

TABLE

| $\log s_d$ | $\log R_{is}$ | $\log \tau$ |
|---|---|---|
| 12.75 | 13.34 | 6.99 |
| 13.00 | 13.65 | 7.67 |
| 13.45 | 13.67 | 8.17 |
| 13.50 | 14.12 | 8.90 |

The radius of the envelope is so large, that we might even expect the formation of a common envelope. Additional luminosity generated by friction may prevent the mass in the envelope from settling down to a disk again.

We suggest that such a process could have happened in Var A in M 33 for which the color index changed from 0.3 to 1.51 (Hubble and Sandage, 1953).

## References:

Allen C.W.: 1973, Astrophysical Quantities, The Athlone Press University of London, third edition.
Bath G.T.: 1979, Nature 282, 274.
Meyer F., Meyer-Hofmeister E.: 1981, Astron. Astrophys. in press.
Hubble E., Sandage A.: 1953, Astroph. J. 118, 353.
Rosino L., Bianchini A.: 1973, Astron. Astrophys. 22, 453.

# THE FREQUENCY OF CEPHEID BINARIES

G. Russo
Capodimonte Astronomical Observatory, Naples, Italy

Abstract: We summarize the methods now available to detect the presence of companions to Cepheids, and then we analyze the results of the application of the recent photometric method proposed by Fernie (1980). If applied to the set of VBLUW observations of southern Cepheids carried out by Pel (1976), this method gives a frequency of about $25°/_o$, which is consistent with the more recent spectroscopic determinations (Gieren, 1981). The importance of the IUE satellite to directly observe the U flux emitted by the B-companion is emphasized.

It is known that a 3-9 $M_o$ star will cross the instability strip several times, and that classical cepheids are at their second-third crossing. If the star is a member of a binary system, the critical orbital period of the binary, in order to have the primary to cross the instability strip two times, is of the order of 200 days (Lloyd Evans, 1968). Therefore, it is difficult to resolve them directly, with photometric or spectroscopic observations. Nevertheless, it is important to know their number, in order to have a statistical test of evolutionary calculations for binary systems: in fact Cepheids evolve from B-stars, which, are well known to show a very high incidence in binary systems (Petrie, 1963; Abt and Levy, 1978). For long time the incidence of binaries among cepheids was thought to be no more than $5°/_o$ (Abt, 1959), but this number has increased in the course of the years: $15°/_o$ according to Lloyd Evans (1968); somewhat more according to Madore (1977) and Pel (1978), and up to $35°/_o$ according to Madore and Fernie (1980). Many methods have been proposed to detect the binary cepheids. The first one of wide applicability is the one by Lloyd Evans (1968), based on the search of variable systemic velocities. However, good quality radial velocity curves are very rare, and only in recent years the method is being profitably used (Gieren, 1981). Direct spectroscopic

23

*Z. Kopal and J. Rahe (eds.), Binary and Multiple Stars as Tracers of Stellar Evolution, 23–26.*
*Copyright © 1982 by D. Reidel Publishing Company.*

evidence of a composite spectrum in the visible has been possible only
in a few cases (Lloyd Evans, 1968) while the detection of the UV emission
of blue companions to cepheids has begun with the IUE satellite (Eichendor
1981; Mariska et al., 1980), but has not yet been widely applied.
Photometric methods, based on the opening of the colour-colour loops
(Madore, 1977) or on the amplitude ratios among various filters (Fernie,
1979), are more suited to measure the frequency of Cepheid binaries, but
cannot be used to ascertain without doubts the binary nature of individual
cepheids.
Recently, Fernie (1980) has proposed a new photometric method, based on
the measure of the shift between the light and colour curve (from now on,
we will use only the U-V as colour curve). The basis of the method is the
following: if the cepheid has a B-companion (from evolutionary considerations, these should be the most frequent ones), the phase of minimum light
$\varphi^{min}(V)$, is unaffected, but $\varphi^{min}(U-V)$ will be altered, because the
colour of the system is changed and, in practice, it will be shifted towards earlier phases, with a shift depending on the luminosity of the
B-star (cf.Fernie, 1980). This method has already been applied by Madore
and Fernie (1980) to UBV photometry. This photometric system, in my
opinion, is not well suited for such a study, because of the low quality
of U observations. As a matter of fact, the VBLUW photometry of southern
Cepheids carried out by Pel (1976) is, on the contrary, especially
suitable for this purpose. In order to apply Fernie's method to VBLUW
photometry, one needs, first of all, a calibration of the phase shift

$$\delta \varphi = \varphi^{min}(V) - \varphi^{min}(U-V)$$

in terms of the magnitude difference between the cepheid and the B-compa_
nion. This has been done by Russo et al. (1981), who have also computed
the parameter $\delta \varphi$ for 88 cepheids taken from Pel (1976) and with good
quality U data. This parameter has been computed by fitting Fourier
series to the U and U-V curves. The resulting distribution of $\delta \varphi$     is
shown in Figure 1.
It can be see that the distribution is both skewed and displaced from
zero with a maximum at about $\delta \varphi = 0.04$. If we assume that the non-gaussian
part of the distribution is due to cepheids with possible companions,
than we can estimate the percentage of binary cepheids to be 25 ($\pm$3)$^\circ/_\circ$
of the total. However, it is not simple to judge if an individual
cepheid with high $\delta \varphi$ is in the tail of the gaussian distribution, or if
it belongs to the non-gaussian part, i.e., if it can safely be considered
as having a companion. Moreover, we can in this way detect only cepheids
with B-companions, which are the most frequent ones, but not the only
ones.
Another interesting question arises from the consideration of the

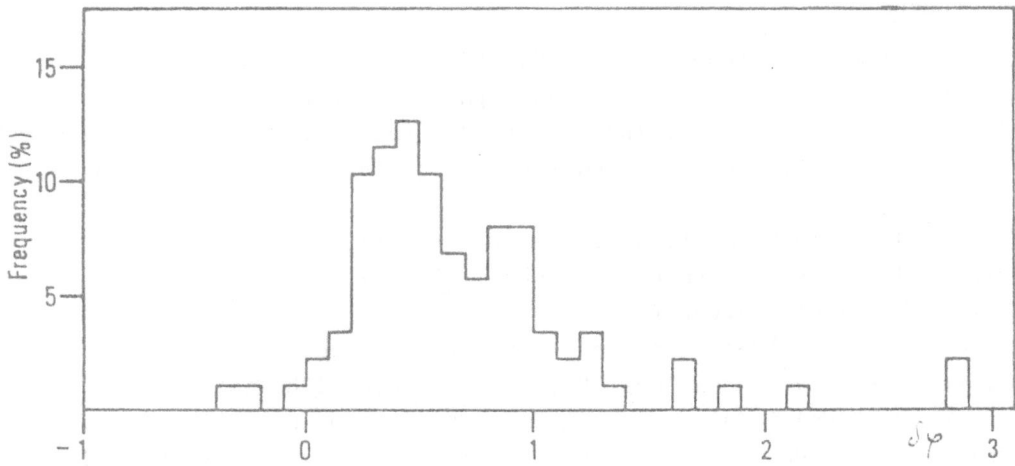

Figure 1: The distribution of $\delta\varphi$ values for 88 cepheids.

separation between the cepheid and its companion. From Kepler's third low, assuming a period of two hundred days, and a total mass of 20 $M_{\odot}$, we obtain a separation of the order of one a. u.. With a radius of $\sim 50$ R$_{\odot}$ for the cepheid and $\sim 3$ R$_{\odot}$ for the main sequence B-companion, the double system, although obviously detached, is a close binary system. This means that one should study the proximity effects on the cepheid due to the presence of the companion, and which should have consequences on the pulsation itself, like the pumping of non-radial oscillations and double mode phenomena.

In conclusion, we can say that:

- the percentage of galactic cepheids which have blue companion is estimated by different authors to be about 25°/$_{o}$, and seems well based;
- the identification of these binary systems is at the beginning, and only a few cases have been ascertained; in this purpose, the IUE satellite (and future UV satellite) can give a substantial contribution;
- the physics of such systems has not yet been explored, and many interesting phenomena will be faced in the near future.

## Acknowledgements

I would like to thank Dr. W. Eichendorf for useful discussion, Dr. W. Gieren for sending a preprint of his paper, and, expecially, Prof. Z. Kopal for very useful suggestions on the prosecution of this work. This research was supported by Consiglio Nazionale delle Ricerche (C.N.R.).

## References

Abt, H.A.: 1959, Astrophys. J. 130, 769

Abt, H.A. and Levy, S.G.: 1978, Astrophys. J. Suppl. 36, 241

Eichendorf, W., Heck, A., Isserstedt, J., Lub, J., Pakull, M., Reipurth, B., and van Genderen, A.M.: 1981, Astron. Astrophys. 93, L5-L6

Fernie, J.D.: 1979, Publ. Astron. Soc. Pac. 91, 67

Fernie, J.D.: 1980, Astron. Astrophys. 87, 227

Gieren, W.: 1981, Astrophys. J., submitted

Lloyd Evans, T.: 1981, Mon. Not. Roy. Astron. Soc. 141, 109

Madore, B.F.: 1977, Mon. Not. Roy. Astron. Soc. 178, 505

Madore, B.F. and Fernie, J.D.: 1980, Pub. Astron. Soc. Pac. 92, 315

Mariska, J.T., Doschek, G.A., and Feldman, U.: 1980, Astrophys. J. Lett. 238, L87- L90

Petrie, R.M.: 1963, in Basic Astronomical Data, p. 64 (Univ. of Chicago Press)

Pel, J.W.: 1976, Astron. Astrophys. Suppl. 24, 413

Pel, J.W.: 1978, Astron. Astrophys. 62, 75

Russo, G. Sollazzo, C., and Coppola, M.: 1981, Astron. Astrophys, in press

# SEARCH FOR PULSATING STARS IN MULTIPLE STELLAR SYSTEMS

E. Antonello[*][x], M. Fracassini[+][x], L.E. Pasinetti[*][+]
and L. Pastori[*]
* Osservatorio Astronomico di Milano - Merate
x Istituto di Astronomia e Geodesia, Milano
+ Dipartimento di Fisica, Università di Milano

The importance of the search for pulsating stars of
the δ Scuti type in multiple stellar systems is emphasized,
in particular for the following objects: the evolutionary
status of these systems, the interconnection between multiplicity and pulsation (Frolov et al., 1980), a verification
of the theory of the turbulent mixing (Vauclair, 1976).
A list of possible pulsating companions in visual and spectroscopic systems with foreseeable period  and amplitude is
given. Some preliminary results of a spectrophotometric
survey begun at the Observatories of Milano - Merate and Bologna - Loiano are also given.

## INTRODUCTION

Catalogues of variable visual binary stars were compiled
by Plaut (1940), Baize (1962), Proust et al. (1981). The
publication of uvbyβ photometry of wide visual binaries
(Oblak, 1978; Oblak and Chareton, 1980) and of some spectroscopic binaries (Hauck and Mermilliod, 1980; Batten et al.,
1978) gives us the opportunity to compile a first list of
possible pulsating (δ Scuti) companions in multiple stellar
systems. The search for pulsating stars in these systems is
important, in particular, for the following subjects:
a. The evolutionary status of the stellar systems, as pointed
   out for some stars such as HD 15165 (BDS 1269, Rucinski,
   1978) and 20 Leo (Fekel and Bopp, 1977).
b. A verification of the theory of the turbulent mixing
   (Vauclair, 1976; Michaud, 1980).
c. The interconnection between multiplicity and pulsation,
   as suggested recently by Frolov et al. (1980).

Surveys of several clusters have shown that
pulsation is a common phenomenon, and it may occur in stars
with different ages (Baglin et al., 1973). But it is not

Z. Kopal and J. Rahe (eds.), Binary and Multiple Stars as Tracers of Stellar Evolution, 27–31.

clear why only about 30% of the stars in the lower part of
the instability strip are variable, and these stars are similar to nonvariable stars (with the exception of Am stars).
The observation of multiple stellar systems such as double
stars (wide visual systems), and particularly those with
both the components in the instability strip, could give some clarification to these problems. Moreover, the survey of
the spectroscopic binaries could give the incidence of variability within this group of stars (see Antonello, 1982) and
verify the suggestion of Frolov et al. (1980).


LIST OF POSSIBLE DELTA SCUTI STARS

    Two lists of possible $\delta$ Scuti stars have been compiled:
one for the wide visual binaries, the other for the spectroscopic binaries in the catalogue of Batten et al. (1978).
These two lists allow to observe two kinds of variable companions coeval with the stellar systems; those without interaction and those with some interaction with other stars of
the system. By means of the uvby$\beta$ photometry of wide visual
binaries (Oblak, 1978; Oblak and Chareton, 1980) and of
spectroscopic binaries (Hauck and Mermilliod, 1980), we have
selected the companions with normal spectral type A, F belonging to the instability strip. Subsequent, by means of the
relations:

$$\log P = -0.33\, M_V + 2.77\,(b-y) - 1.02 \quad (\text{Breger, 1979})$$

$$\log \Delta V = 2.08\, M_V\, \delta c_1 + 0.56\, \log P - 1.09$$

$$(\text{Antonello et al., 1981})$$

we have calculated the expected periods and maximum amplitudes of pulsation (only for visual binaries). These results
are reported in Table 1, which contains 47 companions of
wide visual binaries; 21 spectroscopic binaries are reported
in Table 2 (several visual binaries are also spectroscopic
binaries). The columns contain respectively: (1) and (2) the
star identifications (HD and BD numbers); (3) apparent visual
magnitude; (4) spectral type; (5) and (6) possible period
P and maximum visual amplitude $\Delta V$ of pulsation; (7) separation of the components; (8) remarks.

    The above mentioned subjects show the usefulness of a
joint observational spectrophotometric study for companions
in multiple stellar systems. A cooperation of this kind
between the Observatories of Milano - Merate and Bologna -
Loiano was begun with the following programme stars: HD
43525 (75 Ori A - B), HD 150100 - 150117 (16 - 17 Dra), HD
159541 - 159560 ($\nu 1 - \nu 2$ Dra), HD 173582 - 173607 ($\zeta 1 - \zeta 2$ Lyr)

Table 1. Visual binaries

| HD | BD | $m_v$ | S.T. | P | $\Delta V$ | Sep | R |
|---|---|---|---|---|---|---|---|
| 2358A | +15.00059 | 6.4 | A5 | 0.10 | 0.06 | 96″3 | |
| 2885B | -63.00050 | 4.5 | A2V | 0.03 | 0.01 | 37.7 | |
| 3369B | +32.00101 | 8.6 | A6V | 0.03 | 0.01 | 36.1 | |
| 6288A | + 0.00174 | 6.1 | F0V | 0.04 | 0.02 | 16.6 | |
| 11973A | +22.00288 | 4.8 | F0IV | | | 37.0 | 1,2,S |
| 15695B | + 0.00415 | 7.7 | A7V | 0.04 | 0.02 | 13.5 | |
| 20313A | -79.00091 | 5.7 | F0II | | | 15.2 | S |
| 22077A | -10.00694 | 7.3 | F2 | 0.23 | 0.12 | 80.4 | |
| 23630B | +23.00541 | 8.3 | F0 | 0.06 | 0.04 | 85.6 | |
| 24398B | +31.00666 | 9.9 | A2 | 0.04 | 0.03 | 12.6 | |
| 27934A | +21.00642 | 4.2 | A7V | 0.06 | 0.04 | 39.5 | 3 |
| 27946B | +21.00643 | 5.3 | A7V | 0.06 | 0.04 | 39.5 | |
| 29172B | -10.00958 | 7.7 | A0 | 0.03 | 0.01 | 12.8 | |
| 31203A | -53.00760 | 5.7 | F0IV | | | 12.4 | S |
| 42955B | +14.01211 | 7.9 | F0 | 0.04 | 0.02 | 69.8 | |
| 257937B | +20.01441 | 7.9 | A1V | 0.03 | 0.02 | 112. | |
| 71663A | - 2.02581 | 6.4 | F0 | | | 18.0 | * |
| 76644A | +48.01707 | 3.1 | A7V | | | 10.7 | * |
| 81029A | + 4.02178 | 7.3 | F0 | 0.07 | 0.05 | 21.2 | |
| 83023A | +15.02077 | 6.3 | A1V | 0.04 | 0.03 | 42.4 | |
| 91312A | +41.02101 | 4.8 | A7IV | | | 24.6 | *,3 |
| 286295C | +14.00797 | 10. | A5 | 0.03 | 0.01 | 89.3 | |
| 118349A | -25.09900 | 5.8 | A7III | 0.09 | 0.05 | 10.1 | |
| 118349B | -25.09900 | 6.7 | A7IV | 0.06 | 0.04 | 10.1 | |
| 120641B | -52.06787 | 7.6 | A3 | 0.03 | 0.01 | 18.3 | |
| 120955B | -31.10729 | 8.5 | F0 | 0.03 | 0.01 | 14.9 | |
| 124620A | -56.06215 | 7.2 | A0 | 0.03 | 0.01 | 30.2 | |
| 138268A | -19.04128 | 6.3 | A5V | | | 11.6 | S |
| 138362A | -47.10092 | 7.1 | F0 | 0.04 | 0.02 | 28.3 | |
| 148638A | -60.06560 | 7.9 | A0 | 0.10 | 0.03 | 27.4 | |
| 151431A | + 2.03175 | 6.1 | A2 | | | 23.2 | S |
| 159480B | + 9.03424 | 7.8 | F0IV | 0.03 | 0.02 | 41.3 | |
| 159876A | -15.04621 | 3.5 | F0IV | | | 24.9 | *,2 |
| 161270A | + 2.03390 | 6.1 | A0V | | | 20.8 | S |
| 165910A | +13.03529 | 6.6 | A2V | 0.29 | 0.01 | 42.7 | |
| 174005A | - 6.04913 | 6.5 | A2 | 0.10 | 0.06 | 37.9 | |
| 175638A | + 4.03916 | 4.6 | A5V | 0.05 | 0.03 | 22.3 | |
| 175639B | + 4.03917 | 5.0 | A5n | 0.06 | 0.04 | 22.3 | |
| 178449A | +32.03326 | 5.0 | A7 | | | 128. | *,1 |
| 231195B | +14.03879 | 8.8 | A5 | 0.12 | 0.06 | 80.8 | |
| 188557A | -52.11589 | 7.6 | F0 | 0.04 | 0.02 | 80.5 | |
| 190849A | + 7.04367 | 7.1 | A1V | 0.07 | 0.04 | 65.1 | |
| 191709B | + 0.03937 | 7.9 | A0 | 0.03 | 0.02 | 55.3 | |
| 192461A | - 3.04825 | 7.0 | F0 | 0.05 | 0.03 | 14.2 | |
| 193281A | -29.16981 | 6.6 | A2III | 0.11 | 0.04 | 27.4 | |
| 193281B | -29.16981 | 7.7 | A2IV | 0.08 | 0.05 | 27.4 | |
| 195093B | -19.05830 | 6.7 | A8V | 0.03 | 0.01 | 22.2 | |

Remarks: 1. constant (Breger, 1969)
         2. constant (Millis, 1967)
         3. variable? .(Millis, 1967)
         S. spectroscopic binary
         *. spectroscopic binary with known orbital
            elements

Table 2. Spectroscopic binaries (*)

| HD | BD | m | S.T. | R |
|---|---|---|---|---|
| 1826 | +28.00049 | 6.9 | A5 | |
| 4058 | +46.00146 | 4.9 | A5V | |
| 12111 | +70.00153 | 4.7 | A4V | |
| 13161 | +34.00381 | 3.0 | A5III | 1 |
| 17094 | + 9.00359 | 4.2 | F0IV | 2 |
| 27176 | +21.00618 | 5.6 | A8 | 2 |
| 31109 | - 5.01068 | 4.4 | A9IV | 2 |
| 37507 | - 7.01142 | 4.8 | A4IV | 3 |
| 82610 | -28.07373 | 6.4 | F0 | |
| 104321 | + 7.02502 | 4.6 | A4V | |
| 104350 | +13.02481 | 8.6 | A7 | |
| 107259 | + 0.02926 | 3.9 | A2V | |
| 139319 | +64.01077 | 7.3 | A5 | |
| 151676 | -15.04395 | 6.1 | A3 | |
| 179950 | -25.13866 | 4.9 | F5 | |
| 187949 | -14.05578 | 6.5 | F | |
| 196362 | +25.04299 | 6.2 | A4III | |
| 199603 | -15.05848 | 6.0 | A3 | |
| 205767 | - 8.05701 | 4.7 | A7V | |
| 209278 | -17.06422 | 7.1 | A2V | |
| 217792 | -35.15630 | 5.1 | F0IV | |

(*) the known $\delta$ Scuti stars (12 systems) are excluded.
Remarks: 1. constant (Breger, 1969)
         2. constant (Millis, 1967)
         3. constant (Jorgensen et al., 1971)

and HD 173648 - 173649 ($\varepsilon 1$ -$\varepsilon 2$ Lyr). The photometric observations of the systems $\zeta 1$-$\zeta 2$ Lyr  and $\nu 1$-$\nu 2$ Dra show probable
variations for the former, and no variations for the latter,
according to Gonzalez et al. (1974). Possible small varia-
tions were found in $\zeta^1$, $\zeta^2$ , $\varepsilon^1$, $\varepsilon^2$ Lyr also by Percy (1978).

Acknowledgements: the Italian Consiglio Nazionale delle Ricerche are thanked for the financial contribution, which has
permitted us to attend the present Colloquium.

REFERENCES

Antonello, E.: 1982, this Colloquium, p. 33.
Antonello, E., Fracassini, M. and Pastori, L.: 1981, Astrophys. Space Sci. 78, p. 435.
Baglin, A., Auvergne, M., Valtier, J.C. and Saez, M.: 1980, in "Variability in Stars and Galaxies", Proc. Fifth Europ. Reg. Meeting, Liége, B.3.1.
Baglin, A., Breger, M., Chevalier, C., Hauck, B., Le Contel, J.M., Sareyan, J.P. and Valtier, J.C.: 1973, Astron. Astrophys. 23, p. 221.
Baize, P.: 1962, J.Observateurs 45, p. 117.
Batten, A.H., Fletcher, J.M. and Mann, P.J.: 1978, Publ. Dominion Astrophys.Obs. 15, p. 121.
Breger, M.: 1969, Astrophys.J.Suppl. 19, p. 79.
Breger, M.: 1979, Publ.Astron.Soc.Pacific 91, p. 5.
Fekel, F.C. and Bopp, B.W.: 1977, Publ.Astron.Soc.Pacific 89, p. 216.
Frolov, M.S., Pastukhova, E.N., Mironov, A.V. and Moshkalev, V.G.: 1980, Inf.Bull.Var.Stars, No. 1894.
Gonzalez, S.F., Gomez, T. and Mendoza, E.E.: 1974, Rev.Mex. Astron.Astrofys. 1, p. 119.
Hauck, B. and Mermilliod, M.: 1980, Astron.Astrophys.Suppl. 40, p. 1.
Jorgensen, H.E., Johansen, K.T. and Olsen, E.H.: 1971, Astron. Astrophys. 12, p. 223.
Michaud, G.: 1980, Astron.J. 85, p. 589.
Millis, R.L.: 1967, unpublished thesis.
Oblak, E.: 1978, Astron.Astrophys.Suppl. 34, p. 453.
Oblak, E. and Chareton, M.: 1980, Astron.Astrophys.Suppl. 41, p. 255.
Percy, J.R.: 1978, Publ.Astron.Soc.Pacific 90, p.703.
Plaut, L.: 1940, Bull.Astron. 234, p. 49.
Proust, D., Ochsenbein, F. and Petterson, B.R.: 1981, Astron. Astrophys.Suppl. 44, p. 179.
Rucinski, S.M.: 1978, Acta Astron. 28, p. 545.
Vauclair, G.: 1976, Astron.Astrophys. 50, p. 435.

# AN AD HOC HYPOTHESIS ON THE PULSATION OF DELTA SCUTI STARS

E. Antonello
Osservatorio Astronomico di Milano-Merate, Italy

Taking into account the possible influence of external factors such as the orbital motion in binary systems and the rotational velocity on the pulsation of $\delta$ Scuti stars, it may be possible to explain some properties of these variables.

## INTRODUCTION

It is well known that the incidence of variability within the lower part of the instability strip is of about 30%, and that the luminosity, metallicity and rotational velocity are only some of the factors responsible for the variability (Breger, 1979; see also Kurtz, 1978). Hence it is not clear why, during their evolution, similar stars can pulsate or not. There is a well known P-L-C relation for $\delta$ Scuti stars (Breger, 1979), and also a clear relation between the amplitude of pulsation, the period and luminosity of the low amplitude $\delta$ Scuti stars (Antonello et al., 1981); hence it would be possible to predict the period and the amplitude of any star (variable and non variable) in the lower part of the instability strip. These facts are difficult to explain with internal differences between variable and non variable stars; for classical Am stars that is possible (see e.g. Baglin et al., 1980), but in this case there is also an evident effect on the surface of these stars. Therefore we believe there should be different external conditions.

## HYPOTHESIS

Fitch has shown that the tidal interaction in a close binary could explain some properties (periods) observed in some $\delta$ Scuti stars (e.g. Fitch, 1967, 1980). More recently, Fitch (1980) has suggested that nonradial mode excitation (in stars above the main sequence) will usually occur only

33

*Z. Kopal and J. Rahe (eds.), Binary and Multiple Stars as Tracers of Stellar Evolution, 33–35.*
*Copyright © 1982 by D. Reidel Publishing Company.*

in those stars with significant departures from spherical symmetry; these departures are due either to high rotation or to close companions. Smith (1980) has suggested that non-radial pulsation excites radial pulsation in $\beta$ Cephei stars. Nonradial modes may be common in $\delta$ Scuti stars (e.g. Dziem-bowski, 1980), hence, taking into account Smith's suggestion also for $\delta$ Scuti stars, we can make this hypothesis: the stars in the lower part of the instability strip are pulsa-ting if they have a sufficiently high rotational velocity and/or if they are in an appropriate binary system, that is they need some sort of nonradial hard excitation to pulsate. This should be valid at least for the low amplitude variables.

According to Vauclair's theory (1976), the Am phenomenon requires a slowly rotating star and/or a binary system with synchronous orbital and rotational motions. On the contrary, a variable star should require a high rotational velocity and/or an appropriate binary system without synchronization. Since there is a continuous distribution of possible rotatio-nal velocities and elements of binary systems, there will be stars with intermediate properties.


HYPOTHESIS VERIFICATION

The results of various studies on the $\delta$ Scuti stars do not show a clear confirmation or rejection of this hypothesis. The known binary systems (Batten et al., 1978) with normal spectra (and within the instability strip) in the surveys of Millis (1967; excluded Hyades cluster), Breger (1969), Jorgensen et al. (1971) are eight, and only one was found to be variable. Taking into account the variables in the Catalogue of Bright stars (Hoffleit, 1964), there are only 23 stars indicated as spectroscopic binaries among 92 $\delta$ Scuti stars. As regards the rotational velocity, it has some importance, as shown by Breger (1979), but its behaviour in dwarfs is different from that in giants, and the results are contradictory in various clusters.

However, if we consider the stars in the well studied Hyades cluster (Millis, 1967; Breger, 1970; Horan, 1979), we can see that, also excluding the Am stars from the sample, the rotational velocity has some importance for the discri-mination between variable and nonvariable stars in the sense of our hypothesis (Antonello et al., 1981). Moreover, the different incidence of spectroscopic binaries among variable and nonvariable stars is remarkable, as shown in the Table 1 (the Am stars are excluded, and only stars in the instabili-ty strip are taken into account). These facts seem to support our hypothesis at least for dwarfs; hence only the stars which satisfy our hypothesis could pulsate with the period and the amplitude predicted by the relations above mentioned.

|                          | Variables | Nonvariables |
|--------------------------|-----------|--------------|
| Number of stars          | 7         | 11           |
| S.B.'s with known orbit  | 3         | 1            |
| Total number of S.B.'s   | 6         | 3            |

Table 1.  Hyades cluster.

Acknowledgements: the Italian Consiglio Nazionale delle Ricerche are thanked for the financial contribution, which has permitted us to attend the present Colloquium.

REFERENCES

Antonello, E., Fracassini, M. and Pastori, L.:1981, Astrophys. Space Sc. 78, p. 435.
Baglin, A., Auvergne, M., Valtier, J.C. and Saez, M.:1980, in Variability in Stars and Galaxies, Proc. 5th Europ. Meet. in Astron., Liége, B.3.1.
Batten, A.H., Fletcher, J.M. and Mann, P.J.:1978, Publ.Dom. Astr.Obs. 15, p. 121.
Breger, M.:1969, Astrophys.J.Suppl. 19, p. 79.
Breger, M.:1970, Astrophys.J. 162, p. 597.
Breger, M.:1979, Publ.Astron.Soc.Pacific 91, p. 5.
Dziembowski, W.A.:1980, in H.A. Hill and W.A. Dziembowski (eds.), Nonradial and nonlinear Stellar Pulsation, p. 22.
Fitch, W.S.:1967, Astrophys.J. 148, p. 481.
Fitch, W.S.:1980, in H.A. Hill and W.A. Dziembowski (eds.), Nonradial and Nonlinear Stellar Pulsation, p. 7.
Hoffleit, D.:1964, Catalogue of Bright Stars, Yale University Obs., New Haven, Connecticut.
Horan, S.:1979, Astron.J. 84, p. 1770.
Jorgensen, H.E., Johansen, K.T. and Olsen, E.H.:1971, Astron. Astrophys. 12, p. 223.
Kurtz, D.W.: 1978, Astrophys.J. 221, p. 869.
Millis, R.L.:1967, unpublished thesis.
Smith, M.A.:1980, in H.A. Hill and W.A. Dziembowski (eds.), Nonradial and Nonlinear Stellar Pulsation, p. 60.
Vauclair, G.:1976, Astron.Astrophys. 50, p. 435.

# ON THE INTERNAL STRUCTURE OF MAIN-SEQUENCE STARS

Alvaro Giménez and José M. García-Pelayo
Instituto de Astrofísica de Andalucía. Apartado
2144. Granada. Spain.

We present the main results of a study of the observed internal structure constants, $k_2$, for a wide set of eclipsing binaries. From the analysis of the variations in relative positions of the eclipses and the comparison with different theoretical models, we could deduce that the discrepancy, previously reported by several authors between theory and observations, is no longer supported. Moreover, a strong correlation has been found between the evolution of the parameter $k_2$ and the gravity at the surface of the star, g.

Some forty years ago, Z. Kopal (1940) tried to get observational information about the internal density concentrations of the stars by means of a study of apsidal motion in close binaries. Since that time, this kind of approach has represented one of the few tests for the different theoretical stellar structure models. It is well known that changes in the observed orbital parameters (in particular the position of the periastron) are related to dynamical perturbations arising from rotation, tides and even the presence of a third body. As a consequence, since the perturbing potential can be computed and the orbital movements accurately measured for eclipsing systems, some information about the internal structure of the stars can in principle be obtained.

A detailed agreement between observations and theoretical models is obviously required, but even recent studies (Monet, 1980) show an important systematic deviation between both. On the other hand, the low accuracy of available observational data made impossible any further research on the physical reasons for such a situation. In order to diminish the problem inherent in the accuracy of the apsidal motion determinations, we have developed a new procedure

*Z. Kopal and J. Rahe (eds.), Binary and Multiple Stars as Tracers of Stellar Evolution, 37–46.*
*Copyright © 1982 by D. Reidel Publishing Company.*

already explained in a previous paper (Giménez, 1981). En-
hancing the importance of such a rediscussion of the observations from the theoretical point of view, apsidal motion
periods have been claimed as a test for the validity of different types of opacities (Stothers, 1974).

To keep our data base homogeneous, we have only
used the displacements of minima in eclipsing binaries for
the determination of apsidal motions. A catalogue of 113
candidates was obtained from different sources and all of
them were fully investigated. To avoid false or highly biased determinations, several conditions were imposed to the
systems to consider them as definite candidates. These rec-
trictions are related to the photometric behaviour of the
stars as well as their dynamical and evolutionary status
and have been already discussed (Giménez, 1981). The final
result is a set of 55 eclipsing systems (Giménez and Delgado, 1980) for which a list of some 1500 times of minimum
has been compiled. This list represents in fact the primary
source of information for our study.

The analysis of the final catalogue of 55 eclipsing
binaries with detectable apsidal motion, proceeded in two
directions:
    a) Determination of absolute dimensions, and
    b) Obtention of internal structure constants.
The method applied to the analysis of apsidal motions includes the determination of anomalisitic period and orbital
eccentricity combining frequency and time domains. The equa
tion relating orbital parameters with times of minimum is,

$$T = T_o + P_s E + \frac{\theta P}{2\pi} + \frac{P}{\pi} \sum_{n=1}^{\infty} (-\beta)^n (1 + \sqrt{1-e^2}) \sin n v$$

where $T_o$ is initial epoch, $P_s$ the sidereal period and P the
anomalistic period, while $v$ is the true anomaly, e the eccentricity and $\theta$ the phase of conjunction. $\beta$ is a known
function of the eccentricity (Brouwer and Clemence, 1961).
An example of the method applied can be found in the system
Y Cyg (Giménez and Costa, 1980). Further details on the
equations and procedure will be soon published elsewhere.

To obtain absolute dimensions, as well as estimations of the age, Hejlesen (1980) evolutionary tracks have
been used following the method currently applied by the
group of Copenhagen (Andersen et al., 1979).

The internal structure constants, $k_2$, for each system (weighted mean of both components) were obtained after
correction for relativistic apsidal motion and higher order

terms ($k_3$ and $k_4$). The observed values of log $k_2$ are represented in figure 1 with respect to the logarithm of the mass (also expressed by the weighted mean value). Zero-age main sequence theoretical models are shown for comparison according to the computations by Mathis (1967), Cisneros-Parra (1970), Stothers (1974) and Semeniuk and Paczynski (1968). These models adopted the following opacities and chemical compositions:

| Authors | Opacities | X | Z |
|---|---|---|---|
| Stothers (S) | Carson | 0.730 | 0.020 |
| Mathis (M) | Cox & Stewart | 0.739 | 0.021 |
| Cisneros-Parra (C-P) | " | 0.602 | 0.044 |
| Semeniuk & Paczynski (S-P) | " | 0.602 | 0.044 |

The computations by Semeniuk and Paczyński are only given for 4 and 16 solar masses but include evolved models and we have represented them by dashed lines.

A mere inspection of the diagram shows that almost all the observed values of log $k_2$ are below the ZAMS theoretical models. The few exceptions were re-investigated and it was found that either a very close companion was present (therefore increasing the speed of periastron revolution) or the masses were too low to ignore the effects of convection in tidal evolution. The mean errors of the determination of log $k_2$ range from 0.03 to 0.10 while those of log g are between 0.02 and 0.06. Consequently, it can be confirmed that the systematic deviation of the observed values of the density concentrations from the theoretical models is real and must be due to some physical fact. Actually, we know that given a particular model, the distribution of mass and therefore the value of $k_2$ is a function,

$$k_2 = k_2(m,X,t)$$

where m is the mass, X the chemical composition and t the age of the star. A detailed study of the influence of different values of the chemical composition in the computed ZAMS configurations, lead to the conclusion that, within a reasonable range, changes in the hydrogen and metal content can not explain differences in log $k_2$ larger than 0.1 approximately. Thence, observed deviations in figure 1, for a given mass and theoretical model, must be related to the evolutionary status of the system.

The age t is a difficult parameter to measure accurately enough without involving theoretical evolutionary tracks. For this reason, we have adopted a directly observable element in eclipsing binaries very sensitive to changes

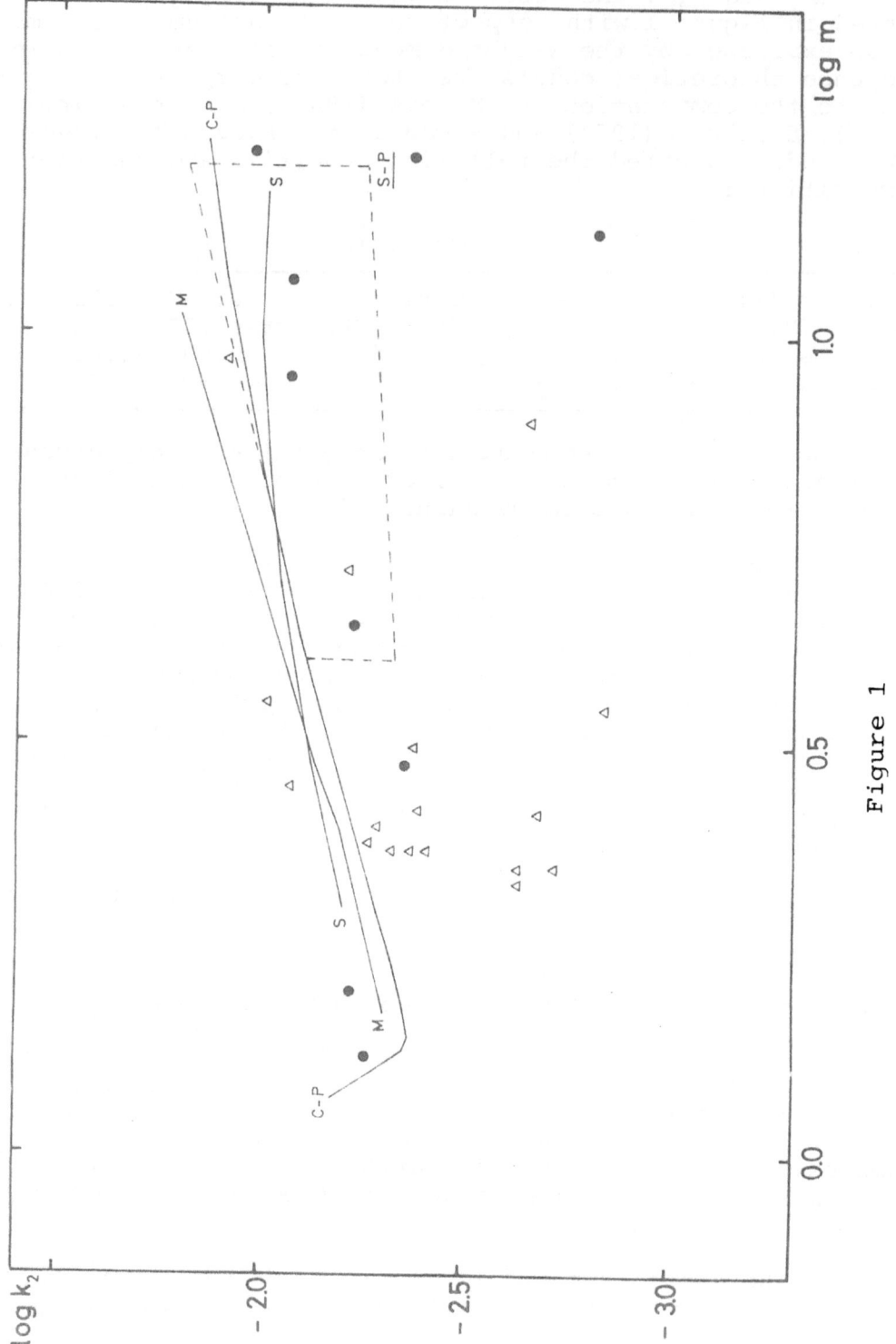

Figure 1

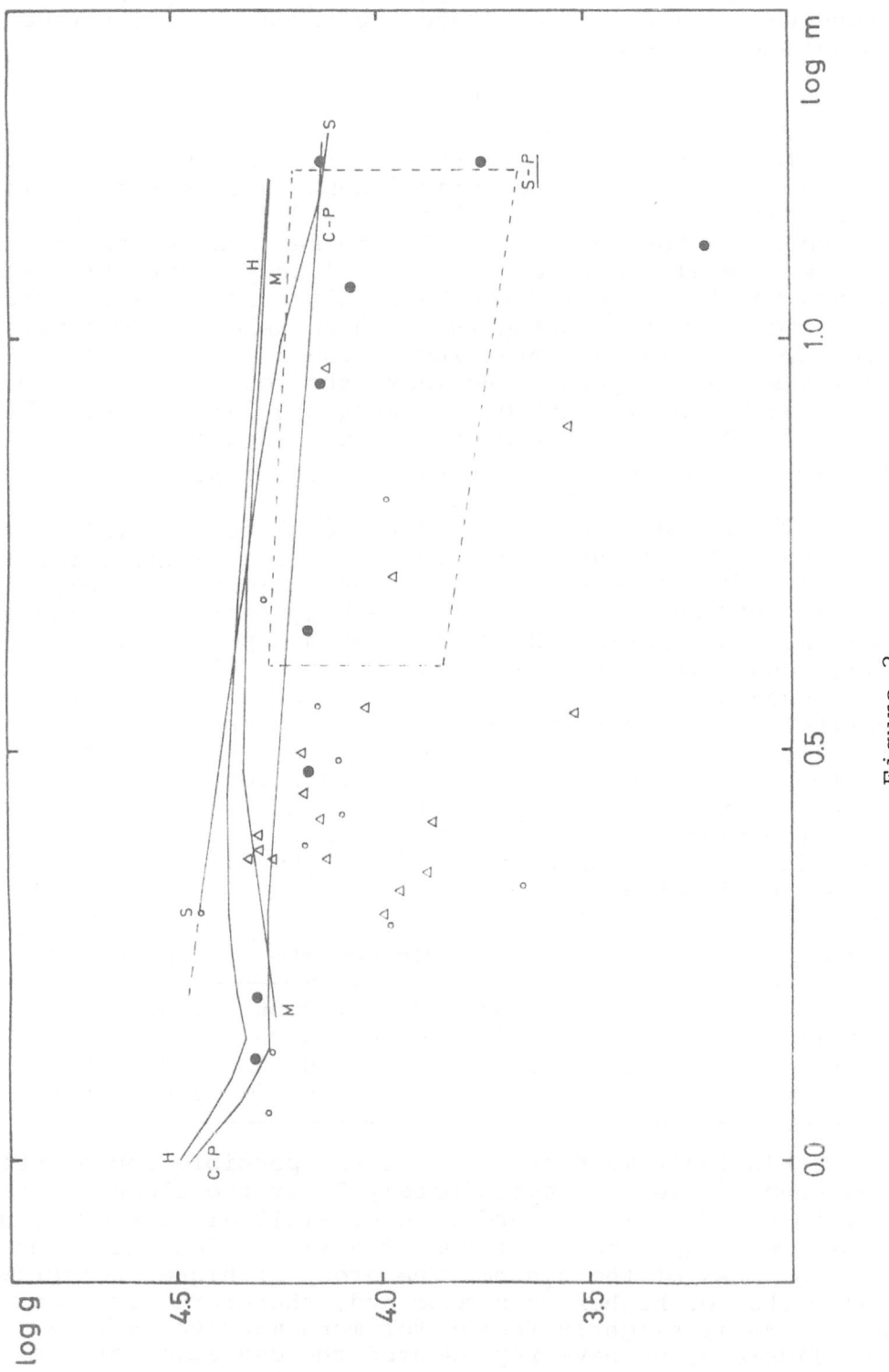

Figure 2

of age: the radius, R. For comparison with $\Delta\log k_2$, where $\Delta$
represents the difference between observed and theoretical
ZAMS values, we have used,

$$\Delta\log g = -2.\Delta\log R$$

        In figure 2, we have represented the observed values
of log g together with theoretical computations like in figure 1. Models by Hejlesen (H) for X = 0.700 and Z = 0.020
are also shown for comparison. It can be clearly seen that
the same general behaviour of figure 1 is present. All the
observed points are below the ZAMS values as it is expected
due to the evolution during the main sequence or consumption
of the hydrogen in the core (increasing radius implies decreasing surface gravity). Besides, the few discordant points
in the log $k_2$-log m plane do not show the same extreme positions in agreement with the above given suggestions that do
not have any influence in the evaluation of R.

        The comparison of the observed minus ZAMS values for
log $k_2$ and log g is shown in figure 3 for a particular set
of models. Filled dots indicate larger weight and similar re
presentations have been made for each type of theoretical
models also separating the systems in groups of different
quality and degree of evolution. The straight line in figure
3, represents the best linear fitting for all the points.
Nevertheless, for our purposes, it is better to make separate fittings only for the main-sequence systems, not including those after consumption of the hydrogen in the core.
The obvious reason is the functional behaviour of the density distribution for stars after the TAMS (non-monotonically
decreasing). The results obtained for the slope of figure 3
by means of least-squares are summarized in the following
table,

| Models | Group ● | Groups ●&o | Mean value |
|---|---|---|---|
| Cisneros-Parra | 1.09 ± 0.04 | 0.95 ± 0.06 | 1.02 ± 0.04 |
| Stothers | 1.04    0.09 | 0.96    0.06 | 1.00    0.06 |
| Mathis | 0.99    0.04 | 0.95    0.06 | 0.97    0.04 |
| Mean value | 1.04    0.04 | 0.95    0.04 | 1.00    0.03 |

        Within their mean errors, all the possible representa-
tions gave a value of approximately 1 for the slope of the
relation between $\Delta\log k_2$ and $\Delta\log g$. Small differences could
be very well explained in terms of a slight dependence on
the total mass of the system. The group of higher weight(●)
is also that of higher mean mass and, therefore, it seems
that $\Delta k_2/\Delta R$ is slightly faster for more massive systems.
Using figure 3, we have represented the deviation of the

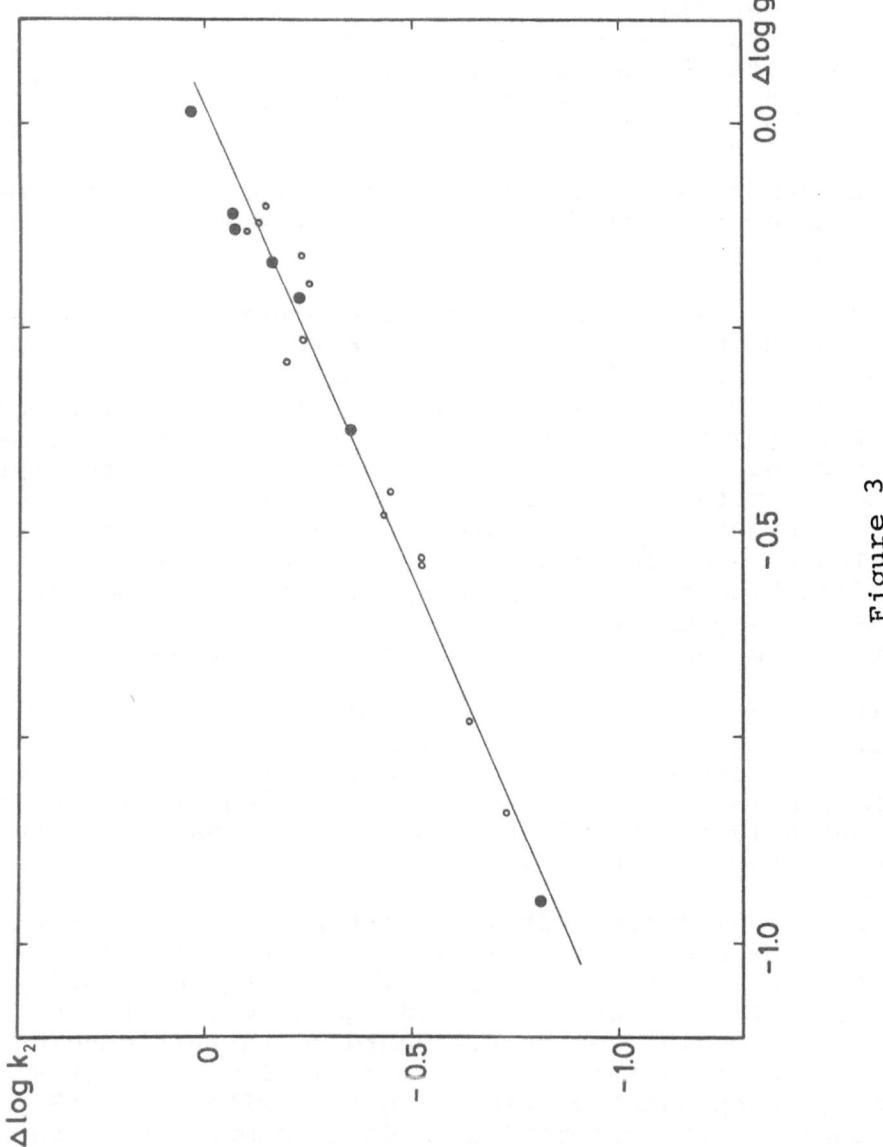

Figure 3

points with respect to the best fitting versus log m and a
significant although small correlation could be detected.
After removing this "effect of mass" the dispersion in figure
3 was considerably diminished.

We have seen that, in general, the observed values of
the internal structure constants in eclipsing binaries are
in agreement with theoretical predictions for the main sequence provided that they evolve in such a way that,

$$\Delta \log k_2 = c.\Delta \log g$$

where $\underline{c}$ is approximately 1. After consumption of the hydrogen in the core, the value of $\underline{c}$ seems to reach a lower slope
close to 0.8.

In order to see if the suggested evolutionary behaviour
of $k_2$ is also shown by theoretical models, we have to integrate the internal density concentrations for evolved configurations using Radau's differential equation. Unfortunately
this is still something to be done for a wide range of masses and ages but some data are available from the literature
for particular cases (Stothers, 1974 or Odell, 1974). For
these few examples, we have found the same linear relation
between $\Delta \log k_2$ and $\Delta \log g$ with a slope close to 1 although
slightly depending on the adopted opacity tables. However,
more high-accuracy observational data are still needed to
be able to select empirically the "best" opacities.

With respect to the zero value of the already discussed
linear fittings, it is clear that for the ZAMS, $\Delta \log k_2 = 0$.
Since small displacements (up and down) of the points in figure 3 are possible by changing the adopted chemical composition (essentially the metal abundance, Z), we have an extra
procedure to check the best values.

As a final remark, it is important to notice that
assuming $\underline{c}$ to be exactly 1, the simple relation found implies that, during the main sequence, the internal density
distributions of the stars evolve in such a way that the
function $k_2.R^2$ remains constant. According to this conclusion, we have plotted the observed systems in the $\log(k_2R^2)$-
log m plane, independent of time, together with the predicted values for the theoretical models computed by Cisneros-Parra and Stothers as shown in figure 4. This diagram
supports the suggestion that Carson's and Cox-Stewart's opacities "bracket" the real values (Söderhjelm, 1976) but we
have to be very cautious in this conclusion. In fact, some

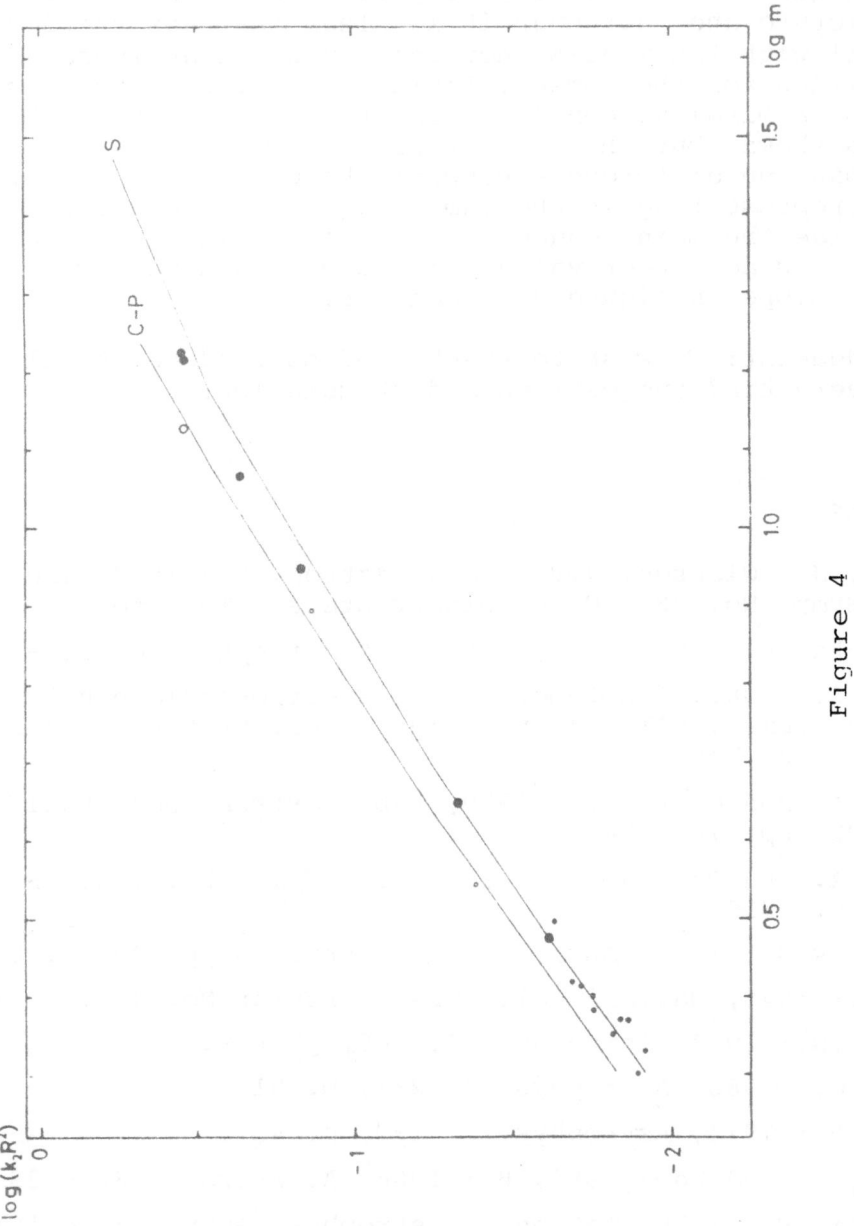

Figure 4

small changes in the chemical compositions could well rever se the picture and, although it is clear that observations agree well with the models, nothing can be established about the opacities for the moment. Again, like other astrophysical tests, a marginal confirmation of the computations by Carson is shown, but the mean errors of $\log(k_2 R^2)$ are much larger than the differences between the C-P and S curves. Finally, representing in the same diagram some observed sys tems outside the main sequence (open circles), the statement c = 1 is no longer valid and the points move upwards since the slope in figure 3 is smaller.

Acknowledgement: We wish to thank Professor Zdenek Kopal for his very kind suggestions and stimulation.

REFERENCES:

Andersen, J., Clausen, J.V. and Nordström, B.: 1979, IAU Symp. No. 88, "Close Binary Stars". Toronto.

Cisneros-Parra, J.U.: 1970, Astron. & Astrophys. 8, 141-147.

Giménez, A.: 1981, "Photometric and spectroscopic binary systems". NATO Advanced Inst., ed. by E.B. Carling and Z. Kopal.

Giménez, A. and Costa, V.: 1980, Publ. Astron. Soc. Pacific 92, pp. 782-784.

Giménez, A. and Delgado, A.: 1980, IAU Comm. 27, I.B.V.S. No. 1815.

Hejlesen, P.M.: 1980, Astron. & Astrophys. Supp. 39, p. 347.

Kopal, Z.: 1940, Harvard Coll. Obs. Circular No. 443, p. 1.

Mathis, J.S.: 1967, Astrophys. J. 149, p. 619.

Monet, D.G.: 1980, Astrophys. J. 237, p. 513.

Odell, A.P.: 1974, Astrophys. J. 192, p. 417.

Semeniuk, I. and Paczynski, B.: 1968, A. Astron. 18, p 33.

Söderhjelm, S.: 1976, Astron. & Astrophys. Supp. 25, p 151.

Stothers, R.: 1974, Astrophys. J. 194, p 651.

# A STUDY OF THE ORBITAL PERIODS OF 22 ECLIPSING BINARIES

T. Panchatsaram and K.D. Abhyankar
Centre of Advanced Study in Astronomy
Osmania University, Hyderabad, India

## 1. INTRODUCTION

Investigation of the variability of the orbital periods of eclipsing binaries is important not only from the evolutionary point of view but also for detecting additional components in them. Systematic study of the period changes in binary stars was started by Plavec (Plavec et al 1960) more than twenty years ago. Work on the same lines for 20 detached systems was reported by Herczeg (1980) at the I.A.U. Symposium No. 88 held in Toronto two years ago. Here we describe the study of 22 systems carried out by us at Hyderabad.

It is a common experience that the large errors in the visual and photographic minima obscure the real nature of period variations. Hence, in order to have accurate and homogeneous data it was decided to use only photoelectric times of minima. Only those systems for which such data was available for at least a decade were considered. Our sample includes all three kinds of systems: detached, semi-detached and contact, having periods up to 4 days. It has been possible to divide them into four groups: (i) Systems with constant periods, (ii) Periodic systems, (iii) Secular systems and (iv) Peculiar systems.

Group (i) contains the four systems: CM Lac, AB And, YY Eri (Panchatsaram and Abhyankar 1981 c) and Z Her (Panchatsaram and Abhyankar 1981 b). The constancy of the period for the detached system CM Lac was already pointed out by Herczeg (1980). The constant period of the RS CVn type system Z Her found by us agrees closely with the period derived by Plavec et al (1961) on the basis of all types of minima. The remaining two binaries of this group are contact systems.

47

*Z. Kopal and J. Rahe (eds.), Binary and Multiple Stars as Tracers of Stellar Evolution, 47–60.*
*Copyright © 1982 by D. Reidel Publishing Company.*

## 2.   PERIODIC SYSTEMS

Group (ii) contains the four systems: U Oph, AK Her,
SW Lac and RT Per.   Sinusoidal variation of the detached
system U Oph was earlier pointed out by Herczeg (1980).
Now Panchatsaram (1981 a) has derived the following light
time orbit for this system:

$a_{12}$ sin i = 1.08 AU, $P_3$ = 27.55 yrs, e = 0, f(m) = 0.0017;

giving a mass of the third body equal to 0.57 to 0.66 solar
masses for i greater than 60 degrees.   In the case of AK
Her  he (Panchatsaram 1980) had obtained a third body
period of 41.55 years with a circular orbit of size

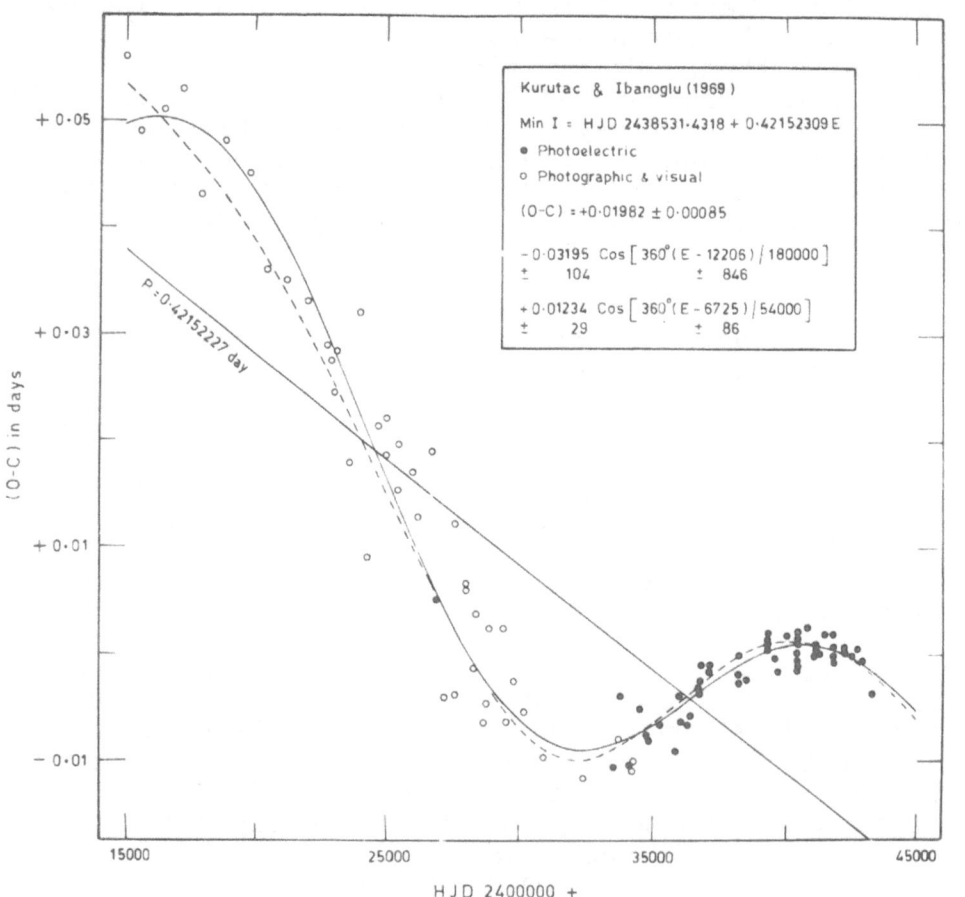

Figure 1.  (O-C) diagram for AK Herculis based on the
           ephemeris of Kurutac and Ibanoglu (1969).

$a_{12} \sin i = 0.92$ AU. However in this case we have tried to see whether the earlier photographic and visual minima fit the postulated orbit, particularly because the residuals are large compared to observational errors. Figure 1 shows the (O-C) diagram for all available primary minima of AK Her based on the period of 0.42152309 day given by Kurutac and Ibanoglu (1969). It is obvious that we have here a double sinusoid pointing to the existence of a third as well as a fourth body. The best fit for a four body solution is shown by the continuous line in the figure. Barker and Herczeg (1979) have however used a different period of 0.42152227 day as indicated by the straight line in Figure 1. On that basis they have obtained an eccentric third body orbit represented by the dashed curve. The two models which follow are given in Table I and shown schematically in Figure 2. It is clear from Figure 1 that the choice between them depends critically on the assumed period of the binary. It should be possible to remove this ambiguity by astrometric measurements in addition to future observations of the minima.

TABLE I - TWO INTERPRETATIONS OF PERIOD CHANGES IN AK Her.

| Period | P = 0.42152227 day | P = 0.42152309 day | |
|---|---|---|---|
| Parameter | 3rd body | 3rd body | 4th body |
| $a_{12} \sin i$ | 2.74 AU | 2.13 AU | 5.52 AU |
| e | 0.3 | 0 | 0 |
| P | 78 yrs | 62 yrs | 207 yrs |
| $f(m)/M_{\odot}$ | 0.00336 | 0.00252 | 0.00392 |
| $m_{12(3)}/M_{\odot}$ | 1.5 | 1.5 | $1.5 + m_3$ |
| $m_{3(4)}/M_{\odot}$ | | | |
| $i = 90^{\circ}$ | 0.21 | 0.20 | 0.24 |
| $i = 60^{\circ}$ | 0.25 | 0.22 | 0.31 |
| $i = 30^{\circ}$ | 0.47 | 0.43 | 0.58 |

Two other binaries: the contact system SW Lac and the
semi-detached system RT Per, show double sinusoids in their
(O-C) diagrams.  They have been interpreted by us (Panchatsaram and Abhyankar 1981 a, Panchatsaram 1981 b) as light

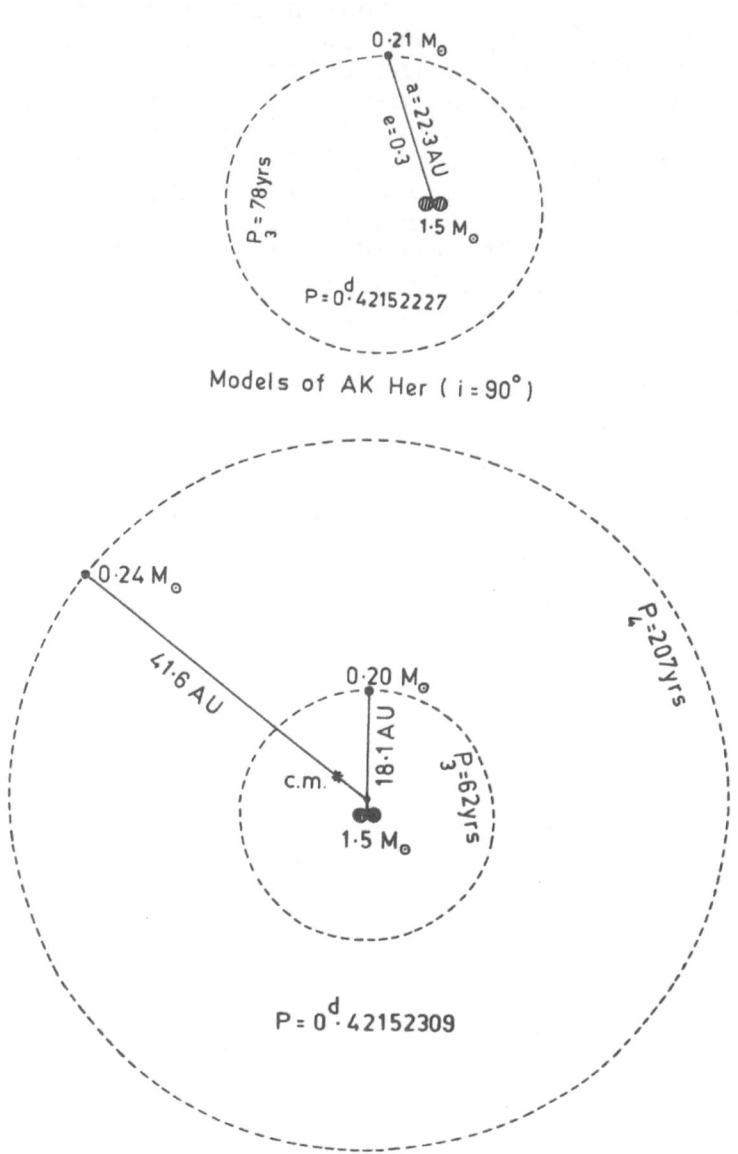

Figure 2.  Two alternate models of the system AK Herculis.

time effects in quadruple systems. The possible third and fourth masses are given in Table II. They are small enough to infer that they might be white dwarfs. Again it should be possible to detect them by astrometric measurements and also by uv observations from satellites.

## TABLE II — POSSIBLE QUADRUPLE SYSTEMS

| Parameter | RT Per | SW Lac |
|---|---|---|
| $a_{12} \sin i$ (AU) | 3.10 | 3.51 |
| $P_3$ (yrs) | 41.86 | 19.67 |
| $f(m)/M_\odot$ | 0.017 | 0.112 |
| $m_3/ M_\odot (i \geqslant 60^\circ)$ | 0.49 – 0.58 | 0.99 – 1.19 |
| $a_{123} \sin i$ (AU) | 3.94 | 7.11 |
| $P_4$ (yrs) | 100 | 70.25 |
| $f(m)/M_\odot$ | 0.006 | 0.073 |
| $m_4/M_\odot (i \geqslant 60^\circ)$ | 0.40 – 0.48 | 1.05 – 1.26 |

## 3. SECULAR SYSTEMS

We, now, come to Group (iii) containing systems which show secular variation of period. Here we have two sub-groups: one with secularly increasing periods and the other with secularly decreasing periods; each group contains five systems. Details of their period studies will be published elsewhere.

Figure 3 shows the (O-C) diagrams for two contact systems V 566 Oph and AH Vir and two semi-detached systems KO Aql and AG Vir, all showing secular increase of period as indicated by the fitted parabolas. Figure 4 shows the (O-C) diagram for the contact system 44 i Boo B. Since it is a member of a visual binary we have removed the effect of its motion in the visual binary orbit on the basis of the elements given by Heintz (1963), in the lower part of the figure.

Figures 5 and 6 show the (O-C) diagrams of three detached systems: RT And, SV Cam and AR Lac, one semi-detached system TV Cas and one contact system U Peg, which

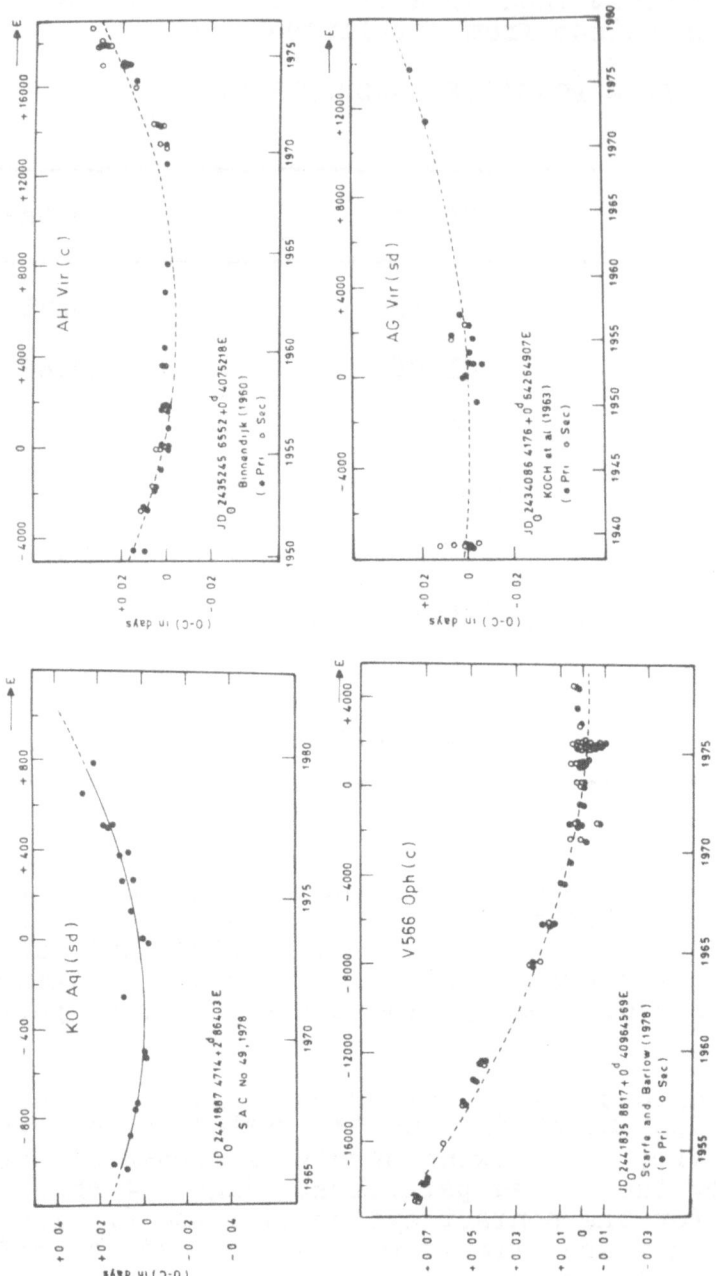

Figure 3. Secular variation of period for KO Aql, AH Vir, V 566 Oph and AG Vir.

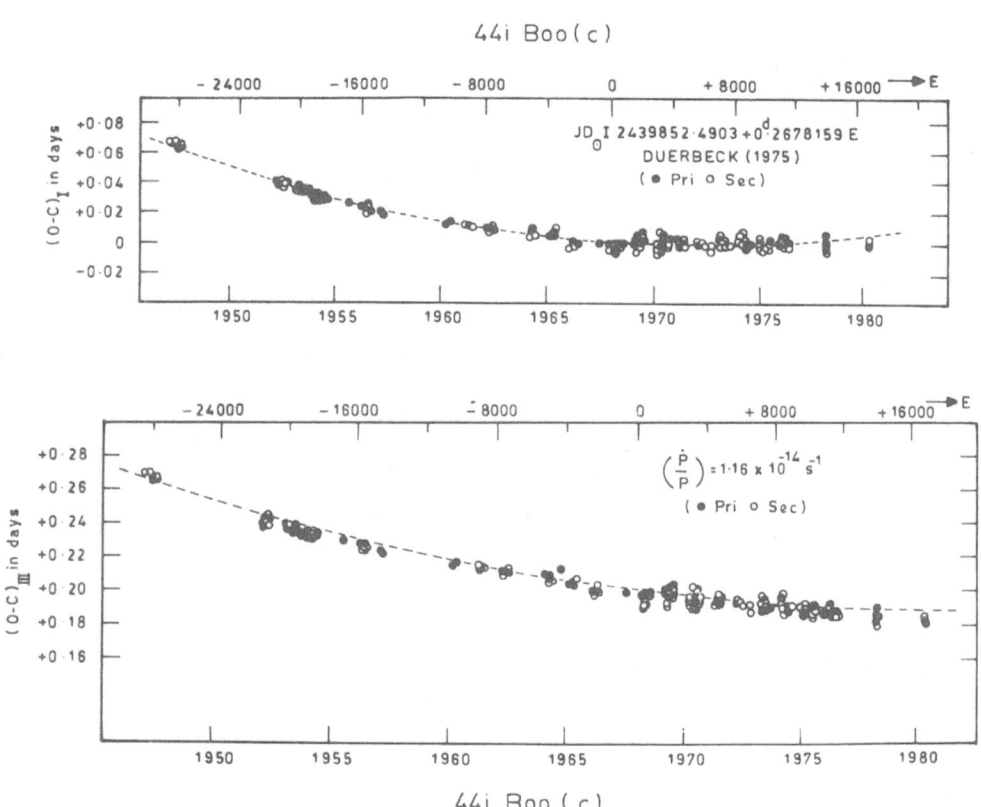

Figure 4. Secular variation of period of 44 i Bootis B;
         lower part indicates variation after removing
         the light time effect of the visual companion
         according to the elements given by Heintz.

exhibit secular decrease of period.  It is to be noted that
the phenomenon of secular variation of period is common to
detached, semi-detached and contact systems which are at
different stages of evolution.  Hence the commonly assumed
 mechanisms of mass-exchange and mass-loss may not account
for the variations of period in all cases.  Light time
effect due to the presence of additional components could
be another common cause for such variations.

    Correct interpretation of the (O-C) diagrams by light
time effect in cases where the sinusoidal variation is not
apparent becomes difficult in the absence of the knowledge
of the true period of the binary.  Even where sinusoidal

variation is seen we can have ambiguity as we have seen in
the case of AK Her.  Hence we should look for another
parameter which is not critically dependent on the period.

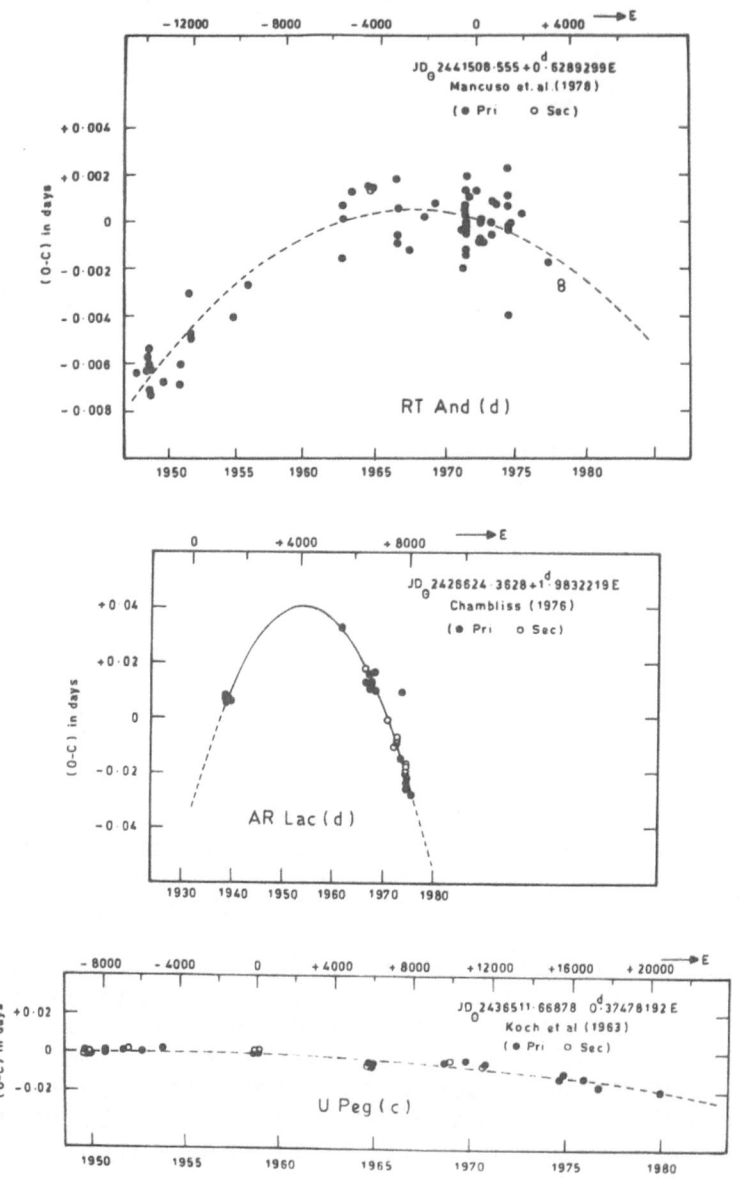

Figure 5.  Secular variation of period in RT And, AR Lac
           and U Peg.

Abhyankar (1981) has tried to identify one such parameter
which could give information about the third body if it
exists.

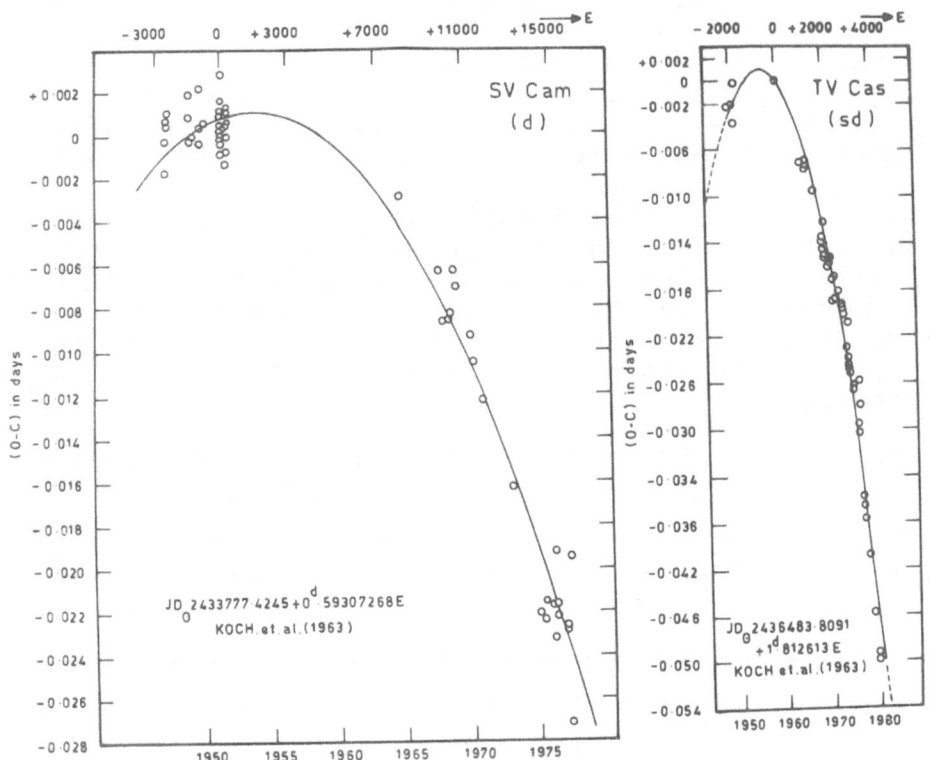

**Figure 6.** Secular variation of period for SV Cam and TV Cas.

First we note that the quadratic representation of
(O-C)'s gives us $(1/P).dP/dt$ which, when multiplied by the
velocity of light c, has the dimensions of acceleration.
The quadratic representation indicates that the system is
experiencing a constant acceleration for considerable length
of time. It is easy to see that $(c/P).dP/dt$ represents
the acceleration of $Z_{12}$, the distance of the centre of mass
of the binary along the line of sight. Measuring this
distance from the sky plane passing through the centre of
mass of the triple system it is found that the acceleration
of $Z_{12}$ remains constant for a large fraction of the period
of the third body near $M \approx v \approx 180$ degrees when $\omega = \pm 90$
degrees for large values of e. The magnitude of the
constant acceleration decreases with increasing eccentricity,
hence the probability of observing a system in this phase
of light time orbit is largest for e = 0.4 to 0.7 which is

also the modal range for the eccentricities of visual
binary orbits. In this case the constant acceleration
lasts for about half the length of period. Hence we can
obtain a tentative value of the mass function from

$$f(m)/M_\odot = (d^2 Z_{12}/dt^2)^3 P_3^4 (1+e)^6 / 64\pi^6$$

where the acceleration is in $AU/yr^2$ and $P_3$ in years, by
putting $e = 0.6$ and $P_3$ equal to twice the observed duration
of constant acceleration. The known mass of the binary can
then give us an estimate of the mass of the third body.

Table III gives the results obtained in the above manner
for the secular systems considered by us. We get reasonably
small values for the mass of the third body to make it normally undectable in the case of 6 systems: SV Cam, U Peg,
AG Vir, RT And, V 566 Oph and 44 i Boo B. For the remaining
four systems the third body interpretation may not be
acceptable.

TABLE III – ESTIMATED MASSES FOR THIRD COMPONENTS
FROM SECULAR PERIOD CHANGES

| System | $(1/P).dP/dt$ $\mathrm{sec}^{-1}$ | $P_3$ years | $f(m)/M_\odot$ | $m_3/M_\odot$ ($i \geqslant 60°$) |
|---|---|---|---|---|
| SV Cam | $7.90 \times 10^{-15}$ | 62 | $1.58 \times 10^{-2}$ | 0.4 to 0.5 |
| AR Lac | $2.52 \times 10^{-14}$ | 76 | 1.16 | 3.5 to 4.6 |
| TV Cas | $1.44 \times 10^{-14}$ | 60 | $8.38 \times 10^{-2}$ | 1.4 to 1.7 |
| KO Aql | $6.12 \times 10^{-14}$ | 28 | $3.05 \times 10^{-1}$ | 2.1 to 2.6 |
| U Peg | $5.14 \times 10^{-15}$ | 64 | $4.94 \times 10^{-3}$ | 0.3 to 0.4 |
| AG Vir | $5.44 \times 10^{-15}$ | 78 | $1.29 \times 10^{-2}$ | 0.5 to 0.6 |
| RT And | $3.28 \times 10^{-15}$ | 66 | $1.45 \times 10^{-3}$ | 0.2 to 0.3 |
| AH Vir | $2.62 \times 10^{-14}$ | 54 | $3.32 \times 10^{-1}$ | 1.6 to 2.0 |
| V 566 Oph | $2.04 \times 10^{-14}$ | 54 | $1.57 \times 10^{-1}$ | 1.1 to 1.3 |
| 44i Boo B (Heintz elements) | $1.16 \times 10^{-14}$ | 68 | $7.27 \times 10^{-2}$ | 0.6 to 0.8 |

We can arrive at the same conclusion by an alternative approach. If we make the other extreme assumption that the third body motion can be represented by a long period circular orbit we can proceed as follows. In this case we can get

$$(a^2/m_3 \sin i) = 4\pi^2/(d^2 Z_{12}/dt^2).$$

Then for a small third mass of the order of one solar mass we can obtain an estimate of 'a' and calculate period $P_3$ by putting the total mass of the system equal to $m_{12} + M_\odot$. Since in this case the acceleration will be continuously changing the observed duration of secular variation of period should come out to be a small fraction of the third body period. From Table IV, which shows such calculations for the ten secular systems considered by us, we again find that the same six systems as before qualify for possible presence of a small third body. Hence we feel that many systems showing secular variation of period might be triple. It is interesting to note that Hilditch et al (1979) have found for SV Cam a third body period of 64 years in an orbit with e = 0.6 and $\omega$ = 90 degrees in agreement with the tentative values given in Table III.

TABLE IV – CIRCULAR THIRD BODY ORBIT REPRESENTATIONS
FOR SECULAR SYSTEMS

| System | a $(m_3 \sin i = 1)$ | $m_{12}+m_3$ $M_\odot$ (i = 90°) | $P_3$ years | Observed fraction | Small 3rd body |
|---|---|---|---|---|---|
| SV Cam | 50 AU | 2.70 $M_\odot$ | 215 | 1/7 | Possible |
| AR Lac | 28 | 3.70 | 77 | 1/2 | ? |
| TV Cas | 37 | 5.49 | 96 | 1/3 | ? |
| KO Aql | 18 | 4.49 | 36 | 1/2.5 | ? |
| U Peg | 62 | 3.40 | 265 | 1/8 | Possible |
| AG Vir | 60 | 3.83 | 239 | 1/6 | Possible |
| RT And | 78 | 3.52 | 365 | 1/11 | Possible |
| AH Vir | 28 | 2.96 | 84 | 1/3 | ? |
| V 566 Oph | 31 | 2.74 | 105 | 1/4 | Possible |
| 44i Boo B (Heintz elements) | 41 | 2.24 | 177 | 1/5 | Possible |

## 4.    PECULIAR SYSTEMS

Finally we come to Group (iv) containing four contact systems which are put under peculiar category.  Most prominent among them is the prototype W UMa itself.  From Figure 7 we see that in addition to showing a secular decrease of period the (O-C) diagram indicates a discontinuity in 1964 which coincides remarkably with the flare observed in that system by Kuhi (1964).  Relation of the

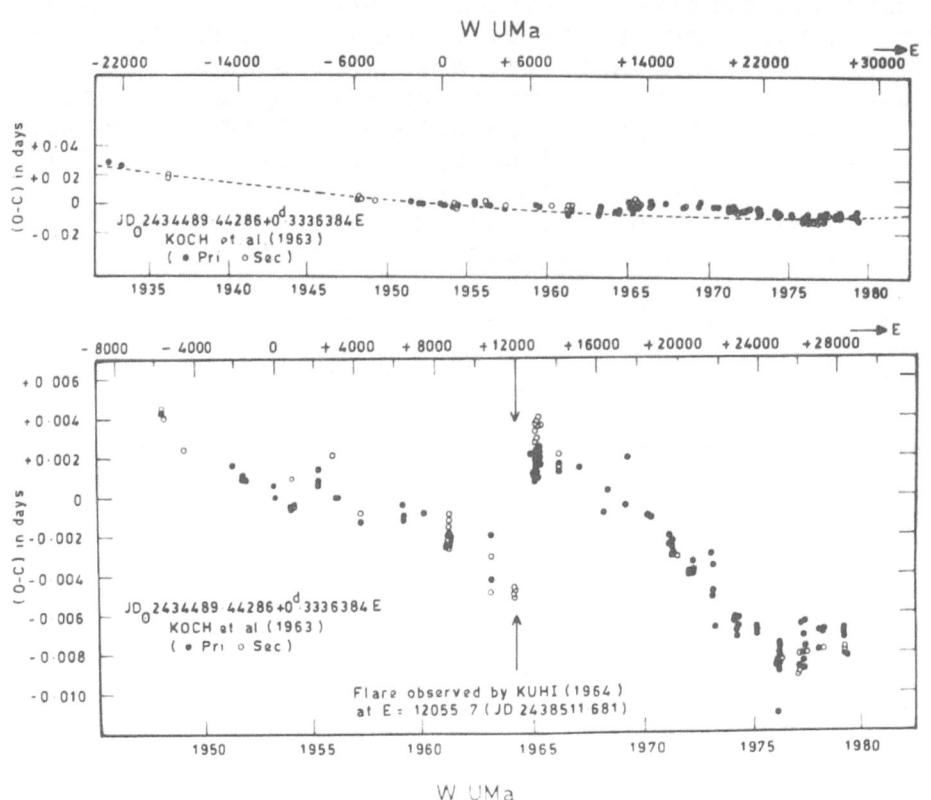

Figure 7. The (O-C) diagram for W UMa, lower part on
          expanded scale to bring out discontinuity.

flare to the period discontinuity is not clear, but we can observe the Kwee (1966) effect of separation of primary and secondary minima which occurs immediately after the flare.  Discontinuities similar to W UMa are also observed in three other systems VW Cep, UV Leo and XY Leo, whose (O-C) diagrams are shown in Figure 8.  However, the existence of third bodies in W UMa and VW Cep cannot be ruled out. The periods used for VW Cep and UV Leo are those due to Koch et al (1963) while that for XY Leo is obtained in the present study.

Concluding, it is commonly assumed that period changes in binary systems are mostly caused by mass exchange between components or loss of mass by the system during the course of its evolution. However, light time effects due to the presence of additional components may be equally important

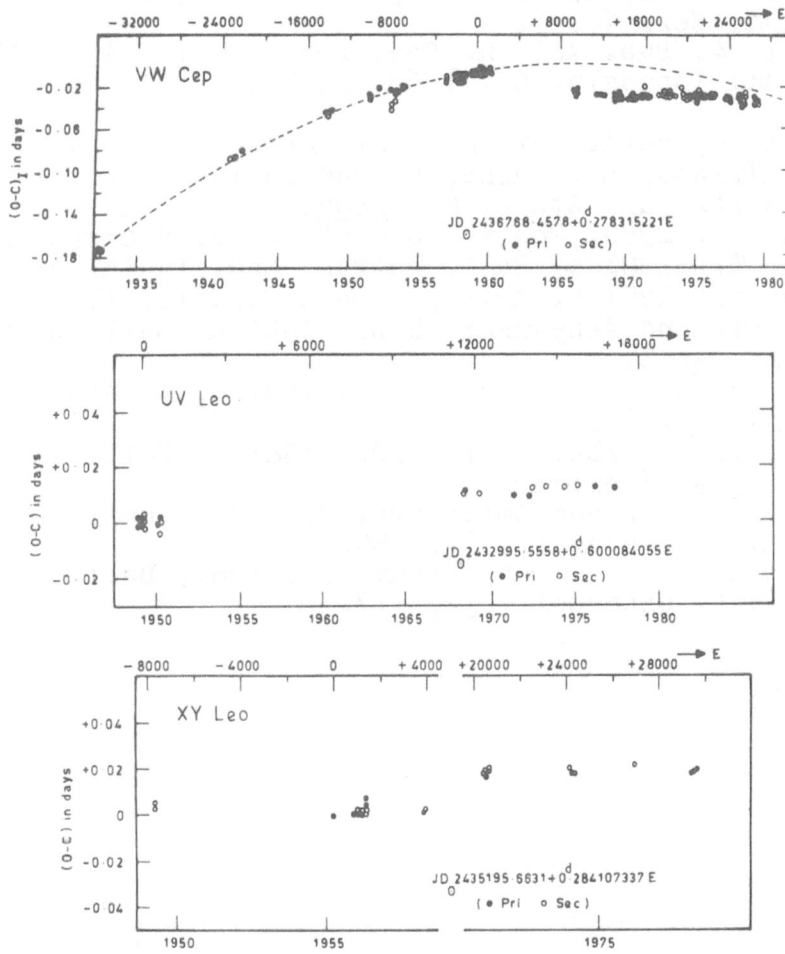

Figure 8. (O-C) diagrams for VW Cep, UV Leo and XY Leo.

in many cases. We would also like to suggest making regular astrometric measurements of all such eclipsing binaries.

# REFERENCES

Abhyankar, K.D.: 1981, Bull. Astron. Soc. India, 9, 99.
Barker, L.A. and Herczeg, T.J.: 1979, Pub. Astron. Soc.
    Pacific, 91, 247.
Binnendijk, L.: 1960, Astron. J., 65, 88 and 358.
Chambliss, C.R.: 1976, Pub. Astron. Soc. Pacific, 88, 762.
Duerbeck, H.W.: 1975, Inform. Bull. Var. Stars, No. 1497.
Heintz, W.D.: 1963, Veroff. der Sternwarter Munchen, 5, 257.
Herczeg, T.J.: 1980, I.A.U. Symposium No. 88, 89.
Hilditch, R.W.; Harland, D.M. and Mclean, B.J.: 1979, Mon.
    Not. Roy. Astron. Soc., 187, 797.
Koch, R.H., Sobieski, S. and Wood, F.B.: 1963, Pub. Univ.
    Pa. Astron. Ser. No. 9.
Kuhi, L.V.: 1964, Pub. Astron. Soc. Pacific, 76, 430.
Kurutac, M. and Ibanoglu, C.: 1969, Inform. Bull. Var. Stars
    No. 369.
Kwee, K.K.: 1966, Bull. Astron. Inst. Nether., 18, 448.
Mancuso, S., Milano, L., Russo, G. and Sollazzo, C.: 1978,
    Inform. Bull. Var. Stars, No. 1409.
Panchatsaram, T.: 1980, Thesis, Osmania Univ. Hyderabad.
Panchatsaram, T.: 1981 a, Bull. Astron. Soc. India, 9, 139.
Panchatsaram, T.: 1981 b, Astrop. Space Sc., 77, 179.
Panchatsaram, T. and Abhyankar, K.D.: 1981 a, Bull. Astron.
    Soc. India, 9, 31.
Panchatsaram, T. and Abhyankar, K.D.: 1981 b, J. Astrop.
    and Astron., 2, 29.
Panchatsaram, T. and Abhyankar, K.D.: 1981 c, Bull. Astron.
    Soc. India, 9, (in press).
Plavec, M., Pekny, Z. and Smetanova, M.: 1960, Bull.
    Astron. Inst. Czechosl., 11, 180.
Plavec, M., Smetanova, M. and Pekny, Z.: 1961, Bull.
    Astron. Inst. Czechosl., 12, 117.

# ON THE GROUP FORMATION OF STARS IN STELLAR ASSOCIATIONS

L.V. Mirzoyan
Byurakan Astrophysical Observatory,
Armenia, U.S.S.R.

The discovery and study of stellar associations by Ambartsumian (1947, 1954 a) proved two fundamental points regarding the star formation process, e.g., Mirzoyan, 1976.  One of them is that the formation process still continues in the present evolution of the Galaxy, and this thesis is not disputed now.  According to the second, stars form as physical groups of multiple stars and star clusters, which become stellar associations during the initial phases of their evolution.

This paper deals with observational data accumulated since the discovery of stellar associations, proving the common origin of the components of stellar systems.  The first evidence in its favor was derived from statistics of double stars, before the discovery of stellar associations.

It has been shown (Ambartsumian, 1937), that a double star can form as a result of a two-star encounter, or the close encounter of three independent stars during their motions.  A double star system can also split as a result of a close encounter with a third star, and a statistical equilibrium between the formation and decay processes of pairs must be established in time.  The available observational data testify to the absence of such a so-called dissociative equilibrium, between the processes of formation and decay of pairs due to close stellar encounters.  The observed portion of wide pairs among single stars in the Galaxy is some ten million times less than the portion expected for the dissociative equilibrium.  It means that at the present time, the number of wide pair decays in the Galaxy is some ten million times larger than the number of formations due to capture by the close passing of triple stars.  It is therefore impossible to explain the formation of all double stars in the Galaxy by the stellar encounter theory.

On the other hand, large values of the angular momentum of double stars relative to their center of gravity, and the absence of strong differences between close and wide pairs, testify against their

61

*Z. Kopal and J. Rahe (eds.), Binary and Multiple Stars as Tracers of Stellar Evolution, 61–66.*
*Copyright © 1982 by D. Reidel Publishing Company.*

formation due to the division of separate stars (Ambartsumian, 1956).

Based on the above mentioned observational facts, Ambartsumian (1947, 1956) concluded that the components of double stars have a common origin.

The formation of multiple stars and clusters by captures is no less difficult to explain. It is feasible then, that the components of a physical system, independent of their number, are formed in common; that is, the stars originate in groups. Also, the results of morphological studies of stellar associations and clusters, strongly confirm this idea of a common origin, since among recently formed young stars (as are the members of stellar associations) a very high portion of double and multiple stars was observed.

In O-associations for example, there are many visual double stars, close pairs and spectroscopic binaries. We can say in this respect, that the Wolf-Rayet stars which usually occur in O-associations, show the same characteristics. According to statistical evidence (Mirzoan, 1949), almost all of them are double, but because of the selective nature of the observations, duplicity was found for only a few of them. This conclusion, obtained in 1949, was recently confirmed (Brutian, 1981) based on new observational data.

In both O-and T-associations, the abundance of multiple stars is very high, especially of dynamically unstable, multiple systems and stellar chains. All these facts indicate that stars are formed in groups.

From this point of view, the existence of Trapezium-type multiple stars in associations, is of particular significance. Ambartsumian (1954 b) was the first to pay attention to the fact that the motions in the systems having a space configuration resembling the famous Orion-Trapezium (Trapezium-type systems [+] ), must differ sharply from the motions in the "ordinary"-type systems, establishing their dynamical instability. He showed that the lifetime of the Trapezium-type systems as such, (keeping the Trapezium-type configuration) must be around $2 \times 10^6$ years if the total energy of the system is negative, and $5 \times 10^5$ years and even less if the total energy is positive.

Statistics on the configuration of multiple stars (Ambartsumian, 1954 b), show that among the real Trapezium-type systems in the Galaxy, those systems prevail in which the main (brightest) members are of spectral types O-B2. It is thus concluded that the real Trapezium-type

---

[+] All multiple systems in which at least three members share mutual distances in the same order of magnitude belong to the Trapezium-type systems (Ambartsumian, 1954 b), independent of the total number of their members.

systems consist almost exclusively of young stars (Ambartsumian, 1954 b). This is confirmed in Table 1, taken from the paper by Salukvadze (1978), and given here with an additional two columns.

TABLE 1

Statistics of Multiple Stars According to the
Index Catalogue of Visual Double Stars

| Spectrum of the main member | Total number of multiple stars | Number of observed Trapezia | Estimation of the number of Pseudo Trapezia | Probable number of Real Trapezia | Relative number of Trapezia (%) |
|---|---|---|---|---|---|
| 0-B2 | 59 | 39 | 5 | 34 | 58 |
| B3-B5-B | 72 | 23 | 6 | 17 | 24 |
| B8-B9 | 1118 | 25 | 11 | 14 | 12 |
| A | 394 | 60 | 35 | 25 | 6 |
| F | 309 | 41 | 28 | 13 | 4 |
| G | 224 | 33 | 20 | 13 | 6 |
| K | 153 | 37 | 14 | 23 | 15 |
| M | 11 | 8 | 1 | 7 | 64 |
| Unknown spectrum | 526 | 146 | 47 | 99 | 19 |

Table 1 gives in consecutive columns: the spectrum of the main member of systems, the total number of such multiple stars, the number of multiple stars having a configuration of the Trapezium-type, the estimated number of Pseudo-Trapezia (the multiple systems observed as Trapezia due to projection of the ordinary-type multiple systems on the sky), the probable number of real Trapezia and the relative number of real Trapezia (in per cent) among all multiple stars with the main members having the same spectral type. Estimates of the number of Pseudo-Trapezia presented in Table 1 have been derived from the probability of such phenomenon for triplets and quartets with $P = 0.09$ as defined by Ambartsumian (1954 b).

Table 1 shows that in fact there are no real Trapezium-type systems among multiple systems in which the main members belong to the spectral classes A, F, G; only a few systems are observed in which the main members are of spectral type B8-B9 and K. This is even more obvious if one takes a slightly higher value than 0.09 (Ambartsumian, 1954 b) for the probability of Pseudo-Trapezia.

Table 1 also shows that the relative number of real Trapezia is very large among multiple systems containing O-B2 type stars. Sharpless (1954) reached the same conclusion after evaluating data on the multiple systems of Trapezium-type in emission nebulae, showing that there is a strong tendency in the brightest components to have spectral classes earlier than O9.

It is an important observational fact that the real Trapezium-type systems, in the overwhelming majority of cases, are found among the multiple systems containing young O-B stars. This fact testifies that the life-time of these systems is really rather short, at least shorter than the ages of the above mentioned stars. One must therefore assume that after this short span, the Trapezia either disintegrate completely or lose some of their members and form stable systems of the ordinary type with less members than earlier.

In Table 1 the high proportion of Trapezium-type systems among multiple stars whose main components belong to the spectral class M is shown. This fact is probably due to the tendency of cool supergiant variables of spectral class M to be members of O-associations (Ambartsumian, 1953 a). However, the total number of multiple stars for this spectral class is too small for serious statistical conclusions.

The fact that Table 1 shows only a small portion of Trapezia whose main members are of other spectral classes can be understood if one takes into account that recently originated young systems could also be among them.

An original evidence in favor of the youth of the Trapezium-type systems comes from the true distribution of triple star configurations derived by Agekian (1954) from their apparent distribution according to Aitken's Catalogue of Double Stars. It shows that the portion of non-stable (Trapezium-type) systems, is greatest for spectral classes O and B. In this regard, it must be noticed that a considerable number of Trapezium-type multiple systems in T-associations has been found and studied recently by Salukvadze (1980 a, b).

Further, the results obtained by Allen and Poveda (1974) on the dynamical evolution of the Trapezium systems, in spite of the author's opinion, are in perfect agreement with Ambartsumian's (1954 b) estimations of the life-time of these systems. Indeed, studying the problem for 30 sextets of the Trapezium-type with different parameters of structure and negative total energy by means of a computer, they found that the probability of keeping their configuration during $10^6$ years is equal to 2/3. This means (Mirzoyan and Mnatsakanian, 1975), that already during $2 \times 10^6$ years, more than half of all studied Trapezium systems must lose their characteristic configuration. In other words, the life-time of the Trapezium systems having negative total energy is in fact $2 \times 10^6$ years.

The origin of associations themselves, which are dynamically unstable and due to this are still in expansion (see, for example, Mirzoyan 1976, 1981) and the presence of Trapezium-type multiple systems in associations, is a confirmation of the findings on group formation of stars. Thus, the results of the morphological study of O-associations by Markarian (1950, 1951) used for the new classification of open star clusters are of great interest.

It is well known that multiplicity is a wide-spread phenomenon among galactic field stars as well. However, multiple star systems in the general galactic field are almost without exception, of an ordinary type, that is they are dynamically stable.

Thus, according to observational data, the general galactic field is rich in multiple stars of an ordinary type, which are dynamically stable, and stellar associations are rich also in Trapezium-type multiple stars, which are dynamicall unstable.

This phenomenon – the abundance of stable multiple systems in the general galactic field and the abundance of non-stable multiple systems in stellar associations – has a natural explanation. The stars are being formed in stellar associations in groups, which are dynamically stable and, especially, dynamically unstable (Trapezium-type systems). The dynamically stable multiple systems are decaying extremely slowly, only as a result of the interaction of the stars in the systems during their close passing; on the other hand, the unstable systems are decaying completely or in part more rapidly than the associations themselves. As a consequence, the general field in the Galaxy enriches itself by dynamically stable multiple systems and practically no unstable multiple systems remain in this field.

We did not discuss here the problem of the nature of protostellar matter, but note that the idea of group formation of stars, which was strongly confirmed by the results of morphological studies of stellar associations, can be better explained in the frame of a high-density "protostar" hypothesis (Ambartsumian, 1953 b).

REFERENCES

Agekian, T.A.: 1954, Astron. Zh, 31, 544.
Allen, C. and Poveda, A.: 1974, The Dynamical Evolution of Trapezium Systems, Instituto de Astronomia, Universidad Nacional Autonoma de Mexico, Mexico, preprint.
Ambartsumian, V.A.: 1937, Astron. Zh. 14, 207.
Ambartsumian, V.A.: 1947, Stellar Evolution and Astrophysics, Ac. Sci. Armenian SSR, Yerevan.
Ambartsumian, V.A.: 1953 a, CR Ac. Sci. Armenian SSR, 16, 73.
Ambartsumian, V.A.: 1953 b, CR Ac. Sci. Armenian SSR, 16, 97.
Ambartsumian, V.A.: 1954 a, Transactions of the IAU, vol. 8, ed. P. Th. Oosterhoff, University Press, Cambridge, p. 665.
Ambartsumian, V.A.: 1954 b, Contr. Byurakan Obs., 15, 3.

Ambartsumian, V.A.:  1956, Vistas in Astronomy, vol. 2, Pergamon
     Press, London-New York,p. 1708.
Brutian, G.A.:  1981, private communication.
Markarian, B.E.:  1950, Contr. Byurakan Obs., 5, 3.
Markarian, B.E.:  1951, Contr. Byurakan Obs., 9, 3.
Mirzoyan, L.V.:  1949, CR Ac. Sci. Armenian SSR, 10, 193.
Mirzoyan, L.V.:  1976, in V.A. Ambartsumian (ed.), Probleme der
     Modernen Kosmogonie, Birkhäuser Verlag, Basel and Stuttgart,
     Chapter II.
Mirzoyan, L.V.:  1981, Stellar Nonstability and Evolution, Ac. Sci.
     Armenian SSR, Yerevan.
Mirzoyan, L.V. and Mnatsakanian, M.A.:  1975, Astrofizika, 11, 551.
Salukvadze, G.N.:  1978, Astrofizika, 14, 57.
Salukvadze, G.N.:  1980 a, Astrofizika, 16, 505.
Salukvadze, G.N.:  1980 b, Astrofizika, 16, 687.
Sharpless, S.:  1954, Astrophys. J., 119, 334.

Comments by J. Dommanget following L.V. Mirzoyan's paper:

In connection with the idea of Prof. V.A. Ambartsumian that
in stellar associations stars are formed in groups, I would like to
mention  that from visual binary orbit considerations, it appears that
stars in the surroundings of the Sun also seem to have been formed in
groups.  Such groups seem to occupy volumes of the order of ten to
twenty parsecs.  This clearly appears from a first study we made in
1967 on the distribution of the orbital poles of some 70 binaries of
which the orbital ascending nodes are well defined from radial velocity observations.  All binaries nearer than approximately 10 parsec.
show some similarity in their orbital plane orientation, while the
orientation for more distant systems appears clearly different.

A new list of binaries for which the orbital poles are well
defined is presently being prepared.  It probably will contain some
140 systems.  From a first look at this material, the above mentioned
phenomenon seems to be confirmed.

# THE AGE OF THE TRIPLE SYSTEM HD 165590

Gustav A. Bakos, Department of Physics, University of
Waterloo, Waterloo, Canada.
Jozef Tremko, Skalnate Pleso Observatory, Tatranska Lomnica,
Czechoslovakia.

ABSTRACT

It has been found that the close spectroscopic pair of this system
consists of a G type primary and a late type secondary, which most
likely is a T Tauri star. As such, the system is semidetached and a
mass transfer takes place between the components.

## 1. INTRODUCTION

The spectroscopic orbit of the triple system HD 165590 has been
discussed by Batten et al. (1979). They found that the A component of
the visual pair is a spectroscopic binary with a period of 0.88 days.
The mass function indicates that the secondary is a star of spectral
type M and the visual companion of spectral type G5. At the time of
periastron passage the visual pair has a separation of 0.4 A.U. while
the separation of the spectroscopic pair is only $3.2 \times 10^6$ km. The
authors have discussed the stability of this system and have concluded
that despite the closeness of the stars the system appears to be
stable. A sensitive indicator of an instability would be a change of
the period of the close pair.

Another point raised by the authors was the age of the system.
Since the strength of the lithium lines is considered to be a good
indicator of the age of a star, they measured the equivalent width of
the lithium lines and by comparing the abundance of lithium with that in
the Hyades and Pleiades they concluded that the age of HD 165590 is
about $5 \times 10^7$ years.

In this presentation we shall address ourselves to the two points
raised by Batten et al., namely to find other indicators for the
stability and the age of this system.

*Z. Kopal and J. Rahe (eds.), Binary and Multiple Stars as Tracers of Stellar Evolution, 67–71.*

## 2. OBSERVATIONS

Since 1977 we have made observations of this system by means of a photoelectric photometer in the visual and the blue region of the spectrum. The internal accuracy of our observations is about ±0.005 mag. As already found by Scarfe (1977) the spectroscopic pair is variable and it shows a shallow (0.06 mag) minimum. Our light curve appears to be variable at other phases as well and the most remarkable phenomenon is the rare appearance of the secondary minimum. A representative light curve exhibiting both minima is shown in Fig. 1a, while the two minima are plotted separately in Fig. 1b. The duration of

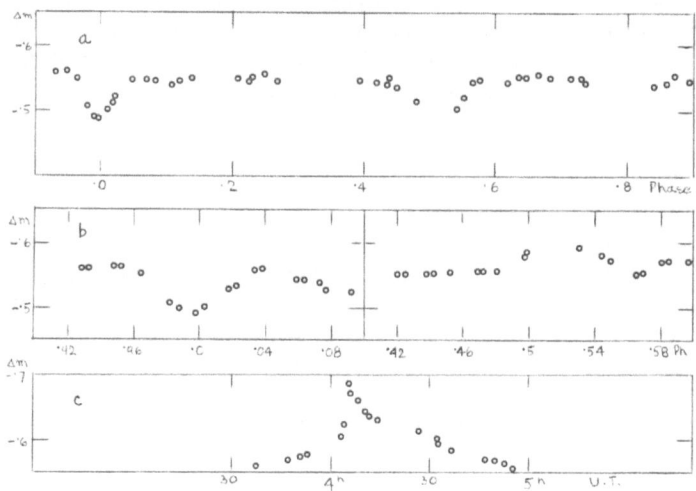

Fig. 1a:   The light curve of HD 165590.
Fig. 1b:   The light curve near a primary and a secondary minimum.
Fig. 1c:   A flare observed on June 16, 1979.

the primary eclipse is about two hours. Between the minima the light curve is subject to variations over short and long time intervals.

From observations made between 1978 and 1981 we derived six epochs of primary minima. These are listed in Table I. For the period of the spectroscopic orbit a value satisfying all observations is P = 0.8794998 days. The 0-C diagram indicates the tendency for the residuals to be slightly positive. However, it would be wrong to conclude that a change

TABLE I.

Photoelectric Epochs of Primary minima of HD 165590.

| J.D. hel. | E | O-C |
|---|---|---|
| 244 3656.6635 | 0 | 0$\overset{d}{.}$0000 |
| 3998.7889 | 389 | 0.0000 |
| 4050.6792 | 448 | -0.0002 |
| 4459.6474 | 913 | +0.0006 |
| 4467.5623 | 922 | 0.0000 |
| 4816.7249 | 1319 | +0.0012 |

of the period is present. The time interval is still too short to be able to detect a change as small as $10^{-10}$ days/cycle, a typical value that would indicate a transfer of one solar mass in $5 \times 10^7$ years. Therefore, from this point of view we are unable to conclude whether the system has remained stable for the duration of its lifetime.


3. ANALYSIS OF THE LIGHT CURVE

3.1. The Primary Minimum

The primary minimum appears to be well defined as far as its depth and duration is concerned. However, a closer inspection indicates that the ascending branch shows a definite brightening at phase 0.02, up to 0.025 mag and duration of about 20 minutes. This is not necessarily a permanent feature of the primary minimum since in a few cases the brightening is hardly observable. It appears that the primary minimum is an occultation and a presence of a hot spot on the surface of the G star facing the M star would explain the brightening at this phase. A symmetrically opposite increase of brightness at phase 0.98 is not obvious.

We have plotted the apparent magnitude of the minimum and the maxima on each side of the minimum (phase 0.9 and 0.1) as a function of time. We have found that the brightness varies by 0.05 mag, being lowest in 1978 and again in 1981, while it reached a maximum in 1980, a variation on a time scale of three years. For the visual pair the periastron passage happened in June 1978. Thus the maximum light of the system does not correspond to the periastron passage of the G5 star.

From the geometry of the system and the constant depth of the minima it follows that the G0 component is not responsible for the changes of brightness. Otherwise the depth of the minima would be variable. Thus, either the M component is intrinsically variable or it might be the G5 star.

## 3.2. The Secondary Minimum

Since the M star represents about 3 per cent of the total
luminosity of the system one would expect to see at the time of the
secondary minimum a drop of brightness by 0.03 mag. In our composite
light curve such a drop is indicated, however, a complete secondary
minimum has not been observed. Instead a brightening sets in at phase
0.5 reaching a maximum at phase 0.52. In fact the largest light
variations occur near the secondary minimum. This is also confirmed by
a very large scatter of the B-V colour around the secondary minima.
There is no  obvious mechanism that could explain this anomaly.

## 3.3. The Light Curve Outside Eclipses

Assuming the mass ratio of 0.6 for the spectroscopic pair, as
derived by Batten et al., this system is definitely detached. Therefore
we would expect to find a constant brightness of the light curve outside
eclipses. However, the light curve indicates considerable variations.
We would like to call attention to the drop of brightness by 0.03 mag
following the primary minimum, near phase 0.10. At other phases we find
variations of a similar magnitude. The time scale is relatively short.

## 3.4. Flaring

A remarkable change in the light curve happened on June 16, 1979
when the star increased its brightness by 0.125 mag during a time
interval of 20 minutes. The shape of the light curve showed a steep
increase followed by a slower decline for about 50 minutes. At the end
of the flare the brightness reached the same level (Fig. 1c) as at the
start of the flaring. In the blue light the amplitude was 0.2 mag. The
flare occured shortly after a secondary minimum. Since the eclipsing
pair was approaching a quadrature it is not obvious which star was
responsible for the flare. By its appearance the flare resembles
closely the brightness variations of UV Cet type stars.

## 4. Discussion

If indeed the flare represents a UV Cet type activity then, it is
associated with the red star and it would represent an increase of
brightness by 2.8 mag. There exists a relation, derived by Parsamian
(1977) between the age, absolute magnitude and the amplitude of a
flare. With the given amplitude and the known absolute magnitude of the
star Parsamian's relation gives the upper limit of the age of the
system, at $2 \times 10^9$ years. Since such an age would be at variance with the
observed abundance of lithium it became obvious that the observed flux
was not the maximum flux to be expected for this type of star or,
alternatively, we have not observed a UV Cet phenomenon. There is
another reason why we should not consider this flare of a UV Cet type.

The total amount of energy released during the flare amounted to $10^{37}$ ergs, which is a much higher energy than the one produced in UV Cet flares. Thus we are lead to conclude that the red star in this system is, most likely, a T Tau star. The types of flares in the considered range of energies are typical for T Tau itself. A frequency of flares in T Tau was found to be one in 30 to 40 hours. Although we do not know the radius of the M Star, radii of T Tau stars are about one to two solar radii. If this were the case for HD 165590 then we would have a semidetached system and a mass transfer mechanism capable of moving mass from one star to another through its inner Lagrangian point. We suggest that the hot spot, observed on the G star might indicate an infall of mass from the M star.


5. Conclusion

    Since there are good reasons to believe that the red star is a T Tau star having a radius of about two solar radii then we have to assume that the star is a young object. This indirectly confirms the findings of Batten et al. If the star overflows its Roche lobe the system is semidetached and a lengthening of the orbital period is to be expected. Finally, the orbital elements derived by Batten et al., especially that of the orbital inclination may have to be revised in order to conform with the changed dimensions of the system.


References

Batten, A.H., Morbey, C.L., Fekel, F.C. and Tomkin, J.: 1979, Publ. Astron. Soc. Pacific 91, 304.

Parsamian, E.S.: 1977, Flare Stars, Symposium Byurakan Obs., Ed. Mirzoyan, L.V., Yerevan.

Scarfe, C.C.,: 1977, Info. Bull. Var. Stars. Comm. 27 I.A.U., No. 1357.

# PROPERTIES OF RELATIVISTIC LINEAR STELLAR MODELS

V. Ureche
University of Cluj-Napoca
Cluj-Napoca, Romania

## 1. INTRODUCTION

In the late stages of stellar evolution, relativistic objects are formed, such as neutron stars or black holes. These relativistic stars possess a strong gravitational field, therefore their structure and their space-time geometry can be described only in the frame of GRT (Zeldovich and Novikov, 1971; Misner, Thorne and Wheeler, 1973). For this purpose, the following four-dimensional interval is used (spherical gravitational field), (Zeldovich and Novikov, 1971).

$$ds^2 = e^{\nu(r)} c^2 dt^2 - e^{\lambda(r)} dr^2 - r^2 (d\theta^2 + \sin^2\theta \, d\phi^2). \qquad (1)$$

If we take the energy-momentum tensor in the form

$$T^{ik} = (\rho c^2 + P)u^i u^k - P\, g^{ik}, \qquad (2)$$

then for a relativistic star in equilibrium, with spherically symmetrical gravitational field, from Einstein's field equations (in the cold matter approximation) we obtain the following equations (Zeldovich and Novikov, 1971; Weinberg, 1975; Hawking and Ellis, 1977)

$$\frac{dM(r)}{dr} = 4\pi r^2 \rho(r) , \quad M(0) = 0$$

$$\frac{dP(r)}{dr} = - \frac{G(\rho + P/c^2)(M(r) + 4\pi r^3 P/c^2)}{r^2(1 - 2GM(r)/c^2 r)} , \quad P(R) = 0 \qquad (3)$$

$$P = P(\rho) , \quad \rho(R) = 0$$

where the notations are usual. For a given state equation $P = P(\rho)$ the system of equations (3) allows the determination of the functions

*Z. Kopal and J. Rahe (eds.), Binary and Multiple Stars as Tracers of Stellar Evolution, 73–78.*
*Copyright © 1982 by D. Reidel Publishing Company.*

$\rho(r)$, $P(r)$ and $M(r)$ which describe the structure of the relativistic stars.

From Einstein's field equations we also obtain the following equations (Tooper, 1964)

$$e^{-\lambda} = \begin{cases} 1 - 2GM(r)/c^2r, & \text{for } r < R \\ 1 - 2GM/c^2r, & \text{for } r \geq R \end{cases}$$

$$e^{\nu} = e^{-\lambda} = 1 - 2GM/c^2r, \qquad \text{for } r \geq R \tag{4}$$

$$\tfrac{1}{2}(\rho c^2 + P)\frac{d\nu}{dr} + \frac{dP}{dr} = 0, \qquad \text{for } r < R$$

$$\lim_{r \to R} \nu(r) = \nu(R)$$

where M is the total mass of the star. The system of equations (4) allows the determination of the functions $\nu(r)$ and $\lambda(r)$, which describe the geometry of the space-time continuum inside and outside the relativistic star.

2.  NON-DIMENSIONAL FORM OF EQUATIONS

For the structural and geometrical research of some concrete models of relativistic stars it is convenient to put the equations (3) and (4) in a non-dimensional form, through the transformations (Ureche, 1980 a,b, 1981)

$$r = a\eta, \qquad \rho = \rho_c, \qquad P = \rho_c c^2 p, \qquad M(r) = M^* m. \tag{5}$$

Taking

$$M^* = 4\pi a^3 \rho_c, \qquad a^2 = c^2/4\pi G \rho_c, \tag{6}$$

from (3), (4) and (5) we obtain

$$\frac{dm}{d\eta} = \eta^2 \varphi, \qquad p = p(\varphi), \qquad m(0) = 0, \qquad \varphi(\eta_s) = 0$$

$$\frac{dp}{d\eta} = -\frac{(\varphi + p)(m + \eta^3 p)}{\eta^2(1 - 2m/\eta)}, \qquad p(\eta_s) = 0 \tag{7}$$

$$e^{-\lambda} = \begin{cases} 1 - 2m/\eta, & \text{for } \eta < \eta_s \\ 1 - 2m_s/\eta, & \text{for } \eta \geq \eta_s, \qquad m_s = m(\eta_s) \end{cases}$$

$$e^{\nu} = e^{-\lambda} = 1 - 2m_s/\eta, \qquad \text{for } \eta \geq \eta_s \tag{8}$$

$$\frac{1}{2}(\gamma + p)\frac{d\nu}{d\eta} + \frac{dp}{d\eta} = 0 \quad , \quad \text{for} \ \eta < \eta_s$$

$$\lim_{\eta \to \eta_s} \nu(\eta) = \nu(\eta_s) \equiv \nu_s \tag{8}$$

where $\eta_s$ = R/a is the value of the non-dimensional coordinate $\eta$ at the surface of the star. The system of equations (7) will determine the physical structure of the relativistic star, while the system (8) will describe the geometry of the space-time continuum inside and outside the relativistic star.

## 3. RELATIVISTIC LINEAR STELLAR MODEL

In the study of the newtonian stars, Stein (1966) has used the linear stellar model, showing that this model is useful for the determination of some representative values of the stellar characteristics, as well as for the construction of some non-homogeneous stellar models (with envelopes).

We have studied the properties of the linear stellar model in the frame of GRT (Ureche, 1980 a). In this case the density distribution is given by the law

$$\varsigma = \varsigma_c(1) - r/R) \tag{9}$$

Using the non-dimensional variables, the equation (9) becomes

$$\gamma = 1 - \eta / \eta_s \tag{10}$$

From the equations (7) and (10) we obtain the differential system

$$\frac{dm}{d\eta} = \eta^2(1 - \eta/\eta_s) \ , \quad m(0) = 0$$

$$\frac{dp}{d\eta} = -\frac{(1 - \eta/\eta_s + p)(m + \eta^3 p)}{\eta^2(1 - 2m/\eta)} \ ; \ p(\eta_s) = 0 \tag{11}$$

where the quantities $a$ and $M^*$ have the expressions

$$a = R \sqrt{R/6R_g} \ , \quad M^* = 2M\sqrt{R^3/6R_g^3} \tag{12}$$

The first equation in (11) is immediately integrated, obtaining

$$m(\eta) = \frac{\eta^3}{3}(1 - \frac{3}{4}\frac{\eta}{\eta_s}) \ , \tag{13}$$

that is formally the same expression as in the case of the newtonian
stars.

For the equation of the hydrostatic equilibrium it cannot be
obtained an exact solution as for the newtonian case. In order to
integrate numerically this second equation from (11), we put it in
another form by the change of variable $\eta = \eta_s y$ . So, the non-dimensional density $\mathcal{Y}$ will be written $\mathcal{Y} = 1 - y$, while the equation of
the hydrostatic equilibrium becomes

$$\frac{dp}{dy} = - \frac{\eta_s^2}{2} \frac{y(p - y + 1)(12p - 3y + 4)}{6 - \eta_s^2 y^2 (4 - 3y)} \quad , \quad p(1) = 0 \quad (14)$$

It is easy to verify that the differential equation (14) has an
unique solution, if $\eta_s^2$ fulfills the restriction $\eta_s^2 < 729/128$. This
solution was effectively obtained by numerical integration, using
the Runge-Kutta algorithm (Gill's variant). The obtained results
are given in tables and graphs (Ureche, 1980 a).

For the maximum mass of linear neutron stars we have obtained
$3.4M_\odot$ , if $P \leq (1/3) \rho c^2$ and $4.7M_\odot$ if $P \leq \rho c^2$.

From (8) and (13), with the change of variable $\eta = \eta_s y$,
it results

$$e^\lambda = \begin{cases} (1 - \frac{1}{6} \eta_s^2 y^2 (4 - 3y))^{-1}, & \text{for } 0 \leq y < 1 \\[2mm] (1 - \frac{1}{6} \frac{\eta_s^2}{y}) - 1 \, , & \text{for } y \geq 1 \end{cases} \quad (15)$$

$$e^\nu = \begin{cases} e^\nu, & \text{where } \nu = \nu(y) \text{ is the solution of the differential} \\ & \text{equation (17) for } 0 \leq y < 1 \\[2mm] = 1 - \frac{1}{6} \frac{\eta_s^2}{y} \, , & \text{for } y \geq 1 \end{cases} \quad (16)$$

$$\frac{d\nu}{dy} = \eta_s^2 \frac{y(12p - 3y + 4)}{3\eta_s^2 y^3 - 4\eta_s^2 y^2 + 6} \quad , \quad \text{for } 0 \leq y < 1$$

$$\nu_s \equiv \nu(1) = \ln(1 - \eta_s^2/6) \quad (17)$$

The determination of the function $e^\nu$ for $y \in (0, 1)$ requires
the integration of the differential equation (17) which contains the
function $p(y)$. This function was tabulated (Ureche, 1980 a), but
the use of these tables for the numerical integration of the differential equation (17) is not convenient. From a practical point
of view it is more convenient to integrate simultaneously the

differential equations (14) and (17).

So, we obtain the following system of differential equations

$$\frac{dp}{dy} = -\frac{\eta^2_s}{2} \frac{y(p - y + 1)(12p - .3y + 4)}{3\eta^2_s y^3 - 4\eta^2_s y^2 + 6} , \quad p(1) = 0$$

$$\frac{d\nu}{dy} = \eta^2_s \frac{y(12p - 3y + 4)}{3\eta^2_s y^3 - 4\eta^2_s y^2 + 6} , \quad \nu(1) = \ln(1 - \frac{1}{6}\eta^2_s)$$

(18)

which has an unique solution, if $\eta^2_s < 729/128$.

Using the numerical methods, the functions describing the geo-metry of the space-time continuum are determined. These functions are given in tables and graphs (Ureche, 1981).

For the relativistic linear stellar model an absolute minimum radius (for which $P \rightarrow \infty$) was determined, namely: $R_{min} = 1.335 R_g$. The corresponding absolute maximum mass is (Brecher and Caporaso, 1977) $M_{max} = 6.22 M_o$. This value can be compared with the value of $8M_o$ for the homogeneous model.

The coefficient of gravitational packing (Zeldovich and Novikov, 1971) has the expression

$$\alpha_1 = 12\sqrt{6} \int_0^1 \frac{y^2(1 - y)\,dy}{\sqrt{3\eta^2_s y^3 - 4\eta^2_s y^2 + 6}} - 1$$

(19)

For the typical values of the parameter $\eta^2_s$ ($\eta^2_s \simeq 3 \div 4$), the values of the coefficient $\alpha_1$ are of the order of $0.3 \div 0.5$. This means that, for relativistic linear stars, the gravitational energy can reach 50% from total energy.

REFERENCES

Brecher, K., Caporaso, G., 1977, Ann. New York Acad. Sci. 302, 471.
Hawking, S.W., Ellis, G.F.R., 1977, The Large Scale Structure of Space-Time (in Russian), MIR, Moscow.
Misner, C.W., Thorne, K.S., Wheeler, J.A., 1977, Gravitation (in Russian), MIR, Moscow.
Stein, R.F., 1966, in "Stellar Evolution (eds. R.F. Stein, A.G.W. Cameron), Plenum Press, New York.
Tooper, R.F., 1964, Astrophys. J., 140, 2, 434.
Ureche, V., 1980 a, Rev. Roum. Phys., 25, 3, 301.

Ureche, V., 1980 b, Abstr. of Contrib. Papers, 9th Internat. Conf.
    on GRG, Jena, July 14-19, 1980, Vol. 2, 299.
Ureche, V., 1981, Rev. Roum. Phys. (in press).
Weinberg, S., 1975, Gravitation and Cosmology (in Russian), MIR,
    Moscow.
Zeldovich, Ya.B., Novikov, I.D., 1971, Stars and Relativity,
    Univ. Chicago Press, Chicago-London.

# PART II

# EVOLUTIONARY TRENDS IN WIDE BINARY SYSTEMS

# EVOLUTIONARY TRENDS IN WIDE BINARIES

Peter van de Kamp
Sproul Observatory
Swarthmore, Pennsylvania

## 1. INTRODUCTION

Zdeněk Kopal has kindly invited me, and I have accepted, to
"instruct the theoreticians on known facts". He also asked me to
express my opinion on the relative evolutionary stages of components.

I am essentially an observing astronomer, occupied with stars in
our immediate neighborhood, say within 10 or at most some 25 parsec,
i.e., the lower main sequence and the white dwarf degenerate branch.

I hope that I may perhaps contribute by surveying and reporting
some relevant data. I shall touch on a number of topics, limited
because of selection and lack of knowledge. My contributions to
binary stars lie in the realm of parallaxes, mass-ratios and masses,
- and for the past half century, perturbations, interpreted as unseen
companions, stellar and otherwise. I shall briefly report on some
results, and I shall be wondering and hoping that some trace of stellar evolution may possibly be present in these results. After having
witnessed for more than half a century my own astronomical evolution,
the time has come for me to become more aware of theoretical, evolutionary and cosmological aspects of the cosmic material, I have been play-
ing with so long.

I dare say that basically I deal with <u>wide</u> binaries, i.e., systems
with sufficient separations which make mass-exchange unlikely. Thus
far, I have not been particularly concerned with evolutionary aspects,
but I welcome the occasion to learn more about them.

The word evolution refers to individual components of binary and
multiple stars, but also, as a consequence, to the changing frequency
of the numerical distribution of the objects. In the cases of stars,
evolution cannot be followed over the required long intervals, except
through theoretical considerations. And the results of these may be
tested through the momentary situation, i.e., the various stages of
evolution, reached by different components, as observed at this time.

*Z. Kopal and J. Rahe (eds.), Binary and Multiple Stars as Tracers of Stellar Evolution, 81–103.*
*Copyright © 1982 by D. Reidel Publishing Company.*

Time-probes in the past, attainable for bright very distant objects
such as galaxies and quasars, cannot be used in the study of the binary and multiple structures, which can hardly be studied beyond a
distance, or past, of a few thousand lightyears at the very most.

My presentation is bound to incomplete, very limited and based
on personal experience and preference. May the choice of the scattered subjects be helpful in our soul-searching attempts, suggested
by the title of this Colloquium.

2. "WIDE" BINARIES, ORBITAL MOTION, COMMON PROPER MOTION

How do we define a wide binary? Traditionally visual binaries
are discovered from "resolving" apparently single stars, with adequate
telescopic power. With the exception of Mizar and Alcor, $\epsilon$ 1 and
$\epsilon$ 2 Lyrae, possibly a few others, double stars are not evident to
the naked-eye , and play a minor role in low-optical surveys. But
with larger, even medium-size telescopes, double stars prove to be a
major, possibly the major portion, of the stellar population. To
separate true physical., i.e., gravitationally bound, from optical
binaries, i.e., chance alignments of stars, at widely different
distances, orbital motion ultimately will tell the story. But common
proper motion, if desired, followed by parallax determination furnishes
an infallible proof for physical connection. For example, the wide
pair G 175-34, proved to have been observed long ago as a double:
Stein 2051. Its large proper motion ($\mu$ = 2''37) and large parallax
(p = 0''192) make this a most interesting object.

Most well determined orbits of visual binaries have semi-axes
major a between 15 and 30 a.u.; hardly any orbits are known with
a $\rangle$ 100 a.u. Provisional orbits have been derived for a number of
visual binaries with a between 50 and 100 a.u., and periods between
350 and 700 years. Van Biesbroeck (1957) compares large double star
orbits, on the average, with planetary orbits in the solar system.
For a total of 163 orbits with sufficiently reliable parallaxes he
finds that the greatest numer have sizes of something like 30 a.u.
Selection effects are obvious of course, but there is a marked decrease beyond 100 a.u. The question arises naturally to what
separation binaries do exist. Using the common proper motion criterion, Van Biesbroeck lists 9 nearby wide binaries with reliable
parallaxes and with projected linear separations ranging from 7500
to 44000 a.u., the latter value being 0.21 parsec or 0.7 lightyear;
the corresponding periods would be of the order of ten million years.
Williams and Vyssotsky (1942) have found separations ranging from
1.000 to 50.000 a.u. Tolbert finds an upper limit of about 40.000
a.u. for the intrinsic separations; two systems ADS 1073 and ADS 15434
were found with exceptionally large separations 140.000 a.u. (0.68 pc)
and 220.000 a.u. (1.06 pc).

For these very wide binaries the question arises whether the
binding energy is sufficiently large to protect the binary against

dissolution, i.e., whether the relative orbit is not, or could become, hyperbolic. Observational orbital tests have been proposed by the author (1961).

The extreme spacings could not be a large fraction of the average observed spacing between stellar systems of about 2 parsec in our immediate neighborhood (within 5 parsec). Huang has remarked that the binding energy of a very wide binary system is so small that the system can easily be disrupted by a stellar encounter. Hence, such a system may prove to be unstable and dissociate before completing a few revolutions. A rough estimate for instability would be a separation larger than about one parsec, i.e., one half of the average spacing of stellar systems. The fact that wide binaries exist, suggests that these may have been created comparatively recently.

The spacings of the above mentioned extreme two cases are still below the range of stability given by Huang. Moreover, these two systems have very massive components, about 15 $M_o$, which would render them more resistant to disruption.

## 4. PROPER MOTION BINARIES

Already referred to in Van Biesbroeck's studies, a simple efficient, infallible method for discovering binaries, and automatically excluding any optical ones, is furnished by the obvious fact that for binary components, the proper motions are virtually the same, save for the generally small, relative orbital motion. Their proper motions may be measured, "absolute", or relative, on a background of "fixed" stars. This leads us to the large and fruitful field of common proper motion stars (historically sometimes referred to as 61-Cygni binaries): visual binaries often with large angular (and linear) separation between the two components, and correspondingly large periods.

The foremost discoverer of these common proper motion binaries is W.J. Luyten, who has stressed that the classical visual double star is not representative of the typical binary in space (Luyten, 1969). The vast majority of stars in space are probably main sequence stars, less luminous than the Sun, but the majority of doubles listed in general catalogues, are more luminous than the Sun, and include substantial numbers of red, yellow and blue giants.

The classical discovery surveys based on apparent magnitude have not, or only the weakest criterion, for parallax. The common proper motion surveys have the strongest possible criterion for parallax, and are likely therefore to be more representative of the true spatial situation. The common proper motion approach is easily extended by means of photography to very faint stars, say 21 pg., down to a common proper motion of 0".2 and a minimum separation of 1".5 or less. Hence the importance of the Luyten surveys and analyses. Statistical considerations yield a (geometric) mean value for the semi-axis major

in a.u., $(a= \frac{12s}{\mu})$ where s is the angular separation, $\mu$ the (common)
annual proper motion in arcseconds.  For s = 1", $\mu$ = 0".2m we find
a = 60 a.u.  The wide doubles thus discovered have semi-major axes
from 50 to 100 a.u., and periods from 300 years upward.

To quote Luyten, there is a frightening amount of observational
selection in the finding of these proper-motion wide pairs.  While
all the selection effects resulting from these observational res-
trictions can be easily recognized, it is very difficult, if not
impossible, to make quantitative allowance for them.  For this reason
Luyten emphasizes that all conclusions must be considered as extremely provisional.

From his proper motion survey with the 48-inch Schmidt telescope,
Luyten ultimately expects up to 100.000 wide common binaries, down to
annual proper motion of 0".05 (statistically nearer than 200 parsec)
separations down to 1" and faint companions down to 23 mag.  This
number exceeds by a wide margin the total number of visually discovered double stars nearer than 200 parsec.  These relatively wide
common proper motion pairs constitute observationally the most common
type of binaries in space.  No spectra are known for these faint stars,
but color estimates b a f g k m have been made from photographic
plates, white dwarfs corresponding to color range b to f, red dwarfs
to color range k to m.

## 5.  LONG FOCUS PHOTOGRAPHIC ASTROMETRY

For a detailed presentation of this subject the reader may
consult the author's "Stellar Paths" (van de Kamp, 1981).

It is now time to first say something about the technique and
possibilities of photographic astrometry applied to individual stars
and stellar systems with long-focus optical instruments, the real
hero's of high accuracy astrometric investigations, to quote my late
friend and colleague Joe Ashbrook.  First employed in the beginning
of the current century for measuring accurate stellar parallaxes, the
same technique provides the geometric and dynamical properties of
double and multiple stellar systems (and of associations and star
clusters) and thus plays an obvious role in the study of origin and
evolution of these systems.  The high accuracy results from the large
scale portrayal of a small portion of the sky, usually less than one
degree across, the high quality of photographic emulsions and of
measuring machine, and the differential nature of measurements made
relative to a number of reference stars within small angular distance
of the central "parallax" star.  The ultimate limiting accuracy appears to be something like 0".002 (about 0.1 micron for a representative instrument).  Parallax determinations yield absolute magnitude;
mass-ratio determinations, together with the space-time dimension of
visual binaries, yield masses.  And presently the same photographic
technique leads to the discovery and subsequent study of perturbations
in stellar paths.

The <u>absolute magnitude</u>

$$M = m + 5 + 5 \log p$$

plotted against <u>spectral class</u> furnishes the well-known H-R diagram. Of particular interest for low luminosity stars, whose spectra may not be known, is the color-absolute magnitude relation.  The illustration for stars nearer than 22 parsec (figure 1) clearly reveals the lower main sequences and the white dwarf or degenerate section of the diagram.  Were it not for scatter due to observational errors, the main sequence - especially - would be rather narrower than the diagram indicates.

For a binary the combined mass, expressed in terms of the Sun's mass, is given by the <u>harmonic relation</u>:

$$M_A + M_B = \frac{a^3}{P^2}$$

where the semi-major axis a is expressed in astronomical units of distance, and P in years.  Or, since the parallax p is a required datum, we write

$$M_A + M_B = \frac{a^3}{p^3} \cdot \frac{1}{P^2}$$

where a and p are expressed in arcseconds.  The cube power of p puts a severe limitation on the attainable accuracy for the combined mass, even for relatively nearby systems.  The separate masses may be found from the observed orbital motion of the two components relative to the uniform motion of the center of mass.

<u>Stellar masses</u> thus found faced with absolute magnitude yield the mass-luminosity relation for the lower main sequence and a number of white dwarfs.  An illustration (Fig. 2) is given for the components of visual binaries nearer than ten parsec with well-determined orbits; the degenerate components of Sirius and Procyon, to a lesser extent $o_2$ Eridani B, form striking exceptions to the general mass-luminosity relation which represents the main sequence. Minor deviations are indicated for a few binaries, such as Zeta Herculis and for 85 Pegasi.

6.  WIDE ECLIPSING BINARIES

At this stage we mention two wide binaries with very massive components that have received special intensive observational attention (van de Kamp, 1978).  They are the two long-period eclipsing

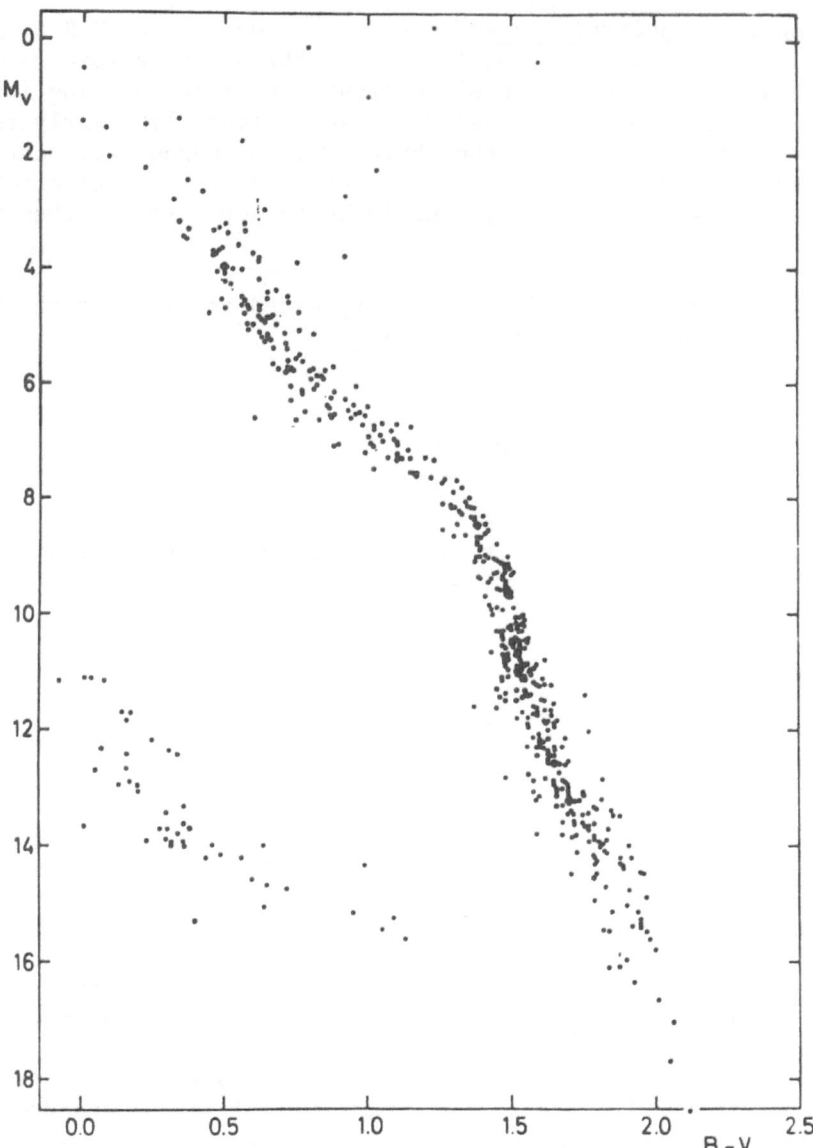

Fig. 1.   Color (B–V)–luminosity ($M_v$) diagram for stars
          nearer than 22 parsec (Gliese).

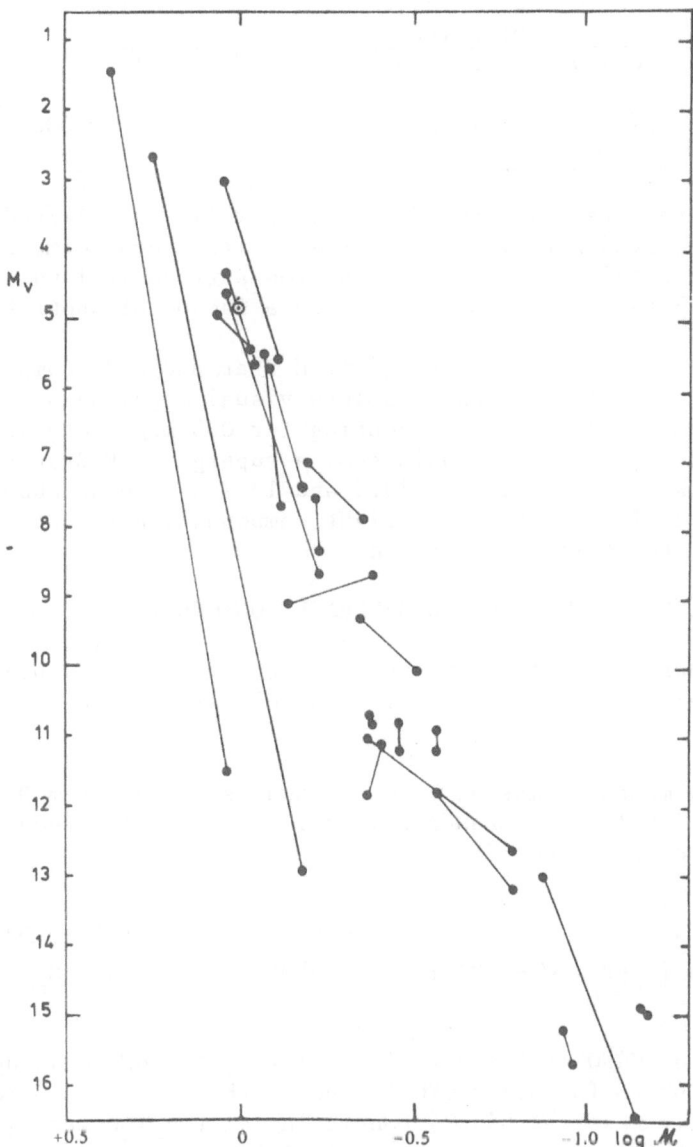

Fig. 2. Mass-luminosity relation for binary components nearer than 10 parsec. The Sun is indicated by ⊙ .

binaries VV Cephei and Epsilon Aurigae.  The relevant data for these
objects are

|  | Period | Semi-major axis of relative orbit | Parallax | |
|---|---|---|---|---|
| VV Cephei | 20.4 yr | 25 a.u. | 0".0014 | $+$ 0".0002 |
| Epsilon Aurigae | 27.08 | 27 a.u. | 0.00172 | 0.00018 |

The very accurate parallaxes are "orbital" parallaxes obtained by
equating the observed astrometric orbit with its linear dimension
obtained spectroscopically which exceeds the astronomical unit, the
base of annual parallax determinations, by a factor of about ten.

VV Cephei consists of a super giant M Star and a B9 companion,
with masses 18.3 and 19.7 $M_o$ and absolute visual magnitudes - 4.3
and - 2.6, respectively, after correcting for 0.3 mag. interstellar
absorption.  Epsilon Aurigae consists of a supergiant F Star, and a
mysterious companion, with masses 15.5 and 13.7 $M_o$ and an absolute
visual magnitude of - 6.7 for the bright component, after correcting
for 0.84 mag. interstellar absorption.

7.  CLASSICAL BINARIES WITH ONE DEGENERATE COMPONENT

In anticipation of reviewing Luyten's studies of wide binaries
with one degenerate component, we list the classical cases of such
binaries:

These, the only measured masses of white dwarfs, range from 0.42 $M_o$
to 0.94 $M_o$, well below the Chandrasekhar limit.  We also record the
case of an unseen white dwarf:

|  |  |  | Parallax | Inferred Mass |
|---|---|---|---|---|
| Zeta Cancri D | 17 yr | $\alpha = 0".191$ | 0".042 | 0.9 $M_o$ |

A recent study at USNO of the most interesting, by now, quadruple
system G 107 - 69/70 (parallax 0".091 reveals the secondary G 107-70
to be a partially resolved binary consisting of two low-mass white
dwarfs, each with masses of approximately 0.5 $M_o$.

8.  PROPER-MOTION BINARIES WITH ONE DEGENERATE COMPONENT:
    STATISTICAL STUDIES

The Luyten proper-motion binaries include particularly interesting pairs.  Six percent of the total studied contain a white dwarf
or degenerate component.  Sixty percent of these pairs are similar to
the classical pair o$_2$ Eridani BC, a white dwarf plus a late main
sequence K or M dwarf.  Luyten considers these as "normal" pairs

Classical Examples of Binaries with One White Dwarf Component

| | Period | Semi-major axis | Parallax | $M_v$ | | Sp | | Masses 1 | Masses 2 |
|---|---|---|---|---|---|---|---|---|---|
| Sirius | 50.09 yr | 19.9 a.u. | 0".377 | 1.4 | 11.2 | A1 | DA | 2.20 M$_\odot$ | 0.94 M$_\odot$ |
| Procyon | 40.65 | 15.9 | 287 | 2.6 | 13.1 | F5 | ... | 1.78 | 0.65 |
| o² Eridani BC | 247.9 | 33.6 | 205 | 11.1 | 12.8 | DA | M4e | 0.42 | 0.20 |
| Stein 2051 | >300 | ~40 | 183 | 12.4 | 13.7 | M5 | DC | 0.22 | 0.48 |

with a white dwarf component. For only one pair, LP 129 - 620/621,
is the white component bolometrically definitely brighter than the
red component: 16.6 pg, f; 20.8 pg, m .

Since virtually no parallaxes are known for any of these proper
motion pairs, Luyten employs the <u>reduced proper motion</u>
$H = m + 5 + 5 \log p$, which has a close relationship to and averages
about 6 units higher than the absolute magnitude $M = m + 5 + 5 \log p$ .
For the average of the components Luyten finds

red stars     :   $H$ = 17.5 pg mag, average color   m   ( k  - m)
white stars :   $H$ = 17.5    "               "       "      a3 (b - f)
$o^2$ Eridani B:   $H$ = 17.8                                a3

A plot of H against color (fig. 3) shows the above average for the
red stars very close to the main sequence, for the white stars well
within the white dwarf area, both established from a study of proper
motion stars in the North Polar Cap.

A similar plot for pairs where either the degenerate component is
later than color f, or the main sequence component earlier than k, or
both, shows that for the great majority, one component is definitely
in the white dwarf area of the diagram. A plot for the dozen or so
pairs for which both components are degenerate (which includes LDS 275,
the first double white dwarf discovered) shows that with one exception
the whiter star is always the brighter of the two, exactly what one
would expect if a single degenerate branch exists. Four m- components
could be degenerate, or because of the crude color estimates, could
be actually very faint main sequence stars.

Common proper motion binaries containing a white dwarf and a
non-degenerate star exist therefore in large numbers. Altogether,
24 of these objects were studied by Wegner (1981); comparative kinematical youth is indicated. Wegner concludes that some white dwarfs
are remnants of fairly massive evolved stars that were originally
above the Chandrasekhar limit of 1,4 $M_o$ for white dwarfs. Studies
by others have also indicated the existence of relatively massive
white dwarf progenitors.

9. SOME THOUGHTS OF EVOLUTION

Cecilia Payne-Gaposchkin and Sergei Gaposchkin (1946) have
studied the spectrum-luminosity relation between components of binaries.
They draw attention to the well-known fact, that associations between
components of practically all physical types are observed, witness
for example the above mentioned VV Cephei and Epsilon Aurigae. They
suggest that the components of one and the same binary may have
different constitutions, and they do not exclude the formation of one
star in the neighborhood of a pre-existing star.

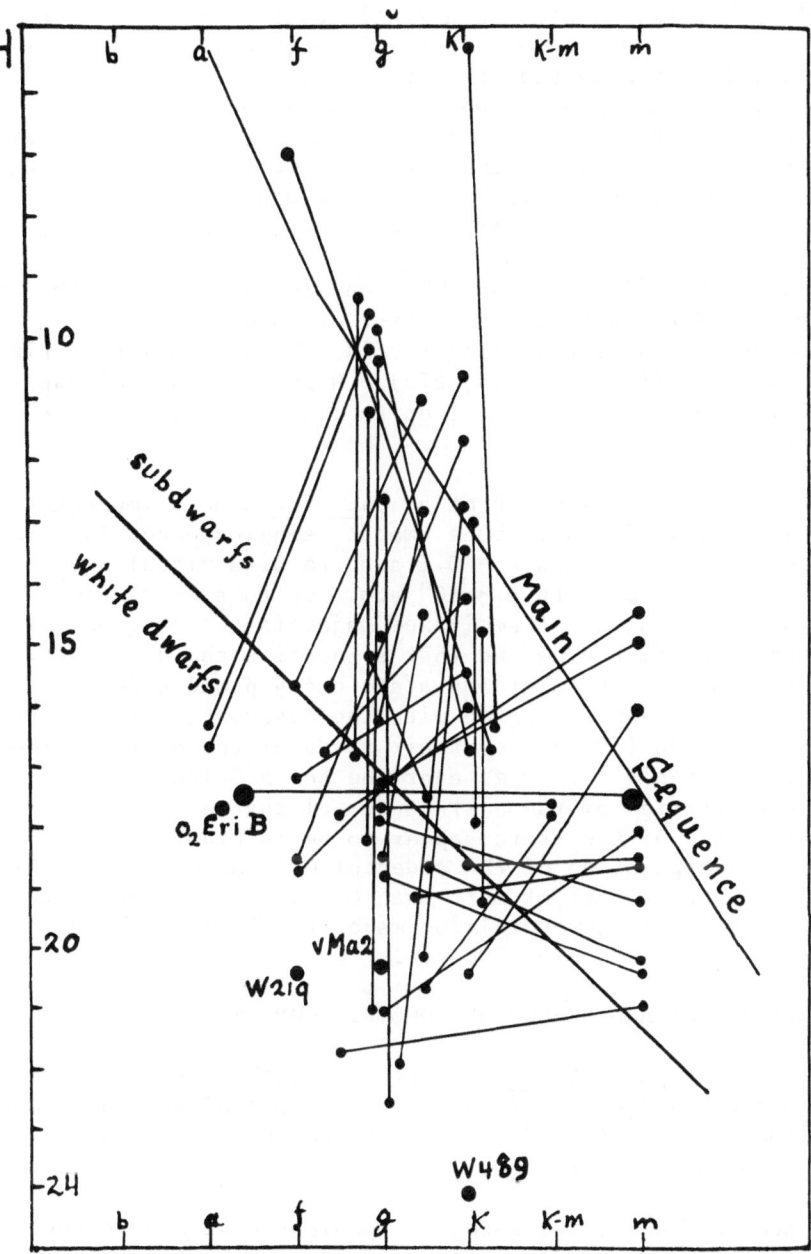

Fig. 3.  Reduced proper motion vs color for proper-
          motion binaries with one degenerate component.

From a study of 94 wide visual binaries Tolbert (1964) finds confirmation for the theories of common origin and evolution. Each component apparently evolved as a single star without close-binary-type interactions. He assumed that all secondaries of luminosity class V (IV, III or A m) had absolute magnitudes on the main sequence; differential colors and magnitudes were used for the primaries. The diagram for the V-primaries is in agreement with evolution, upward and to the right in the H-R diagram with minor exceptions of no consequence. Clearly, these binaries cover a wide range of ages, up to binaries with early-type primaries and late-type secondaries.

We review briefly a recent study on wide binaries in the solar neighborhood and the statistical effect of encounters, by John M. Retterer and Ivan R. King (1981). They recall that single stars may be rare (Abt and Levy, 273). High frequency of duplicity exists in the solar neighborhood, up to 1.3 (i.e., 100 "stars" contain 230 components; Abt, 1978).

The observed distribution of a, <u>one of the most important binary characteristics</u>, is found to be a smooth, single-peaked function from spectroscopic binaries with a = 0.1 a.u. to wide visual and common-proper-motion binaries with a $\sim 10^4$ a.u. Up to a = 2 x $10^4$ a.u. = 0.1 parsec, the distribution of semi-axes major is the original distribution, undisturbed, but beyond that separation the distribution is determined completely by a steady-state decay process due to stellar encounters. This seems to agree with the observational results of Bahcall and Soneira (1981) from their study of the distribution of stars brighter than V = 16 mag. near the North Galactic Pole. The clustering properties of these stars reveal that a significant fraction ( $\sim 14\%$) of the stars appear to be in binaries or triples with a typical separation of the order of 0.1 parsec. Very few binaries have separations of more than 0.1 parsec. Compare this with the separation for the overwhelming majority of visual binaries, which is less than 0.001 parsec or 200 a.u.

10. PERTURBATIONS AND UNSEEN ASTROMETRIC COMPANIONS

The study of <u>perturbations</u> in stellar paths, systematically begun half a century ago has yielded over a score of well determined perturbation orbits. All of these announce the existence of a new hitherto unknown binary system.

For some of these, the unseen companions have been detected visually on the basis of prediction and for many more it is clear that the systems are "normal", i.e., that the companions not seen yet, will not yield any new, shocking information. But a limited number of discoveries made thus far, appear to reveal companions with masses below 0.06 $M_o$, i.e., of a sub-stellar nature; even Jupiter-like companions are indicated in one case.

Perturbations therefore should increasingly contribute to our knowledge of binary systems and unavoidably, this additional information will have bearing on stellar evolution.

The principal orbital data for dynamical interpretation are the semi-axis major (scale) $\alpha$ expressed in astronomical units and the period of revolution P expressed in years. The harmonic relation for the now unresolved astrometric binary is replaced by the mass-function

$$\frac{\alpha^3}{P^2} = (M_A + M_B)(B - \beta)^3$$

where B and $\beta$ are the fractional values mass and luminosity respectively of the companion, relative to the total system.

Or, we may write

$$M_B = \alpha P^{-\frac{2}{3}} (M_A + M_B)^{\frac{2}{3}} + \beta (M_A + M_B)$$

Interpretation of this formula requires adopted values for $M_A$ and for $\beta$ .

For details, the reader is referred to "Stellar Paths", Chapters 13ff.

For the case of a small value of the <u>orbital constant</u> $\alpha P^{-\frac{2}{3}}$, and of no visual evidence for the companion, we may write to a high degree of approximation : $M_B = \alpha P^{-2/3} - M M_A^{2/3}$ .

The first pre-photographic discoveries of the perturbations in the proper motions of Sirius and of Procyon, announced by Bessel in 1844 were followed by visual detection of the faint companions in 1862 and 1896, respectively. The first photographic discovery of a perturbation of the red dwarf Ross 614 (1936) was followed by the visual detection of its companion in 1955. The fainter component of Ross 614 has the smallest known well-determined mass of a visible star, namely 0.06 $M_o$. The second visual detection of an unseen companion followed the photographic study of the perturbation of the G 5 main sequence star VW Cephei (1975).

As long as the companion ascribed to a perturbation is not seen, no final value for its mass can be established; however, generally its mass can be evaluated within fairly narrow limits. Likely locations in the mass-luminosity diagram (Lippincott) are not at all inconsistent with the lower part of the main sequence for components of visual binaries. Exceptions are the white dwarf component of Zeta Cancri C and half a dozen <u>substellar</u> objects. Four of these have masses ranging from about 0.00 5 to 0.02 $M_o$; two appear to be of planetary

nature with masses somewhat below that of Jupiter. Whether these
substellar and planetary companions lie on an extension of zero-age
main sequence toward very low luminosities and very low masses, is not
excluded but remains to be "seen".

A few illustrations of well-established perturbations are given
on Figure 4. For details, consult "Stellar Paths" (van de Kamp, 1981).

A special example of a multiple (in fact double) perturbation is
that of the nearly red dwarf Barnard star, at a distance of 6 lightyears. The pattern of yearly mean residuals may be attributed to
two component perturbations with circular orbits, (Fig. 5) periods of
13.5 and 19.0 years and radii of about 0".01 each. The reality of
these small perturbations is supported by the (Fig. 6) instrumental
profile of the Sproul refractor based on measurements of several stars
without detectable perturbations. An instrumental stability within
0.1 micron or 0".002 in the focal plane is found over the past three
decades.

Interpretation of the perturbation of Barnard's star leads to
masses of about 2/3 that of Jupiter for each component. The orbits
of the components are not far from being co-planar, and could be co-
revolving.

## 11. MULTIPLICITY AMONG BINARIES

The degree of multiplicity beyond binary is high. Finsen and
Worley find that 18% of visual binaries with calculated orbits have
third components while 6% contain four to six components. Similar
results are found from spectroscopic studies by Petrie, by Batten, and
from other samples of binaries. In several of the wide binaries, one
or two of the components are found to be double, and the resulting
system may be quadruple or even more (Sigma Coronae Borealis, Castor,
Zeta Cancri, Xi Ursae Majoris, G 107-69/70).

In the mass-luminosity diagram, no marked difference is shown
within the available accuracy in the behaviour of components of
binaries, whether these are members of multiple systems or not (Fig. 7).

Note that our present knowledge of individual masses and luminosities of white dwarfs is furnished by two binaries (Sirius and
Procyon), by one binary which is part of a triple system (o$^2$ Eridani)
and three binaries which are part of a quadruple system (Zeta Cancri,
Stein 2051, G 107 - 69,70).

A further note: The several papers on triple near collisions,
which I heard recently at Cortina d'Ampezzo, referring among others
to my old friend Carl Siegel, and remarks and discussions with
others, revived and renewed my interest in the general subject of
the origin of binary stars and their evolution. A near collision of
two stars might not result in the formation of a binary; the chances

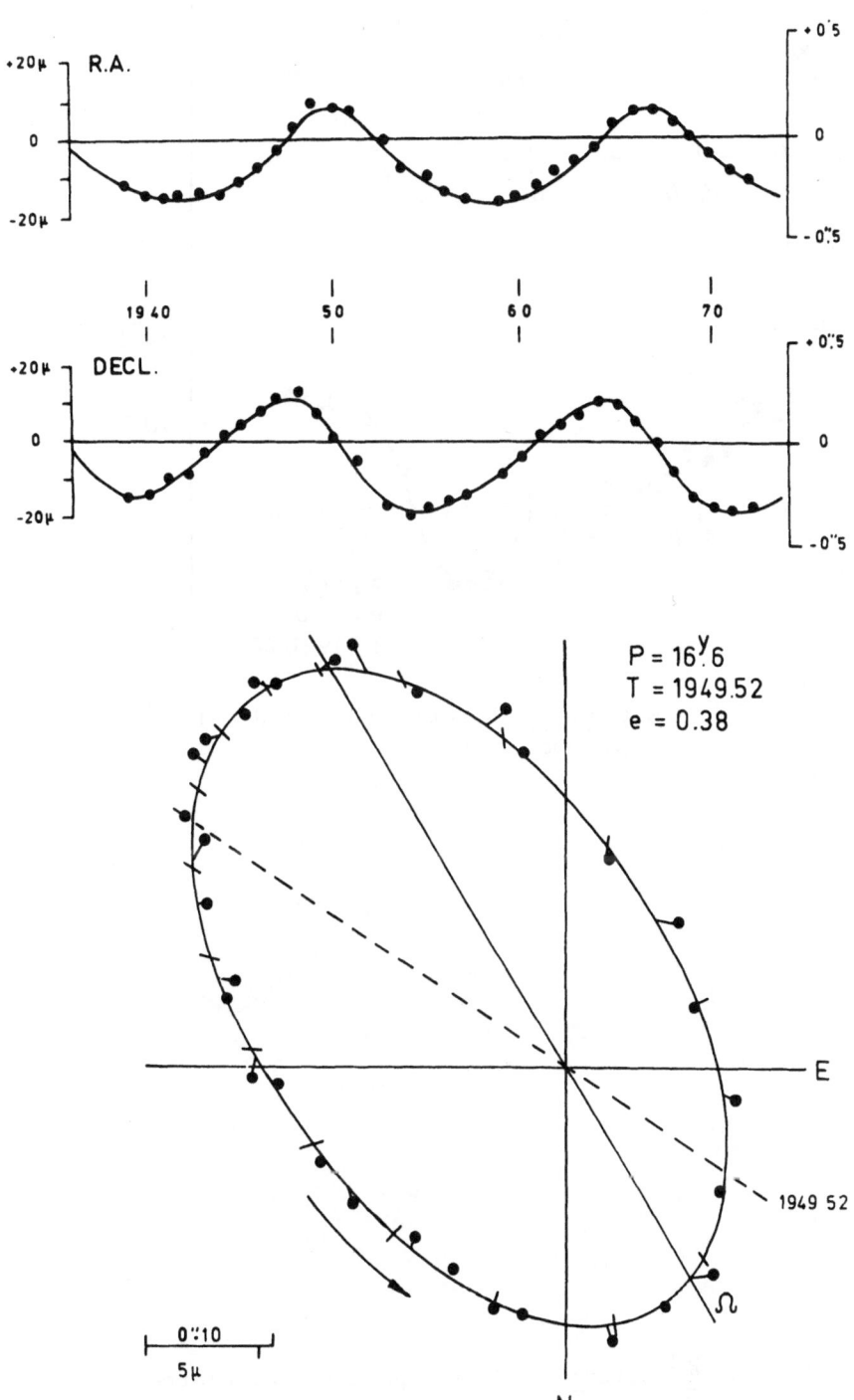

Fig. 4.   <u>Ross 614</u>.  Normal points, calculated displacement curves and
photocentric orbit.   Sproul Observatory.

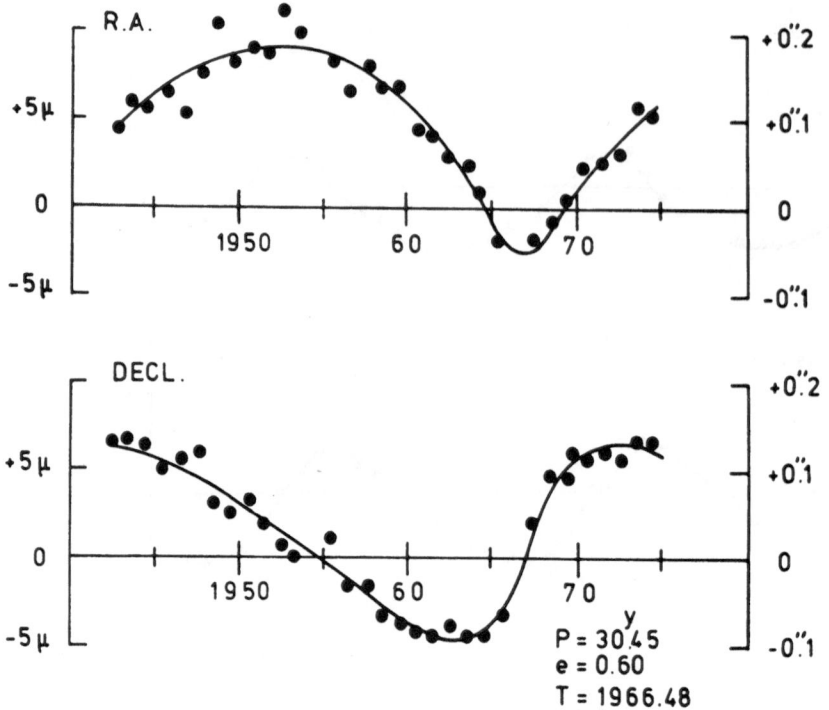

Fig. 4 (cont.).  VW Cephei.  Normal points and calculated
                 displacement curves.

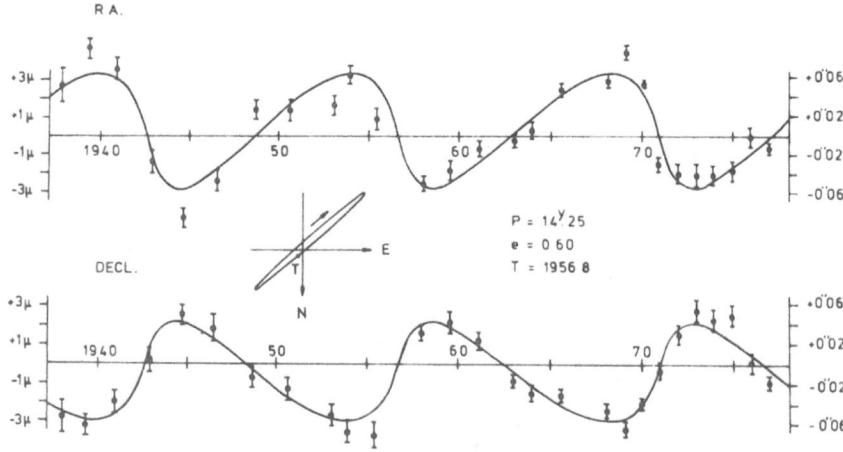

Fig. 4 (cont.). $\chi^2$ Orionis.  Normal points. Calculated
                 displacement curves.  Photocentric orbit.
                 Sproul Observatory.

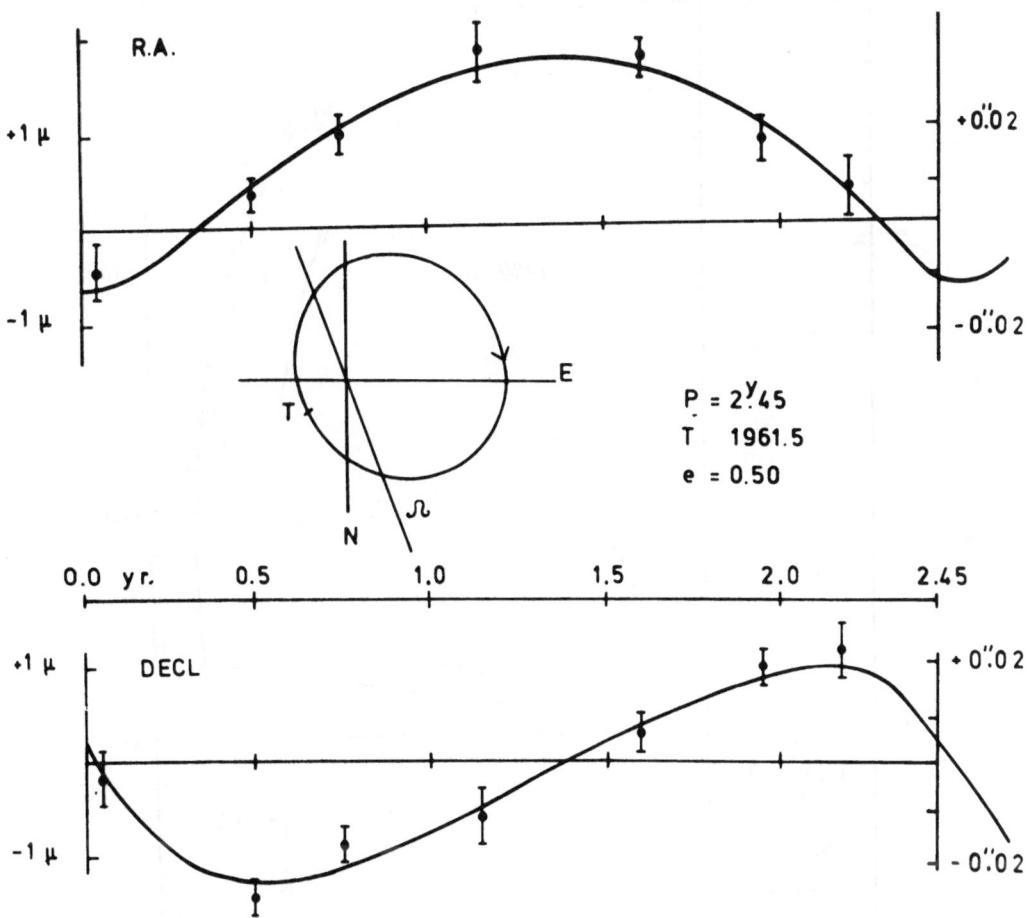

Fig. 4 (cont.). Wolf 1062. Normal points. Calculated
displacement curves. Photocentric orbit.
Sproul Observatory.

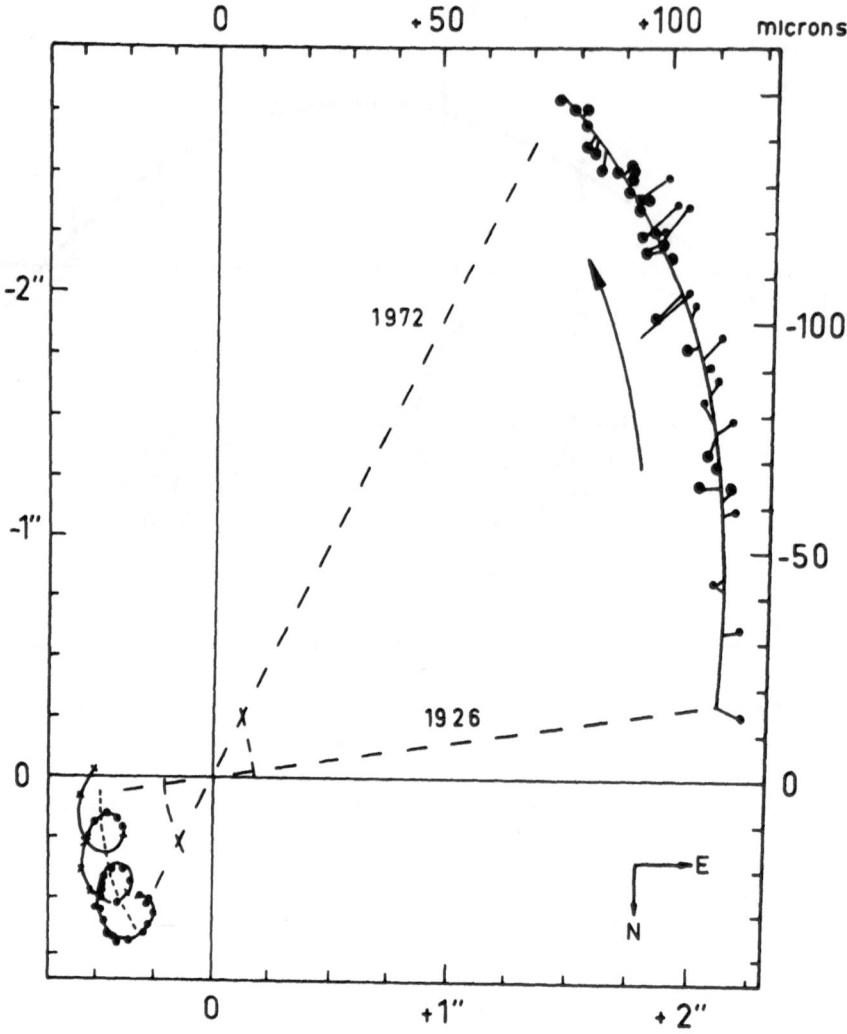

Fig. 4 (cont.). <u>BD +66°34 A.</u> Normal points. Perturbed orbital motion of A with respect to center of mass of A and B. Sproul Observatory.

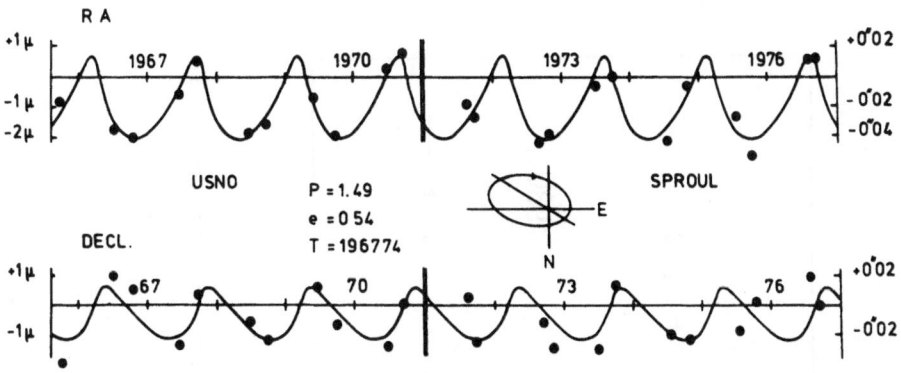

Fig. 4 (cont.).  <u>G 24-16.</u>  Normal points, calculated displacement
curves and photocentric orbit.  USNO and Sproul
Observatory.

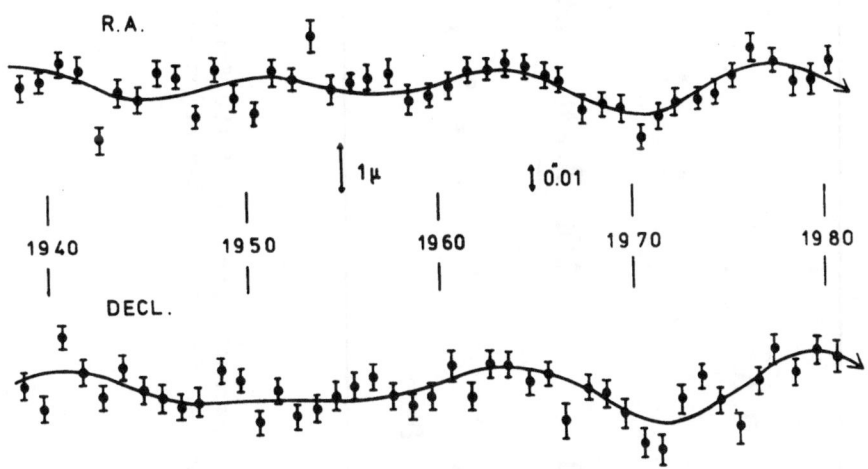

Fig. 5.  Barnard's star.  Yearly normal points and calculated
orbital displacement curves in RA and Decl over the
interval 1938-1980 from Sproul photographs obtained
on 1165 nights.

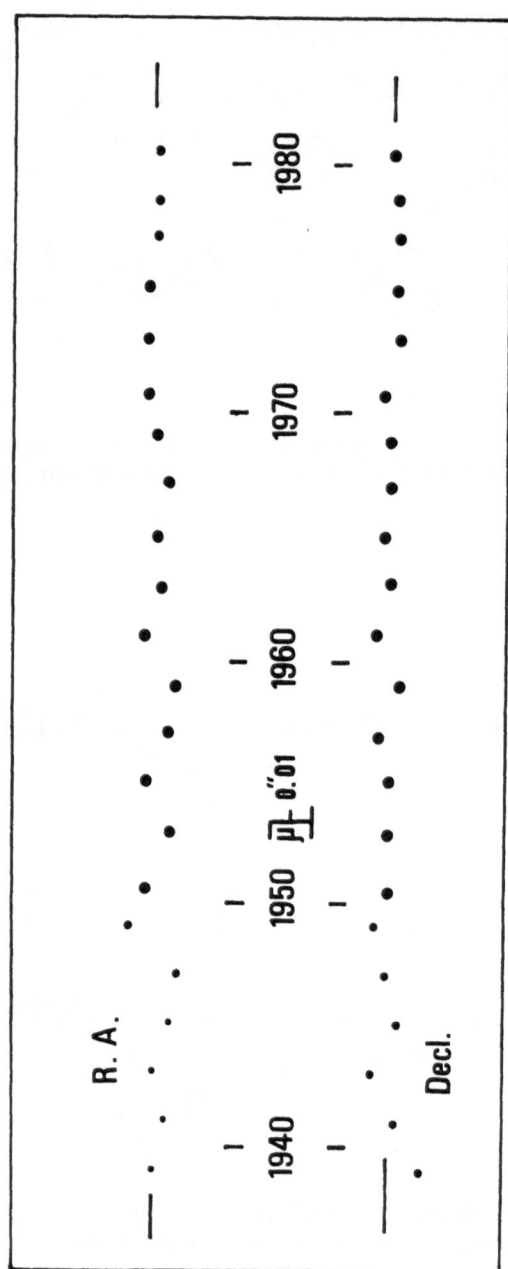

Fig. 6.   Instrumental profile of the Sproul 61 cm refractor, based on 8 photographic series. The average deviation amounts to 0.1 $\mu$ = 0".002.

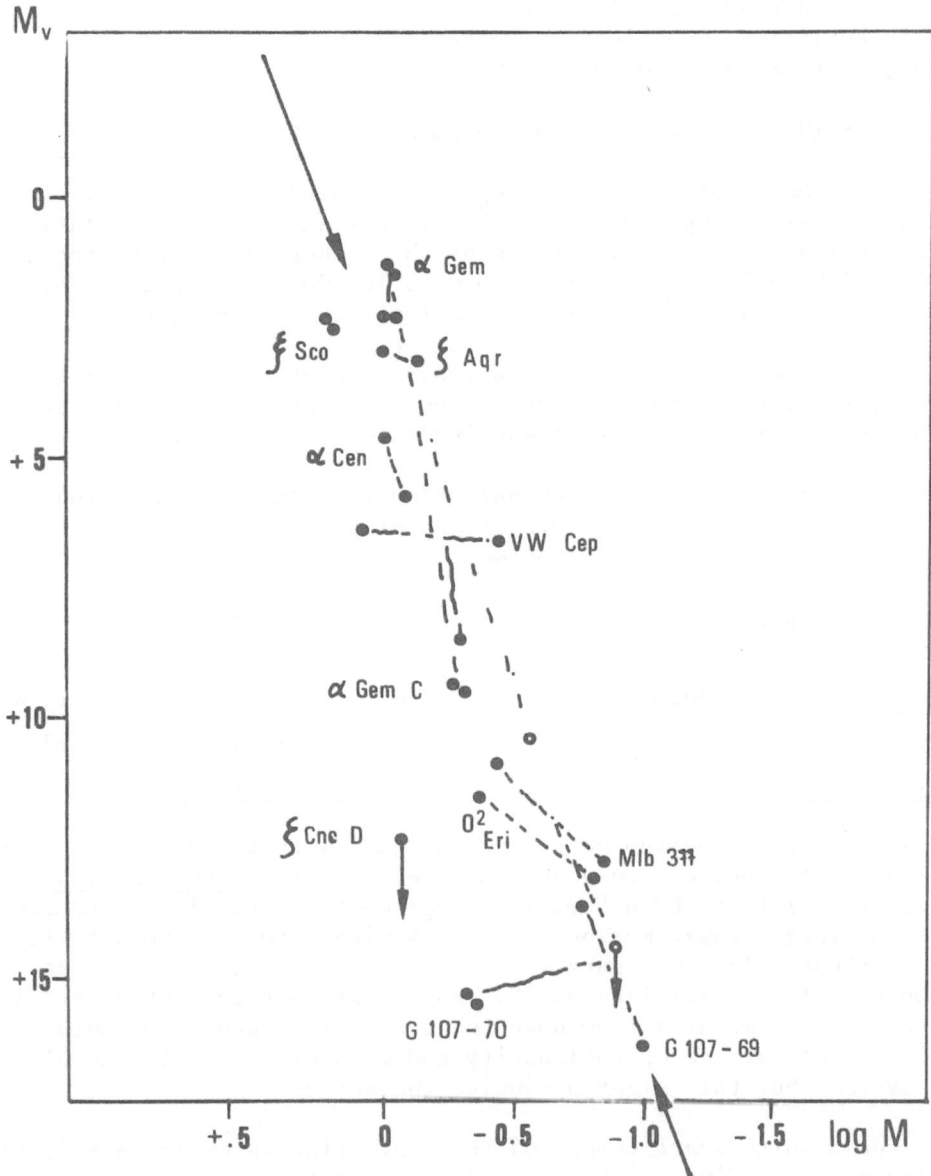

Fig. 7. Masses in multiple stars. The full drawn arrows indicate
the general mass-luminosity relation.

would seem to be better for a binary being left after a near triple collision.  And what about near quadruple etc. collisions and how scarce they likely would be.

Embroidering a little further on this subject.  Could a near triple collision of a binary with two single stars result in a triple star or in two double stars?  And so on.

## 12.  DOES THE SUN HAVE A DISTANT STELLAR COMPANION?

From the observational standpoint the possible existence of a distant stellar companion of the Sun is not excluded.  Calculations were carried out twenty years ago by the author (1973) for circular orbits with radii ranging from 1.000 to 100.000 a.u.  and a range in absolute magnitude of + 15 to +30 for such a companion.

For a companion of small mass, say a faint M star or substellar object, the corresponding range for period, annual proper motion, annual parallax, and apparent magnitude would be as follows:

| Orbital Radius | | Period | Annual Proper Motion | Annual Parallax | Apparent Magnitude for | |
|---|---|---|---|---|---|---|
| a.u. | Parsecs | in yrs | | | abs.mg + 15 | abs.mg + 30 |
| 1.000 | 0.005 | 30.000 | 41" | 210" | - 1.6 | + 13.4 |
| 100.000 | 0.5 | 30.000.000 | 0".041 | 2".1 | + 8.4 | + 23.4 |

The annual proper motion decreases rapidly, with the 3/2 power of the distance.  The object would be characterized by a large parallax but at greater distance by a "small" proper motion.  Parallax measurements of large proper motion stars, say with $\mu > 0".5$ have not yielded yet an abnormally large parallax of say 10" or more; the largest known parallax is still below 1".  Who knows what the future still may bring.  We might narrow down the search for such a companion, by preconceived notions of coplanarity and corevolution with the planetary system, but this might prejudice the search.

For a white-dwarf companion the above figures would be slightly different, depending on the adopted mass for the companion; at the same distance the proper motion would increase somewhat.

REFERENCES

Abt, H.A. 1978. In Protostars and Planets. Univ. of Arizona Press,
    p. 323
Abt, H.A., Levy, S.G. 1981. Ap.J. Suppl. $\underline{30}$, 273
Bahcall, J.N., Soneira, R.M. 1981. Ap.J. $\underline{246}$, 246
Biesbroeck, van, G. 1957. Journ. Roy. Astron. Soc. Canada $\underline{51}$, 35
Kamp, van de, P. 1961. PASP $\underline{73}$, 389
Kamp, van de, P. 1961. PASP $\underline{73}$, 404
Kamp, van de, P. 1978. Sky and Telescope $\underline{56}$, 397
Kamp, van de, P. 1981. Stellar Paths. D. Reidel Publ. Co., Dordrecht
King, D.S. 1981. Ap.J., in press
Luyten, W.J. 1969. Minnesota Publ. XVIII
Payne-Gaposchkin, C., Gaposchkin, S. 1946. Astron. J. $\underline{52}$, 29
Tolbert, Ch.R. 1964. Ap.J. $\underline{139}$, 1105
Wegner, G. 1981. Astron. J. $\underline{86}$, 264
Williams, E.T.R., Vyssotsky, A.N. 1942. PASP $\underline{54}$, 260

# A NOTE ABOUT MULTIPLE SYSTEMS OF DENSE MOLECULAR CLOUDS

V. Vanýsek
Department of Astronomy and Astrophysics, Charles University,
Prague, Czechoslovakia

ABSTRACT

The space distribution of some small dense clouds with point-like IR sources, resembles the clustering of young OB stars. It can be assumed that such objects contain heavy obscured high-luminosity stars on the ZAMS. From the comparison of infrared and radio data it follows that in typical cases, only one B star is the source of the radiation of the cloud. The total mass of the cloud is of the order of one solar mass. If the individual fragments of the cloud are gravitationally unstable, then in the later stage of the evolution only low-mass stars are formed. One can therefore expect that young OB stars are most frequently accompanied by low-mass pre-Main Sequence stars.

Observations of molecular clouds in the radio and infrared spectral region have lead to discoveries of multiple systems of IR sources associated with the dense cores of molecular complexes and accompanied by H II regions, OH and $H_2O$ masers. A few examples can be mentioned: OMC-1, OMC-2, CRL 437, CRL 2591, W3, IRS 5,6,7, Barnard 42 ( $\varrho$ Oph molecular cloud), Mon R2, NGC 7538, S 106, S 140, S 255 and Cep A OB association.

Most of such infrared sources tend to form multiple systems with the average distance of the order 0.1 pc between the individual components and resemble the Trapezium-like clustering of OB stars. The co-existence of OB associations and dense molecular clouds with IR sources can be demonstrated by OMC-1 which is evidently associated with Trapezium stars in the Orion region (Wynn-Williams and Becklin 1974).

The typical bolometric luminosities of such infrared sources are $10^4$ to $10^5$ $L_o$ supporting the idea that these objects are O and B stars in an early evolutionary stage on/or just reaching the ZAMS, obscured in heavy absorbing clouds. The occurrence of recombination lines in the infrared spectra and continuum radiation at GHz frequencies clearly indicates the presence of ionized gas which would be located in

*Z. Kopal and J. Rahe (eds.), Binary and Multiple Stars as Tracers of Stellar Evolution, 105–108.*
*Copyright © 1982 by D. Reidel Publishing Company.*

a fairly dusty H II region inside the cloud core.

The assumption that the underlying energy source within the cloud can be a hot Main-Sequence star is also supported by the silicate absorption feature at 10 $\mu$m which was detected in several infrared sources (IRS). The compact IRS possess some variety of observable properties; however, the pre-Main-Sequence evolution of wide binaries or common-proper-motion pairs (CPM) should preferably be related to the evolutionary processes in multiple systems of small dense clouds with point-like IRS.

It is reasonable to guess that almost the entire radiation of the star is absorbed by the dust and reradiated in the infrared with a maximum near $\lambda \geq 6$ $\mu$m, i.e. corresponding to a black-body temperature $\leqq 450$ K. Since the continuum radiation at 5 GHz is due to photoionization inside the dense core, a comparison of the infrared luminosity and radio data can help to determine the spectral type of the obscured star and the optical depth for the Lyman-continuum of a dusty H II region. Due to the small angular dimension of compact molecular clouds, the useful observational data in the GHz spectral range can only be provided by VLA observations.

Multiple systems of IRS suitable for such a study are for instance, Cepheus A, Sharpless 140 and 255, which were observed by Beichman et al. (1979) with a high VLA angular resolution at 5 GHz, and in the infrared up to $\lambda = 25 \mu$m. The observed properties of the above mentioned IRS are used for an estimate of the spectral type of the embedded star. Table 1 gives the distance and apparent size of the studied objects, the total luminosity $\int L_\lambda \cdot d\lambda$ derived from the 25 $\mu$m flux and the Lyman-continuum photon flux $N_{Lyc}$. All these data are adopted and revised from the paper by Beichman et al. Since Beichman's measurements provide an integral flux for all three IRS in Cep A only, it was assumed that the total luminosity of the individual source is approximately proportional to its size at $\lambda = 6$ cm. The Lyman photon flux $N_{Lyc}$ is derived by Rubin's method (Rubin 1968). For the $\mathcal{T}$ Lyc estimate and the determination of the star type the methods of Panagia (1973) and Natta and Panagia (1976), respectively, were applied. $\mathcal{T}$ Lyc is the optical depth for the Lyman-continuum of the dusty H II region inside the dust core.

The mass of gas and dust $M_{gd}$ in the cloud are estimated under the following assumptions: a) The upper limit of the average column densities of $^{13}CO$ is $10^{17}$ cm$^2$. b) The abundance of isotopic species is terrestrial. c) The abundance ratio of $CO/H_2$ is typical for a compact cloud: 6 x $10^{-5}$ (Thaddeus 1977). d) The contribution of dust to the total mass is considerably less than 0.1 of the gas mass. Although the results depend on the models of high-luminosity ZAMS stars, the obscured stars always tend to be single B0 stars, with the exception of S 255 where there are two possibilities: either a single O 9.5 star or a binary of types B 0.5 + B 0. The narrowness of the star type range is due to the narrow range of allowed combinations of total luminosity, $N_{Lyc}$ and reasonable upper limits for $\tau_{Lyc}$.

Table 1

| IRS | R | Size at 6 cm | $\int L_\lambda \cdot d\lambda$ | $N_{Lyc}$ | Star type | $\tilde{\tau}_{Lyc}$ | $M_{gd}$ |
|-----|---|---|---|---|---|---|---|
|     | kpc | arcsec | $10^4 L$ | $10^{44} s^{-1}$ |  |  | $M_0$ |
| Cep A | 0.73 | | | | | | |
| 1 | | 4x1 | 2.5 | 7 | B0 | 7.6 | 0.45 |
| 2 | | 2x1 | 1.5 | 3 | B0 | 8.0 | 0.30 |
| 3 | | 2x1 | (1.0) | (2) | (B0) | (9.2) | (0.3) |
| S 140 | 0.9 | | | | | | |
| 1 | | 2x1 | 1.4 | 4 | B 0.5 | 4.8 | 0.35 |
| S 255 | 2.5 | 3 | 2.3 | 33 | O 9.5 or B 0.5 + B 0 | 7.0 5.5 | 3 |

The results, of course, must be regarded as approximate and preliminary. The aim of this paper is to show the possible perspective of
such studies. Nevertheless, it seems to be evident that the presence
of only late single O or early B stars in IRS associated with compact
molecular clouds, is more likely than the occurrence of a hot binary
with equal mass of the components.

The mass of the clouds surrounding the IRS may be somewhat under-
estimated but the correct value should be about $1 \le M_o \le 3$. Therefore,
a further collapse in the cloud would result in a low-mass star.
The minimal critical mass $M_c$ of a collapsing fragment is according to
Silk (1978), $M_c \sim (T^6 M_0)/(100 K \tau)$ where T is the temperature of the
dust grains (identical with the gas temperature, i.e. $T \le 100$ K) and
$\tau$ the visual optical depth of the fragment. Assuming that
$\tau \sim 5 \tau_{Lyc} \sim 50$, the minimal value of $M_c$ is about $0.02 M_o$. The
mass distribution of secondaries can therefore be expected to be in
the range $0.02 < M_0 < 3$. This conclusion is in agreement with the fact
that long-period binaries possess a van Rhijn distribution of masses
(Abt and Levy, 1976).

Since the free-fall time and pre-Main-Sequence phase of a star
with $1 M_o$ is about $5 \times 10^2$ longer than that of a typical B0 star, one
can expect that a relatively large number of OB stars on the main-
sequence are accompanied (in wide binaries or CPM systems) by low-
luminosity stars. They are still in the pre-main sequence stage as,
e.g., the T Tau or YY Ori stars.

This paper is part of a Research Project supported by DFG Grant Ra 137/9 in which the author participated during his stay at the Remeis Observatory Bamberg, Astronomical Institute of the University Erlangen-Nürnberg.

REFERENCES:

Abt, H.A. and Levy, S.G.:  1976', Astrophys. J. Suppl. 30, 273.
Beichman, C.A., Becklin, E.E. and Wynn-Williams, C.G.:  1979, Astrophys. J. 232, L 48.
Natta, A. and Panagia, N.:  1976, Astron. and Astrophys. 50, 191.
Panagia, N.:  1973, Astron. J., 78, 929.
Rubin, R.H.:  1968, Astrophys. J. 154, 391.
Silk, J.:  1978, in Protostars and Protoplanets, ed. T. Gehrels, Univ. of Arizona Press, Tucson, p. 172-188.
Thaddeus, P.:  1977, in Star Formation (IAU Symposium No. 75) p. 37-67.
Wynn-Williams, C.G. and Becklin, E.E.:  1974, Publ. Astr. Soc. Pacific 86, 5.

# TRAPEZIUM TYPE MULTIPLE SYSTEMS AND FORMATION OF STARS

G.N. Salukvadze
Abastumani Astrophysical Observatory,
U.S.S.R.

ABSTRACT
    By a comparison of Trapezium-type multiple star systems in the
Abastumani Catalogue and the famous Catalogue of Stellar Associations
and Clusters, it is confirmed that the great number of Trapezium-type
multiple stars, which belong mostly to spectral classes O-B2, are
found in associations and clusters.

    In 13 T-associations, 120 Trapezium-type multiple systems and 182
common multiple and double stars could be detected.

    Based on present-day observational data and the author's photo-
graphic observations, the kinematic of Trapezium-type multiple stars
of spectral class O-B2 is studied.

    In studying the problems of stellar formation and evolution,
special consideration is given to the investigation of unstable
stellar groups and stars at non-stationary states.

    The unstable stellar systems in the galaxy are O-associations,
Trapezium-type multiple systems and O-type galactic stellar clusters.

    In stellar associations, Trapezium-type multiple systems are
observed along with galactic stellar clusters.  These are the multiple
systems, in which, at any rate, three components can be distinguished.
The distances between these components are of the same order.
Prof. V.A. Ambartsumian regards the greater part of Trapezium-type
multiple systems to have a positive energy.  This means that such
systems must expand.  The calculations show that the expansion time is
of the order of $2 \cdot 10^6$ years if the system has a negative energy and
$\sim 10^5$ -$10^6$ years if it is positive.  Therefore, in both cases
Trapezium-type multiple systems are the youngest objects in associations
(Ambartsumian, 1960).

    The presence of a great number of Trapezium-type multiple systems
in some stellar association or galactic cluster indicates that the

*Z. Kopal and J. Rahe (eds.), Binary and Multiple Stars as Tracers of Stellar Evolution, 109–113.*
*Copyright © 1982 by D. Reidel Publishing Company.*

stellar formation process is either still going on in them, or that it has just ceased.

In the works by Profs. V.A. Ambartsumian and B.E. Markarian (1949, 1950, 1951), the problems concerning the relation of Trapezium-type systems with associations and clusters, are considered; the findings derived need some confirmation, as they rely on the highly scarce statistical material.

In Salukvadze (1978) we reported on the Catalogue of Trapezium-type multiple systems compiled by us on the basis of the Index Catalogue of Visual Double Stars. This Catalogue was used for checking the presence of the above mentioned relations.

The Abastumani Catalogue of Trapezium-type multiple stars was compared with the known card catalogues of associations and clusters (Alter et al., 1970). The results of this comparison are given in Table 1.

Table 1

| Spectral Types | Number of Real Trapezia | Number of Trapezia Belonging to | | Percentage of Trapezia Belonging to | |
|---|---|---|---|---|---|
| | | Associ-ations | Clusters | Associ-ations | Clusters |
| O-B2 | 33 | 27 | 21 | 82 | 64 |
| B3-B5+B | 16 | 12 | 6 | 89 | 39 |
| B8-B9 | 13 | 8 | 2 | 61 | 15 |
| Unknown Spectral Type | 96 | 32 | 10 | 34 | 13 |

The first two lines of the Table show that out of the total number of real Trapezia, whose primary stars are of the O-B2 spectral-type, 82% are members of associations and 64% members of clusters. At the same time, the majority of Trapezia are evidently members of both objects mentioned.

The other lines of Table 1 show that among the Trapezia with the primaries of B3-B9 and of unknown spectral types, there is an ap-preciable quantity of Trapezia belonging to associations; that of cluster members decreases sharply.

Based on the above statement, the following conclusions can be drawn:

a)  Belonging of Trapezia to associations is their common
    feature;
b)  A high percentage of O-B2 spectral type Trapezia represents
    the nuclei  of galactic clusters;
c)  In many cases the Trapezia of O-B2 spectral type belong to
    associations and clusters simultaneously.

Belonging of a great majority of Trapezia to associations, established by us, confirms Prof. Ambartsumian's finding that inside
the association not all stars are formed together, but that they
originate in separate groups, clusters and Trapezium-type systems
(Ambartsumian, 1954).

The presence of Trapezium-type multiple systems in associations
is therefore a confirmation of the findings on group formation of
stars.  From this point of view, the search of Trapezia in T-associations is of great interest.

The high percentage of double and multiple stars among the objects in T-associations was first studied by Prof. V.A. Ambartsumian.
Further, at the beginning of the sixties, many investigators of
variable stars (Kukarkin, Herbil, et al.) pointed to the necessity
of studying stellar duplicity in T-associations.  Just then the first
lists of wide pairs in T-associations appeared (Badalian, 1962;
Baize, 1962; Perova, 1963, Zakirov, 1975).

We set a task to search for the Trapezium-type multiple stars
in T-associations involving the groupings of T Tau variable stars
from Kholopov's (1970) list up to a distance of 500 pc.  There are
19 such associations, but we succeeded in searching only 12.

The method and the criteria for excluding optical systems is
described in Salukvadze (1980).  The result is that about 85% of
Trapezia are physical systems.

Assuming that Trapezium-type systems are unstable systems,
V.A. Ambartsumian suggested a hypothesis for the formation of stars
from a so-called "protostar" of a rather high density; it is split
by some presently unknown mechanism.  As a consequence, fast expanding Trapezium-type systems are formed.  From this we can conclude that
kinematic behavior of Trapezium-type multiple systems provides
valuable information on the initial conditions under which stars
are formed.

Based on the present-day observational data including our own
photographic observations, we have analyzed the stability of the
Trapezium-type systems.  Twenty-six such systems were considered
with O-B2 spectral type primaries.  From these, 13 are fairly well
observed.

Table 2

| No. | ADS | Components | dD/dt (") | Over 100 years (") | Associations, Clusters | Distances in pc | Vt km·sec⁻¹ |
|---|---|---|---|---|---|---|---|
| 1 | 719 | AB | 0.012 | 0.057 | NGC 281 | 1100 | 3.0 |
| | | AC | 0.016 | 0.314 | | | 16.4 |
| | | AD | 0.061 | 0.490 | | | 23.2 |
| | | AE | 0.322 | 0.986 | | | 51.4 |
| 2 | 2783 | AB | 0.041 | 0.276 | NGC 1444 | 800 | 10.4 |
| 3 | 2843 | AB | 0.031 | 0.295 | Per OB2 | 330 | 4.6 |
| | | AC | 0.044 | 0.510 | | | 8.0 |
| 4 | 3709 | AB | 0.112 | 4.036 | | | |
| | | AC | 0.040 | 0.971 | | | |
| 5 | 4241 | AC | 0.035 | 0.325 | Orion OB1 | 460 | 7.1 |
| | | AD | 0.014 | 0.064 | | | 1.4 |
| 6 | 4728 | AB | 0.014 | 0.165 | NGC 2169 | 850 | 5.8 |
| | | AE | 0.041 | 0.204 | | | 8.2 |
| 7 | 5322 | AB | 0.013 | 0.102 | Mon OB1 | 760 | 3.8 |
| | | AC | 0.028 | 0.056 | NGC 2264 | | 2.0 |
| 8 | 5977 | AB | 0.048 | 0.430 | NGC 2362 | 1500 | 30.6 |
| | | AC | 0.067 | 0.361 | | | 25.6 |
| 9 | 13374 | AB | 0.060 | 0.299 | Cyg OB3, NGC 6871 | 1580 | 22.4 |
| | | AC | 0.108 | 3.499 | | | 262.0 |
| | | AD | 0.055 | 0.270 | | | 20.2 |
| 10 | 14526 | AB | 0.017 | 0.062 | Cyg OB2 | 300 | 0.9 |
| 11 | 14831 | AB | 0.084 | 0.175 | Cyg OB4 | 345 | 2.9 |
| | | AC | 0.042 | 0.484 | | | 7.9 |
| 12 | 15184 | AC | 0.013 | 0.119 | Cep OB2, IC 1396 | 700 | 4.0 |
| 13 | 16381 | AB | 0.030 | 0.098 | Lac OB1 | 600 | 2.8 |

For each of these 13 Trapezia, the measured distances between
the primary and its components are plotted versus time.  The observational data used in most cases cover a time interval of more than 100
years.  The observations are performed at different times by different
authors.  Various observers were assigned different weights depending
on the internal agreement of their measurements.  Then the observations
were treated by the least square method.

Having carefully inspected all the plots and considered the
results of calculations, it is found that out of 13 studied Trapezia,
six show expansion for all components.

Table 2 gives the variations of the distances for 13 Trapezia
under investigation in a time span of 100 years.

As the sampling of these 13 Trapezia was based only on the
availability of a considerable number of measurements over a long
period of time, i.e., as the criteria of the sampling was not related
to the character of the observational motions, the derived predominance of positive values in the distance measurements over negative
indicates that the Trapezia components showing relative motions are
not field stars.  If the velocities are equal to zero, then one can
hardly find systematic errors, which seems to imply that the component
always moves away from the primary.

Table 2 also shows the results of computed tangential velocities
over 100 years with the rms errors.

Hence, the results of our investigations, based on the treatment
of observational data, confirm the suggestion that a considerable
number of Trapezium-type multiple systems are unstable systems.  The
latter statement favors the hypothesis for the formation of stars from
superdense protostars.

REFERENCES

Alter, G., Balzs, B. and Ruprecht, J.:  Catalogue of Star Clusters and
     Associations, Budapest, 1970.
Ambartsumian, V.A. and Markarian B.E.:  Soobsh. Bjurakanskoi Obs. 2,
     3, 1949.
Ambartsumian, V.A.:  Soobsh. Bjurakanskoi Obs. vol. 15, No. 1, 1954.
Ambartsumian, V.A.:  Nauchnie Trudi, Vol. 2, Erevan, 1960.
Badalian, G.S.:  Soobsh. Bjurakanskoi Obs. Vol. 31, 1962.
Baize, M.P.:  TO, vol. 45, No. 6, 1962.
Kholopov, P.N.:  Eruptive Stars, Moscow, 1970.
Markarian, B.E.:  Soobsh. Bjurakanskoi Obs., Vol. 5, No. 3, 1950.
Markarian, B.E.:  Soobsh. Bjurakanskoi Obs., Vol. 9, No. 3, 1951.
Perova, N.B.:  Peremenye Zviozdy, Vol. 14, No. 5, 1963.
Salukvadze, G.N.:  Astrophysika, Vol. 14, No. 1, 1978.
Salukvadze, G.N.:  Astrophysika, Vol. 16, No. 3, 1980.
Zakirov, M.M.:  in "Investigation of Extremely Young Stellar
     Complexes", Tashkent, 1975.

# A HYPOTHESIS ON THE RICHNESS OF A GALAXY IN BINARY STELLAR SYSTEMS

Hans Zinnecker

Max-Planck-Institut für Physik und Astrophysik
Institut für Extraterrestrische Physik
D-8046 Garching, W.-Germany

## I. INDRODUCTION

This contribution is concerned with the origin of binary stellar systems (cf. Abt 1977, Huang 1977). The hierarchical fragmentation scheme of rotating interstellar clouds, proposed by Bodenheimer (1978), is combined with the role of magnetic fields in the early stages of star formation (Mouschovias 1978, Dorfi 1981). The possible influence of the local strength of the mean interstellar magnetic field on the local fraction of binary systems in a galaxy is stressed.

## II. THE BASIC IDEA

The hierarchical scheme of Bodenheimer (1978) is conceived to solve the angular momentum problem of star formation in several steps by conversion of the spin angular momentum of a fragment into orbital angular momentum of two or more subfragments, without transport of angular momentum. It is based on successive 'ring formation and ring fragmentation' during hydrodynamic collapse (Larson 1972, Tohline 1980, Norman and Wilson 1978). The effect of magnetic fields is not incorporated in the scheme. Magnetic fields essentially cause a magnetic braking of the rotation of diffuse interstellar clouds up to a gas density at which ambipolar diffusion (Mestel and Spitzer 1956) starts to decouple the magnetic field from the bulk of the gaseous matter (e.g. Mestel 1977, Spitzer 1978). This is believed to take place at a gas density of about $10^{-19}$ g/cm$^3$, where the fractional degree of ionization becomes low enough, i.e. of order $10^{-8}$ (cf. Nakano 1979, Elmegreen 1979). Up to this density the torques of the frozen-in magnetic field help to solve the angular momentum problem of star formation (Mestel and Paris 1979, Mouschovias and Paleologou 1980). After the decoupling the residual angular momentum problem may be solved along the lines of hierarchical fragmentation suggested by Bodenheimer (1978) which is supported by the recent observations of Fekel (1981).

It is the basic idea of this note to point out the hypothesis that the fraction of binary systems in a galaxy may depend on how much of the angular momentum problem of star formation can be solved by magnetic

115

*Z. Kopal and J. Rahe (eds.), Binary and Multiple Stars as Tracers of Stellar Evolution, 115–117.*
*Copyright © 1982 by D. Reidel Publishing Company.*

torques, and how much of the angular momentum problem is left over to the cascade process of Bodenheimer (1978). Of course, the issue that I have raised should actually be discussed in terms of mechanisms promoting binary star formation versus mechanisms promoting single star formation. However, as long as astrophysicists do not have a theory of turbulence and angular momentum transport, it is very difficult to assess the importance of the mechanisms promoting single star formation (cf. Tscharnuter 1980). Here I am dealing only with the first half of the problem; the conclusion will be tentative to the extent that I shall neglect the second half of the problem.

The amount of magnetic braking depends on the critical gas density $\rho_{crit}$ above which magnetic braking becomes inefficient due to the ambipolar diffusion of the magnetic field; $\rho_{crit}$ is determined by the requirement that the braking timescale is equal to the ambipolar diffusion time-scale. According to equ.(12) in Mouschovias (1978) this requirement leads to

$$\rho_{crit} \propto M_{cl} \; n_{ion}^{3/2} \; \bar{B}^{-3/2}, \tag{1}$$

where $M_{cl}$ is the cloud mass, $n_{ion}$ is the number density of charged particles, and $\bar{B}$ is the mean interstellar magnetic field. From the above formula one may infer a systematic tendency that the higher the mean interstellar magnetic field the more magnetic braking occurs (on the implicit assumption that the other parameters remain constant). For this reason it is interesting theoretically to know the local strength of the mean interstellar magnetic field. The following model of the local strength of the mean interstellar magnetic field is presented: Imagine a uniform primordial seed magnetic field which gets amplified during the nonhomologous collapse of the proto-galaxy in a magneto-hydrokinematic manner (the energy density in the galactic magnetic field is always much less than the gravitational energy density of the galaxy). From conservation of mass and magnetic flux it is found (e.g. Mestel 1965) that the local mean magnetic field strength $\bar{B}(r)$ is proportional to the 2/3th power of the local mean interstellar gas density $\bar{\rho}(r)$ in a galaxy, i.e.

$$\bar{B}(r) \propto [\bar{\rho}(r)]^{2/3}. \tag{2}$$

r denotes the galacto-centric distance (either in a spiral or in an elliptical galaxy). Equs.(1) and (2) combine to the simple relation

$$\rho_{crit} \propto \bar{\rho}(r) \tag{3}$$

which says how the local effect of the magnetic field in star formation is related to the global density structure of a galaxy (the gas density in a galaxy is roughly proportional to the total matter density).

## III. PREDICTION

From the foregoing considerations I predict that there is a gradient of the percentage of binary stellar systems in a galaxy, in the sense that the central regions will not be as rich in binary systems as the outer regions. It might be that the dark halos of galaxies consist of a very large number of very low-mass binary systems (cf. Zinnecker 1981 a,b).

REFERENCES

Abt H. (1977): Revista Mexicana de Astronomia y Astrofisica Vol.3 (IAU-Colloqu. No. 33 Special Issue)

Bodenheimer P. (1978): Ap. J. 224, 488.

Dorfi E. (1981): Ph.D. Thesis (Univ. Wien)

Elmegreen B.G. (1979): Ap. J. 232, 729.

Fekel F. (1981): Ap.J. 246, 879.

Huang S.S. (1977): Revista Mexicana de Astronomia y Astrofisica Vol. 3 (IAU-Colloqu. No. 33 Special Issue)

Larson R.B. (1972): M.N.R.A.S. 156, 437.

Mestel L. (1965): Quart. J. Roy. Astr. Soc. 6, 265.

Mestel L. (1977): IAU-Symp. No. 75 (de Jong T. and Maeder A., eds.), p.213.

Mestel L. and Spitzer L. (1956): M.N.R.A.S. 116, 503.

Mestel L. and Paris R.B. (1979): M.N.R.A.S. 187, 337.

Mouschovias T. (1978): in Protostar and Planets (Gehrels T., ed.),p.209.

Mouschovias T. and Paleologou (1980): The Moon and the Planets 22, 31.

Nakano T. (1979): P.A.S.J. 31, 697.

Norman M.L. and Wilson J.R. (1978): Ap. J. 224, 497.

Spitzer L. (1978): Physical Processes in the Interstellar Medium (Wiley, New York)

Tohline J.E. (1980): Ap. J. 236, 160.

Tscharnuter W.M. (1980): IAU-Symp. No. 93 (Kyoto).

Zinnecker H. (1981a): Ph.D. Thesis (TU München)

Zinnecker H. (1981b): in preparation

# A MASS-ECCENTRICITY CORRELATION IN SPECTROSCOPIC AND VISUAL BINARY ORBITS

J. DOMMANGET

Royal Observatory, Belgium

Looking for a possible explanation for the debated existence of a correlation between period and eccentricity in binary orbits (of which a diagramme is given in fig. 1.) we made the assumption that the orbital evolution of the binaries could be the consequence of a substantial mass-loss of their components even when these components are late type main sequence stars.

This led first to the consideration of various classes of the areal constant. But it immediately appeared much better to consider for each of such classes, the total mass of each system instead of its orbital eccentricity and thus to consider the mass-period diagramme rather than the period-eccentricity diagramme. Eleven such diagrammes were considered. Figure 2.- gives only the five most typical ones established already in 1963 on the basis of some 212 visual and 98 spectroscopic selected binaries.

One can easily recognize on each diagramme, the existence of an upper limit to the masses of the binaries as a function of the period. It is important to mention that this limit is not due to any selection effects : binaries above this limit should have been discovered and observed more easily than those situated below and would have led to an orbit generally under better conditions.

In order to assure a better definition of this limit, all diagrammes have been piled up after an appropriate shift in abscissae. The final diagramme is given by figure 3.-, where :

$$X = \log P - 3 \log C,$$

C being the areal constant.

The first two diagrammes are the ones obtained with the 1963 material. The other two have been obtained recently on the basis of some 550 visual pairs. No change is observed !

*Z. Kopal and J. Rahe (eds.), Binary and Multiple Stars as Tracers of Stellar Evolution, 119–122.*
*Copyright © 1982 by D. Reidel Publishing Company.*

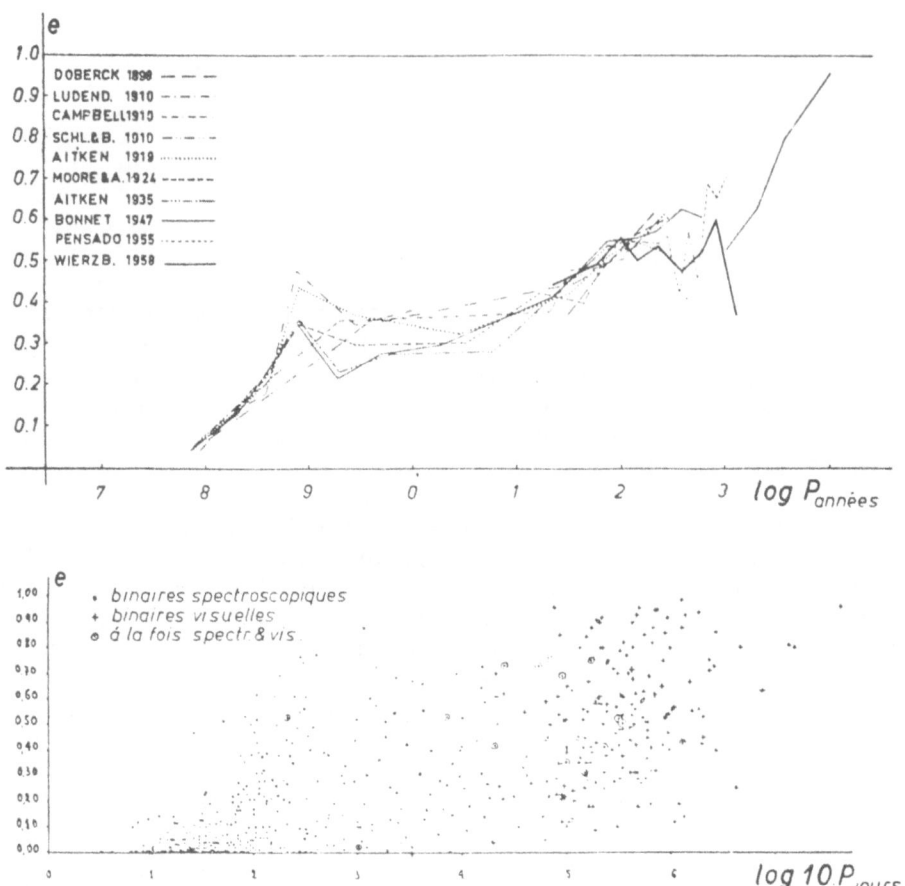

Fig.1.- Period-eccentricity correlation.- Representative curves proposed by various authors (above) and statistical material considered by
R. Bonnet (below) - J. Dommanget, 1963.


        The upper limit is well represented by the equation :

$$e^{2,8} \, \mathfrak{M}_{AB} = 3,60$$

        Any theory on binary formation and evolution should be able
to explain this limit.

        Details on this research will be found in a paper given
(J. Dommanget, 1981) at a recent I.A.U. Colloquium (n° 59) held at
Trieste in septembre last year.

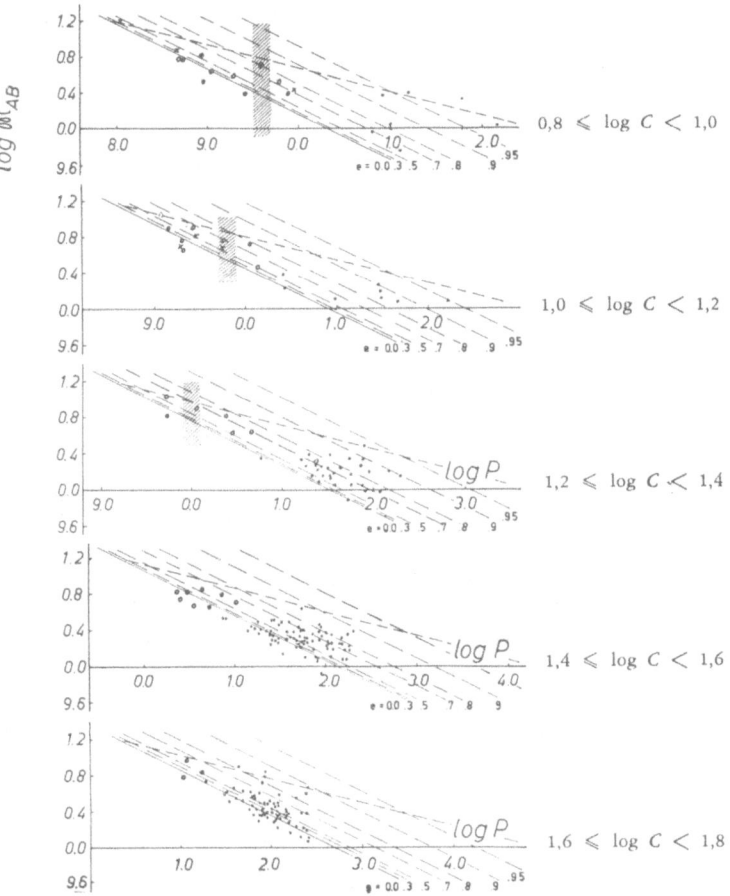

Fig.2.- The five most typical of the eleven diagrammes (log P, log $\mathfrak{M}_{AB}$) established by the author in 1963 and showing on each of them, the existence of an upper limit to the masses of the binaries as a function of the period.

## BIBLIOGRAPHY

J. DOMMANGET, 1963.- Recherches sur l'évolution des étoiles doubles, par voie statistique et par application de la mécanique des masses variables, - Annales de l'Observatoire Royal de Belgique, 3ème Série, tome IX, fasc. 5;

J. DOMMANGET, 1964.- Les étoiles doubles et l'évolution stellaire, - Ciel et Terre, 80, p. 315.- Communication de l'Observatoire Royal de Belgique, n° 232.

J. DOMMANGET, 1981.- Is this diagramme an argument for binary orbital evolution due to mass-loss ? - I.A.U. Colloquium n° 59, Effects of Mass-Loss on Stellar evolution, p. 507. - Communication de l'Observatoire Royal de Belgique, B, 119.

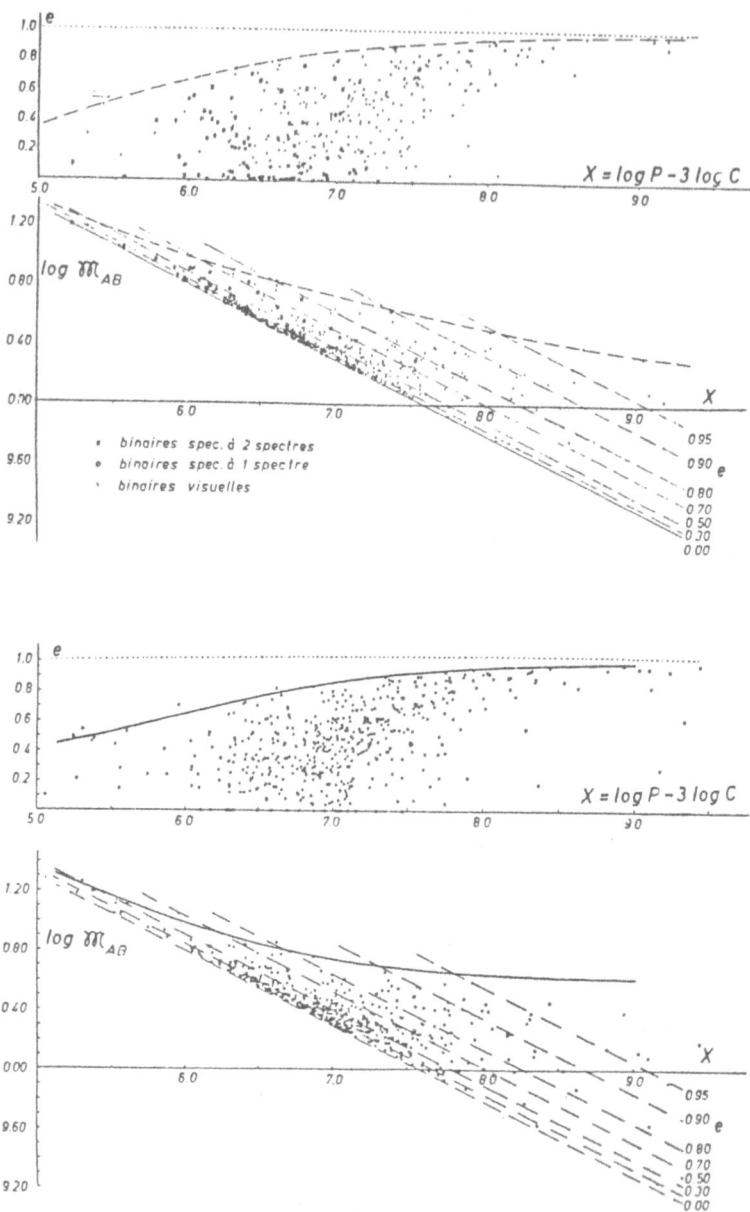

Fig.3.- Diagrammes (X,e) and (X, log$\mathfrak{M}_{AB}$) established by the author in 1963 (above) and with a more recent material (below) (J. Dommanget 1981). The curved lines are well represented by the equation :
$$e^{2.8}.\mathfrak{M}_{AB} = 3.60.$$

# PHOTOMETRIC INVESTIGATION OF VISUAL BINARIES WITH A COMPONENT ABOVE THE MAIN SEQUENCE

O. G. Franz
Lowell Observatory, Flagstaff, Arizona 86002, U.S.A.

*Abstract.* Visual binaries containing a component above the main sequence in combination with a main-sequence star provide an excellent opportunity for determining absolute magnitudes of evolved stars. Some aspects of this technique are discussed, and preliminary results are presented.

## 1. INTRODUCTION

Studies of the photometric and spectroscopic properties of visual binaries containing a non-main-sequence star in combination with a main-sequence component have long provided a useful method for determining the luminosities of evolved stars. Well-known examples of such investigations include the pioneering work by Leonard (1921, 1923) and the more recent studies by Bidelman (1958), Eggen (1956, 1965), Slettebak (1963), Stephenson (1960), and Stephenson and Sanwal (1969), the latter representing a comprehensive investigation of the masses of stars above the main sequence.

One difficulty usually encountered by these investigators was the lack of accurate magnitudes or magnitude differences for the components of visual binaries. Good magnitudes of the individual components of double stars available through the work of Johnson (1953) and Eggen (1963) generally remained limited to pairs of angular separation greater than 10 arcsec.

For some closer pairs, magnitude differences measured by visual or photographic techniques were available and were generally taken from the catalog of Wallenquist (1954). In many instances, however, the only magnitude information in existence consisted of the visual estimates listed in the "Aitken Catalog" (1932) and in the *Index Catalogue of Visual Double Stars* (Jeffers *et al.*, 1963). In the case of small magnitude differences, the absence of accurate values should have had little effect on the resulting luminosity determinations of non-main-sequence stars. Large magnitude differences, however, whether estimated or determined by visual techniques, could have errors large enough to have

123

*Z. Kopal and J. Rahe (eds.), Binary and Multiple Stars as Tracers of Stellar Evolution, 123–128.*

caused serious errors in the absolute magnitudes obtained for evolved
stars by main-sequence fitting of their dwarf companions. When planning
a program of double-star photometry by photoelectric area-scanning, I
therefore included in the observing program a representative sample of
pairs of this type in order to examine this problem and to provide
improved photometric data for its partial solution.

## 2.  THE OBSERVATIONS

While a prototype area scanner developed by Rakos (1965) was used for
initial exploratory observations (Franz, 1966), the observational program
itself was carried out with an area scanner designed and constructed at
the Lowell Observatory.  This instrument and the associated data acquisition system in its original configuration have been described in some
detail elsewhere (Franz, 1970).

The observing program, started in 1969, contains about 350 objects of
which nearly 300 either have been completed or have at least some observations.  Among these are more than 100 pairs either previously known or
suspected on the basis of the new photometry to contain at least one
non-main-sequence component.  Some are newly detected variables or variable suspects found as a result of an effort made throughout the program
to observe each object on several nights during several observing seasons

## 3.  RESULTS AND DISCUSSION

For 15 of the pairs previously known and studied as above-the-main
sequence objects and sufficiently observed in this investigation, MK clas
sifications given by Stephenson (1960) or Stephenson and Sanwal (1969)
show that each pair consists of an evolved star and a main-sequence companion.  By fitting their dwarf companions to a standard main sequence
according to their MK types, absolute magnitudes and colors for the
evolved stars are readily obtained from the observed UBV magnitude differences.  Figure 1 shows the Hertzsprung-Russell diagram obtained in
this manner for the 15 pairs with the use of the standard main sequence
according to Allen (1973).  Solid dots denote MK classifications by
Stephenson (1960) or Stephenson and Sanwal (1969), while crosses indicate
classifications adopted by them from the work of others.  Also shown for
each evolved star is its assigned luminosity class.  The marked lines
represent schematically the standard branches of luminosity classes Ib
through IV according to Allen (1973).  Inspection of Figure 1 shows at
once that stars of luminosity class III according to Stephenson (1960) or
Stephenson and Sanwal (1969), while forming a well-defined giant branch,
*all* lie on the average nearly one magnitude below the standard sequence
of luminosity class III giants.  Before further examining this discrepanc;
attention is drawn also to the pairs containing a component of luminosity
class IV whose absolute magnitudes are in satisfactory agreement with the
mean sequence according to Allen (1973).  Also shown in Figure 1 is the
position of the Ba-star component of ADS 8448 according to the MK classification and photometry by Culver *et al*. (1977) and that of two M-type
supergiants.  One of them, namely ADS 10074A (Antares), has recently been
classified by White (1981) as M1Iab.

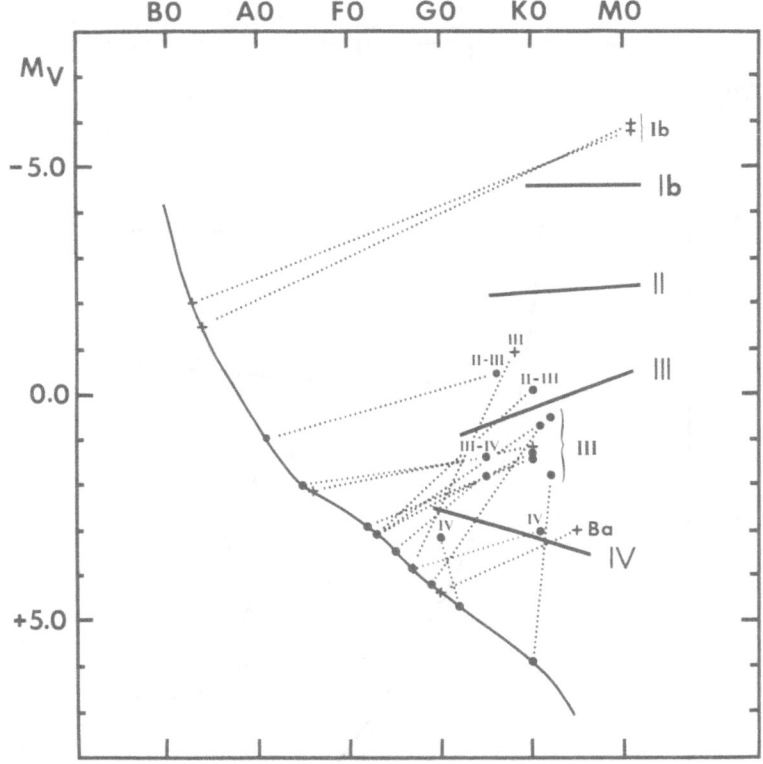

Figure 1.    Hertzsprung-Russell diagram of 16 visual
binaries  consisting  of a main-sequence star and an
evolved component.

A color (B-V) versus absolute magnitude ($M_V$) diagram is shown in Figure 2.
Note that the supergiant ADS 14864A (HR 8164), whose classification is
M1.1Ib according to White and Wing (1978), thus virtually the same as
that of Antares A, is much bluer than the latter.  This is probably due
to the presence in its spectrum of a B-type component (Bidelman, 1954)
which would have significant effect upon the (B-V) color but little on
the V magnitude.  The systematic difference between the newly determined
absolute magnitudes of class III giants and the standard giant branch is
again readily apparent.

While there could be numerous causes for this discrepancy, one most disturbing would be a difference resulting from the use of two principal
calibration techniques, namely luminosity calibrations based upon trigonometric and statistical parallaxes on the one hand, and main-sequence
fitting on the other.  Fortunately, this problem has only most recently
been resolved through the work of Abt (1981) on MK classifications of
visual multiple stars.  Since only two of the 15 pairs have new classifications by Abt (1981), a star-by-star comparison with previous classifications is not possible.  However, in examining systematic and random
classification errors, Abt (1981) finds that a comparison of his

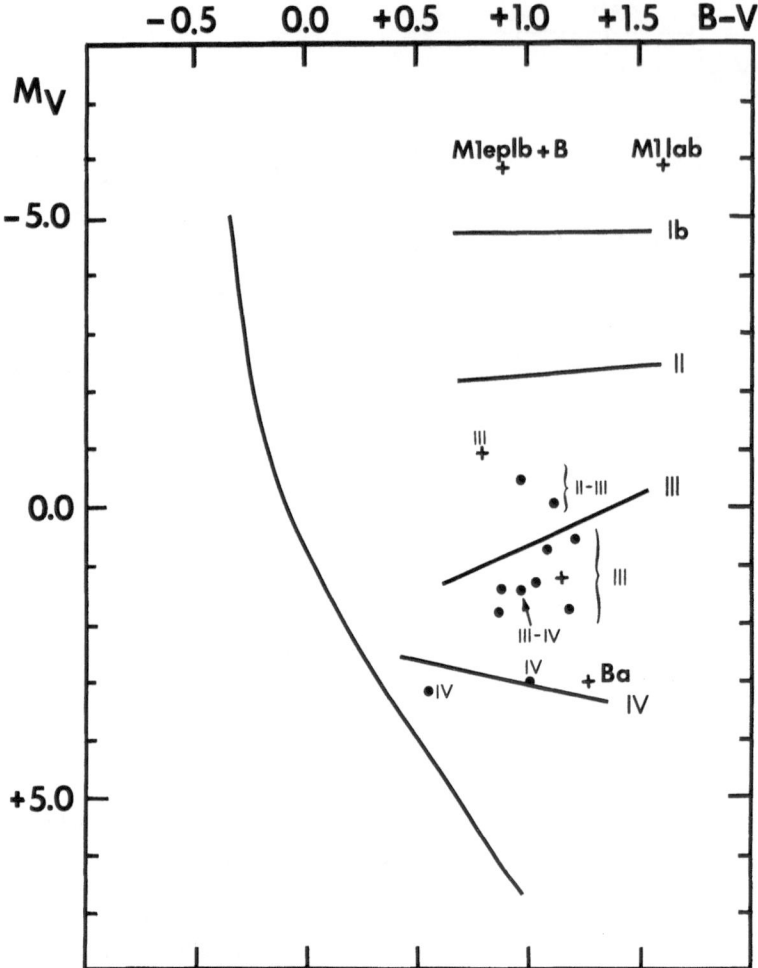

Figure 2. Color (B-V) versus absolute magnitude ($M_V$) for 16 visual pairs containing a main-sequence component and an evolved star.

classifications with those of Stephenson (1960) and Stephenson and Sanwal (1969) "shows the only fairly convincing systematic difference" in the sense that he assigns luminosity classes that are on the average 0.4 of a luminosity class lower than those obtained by the other investigators. For example, the 31 stars which are contained in both sets of classifications, have non-composite spectra, and have been assigned class III by Stephenson (1960) or Stephenson and Sanwal (1969), have been classified by Abt (1981) as shown in Table I. Clearly, there exists a systematic difference in classifications that can fully account for the discrepancies seen in Figures 1 and 2 between newly determined absolute magnitudes and the published standard sequence for class III giants.

## TABLE I.

### LUMINOSITY CLASSES BY ABT (1981)
### FOR 31 STARS PREVIOUSLY ASSIGNED L. C. III
### BY STEPHENSON (1960) OR STEPHENSON AND SANWAL (1969)

| No. of Stars | Abt L. C. |
|:---:|:---:|
| 1 | II |
| 17 | III |
| 2 | IIIb |
| 1 | III - IV |
| 7 | IV |
| 3 | V |

One may therefore conclude that modern MK classifications made against current MK standards combined with accurate magnitude differences obtained by photoelectric area scanning will produce reliable absolute magnitudes for evolved stars that are members of visual binary systems. To this end, a program is now under way, in collaboration with R. B. Culver, to complete the photometry and to carry out MK classifications of 125 pairs known or suspected to contain a non-main-sequence component. Progress and ultimate completion of the work should add substantially to our knowledge of what has been so aptly chosen by its organizers as the subject of this colloquium.

## Acknowledgements

The area-scanning photometer used in this research was developed and assembled with the support of National Science Foundation grants GP-6983 and GP-20090.

## References

Abt, H.A.: 1981, Astrophys. J. Suppl. 45, pp. 437-456.
Aitken, R.G.: 1932, *New General Catalogue of Double Stars*, Carnegie Institution of Washington.
Allen, C.W.: 1973, *Astrophysical Quantities*, University of London, The Athlone Press.
Bidelman, W.P.: 1954, Astrophys. J. Suppl. 1, pp. 175-267.
Bidelman, W.P.: 1958, Publ. Astron. Soc. Pacific 70, pp. 168-179.
Culver, R.B., Ianna, P.A., and Franz, O.G.: 1977, Publ. Astron. Soc. Pacific 89, pp. 397-399.
Eggen, O.J.: 1956, Astron. J. 61, pp. 361-380.
Eggen, O.J.: 1963, Astron. J. 68, pp. 483-514.

Eggen, O.J.: 1965, Astron. J. 70, pp. 19-93.
Franz, O.G.: 1966, Lowell Obs. Bull. 6, pp. 251-256.
Franz, O.G.: 1970, Lowell Obs. Bull. 7, pp. 191-197.
Jeffers, H.M., van den Bos, W.H., and Greeby, F.M.: 1963, Publ. Lick
    Obs. 21.
Johnson, H.L.: 1953, Astrophys. J. 117, pp. 361-365.
Leonard, F.C.: 1921, Publ. Astron. Soc. Pacific 33, pp. 213-214
    (abstract).
Leonard, F.C.: 1923, Lick. Obs. Bull. 10, pp. 169-194.
Rakos, K.D.: 1965, Applied Optics 4, pp. 1453-1456.
Slettebak, A.: 1963, Astrophys. J. 138, pp. 118-139.
Stephenson, C.B.: 1960, Astron. J. 65, pp. 60-79.
Stephenson, C.B., and Sanwal, N.B.: 1969, Astron. J. 74, pp. 689-704.
Wallenquist, Å.: 1954, Ann. Uppsala Astron. Obs. 4, No. 2, pp. 1-79.
White, N.M.: 1981, private communication.
White, N.M., and Wing, R.F.: 1978, Astrophys. J. 222, pp. 209-219.

# ON THE DISTRIBUTION OF YOUNG SPECTROSCOPIC BINARY STARS OVER THE MAJOR SEMIAXES OF THEIR ORBITS

E.I. Popova
A.V. Tutukov
B.M. Shustov
L.R. Yungelson
Astronomical Council, USSR Ac. Sci.

About 60% of stars of the disc population in our Galaxy are close binary systems (CBS). Half of the known CBS are spectroscopic binary stars (Kraitcheva et al., 1978).

To know the distribution of a correlation between the masses of CBS components and semiaxes of their orbits is necessary for the investigation of the origin and evolution of CBS. For such statistical investigations, a catalogue of CBS was compiled at the Astronomical Council. The catalogue is based on the 6th Batten catalogue (Batten, 1967), its extensions (Pedoussant and Ginestet, 1971; Pedoussant and Carquillat, 1973) and data published up to the end of 1980 (Popova et al., 1981). Now it is recorded on magnetic tape and contains data on 1041 spectroscopic binaries; 333 of them are stars with two visible spectra. The latter are mostly systems prior to mass exchange and the distribution of physical parameters in these systems reflects the distribution and presumably conditions at the time of formation. Using some assumptions, we can obtain for spectroscopic binaries masses of the components $M_1$ and $M_2$ (or the ratio $q = M_1/M_2$) and semiaxes of their orbits. Masses of components with the known sin i were obtained by the usual technique; when sin i was not known, masses were estimated from the spectra. We shall discuss here the distribution of CBS in the M-a plane.

In fig. 1 the distribution of 333 spectroscopic binaries with two visible spectra in lg $M_1$ - lg a is shown. Crosses mark pairs with $M_v > 7^m$, dots - brighter pairs. The bottom line $a_{min}$ corresponds to the ZAMS contact systems with equal-mass components. The a $\lesssim$ 10 $R_o$, $M_1 \gtrsim 1.5 M_o$ region is almost empty. The existence of this empty area was noted by Svechnikov (1969) and confirmed by Kraitcheva et al. (1978). Note that the upper boundary of the empty area has a slope d lg a / d lg $M_1 \sim 0.3$. Three stars that lie in this region (GK Cep, UZ Pup and V Sge) display mass exchange and are probably evolved (Batten, 1967). There are less massive stars with a $\lesssim$ 10 $R_o$. It is evident that among these stars there are practically no bright ones. Taking into account the observational selection due to the

*Z. Kopal and J. Rahe (eds.), Binary and Multiple Stars as Tracers of Stellar Evolution, 129–131.*
*Copyright © 1982 by D. Reidel Publishing Company.*

luminosities of the components, their radial velocities, inclinations
of orbits, etc., we could derive the conclusion that the specific
density (i.e., the number of pairs per unit interval lg a) is 30-60
times lower than the one for wider pairs with the same masses
(Popova et al., 1981).

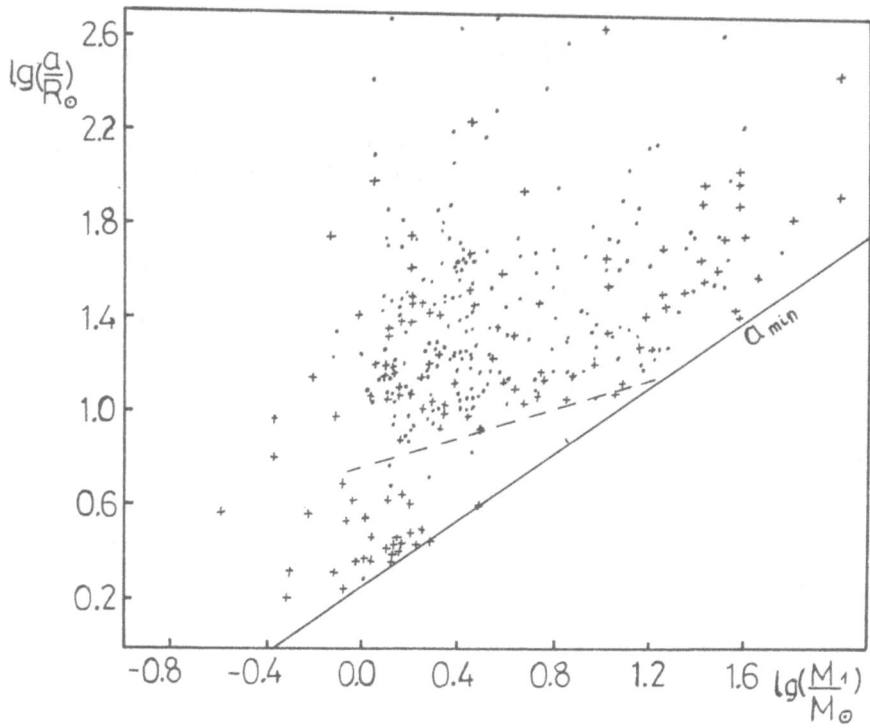

Figure 1.   Distribution of CBS with two visible
spectra in the lg $M_1$ - lg a plane.

The main question is:  What are the reasons for the existence
of this avoidance zone with a $\lesssim$ 10 R ?  The lack of CBS with
a $\lesssim$ 10 R is of genetic nature and the explanation of this feature
is very important for the CBS formation theory.

Analysis of the formation of a single star in the accretion
regime shows that the upper limit of an accreting star is about 30 R ;
the mass of a star at this moment reaches  $\sim$ 2 M .  This estimate is
in good agreement with the results of calculations of stellar evolution
near the Hayashi boundary (Larson, 1969).  There is observational
evidence for these estimates.  Cohen and Kuhi (1979) on the basis of

observations of young stellar aggregates showed that probably no object with a mass higher than 2-3 $M_0$ can be on the Hayashi tracks. Theoretical models of CBS formation in an accretion regime show that components in systems with a $\lesssim 10$ $R_0$ coalesce when stars are fully convective (on the Hayashi tracks). Resulting rapidly rotating stars brake probably due to magnetic stellar wind. Binary stars with a $\gtrsim 10$ $R_0$ remain double systems. Note that the theoretical relation a $\sim M^{1/3}$ for CBS forming in the accretion process from an initially uniform, rigidly rotating cloud is in agreement with the observational feature (see fig. 1). Details will be described in a subsequent paper.

Stars in the zone $M_1 \lesssim 1.5$ $M_0$ and a $\lesssim 10$ $R_0$ appear to be pairs closed due to intensive loss of angular momentum. Indeed in the zone $M_1 \lesssim 1.5$ $M_0$, there are stars with high chromospheric activity (RS CVn type stars) for which X-ray observations reveal strong stellar wind spots and probably magnetic fields (Hall, 1976). It is known that a magnetic stellar wind can be an effective mechanism of angular momentum loss.

Thus, we conclude that CBS with a $\lesssim 10$ $R_0$ cannot form, and that components of CBS with $M_1 \lesssim 1.5$ $M_0$ can come into contact (W UMa stars) only in the process of angular momentum loss due to the magnetic stellar wind. The proposed scenario of CBS formation is preliminary and the solution of such complicated problems as a collapse with fragmentation, evolution with rotation, and magnetic stellar wind is needed to develop this scenario.

REFERENCES

Batten, A., 1967, Publ. Dominion Astrophys. Observ. 13, No. 8.
Cohen, M., Kuhi, L.V., 1979, Astrophys. J., 227, L 105.
Kraitcheva, Z.T., Popova, E.I., Tutukov, A.V., Yungelson, L.R., 1978, Astron. Zh. 55, 1176 (English translation in Soviet Astronomy).
Larson, R.B., 1969, Monthly Not. Roy. Astron. Soc., 1969, 145, 271.
Pedoussant, A., Ginestet, N., 1971, Astron. and Astrophys. Suppl. Ser., 4, 253.
Pedoussant, A., Carquillat, J.M., 1973, Astron. and Astrophys. Suppl. Ser., 10, 105.
Popova, E.I., Tutukov, A.V., Yungelson, L.R., 1981, Astron. Zh., in press.
Svechnikov, M.A., 1969, Catalogue of Orbital Elements, Masses and Luminosities of Close Binary Stars, Ural Univ. Press, Sverdlovsk.
Hall, D.S., in: Multiply Periodic Variable Stars, Proceed. IAU Coll. No. 29, Budapest, 1976, p. 287.

# OBSERVATIONAL SELECTION IN SPECTROSCOPIC BINARY ECCENTRICITIES

M.S. Staniucha
Warsaw University Observatory and
N. Copernicus Astronomical Center, Warsaw, Poland

ABSTRACT

The discovery-and-identification probability for different shaped orbits of spectroscopic binary stars is estimated. The eccentricity distribution observed in the sample of $\sim$1000 binaries with known orbits and appearing as strongly peaked toward e=0 is corrected for observational selection effects. The resulting e-distribution seems to be flat for e in the range $\sim$0.05-0.6 with some excess of circular (or almost circular) orbits and a deficiency of orbits with e $\gg$ 0.6.

## 1. INTRODUCTION

Statistical analysis of the data compiled in the "Seventh Catalogue of the Orbital Elements of Spectroscopic Binary Systems" (Batten et al. 1978, hereinafter referred to as the Catalogue) demonstrates the serious bias of all these data mostly by observational selection effects (Staniucha 1979). The observed distribution of orbital eccentricity for 978 binaries is strongly peaked toward e=0 (cf. Fig.4 in Staniucha 1979). Almost 28% of all systems have circular orbits and 55% have e $<$ 0.1. A similar feature may be found probably for all close binaries, but the problem is whether it is real.

An attempt at extracting the real e-distribution from the observed one was made thirty years ago by Scott (1951). She analysed the selective identifiability of spectroscopic binaries and although she derived several very useful formulae, the rectification of the observed distribution did not give any convincing result.

The purpose of the present paper is to answer two questions: (1) How do the observational selection effects contribute to the observed eccentricity distribution? and then (2) What does the real distribution look like? From this view we estimate the probability

133

Z. Kopal and J. Rahe (eds.), Binary and Multiple Stars as Tracers of Stellar Evolution, 133–143.
Copyright © 1982 by D. Reidel Publishing Company.

of the spectroscopic detection-and-identification of orbits with
different elements (Sec. 2). These estimates are then used (Sec. 3)
to obtain the real eccentricity distribution from the observed one.
Concluding remarks and discussion are given in Section 4.


## 2. OBSERVATIONAL SELECTION EFFECTS

For any observed spectroscopic binary system two conditions have
to be fulfiled to place that system and its orbit in the Catalogue:
(1) detection of its radial velocity variations, and (2) determination
of the period for this variability and of the other orbital elements.
Both steps may end with success only with a certain probability, the
value of which depends on the parameters of the binary itself, on the
available instruments, weather conditions and on the number of the
observed spectra.

To estimate the probability, each step is accomplished with, we
adopt (after Scott 1951) the following approach. First, we assume the
moments of observations chosen at random in time and thus uniformly
and independently distributed throughout the orbital period P. This
assumption will be sufficiently good if we have to do with periods of
reasonable length. We disregard therefore all difficulties with
periods of a day, a month, a year etc., as well as with very long
periods and even with the very short ones (comparable in length with
the time of single observation). Second, we adopt the errors in
radial velocity without regard to their origin as following a normal
distribution with zero-mean and known standard deviation $\sigma$. Then,
assuming the star as having variable radial velocity whenever

$$\frac{1}{\sigma^2} \sum_{j=1}^{n} \left( v_j - \bar{v} \right)^2 \geqslant \chi_\alpha^2 \qquad (1)$$

(where $\bar{V}$ is the mean of the observed radial velocities, $V_j$, n is the
number of observations and $\chi_\alpha^2$ is the value of the classical $\chi^2$ with
the significance level $\alpha$ and n-1 degrees of freedom), we can obtain
the desired formula for the probability of variable radial velocity
detection in the form $\beta = \beta(x, e, \omega, n)$ (Scott 1951, gives the
formula as well as other details). Here $x = (\sigma/K)(1-e^2)^{-1/2}$, where K
is the semi-amplitude of radial velocity and the second factor was
introduced to release x from implicit dependence on eccentricity.

The problem of the probability, we accomplish the second step
with, cannot be solved in an analytic way with mathematical strict-
ness. Since we know quite many stars showing radial velocity variations, but having no derived orbit or even no period found for those
variations (few examples are seen in the data of Abt and Levy 1976,
1978), we decided not to neglect this point and to give at least some,
more or less rough estimate for that probability.

Suppose, a star's radial velocities are equal to $V(t)$ and $V(t^*)$ at times $t$ and $t^*$ respectively. The difference between them will be observable only if

$$\left| V(t) - V(t^*) \right| \geqslant V_{lim}, \tag{2}$$

where $V_{lim}$ is the minimum change in radial velocity still detectable, i.e. it has practically the same meaning as the above introduced $\sigma$, thus we are using both parameters equivalently throughout the paper. The longer is the fraction of the orbital period when Eq.(2) is fulfiled (with the value of $t^*$ fixed and $t$ running from zero to P), the larger is the probability we will be able to obtain a clear velocity curve and then good estimates of all orbital elements. Defining therefore

$$\beta^*(t^*) = \int\limits_{|V(t)-V(t^*)| \geqslant V_{lim}} dt \cdot \left( \int\limits_{0}^{P} dt \right)^{-1} \tag{3}$$

and then integrating $\beta^*$ over $t^*$ from 0 to P we obtain an estimation for the probability of the orbital elements determination:

$$\beta_2(x, e, \omega) = \frac{1}{P} \int\limits_{0}^{P} \beta^*(t^*) \, dt^*, \tag{4}$$

where the notation is the same as previously used one. Also the other expressions were tested for the function $\beta^*$ (e.g. dt in Eq. 3 was weighted by the relative radial velocity changes) but since all of them show practically the same behaviour and give the same final result, we chose the simplest form of Eq.(3). Few examples of the $\beta_2$-function behaviour are shown in Fig. 1. For reasons of symmetry only the values of $\omega$ from the first quadrant were presented. The dependence on x though not given is also very strong: $\beta_2$ steeply decreases with increasing x (e.g. $\beta_2$ = 1.0, 0.5, 0.25 and 0.13 for x = 0.0, 0.5, 1.0 and 1.5 respectively when e=0.8 and $\omega$ =45$^{\circ}$).

We should notice however that $\beta_2$ gives in fact the probability of the determination of period, since in a predominant majority of cases we can always estimate all the other orbital parameters once we know the value of period. Therefore the evaluation of the accuracy, the elements are determined with, becomes essential. A good measure for this accuracy is given by

$$\beta_3 = \beta_3 \left( K', e', \omega' \mid K, e, \omega \right), \tag{5}$$

which is the probability density of the estimates $K'$, $e'$, $\omega'$ of elements with their true values $K$, $e$, $\omega$. Analytic treatment of $\beta_3$ meets many difficulties, however some numerical results in that subject were already obtained. Namely, Scott (1951), using computer

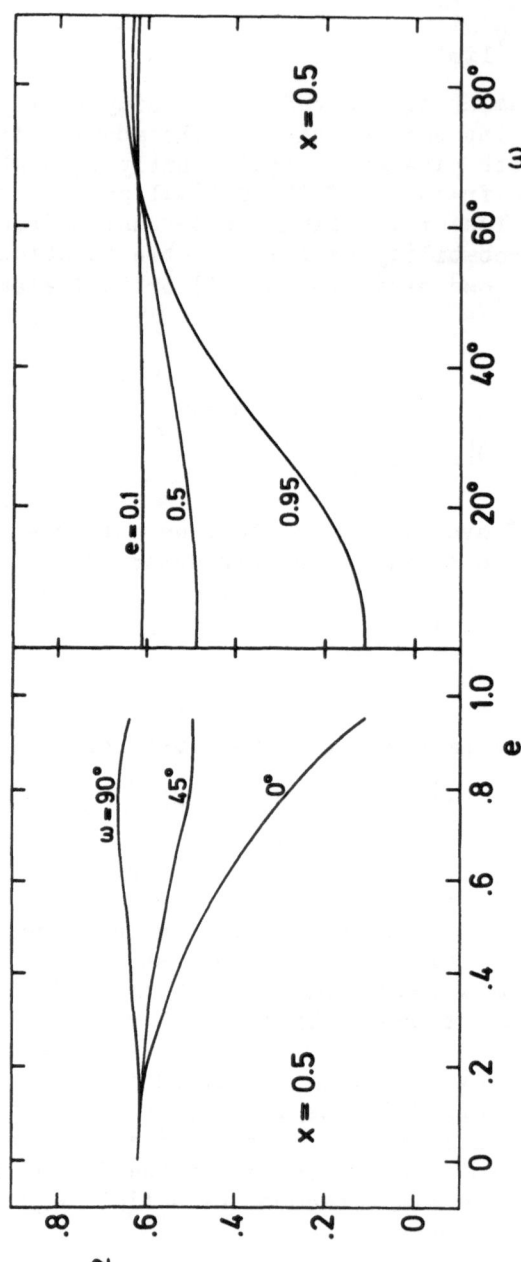

Figure 1. Dependence of the probability $\beta_2$ on the eccentricity e and the longitude of periastron $\omega$ for x=0.5.

generated observations, estimated the $\beta_3$-function for the case of $K/\sigma = 2$, i.e. $x = 0.5/(1-e^2)^{1/2}$. Although the estimation is a little rough (the elements were determined by a visual comparison of the data with a grid of synthetic velocity curves), it shows (cf. Fig. 9 in her paper) that even in such extreme case - where the semi-amplitude K is only twice the standard error $\sigma$ - the circular orbits might be distorted in very small percentage of binaries only (the probability density highly peaks at e=0 and e'=0). As we are going to the higher eccentricities the probability density function becomes wider and flatter, but since at the same time the x value increases, the probability, that we detect such a binary and then we determine its elements, very fast becomes small enough to allow us to argue that the neglection of all effects connected with $\beta_3$ will not enter any serious bias into the present investigation. We do not have perhaps to add, that the derivation of the formula for $\beta_3$ might be of great importance in the rectification of distributions for other orbital elements (like e.g. $\omega$).

With the $\beta_1$, $\beta_2$ and $\beta_3$ functions already defined we can answer the first question of Introduction: the influence of the observational selection effects on the distribution of the elements of spectroscopic binaries can be represented by the formula:

$$\mathcal{O}\left(K', e', \omega'\right) = \frac{\iiint \beta_1 \beta_2 \beta_3 \, \mathcal{R}(K, e, \omega) \, dK \, de \, d\omega}{\iiint \beta_1 \beta_2 \, \mathcal{R}(K, e, \omega) \, dK \, de \, d\omega}, \quad (6)$$

where the observed and the real distributions are denoted by $\mathcal{O}$ and $\mathcal{R}$ respectively and the range of integration is over the extreme limits of each variable. Moreover, if we put

$$\beta_3 = \delta\left(K' - K\right) \cdot \delta\left(e' - e\right) \cdot \delta\left(\omega' - \omega\right), \quad (7)$$

we obtain

$$\mathcal{O}\left(K, e, \omega\right) = \beta_1 \cdot \beta_2 \cdot \mathcal{R}\left(K, e, \omega\right), \quad (8)$$

where the constant factor was omitted.

## 3. INCOMPLETENESS CALCULATIONS

To apply the procedure from the previous section to the correction of the observed e-distribution for the observational selection effects we need, apart from e and $\omega$ listed in the Catalogue, the values of $V_{lim}$ $(\equiv \sigma)$ for every system in our sample. Abt and Levy (1976, 1978) assumed fixed values of $V_{lim}$ for stars with similar spectral types. In general, $V_{lim}$ should be a function of the star's temperature, with a minimum for middle spectral types, being a compromise between broad lines in hot stars and the predominance of many metallic lines in cold ones. The good approximation for the $V_{lim}$ values can be given by the minimum semi-amplitudes of radial velocity, $K_{min}$ found for different spectral type ranges in the Catalogue. Fig. 2 presents log $K_{min}$ -spectral type relation for single ( SB1 ) and double line ( SB2 ) systems treated separately. As we might expect, $K_{min}$ values

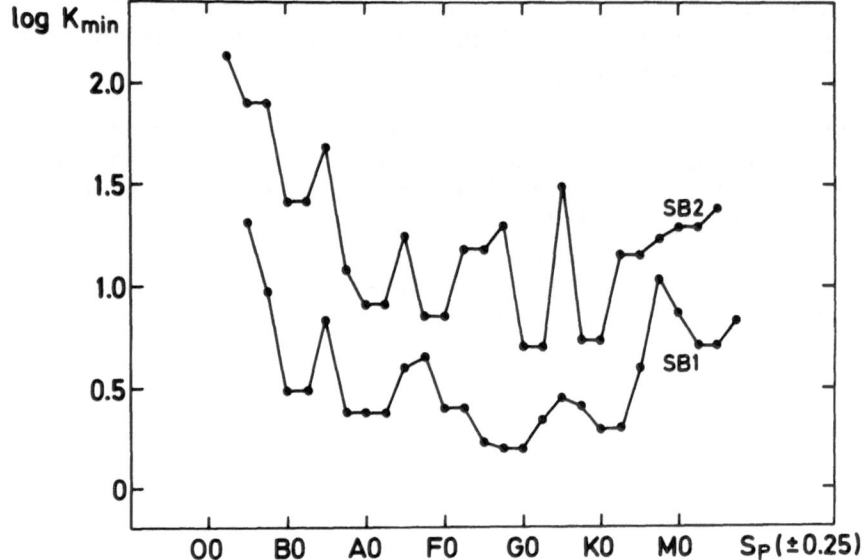

Figure 2. Common logarithm of the minimum semi-amplitude of radial velocity as a function of the spectral type of primary. Spectral type bins overlap each other by one-half.

are in the average about eight times larger for SB2 systems than for SB1s, since the detection of two separate lines needs relatively larger values of K compared to the single line changing its position. Very similar results have been recently found by Kraitcheva et al. (1979), though they used the mass of primary component instead of spectral type as a parameter.

Two curves giving lower boundaries were fitted to the points in Fig. 2 and the values of $V_{lim}$ and x were computed for almost all (i.e. 933) binaries in the Catalogue. Only systems with unknown spectral types or those with Wolf-Rayet star or subdwarf as primary were omitted. Also all binaries originally observed as visual ones were rejected, since the probability of their detection is slightly different (this was confirmed by our calculations). Probabilities $\beta_1(x, e, \omega, n)$ and $\beta_2(x, e, \omega)$ were then determined with the number of observed spectra, n, usually not given in catalogues, assumed to be uniformely distributed between 2 and 15 (several runs have been made to avoid some casuality which might arise from the random sampling of n).

The correction for the incompleteness due to observational selection was made for each 0.1-wide eccentricity bin separately. Circular orbits were also taken as a separate group. The real number of systems in each bin, $N_r$ was obtained by simple summation:

$$N_r = \sum_{j=1}^{N} \frac{1}{\beta_1 \cdot \beta_2} , \qquad (9)$$

where N is the observed number of binaries in that e-bin. The errors, $\delta N_r$ were computed from:

$$(\delta N_r)^2 = \sum_{j=1}^{N} \frac{\left(\frac{\delta \beta_1}{\beta_1}\right)^2 + \left(\frac{\delta \beta_2}{\beta_2}\right)^2}{(\beta_1 \beta_2)^2} + (N_r \delta N/N)^2, \qquad (10)$$

where $\delta N$ was taken according to the Poisson distribution as square root of N. Different values were tried for $\delta \beta_1$ and $\delta \beta_2$ which are not known, but since the sensitivity of the resulting $\delta N_r$-error was rather weak to both quantities they were finally fixed and both taken equal to 0.05.

The eccentricity distribution corrected for the observational selection effects can be characterised as follows:
(1) Large fraction (about one fourth) of all systems has circular orbits. No significant difference in that group of binaries was therefore noticed.
(2) The distribution for eccentric orbits seems to be uniform for e in the range ∼0.05-0.6 (see Fig. 3). Some excess of very small eccentricities as well as a clear deficiency of highly eccentric

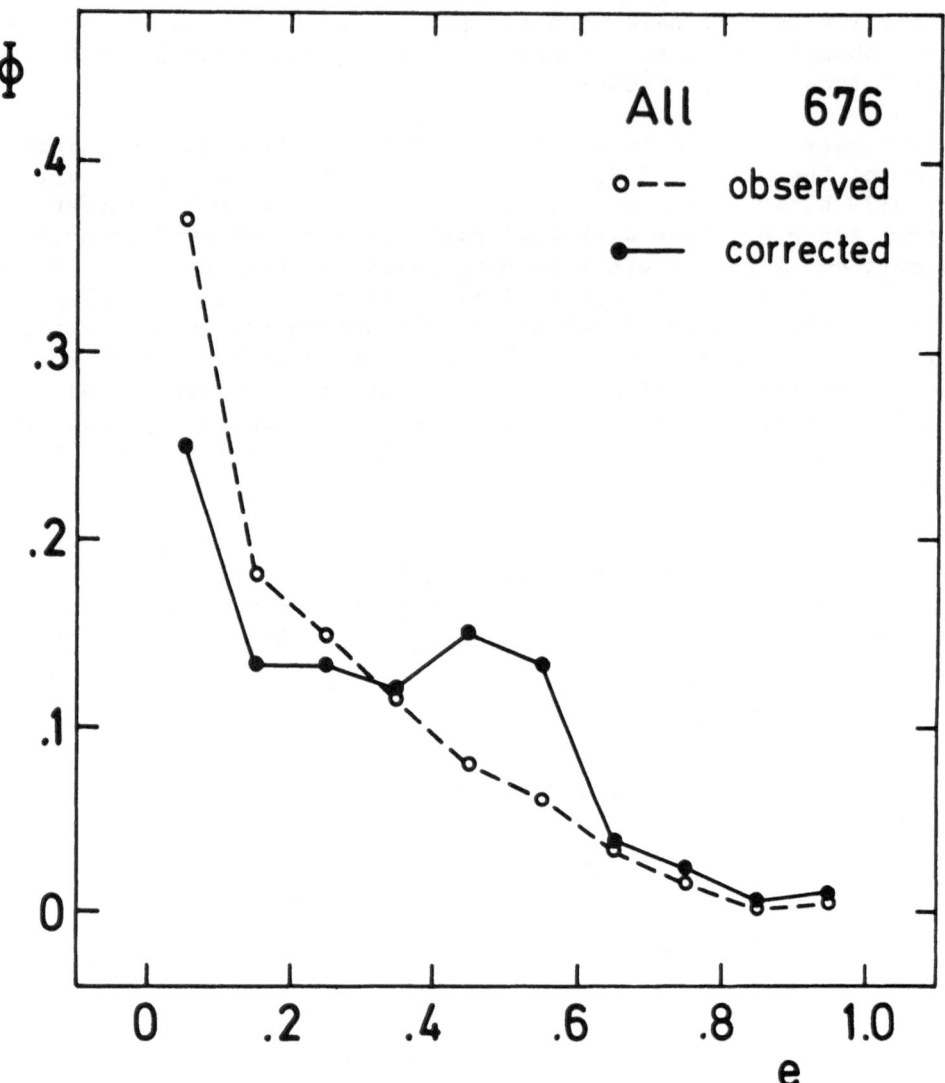

Figure 3. Observed and corrected eccentricity distributions for all
non-circular orbits. The number of systems in the observed sample
is given in the upper right corner.

orbits is seen. Although in Fig. 3 the error bars were not plotted to avoid ambiguity, we should stress the statistical significance of both jumps: at e $\cong$ 0.6 and e close to zero, the latter being even more pronounced when the circular orbits are added to the sample. The flat part, on the other hand, fits well within the one sigma-wide band marked by the data points.

(3) If the samples of SB1 and SB2 systems are treated separately the e-distribution in each sample remains almost the same as for all systems together (cf. Fig. 4).

Some additional attention may be called to the systems with high quality orbits, i.e. those with quality classes a and b according to Batten et al. (1978). The corrected e-distribution in this group is intermediate between the two distributions presented in Fig. 3, but since these are the systems having (to some extent by assumption) the highest probabilities of detection and identification, their useful-ness in this case is restricted to supporting our results for the total sample.

## 4. CONCLUDING REMARKS

The results from our rectification procedure show that the certain number of binaries (at least 25%) have circular or almost circular orbits, while in the group of remaining systems the eccentri-city is distributed more or less uniformly, with the exception of very eccentric orbits (e $\gtrsim$ 0.6) which are evidently unfrequent. Every trial to explain such result must be of necessity only speculative, since our knowledge on the mechanisms of binary star creation and/or evolution is still very uncertain.

If we assume after Popova et al. (1981, this meeting) that most of SB2s are "systems prior to mass exchange"(what probably makes the period-eccentricity correlation existence due to observational selec-tion effects) we arrive at the conclusion that the eccentricity dis-tribution for SB2s, shown in Fig. 4, is (1) very close to the primor-dial one, or (2) defined mainly by the dynamical evolution effects (of what sort?), or (3) both. To clarify this point some support from a theory is needed.

If the orbital distribution in phase space is a function of binary energy only, then the expected eccentricity distribution is $\phi$(e) = 2e (Ambartsumian 1937). Either numerical or theoretical investigations show that it occurs especially in star clusters which are relaxed (see e.g. Heggie 1975). But it has no place for the field stars (which make about 96% of all Catalogue entries), since any significant encounter between them is almost impossible. On the other hand, the escape rates of binaries in clusters as well as the distri-bution of eccentricity among escapers are indeed unknown. Binaries which result from the dynamical decay of small (five bodies) unstable stellar systems were investigated by Harrington (1976). He found  the

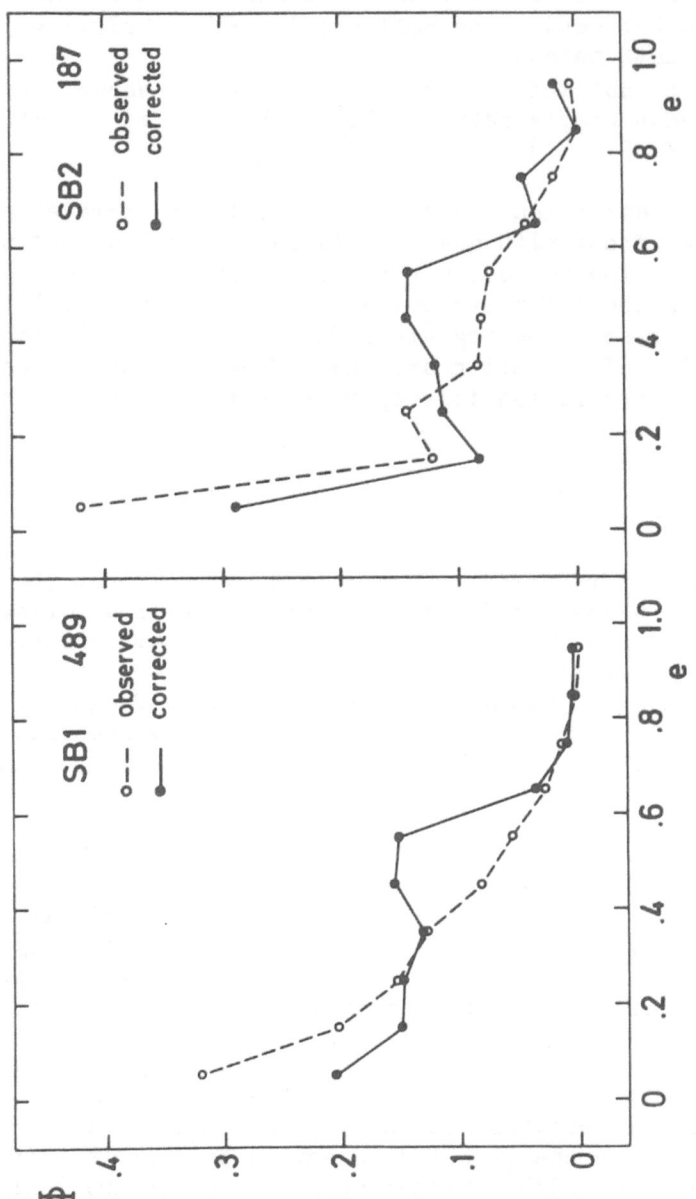

Figure 4. The same as in Fig. 3 but for SB1s (left panel) and SB2s (right panel) separately.

resulting eccentricity distribution very close to the linear Ambartsumian function, but a somewhat flatter at high $(\geqslant 0.6)$ eccentricities. Also a fission of single rotating protostar was simulated (Lucy 1977), but no information on the resulting eccentricities was obtained.

Therefore, as one can see , no prediction to the eccentricity distribution for field (close) binaries can be given. Also the knowledge of mechanisms, the dynamics of binary stars is governed by, is rather poor. Further mainly theoretical developement in those subjects might put a new light on the real e-distribution presently obtained.

Acknowledgements. Many thanks are due to Dr. S. Ruciński for suggesting the subject of this work and for discussions and advice. All computations reported in this paper have been performed on the PDP 11/45 computer donated to the Copernicus Center by the U.S. National Academy of Sciences.

REFERENCES

Abt, H.A., and Levy, S.G.: 1976, Ap. J. Suppl. 30, pp. 273-306.
Abt, H.A., and Levy, S.G.: 1978, Ap. J. Suppl. 36, pp. 241-274.
Ambartsumian, V.A.: 1937, Astron. Zhu. 14, pp. 207-225.
Batten, A.H., Fletcher, J.M., and Mann, P.J.: 1978, Publ. Dom.
    Astrophys. Obs. 15, pp. 121-295.
Harrington, R.S.: 1976, A. J. 80, pp. 1081-1086.
Heggie, D.C.: 1975, Mon. Not. R. astr. Soc. 173, pp. 729-787.
Kraitcheva, Z.T., Popova, E.I., Tutukov, A.V., and Yungelson, L.R.:
    1979, Astron. Zhu. 56, pp. 520-531.
Lucy, L.B.: 1977, A. J. 82, pp. 1013-1024.
Popova, E.I., Tutukov, A.V., Shustov, B.M., and Yungelson, L.R.: 1981,
    this meeting.
Scott, E.L.: 1951, in "Proceedings of Second Berkeley Symposium on
    Mathematical Statistics and Probability", ed. J. Neyman (Univ.
    of California Press, Berkeley), pp. 417-435.
Staniucha, M.: 1979, Acta Astr. 29, pp. 587-608.

DOUBLE STARS:   THE CLOSEST AND THE MOST DISTANT SYSTEMS.

Tibor J. Herczeg
University of Oklahoma

Review and comments on two extremely different samples of binary and multiple stars:  systems within 10 parsecs of the sun and systems discovered in galaxies of the Local Group.

1.  BINARIES AMONG NEARBY STARS (r $\leq$ 10 pc)

Binary and multiple stars within 10 pc - there are 49 binaries and 17 multiple systems known to this limit - furnish us with unique information about multiplicity among the low luminosity, least massive stars, mainly late main-sequence objects and degenerate dwarfs.  Systems with $\pi$ > 0.''100 are listed in a review article by Worley (1968) and, with a few variants, in the forthcoming new edition of Vol. 6 of the Landolt-Börnstein Tables (H. H. Voigt, editor; see section 6.1.0.5).  The latter list is based primarily on the Royal Greenwich Observatory Catalogue (1970) and Gliese's catalogue (1969) of nearby stars.  Not always is the recognition of duplicity definitive:  in one case, all the published information is a laconic note "SB" in Wilson's radial velocity catalogue, in another, an astrometric orbit was published in 1968 but the solution did not appear in a recent review of the topic by the same author.  There can be, of course, no serious question about the duplicity of most stars in these lists; orbits have been calculated for just about half of the systems.

The sample is certainly not complete, as can be judged from the following tabulation - the three zones have equal volumes.

|                     | 0 - 6.6 pc | 6.6 - 8.3 pc | 8.3 - 9.5 pc |
|---------------------|------------|--------------|--------------|
| Singles (incl. sun) | 47         | 27           | 32           |
| Doubles             | 22(44)     | 11(22)       | 9(18)        |
| Multiples           | 7(23)      | 6(19)        | 1(3)         |

*Z. Kopal and J. Rahe (eds.), Binary and Multiple Stars as Tracers of Stellar Evolution, 145–151.*
*Copyright © 1982 by D. Reidel Publishing Company.*

The multiplicity of the components indicated in the first two zones is
60 percent. The third zone is clearly affected by the additional incompleteness of the double star discoveries.

Among the nearby double stars, the great majority are the visual
pairs: 41 systems, if we count the astrometric binaries, too. There
are 7 spectroscopic binaries and one cpm pair (Gliese 49,51); in this
system the projected distance between the components is 5900 AU. Due
to the meagerness of the sample, obtaining a distribution function for
the periods or major axes is not very promising. The relatively large
number of separations between 100 AU and 1000 AU, nearly 25%, would
make the semi-logarithmic distribution N - log a almost a constant
with a flat maximum around the bin a = 4 to 8 AU. The resulting
periods, about 10 to 25 years, correspond roughly to the "peak" in
Abt and Levy's frequency curve for solar type stars, see Abt (1979),
Fig. 6.

This sample is, however, entirely different, here the dM-stars
dominate. The spectral types of the primaries are distributed as follows:

$$A-F : 2, \ G-K : 13, \ M0-M7 : 26.$$

A quirk of the spectral distribution makes the dM4 components particularly frequent, 14 out of the 26 M-type primaries - a fluctuation or
perhaps a classification bias? Also, probably just a fluctuation in
the very small sample but worth mentioning: the solar type stars
(F6-G5) show the lowest binary frequency, 4 doubles against 11 singles.

We found 5 white dwarf components in the list of doubles, roughly
10 percent. A particularly important feature of the 10 pc survey is
the comparatively high frequency of astrometric binaries and "unseen
companions". A somewhat more extensive statistics by van de Kamp (1976)
tentatively ascribes these objects the respectable number density of
$0.21\pm0.06$ pc$^{-3}$. A detailed discussion of individual cases has also
been given by him in 1975; in a few cases, however, there is no complete agreement, see for instance Heintz (1978). Several of these companions seem to have masses in the range $0.01-0.06$ $M_\odot$ and at least some
of them are well below Kumar's mass limit for main-sequence evolution.
They are probably not "planets" and thus we recognize a new class of
astronomical objects, nearly impossible to detect but being a component
in binary systems. Their integral properties are easy to guess at:
essentially cold hydrogen-helium configurations, 10 to 40 times more
massive than Jupiter, sizes perhaps that of Uranus or Neptune, mean
densities 500-600 gcm$^{-3}$. Structure and evolution of these bodies are,
on the other hand, problems not yet considered.

Finally, how close to each other are the components of the closest
binaries in this sample of nearby stars? Six pairs may have semi-major
axes under 1 AU, five spectroscopic binaries and the astrometric pair

G24-16; in some cases the size of the orbit has to be estimated from the dispersion of the observed radial velocities. Even these pairs are not actually close binary systems, in terms of interaction between the components. An exception is perhaps Gliese 268, a dM5e star with a radial velocity range of 110 $kms^{-1}$! Checking this number and calculating an orbit would certainly be a rewarding task; the period can hardly be longer than 1-2 days, the mean distance between the components of the order of 1 to 2 x $10^6$ km. Due to the smallness of the radii of late M-type or degenerate components, eclipses would still have a very low probability and no ellipticity effects are expected (although the star is suspected variable).

Given this distribution of the separations, it is not surprising that no eclipsing system has been discovered within 10 pc. (In the distance range 10 pc to 20 pc we do find a few very interesting objects: i Boo, CM Dra, YY Gem.) However small this sample of very late type dwarfs may be, it supports through the complete absence of contact type binaries the existence of a remarkably sharp cut-off for W UMa type systems at the spectral type K5.

The nearest multiple stars comprise 14 triplets, 2 quadruplets and 1 quintuple system. All combinations except eclipsing systems, are represented: visual, spectroscopic and astrometric binaries as well as cpm pairs. Some of these multiples are among the best known stellar objects, such as α Cen AB + Proxima, 40 Eri with its white dwarf component, the 61 Cyg system - although here the astrometric duplicity needs further confirmation. The highest multiplicity is that of the visual triple Gliese 644 having common proper motion with Gliese 643 which itself is a spectroscopic binary.

2. BINARY SYSTEMS IN NEARBY GALAXIES (r < $10^6$ pc)

The sample of nearby binaries was dominated by visual pairs of low luminosity stars. In the galaxies of the Local Group we have, of course, no choice of discovering visual double stars and the objects we can reach the easiest way are all massive, high luminosity eclipsing systems; we can, in fact, select the very brightest systems in a galaxy. Spectroscopic studies may help, at the limits of our present observational techniques, in the case of LMC or SMC objects, but our best - and, for a long time, only - method of detection is the search for eclipses.

Nevertheless, there are improvements in our techniques of binary detection. The appearance of novae in a system indicates the presence of cataclysmic binaries, a well defined type of evolved, interacting system. And in recent years we were able to identify the majority of X-ray sources, well observable in nearby galaxies, with interacting binaries having as one component a compact object.

Five galaxies have been searched extensively for variable stars, among them eclipsing variables. Work on M31, M33 and the dwarf spheroidal system in Draco was done almost exclusively at the observatories on Mt. Wilson and Palomar Mountain; for the Magellanic Clouds, most of the early discoveries were made at the Southern station of the Harvard Observatory.

In the Clouds, almost 80 eclipsing systems have been recognized; recent lists of them were given by Payne-Gaposchkin and Gaposchkin (1966) and by Hodge and Wright (1967,1977). Although typical (evolved) Algol systems with B-type primaries are within easy reach of the surveys, they seem to represent only about 20% of the observed light curves; most systems have been classified as "β Lyrae type" - some of them may also represent semi-detached systems after the first phase of mass exchange. The brightest system has its $M_{pg}$ close to -5, the light curve is shown in Fig. 1 (P = 4.34 days). Most systems have $M_{pg}$ between -2 and -3.

The Clouds contain 7 or 8 X-ray sources, two of them identified with massive, early type spectroscopic systems (SMC X-1, LMC X-4). The absolute X-ray luminosities of the Magellanic Cloud sources are quite well determined; SMC X-1 and LMC X-4 are the two brightest sources known, close to $10^{39}$ ergs$^{-1}$, Eddington's limit.

Another comparatively near system, the dwarf spheroidal in Draco, offers a striking contrast in its type and population as well, concerning binaries. No eclipsing systems have been found, in spite of a careful search of the central region (Baade, Swope 1961). The complete absence of giant binary systems corresponds exactly with what we find in globular clusters, representing a very similar stellar population.

In M33, there is only one known variable which "appears to be an eclipsing star", at about $m_{pg} \sim 19$ (Hubble 1926). This seems rather surprising since in this Sc spiral one would expect bright eclipsing

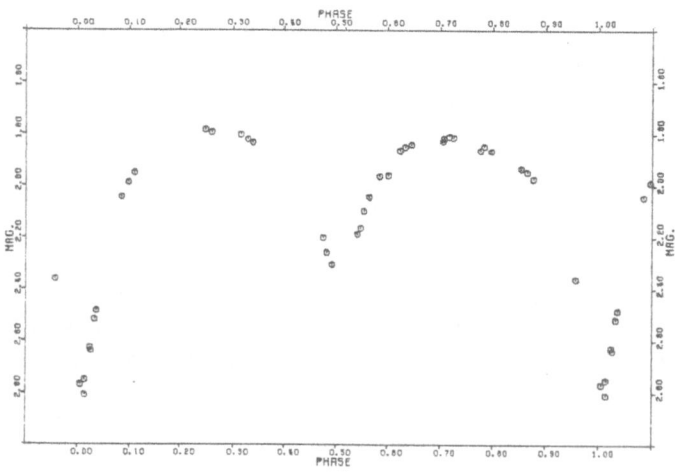

Fig. 1. Photoelectric U light curve of H.V. 2241 (in the LMC). Observations by Herczeg, ESO 1 m telescope.

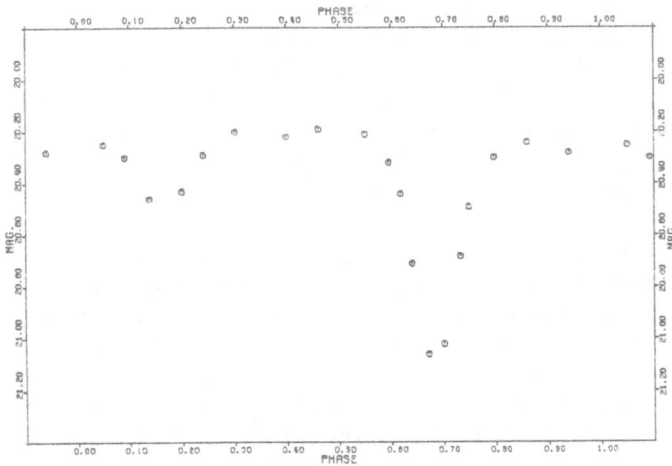

Fig. 2. Photographic
light curve of "star G"
(no. 75a in Field II).
Gaposchkin, AJ 67, 358
(1962).

systems in numbers comparable to those in the Magellanic Clouds. A recent survey (van den Bergh et al. 1975) found no binaries among 38 new
variables. However, this search was based on 67 Palomar Schmidt plates
taken over an interval of nearly nine years, clearly not favorable to
the discovery of variable objects with 5-15 days periods, as massive
binary systems frequently have. Another problem is that the limiting
magnitudes of the 48 in. Schmidt (B ∿ 20, V ∿ 21) exclude the bulk of
bright eclipsing systems which would, at the distance modulus of 24.6,
show up just beyond these limits. A "denser" and somewhat deeper survey is required to find these binaries.

The Andromeda galaxy offers the most interesting picture and the
richest harvest as far as binaries go. Novae are observed by the hundreds in this galaxy, ever since Hubble's early investigations. A systematic search to detect variable stars was started by Baade using the
200-inch Hale telescope, and carried out by him, Miss Swope and Gaposchkin (see references 1963-1965). The search covered only a fraction of the area of M31, four circular fields of 15' diameter but even
this limited program revealed almost as many eclipsing binaries as we
know in the Magellanic Clouds.

Baade selected four fields south preceding the nucleus. Data of
the fields and number of variables discovered is given below.

| Field | Dist. from nucleus | Description | Variables | Eclipsing | Author |
|-------|--------------------|-------------|-----------|-----------|--------|
| I     | 15'                | Edge of main body | 109 | 2 | B,S |
| II    | 35'                | "Mixed",between arms | 223 | 9(+10ecl?) | G |
| III   | 50'                | Rich spiral arm | 334 | 36 | B,S |
| IV    | 96'                | Faint arm, 20 kpc | 54 | 10 | B,S |

Most eclipsing systems are noted as "blue". Periods are given for 34 stars; β Lyrae type light curves dominate, except in Field III. The periods range from 2.3 days to about 960 days, but the range 4 to 10 days contains most of them. A well observed light curve is reproduced in Fig. 2 (P = $4\overset{d}{.}805$). The brightest system is V60 in Field III, in maximum m(pg) $\sim$ 18.9. The light curve resembles that of a massive contact system (P = 7.33 days), suggesting a young object.

The low frequency of massive binaries in Field I is puzzling. This field is in the main body of M31 but well outside the nuclear region. On Baade's identification figure much structure and dust lanes are clearly visible. This is the region, however, where in the spectrum of the galaxy, features of an earlier F-type spectrum fade out and features of the late giant spectrum become dominant: there is a change in the stellar population. This may have some importance for the binary counts. Numerous novae in this field (Baade found seven) suggest that evolved binary systems of lower mass and luminosity may be frequent.

A comparison of these systems with galactic objects is not quite trivial since we may not know the brightest eclipsing systems in our galaxy. We know however several massive systems of early spectral type and presumably very high luminosity; the light curves are mostly of "β Lyrae type". To mention a few: UW CMa, AO Cas, V380 Cyg, V382 Cyg, V Pup, V453 Sco, RY Scu, possibly β Lyrae itself. Transferred to M31, these systems would appear very much like one or other star in Baade's list.

Our knowledge of massive, luminous binary systems in M31 widened dramatically with the X-ray observations of HEAO 2 (Einstein Observatory), as reported in the memorable Nov. 15, 1979, issue of Astrophys. J. Letters (van Speybroeck et al.) While M31 was just a diffuse source for UHURU, now 69 sources have been found in it, one possibly coincident with the nucleus, 7 in globular clusters, 8 further cluster "candidates", 17 rather strongly concentrated in the inner bulge and 36 Population I sources distributed over the spiral arms. These latter sources almost certainly combine a massive early primary with a compact secondary, of the type of Cen X-3 in our galaxy or SMC X-1 in the Small Cloud. We may assume that the bulge-type sources are also binaries, as indirect evidence strongly suggests in the galaxy. Unlike the eclipsing systems in the M31 fields, representing but a small fraction of all objects of the kind, the census of the X-ray sources is nearly complete to the sensitivity limit.

The authors point out that M31 and the galaxy show some conspicuous systematic differences in their "X-ray population". This may be compared to systematic differences between the galactic X-ray sources and those in the Magellanic Clouds, recognized earlier (Clark et al. 1978).

Massive X-ray binaries do not show conspicuous light variations, and any correlation between Baade's lists and the HEAO 2 observations is unlikely. A comparison is further hindered by the circumstance that the X-ray observations neglected the outer regions of the south precedent half of M31 and only Field I was covered in both programs.

ACKNOWLEDGEMENTS

I am thankful to Ms. Linda Barker for her kind help in collecting and tabulating the data of nearby binaries, and to Ms. Jaquine Littell for preparing the manuscript.

REFERENCES

Abt, H. A., 1979:  A J 84, 1591
Baade, W., Swope, H. H., 1961:  AJ 66, 300
Baade, W., Swope, H. H., 1963:  AJ 68, 435
Baade, W., Swope, H. H., 1965:  AJ 70, 212
Clark, C., Doxsey, R., Li, F., Jernigan, J. G., van Paradijs, J., 1978:
    ApJ 221, L37
Gaposchkin, S., 1962:  AJ 67, 334
Gliese, W., 1969:  Veröff. ARI Heidelberg, No. 22
Heintz, W. D., 1978:  ApJ 220, 931
Hodge, P. W., Wright, F. W., 1967:  The Large Magellanic Cloud, Wash-
    ington; 1977:  The Small Magellanic Cloud, Seattle-London
Hubble, E., 1926:  ApJ 63, 236
Payne-Gaposchkin, C. H., Gaposchkin, S., 1966:  Smithsonian Contrib.
    9, 1
van de Kamp, P., 1975:  Ann. Rev. Astron. Astroph. 13, 239
van de Kamp, P., 1976:  Roy. Greenwich Obs. Bull. No. 182, p. 7
van den Bergh, S., Herbst, E., Kowal, C. T., 1975:  ApJ Suppl 29, 303
van Speybroeck, L., Epstein, A., Forman, W., Giacconi, R., Jones, C.,
    Liller, W., Smarr, L., 1979:  ApJ 234, L45
Woolley, Sir Richard, Epps, E. A., Penston, M. J., Pocock, S. B.,
    1970:  Royal Obs. Ann. No. 5
Worley, C. E., 1968:  in "Low-luminosity stars" (ed. S. Kumar), U. of
    Virginia

# THE EVOLUTIONARY STATE OF ZETA AURIGAE

Robert D. Chapman
NASA - Goddard Space Flight Center
Greenbelt, Maryland 20771  USA

Ultraviolet studies, originally undertaken to ascertain the state of the atmosphere of the K-supergiant component of the zeta Aurigae system, have been sidetracked by the discovery of significant accretion effects. An analysis of the phase dependence of the profiles of resonance lines in Mg II and C IV has led to a qualitative model of the wind flow from the K star. At the position of the $B$ star, the flow velocity is about 100 km/sec and the density is $3 \times 10^{-6}$ cm$^{-3}$, leading to a mass loss rate of $2 \times 10^{-8}$ solar masses per year. This wind interacts with the B star in a shock, which will be described, leading to accretion on the B star at a rate of $4 \times 10^{-10}$ solar masses per year.

## INTRODUCTION

Zeta Aurigae--the prototype for a group of eclipsing binary systems-- consists of a K-type supergiant star (R = 200 $R_\odot$, M = 8.3 $M_\odot$) and a smaller, B-type companion (R = 4 $R_\odot$, M = 5.6 $M_\odot$). The system is interesting because, during partial eclipse phases, the light of the hot companion passes through the extended atmosphere of the supergiant, (see Wilson, 1960; Wright, 1970). The resulting atmospheric eclipse permits us to probe the stratification of the atmospheric layers of the K supergiant.

Spectra of zeta Aurigae have been obtained with the International Ultraviolet Explorer before, during and after the 1979 - 1980 eclipse. The high dispersion spectra have been obtained in both the short (1150 to 1950 A) and the long (1900 to 3200 A) wavelength regions. Preliminary descriptions of the spectra have been published by Chapman (1980, 1981a). More detailed discussions of the initial analysis of the spectra are given by Chapman (1981b) and Stencel and Chapman (1981).

## NATURE OF THE WIND NEAR THE B STAR

The resonance doublet of Mg II is the only pair of lines in the zeta

153

*Z. Kopal and J. Rahe (eds.), Binary and Multiple Stars as Tracers of Stellar Evolution, 153–156.*
*Copyright © 1982 by D. Reidel Publishing Company.*

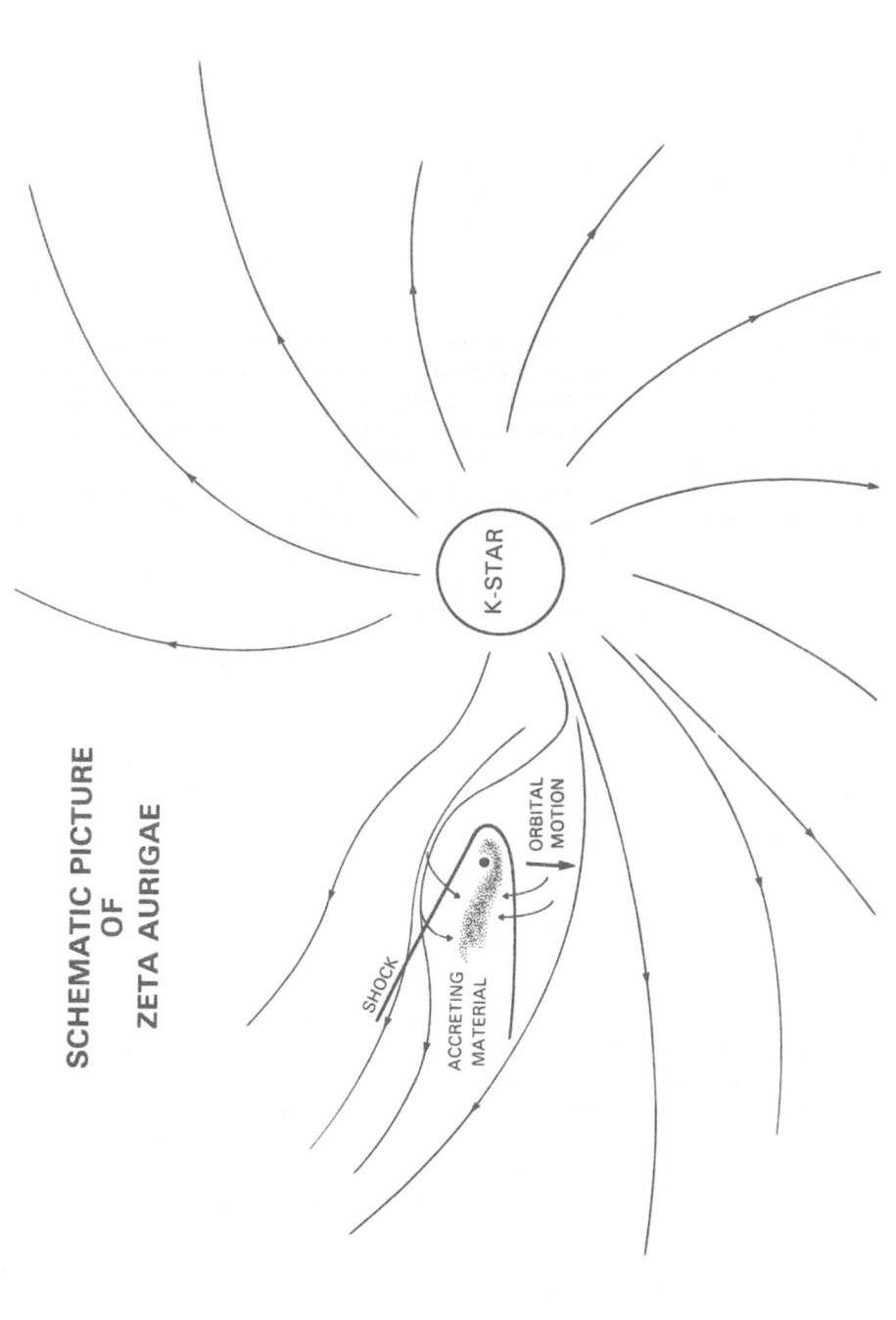

SCHEMATIC PICTURE
OF
ZETA AURIGAE

K-STAR

ORBITAL
MOTION

SHOCK

ACCRETING
MATERIAL

Figure 1.  A schematic model of the zeta Aurigae system.  The C IV absorption occurs in the accreting material, (dotted).

Aurigae spectra to show P Cygni profiles. We conclude from our analysis
(Chapman, 1981b) that the lines originate entirely in the circumstellar
material. Fitting Theoretical P Cygni profiles (Castor and Lamers, 1979)
to the observed profiles leads us to conclude that the wind speed is
approximately 100 km/sec and the matter density at the B star is about
$3 \times 10^{-6}$ cm$^{-3}$. The latter number was derived assuming that the wind
plasma is not accelerated after it leaves the immediate vicinity of the
K star.

The profiles of the C IV resonance lines are a strong function of
orbital phase. Near primary eclipse, before atmospheric effects become
significant, the lines are relatively weak, with a possible double profile. After primary eclipse, the profiles become significantly broader.
Their shape indicates the presence of large amounts of turbulent material. This type of profile has persisted through observations made at
and somewhat after secondary eclipse.


THE MODEL

The model of the zeta Aurigae system derived from the observations is
described in detail by Chapman (1981b) and Stencel and Chapman (1981).
The basic features of the model are summarized in Figure 1. The wind
from the K star flows past the B star at roughly Mach 8.5, giving rise
to a bow shock. Material passing through the shock is heated to high
temperatures (250,000 K) as its kinetic energy is converted to internal
energy. The material then flows into a dense, slowly moving column
which accretes onto the B star. Using the Bondi and Hoyle (1944)
accretion model, we find that the B star accretes $4 \times 10^{-10}$ solar masses
per year, or roughly 0.5 gm per cm$^2$ of surface each year. According to
models calculated by Kurusz (1979), the mass per unit area above optical
depth 2.5 in a B-star atmosphere is roughly 5.5 gm/cm$^2$. Thus the B star
accretes a new photosphere every decade, and the material will probably
remain there, since B stars have radiatively dominated photospheres.
The B-star atmosphere is almost certainly composed primarily of material
processed within the K star. The so-called dredge-up phases of red
giant evolution (Becker and Iben, 1979), will affect the abundances in
the surface layers of the K star and therefore in the wind and in the
B-star atmosphere. The isotopes $^4$He and $^{14}$N should be enhanced while
$^1$H, $^{12}$C and $^{16}$O should be depleted. A careful study of abundances using
the ultraviolet spectrum is in order.


REFERENCES

Becker, S. A. and Iben, I., Jr.: 1977, Astrophys. J. 232, pp831-853.
Bondi, H. and Hoyle, F.: 1944, MNRAS 104, pp.273-280.
Castor, J. H. and Lamers, H. J. G. L. M.: 1979, Astrophys. J. Suppl.
    39, pp481-512.

Chapman, R. D.: 1980, Nature 286, pp580-581.
Chapman, R. D.: 1981a, in The Universe at Ultraviolet Wavelengths: The
    First Two Years of IUE. ed. R. D. Chapman. NASA CP-2171.
    Washington, DC.:GPO.
Chapman, R. D.: 1981b, Astrophys. J., Sept. 15, 1981 issue.
Kurusz, R. L.: 1979, Astrophys. J. Suppl. 40, pp1-340.
Stencel, R. E. and Chapman, R. D.: 1981, Astrophys. J., Dec. 15, 1981
    issue
Wilson, O. C.: 1960, in Stellar Atmospheres. ed. J. L. Greenstein.
    Chicago: University of Chicago Press
Wright, K. O.: 1970, Vistas in Astronomy, 12, pp.147-170.

# PART III

## EVOLUTIONARY PROCESSES IN CLOSE BINARY SYSTEMS

EVOLUTION OF CLOSE BINARY STARS:   OBSERVATIONAL ASPECTS

Mirek J. Plavec
Department of Astronomy
University of California, Los Angeles, CA 90024

ABSTRACT

Detached close binary systems define the main sequence band satisfactorily, but very little is known  about the masses of giants  and supergiants. High-dispersion IUE observations  promise an  improvement,
since blue companions are now frequently found to late-type supergiants.
The interesting cases of μ Sagittarii and in particular of ε Aurigae are
discussed in more detail. The barium star abundance anomaly appears now
to be due to mass transfer in interacting systems.   The symbiotic stars
are another type of binary systems containing late-type  giants; several
possible models for the hotter star and for the type of  interaction are
discussed.  The W Serpentis stars appear to be Algols in the rapid phase
of mass transfer, but a possible link relating them to the symbiotics is
also  indicated.   Evidence of  hot circumstellar  plasmas has now  been
found in several ordinary Algols; there  may exist a  smooth transition
between very quiescent Algols  and the  W Serpentis stars.   β Lyrae  is
discussed in the light of new spectrophotometric results.

INTRODUCTION

By its format and title, this Colloquium closely resembles the Colloquium On the Evolution of Double Stars held at Uccle 15 years  ago, in
September 1966 (Dommanget, 1967). That was a memorable colloquium, since
the evolution  in binary stars was, for  the first time, the topic  of a
whole  meeting.   Since then,  our field  has expanded  tremendously. We
held two large-scale Symposia discussing the evolution of close binaries
only (Eggleton,  Mitton  and Whelan,  1976; Plavec,  Popper  and Ulrich,
1980), in  addition to several other meetings on a slightly  lower scale.
After the most recent Symposium,  held in  Toronto in 1979,  I concluded
that in the future it would  no longer be possible to  cover adequately,
in one  full Symposium,  the  whole field  of close  binaries.   Thus the
goals set for this Colloquium  are in no way small.   By coincidence,  I
have been entrusted with the same type of introductory talk at this Col-

*Z. Kopal and J. Rahe (eds.), Binary and Multiple Stars as Tracers of Stellar Evolution, 159–181.*
*Copyright © 1982 by D. Reidel Publishing Company.*

loquium as I gave fifteen years ago,  and this gives me a  good opportunity to compare.

        The topic of my Uccle talk, as well as the topic of the subsequent
extremely important  contributions by  Paczyński and  by Kippenhahn and
Weigert, was practically entirely the evolution leading from  two binary
components on  the Main Sequence  to a  semi-detached Algol  system.   I
think only Paczynski went  beyond this framework and suggested  that the
Wolf-Rayet stars may be products  of a similar process of  mass transfer
between the components. Thanks to Kippenhahn and Weigert and  to Paczynski, we heard for the first time about actual model sequences describing
this process; naturally, those calculations were based on  the "conservative" assumptions, namely that both  the total  mass of the  system and
its orbital angular momentum remain preserved.  Nevertheless, I remember
vividly the remark made in the discussion by Kruszewski, who declared in
a rather prophetic and (therefore?)  tragic voice:  "... The question of
rate of mass loss looks hopeless  from both the theoretical and  the observational points of view...A  question of first importance ...  is the
ratio  of the matter lost from  the system to the matter  transferred to
the opposite component... The accuracy of magnitude estimate that we can
get from spectroscopic observations tells us nothing about  this ratio."
(Dommanget, 1967, p. 124).  After fifteen  years, this dilemma  is still
plaguing us, and a good part of my  talk will be devoted to  the problem
whether the spectroscopic observations can tell us something or no thing
at all.

        Concerning the scope of the topics discussed at Uccle, it would be
wrong to assume that at that time in the past, the field of close binary
star evolution  was really so  narrow as  to include only  the incipient
concepts of the formation of the Algol systems.  Very little was said at
Uccle about two extremely important types of binary stars  the investigation of which was at that time just about to start the fantastic explosion of activity  and knowledge  that transformed binary  star astronomy
from "arcane art",  to use the term coined  by R. P. Kraft, into  one of
the forefront fields in astrophysics:  I mean the X-ray binaries and the
cataclysmic variables.

        Accretion as  the mechanism  powering the  galactic  compact X-ray
sources emerged at  about that time, perhaps symbolically  introduced to
the wider astronomical community by  the famous remark by Ginzburg  at a
Radio Astronomy Symposium (van Woerden, 1967, p. 411) to the effect that
"We have such a large amount of gravitational energy available in such a
binary source:  we must use it!  of course!".  Soon after, Trimble and
Thorne (1969) opened the search  for black holes in binary  systems; although this venture has so far been  much less fruitful than it  was originally hoped for, their paper is still a landmark.   The evidence that
binary nature is essential for  the existence  of novae and  dwarf novae
developed gradually, but by the time of the Uccle Colloquium it  was already firmly established by the work of Kraft (1963) and others.   There
is no doubt that  the X-ray  binaries, cataclysmic variables,  and other
binary systems remain in  the forefront of interest today.   And I think

we can add to them another important class of binary stars, namely the
RS Canum Venaticorum systems. Their unusual photometric properties,
their X-ray and radio emission, and their obvious relation to chromospheric activity of G-K type stars attracted many astrophysicists who
were never before interested in binary stars. It is really impossible
to cover these three important groups in one talk, and it would make no
sense to attempt it. There have been so many good reviews, talks, and
conferences on them in the recent few years that I have nothing of value
to add. I want to concentrate on binary systems in the earlier stages
of evolution of both components. They may not generate such excitement
and so conspicuous phenomena, but they represent stages of evolution
through which all of the exciting objects had to pass; and since we are
here to trace stellar evolution in all its twists and turns, they deserve proper attention.

You will have noticed that the there exists a subtle difference
between the title of the whole Colloquium, Binaries as Tracers of Stellar Evolution, and the title of my talk, Evolution of Close Binaries.
It is true that close binary stars, in particular their eclipsing variety, are the most important tracers of stellar evolution, since they can
provide the most complete set of parameters characterizing the evolutionary state of each component, if circumstances are favorable. However, quite often they mark a detour from the proper track of the normal
stellar evolution: they lead us along a track which they themselves
laid differently. Since a large fraction of stars are actually members
of close binary systems, it is naturally quite justified to study their
evolution as an important alternative to the single star evolution.
Nevertheless, it is quite proper to say first a few words on how close
binary stars contribute to the knowledge of single star evolution.


DETACHED BINARY SYSTEMS AS TRACERS OF STELLAR EVOLUTION

Tracing stellar evolution means plotting the evolutionary tracks
point by point. A star of a given mass is described by a number of parameters, such as effective temperature, luminosity, radius, chemical
composition, rotation, atmospheric structure, possibly also stellar wind
and/or a circumstellar envelope. Combined photometry, spectrophotometry, and radial velocity studies can give us practically all this information if the star is member of an eclipsing system and circumstances
are favorable.

We often hear it said that eclipses are a real miracle, a royal
road to knowledge. This all is true, but purely physically, the eclipses are a simple consequence of the fact that the orbital planes of
close binary stars are oriented at random. What should be considered as
a truly remarkable fact, one that is not a priori obvious and easy to
anticipate, is that binary stars tend to come as pairs of stars of nearly equal masses. Statistical studies, whether they find bi-modal or
unimodal distributions of mass ratios, agree that there exists a strong

trend towards mass ratios close to one (see, e.g., Trimble, 1974). Close binary stars have become the most important tracers of stellar evolution mainly because of this property. Otherwise, the strong positive dependence of radius, effective temperature and in particular of luminosity on stellar mass would make eclipses shallow and secondary spectra undetectable at any wavelength. This is in particular true about the main sequence band.

Thanks to favorable mass ratios, a large part of the main sequence is now well described empirically by means of the components of eclipsing binaries. Popper (1980) whose criteria are unusually severe, lists 36 reliable systems which cover satisfactorily the range of spectral types between B6 and G2. Then there is a gap between G2 and the two well-determined pairs of early M type stars, YY Gem and CM Dra. This gap is unlikely to be filled. Eclipsing binaries in this region tend to be either of the contact (W UMa) type, or of the probably mildly evolved type (RS CVn).

Popper noticed a somewhat similar difficulty with eclipsing stars earlier than about B6. The difficulty seems to be primarily technical. Proximity effects distort the light curves and shallow and blended spectral lines adversely affect the radial velocity work. As a consequence, it is difficult to distinguish between the detached, semidetached, and contact systems among the early-type binaries. I encountered this difficulty when I attempted to introduce two-dimensional classification of eclipsing binaries (Plavec, 1964). Hot and luminous early-type stars have extensive and dynamical outer atmospheres; thus it may well be that the difficulty is not merely technical but represents an inherent property.

DETACHED SYSTEMS WITH GIANT AND SUPERGIANT COMPONENTS: A NEW ERA BEGINS

As soon as the more massive star of the pair leaves the main sequence, differential evolution will quickly create a large gap in the H-R diagram between the two components, even if their masses are very similar. Now the less massive star, still sitting on the main sequence, will be associated with a late-type giant or supergiant. For stars more massive than about 4 $M_\odot$, i.e. practically for all B stars, the evolutionary track in the H-R diagram is practically horizontal all the way from the main sequence to the red giant tip. The luminosity does not change markedly, while the peak in the spectral energy distribution shifts to longer wavelengths. The giant or supergiant now, as a rule, dominates the visual region of the spectrum. But the other star, although somewhat less massive and therefore also less luminous, will make a strong showing in the ultraviolet. Until recently, the knowledge of this fact was of little comfort to astronomers, and eclipsing binaries with one component away from the main sequence were no good tracers of stellar evolution. Visual binary stars were no better in this respect, although for a different reason: giants and supergiants are rare anim-

als in solar vicinity.   As a consequence of this conspiracy, the masses
of giants and supergiants are   still very poorly known, and   many important studies of the  various peculiar  and exciting objects   suffer from
this lack of knowledge.

Among the systems consisting of  a giant or supergiant and   a main
sequence star, the eclipsing binaries 31 Cyg, 32 Cyg, $\zeta$ Aur, and   VV Cep
became famous, but for a different reason.   They exhibit atmospheric eclipses when the hotter, much smaller star traverses behind the  very extended atmosphere of the cool  supergiant (a K supergiant in  the first
three cases, an  M supergiant in VV Cep).   The systems  are essentially
detached because of the large separations between the components, as indicated by their long periods, between 3 and 20 years.   Therefore, they
are important tracers of single star evolution, and should enable  us to
obtain the mass and other parameters of the supergiant component.   Complete orbital parameters and hence also masses were derived  from radial
velocities obtained  from the  optical spectra, although  with difficulties, since the lines of the blue  star are as a rule  severely blended.
From the published orbits, as  reviewed e.g.  by Wilson (1960)  and more
recently by Wright (1970), it transpires that  in $\zeta$ Aur, 31 Cyg,  and 32
Cyg the supergiant is about twice as  massive than its blue mate,  so it
agrees with single star evolution that the blue components  are probably
still on the main sequence.  In VV Cep, the M supergiant appears to have
a mass  only equal  to its blue  companion, or  even  slightly smaller.
Small discrepancy in this direction can perhaps be explained in terms of
mass loss from the supergiant.  It should be remembered that in spite of
truly heroic efforts, in particular by Wright (1977), the orbital parameters and hence the masses in VV Cephei are poorly known.  No absorption
feature can be safely attributed to the hotter component alone,  and the
orbit of the hotter star is based on a detailed reconstruction of a complex  emission profile of H$\alpha$, of  which one component is supposed  to be
associated with the hotter star; however, it is not clear if  its radial
velocities are identical with those of the photosphere of the  hot star,
even if it could be safely identified, isolated, and measured.

A new epoch came with the  advent of the IUE satellite.   When the
high-dispersion mode of the spectrograph can be used, we have the opportunity to measure radial velocities  of the hot component; and  both the
low-dispersion and high-dispersion modes enable us to study the spectral
energy distribution and the line profiles.   As in the optical region, a
careful study is needed in  each individual  case in  order to isolate
clean lines of the hotter star.  This may not be possible at all in certain cases.   Thus it seems, according to Stencel et al. (1980), that in
32 Cyg the B star is moving  rather deep inside the stellar  wind structure of the K4 supergiant, and that a hot turbulent region surrounds the
B star.   Yet I  am convinced that clean lines  can be found, if not  in
this system,  then in others.   So far,  everyone has been excited about
eclipse studies and about winds and interactions.  I would like to point
out the  importance of the "old-fashioned" approach.  If our  good luck
lasts and  the IUE satellite remains operative  for a few  more years,
there is good hope for improving orbital data.

Nor is it necessary to attach our hopes only to the ζ Aurigae
stars. A number of supergiants are now known or strongly suspected to
be accompanied by blue components. Independently of the far ultraviolet
observations, multicolor photometric studies indicated a large incidence
of blue companions in the Cepheids. From an extensive photometry in the
Walraven five-color system, Pel (1978) concluded that among the southern
Cepheids he studied, at least 25% are members of close binary systems.
Madore and Fernie (1980) use the differential color effect a potential
blue component will have on the minimum phase of the light and color
curves of Cepheids, and conclude that (35 ± 5)% of them have blue compan
ions. Parsons (1981b) examined 50 supergiants of spectral types F and
G, and concluded that at least 17 among them are double, and at least 10
of these have hot companions. All these numbers agree well with the
statistical conclusions by Abt and Levy (1978) on the incidence of binary stars among B type stars. Since binary star components tend to
have similar masses, and since the giants and supergiants examined have
evolved from main-sequence B stars, Abt and Levy's statistics have a direct bearing on the supergiant surveys.

An extension of the supergiant survey to supergiants of an earlier
spectral type than A will certainly reveal additional binaries. Observationally, the task becomes more and more difficult as the supergiant
will also dominate the ultraviolet. A good example is the discovery of
a hot companion to the luminous B8Ia supergiant μ Sagittarii. The hotter star, of spectral type near B0 V, does not contribute significantly
to the total flux of the system except at wavelengths shorter than about
150 nm; and its character can actually be established with some degree
of confidence only thanks to the eclipses. That an eclipse occurs in
spectroscopic binary system of μ Sgr has been known since 1938. But
this must be the shallower eclipse, since it occurs at the conjunction
with the B8 star behind. When R. Polidan discovered lines of P V in the
Copernicus spectrum of the star, obtained in our joint project, it was
clear that another and deeper eclipse must occur when the hotter component is eclipsed by the B8 supergiant. I predicted this primary eclipse
for September 1979 (Plavec, 1979), and combined observations by several
people (Guinan and Dorren, Kondo, Plavec and Polidan) confirmed the prediction. The duration of the eclipse is probably several weeks, but far
from safely determined. The system is not easy to study since its period, 180 days, is not only long but is fairly close to half a year.
Only one primary eclipse can be observed per year, in August-September;
the other occurs at a time when the sun is too close to the star in the
sky. By subtracting the IUE spectra, we were able to obtain the spectral
energy distribution of the hotter component (Plavec, 1981a; Plavec and
Weiland, 1980), which clearly suggests a spectral type near B0; but the
effective temperature remains uncertain within wide limits, probably
mainly because of the quasi-periodic fluctuations of the light of the B8
supergiant, discovered by Dorren, Guinan, and Sion (1981). Our estimates vary between 18,000 - 40,000 K, but we are reasonably sure that the
correct value will be nearer the lower limit of this interval. The radial velocity curve of the B8 supergiant is well determined and gives a

large  mass function,  f(m) = 2.67.   If the  mass of the supergiant lies
between  10 and 20 solar masses,  as is reasonable to suppose,  then the
hotter component  must have 8-13 solar masses.   Since it  is about 2.5$^m$
fainter in  V than the  B8 supergiant,  it is  probably  a main-sequence
star,  and  the two  components have  evolved  essentially independently:
the system is still detached.   But there exists interaction between the
two components in the form of a strong stellar wind blowing from the luminous supergiant.   Additional  absorption  lines in  the spectrum have
been found both by Polidan in the  <u>Copernicus</u> spectra,  and by us  in the
high-dispersion IUE spectra.   They are due mostly to Fe II and have the
character of shell lines.   Thus we may observe a kind of an atmospheric
eclipse  preceding the bodily  eclipse of  the hotter star.   The system
promises to yield valuable  information on the structure of  the stellar
wind  from a supergiant that is  much hotter than those in  which atmospheric eclipses were studied in the past:   thus I believe that this has
been a significant discovery.

     Similar direct discoveries of hotter companions are  becoming more
and more frequent.   Mariska, Doschek and Feldman (1980) report the discovery of  components of  spectral types not  far from  A0 V in  the two
classical Cepheids $\eta$ Aql and T Mon.   Parsons (1981a) announced that V810
Cen (HR 4511 = HD  101947), which is probably another  classical Cepheid
but with quite  a large period of 125  days, is associated with a  hot B
star (actually seen already by Bohm-Vitense and Dettman, 1980);   the hot
star seems  to have a  stellar wind  indicating a supergiant,  while its
continuum flux suggests  a less luminous star,  perhaps  luminosity class
III.

$\varepsilon$ AURIGAE:   ENIGMA OF THE QUARTER CENTURY (OR OF 27 YEARS)

     Before I leave the realm of the supergiants, I would like  to talk
about one of the  most mysterious eclipsing binaries,  namely  $\varepsilon$ Aurigae.
Since the term "Enigma of the  Century" has already  been requisitioned
for SS 433, I must call $\varepsilon$ Aur only an  enigma of a quarter century.   In
fact,  the enigma  always comes  only every 27  years, when we  get an
eclipse  of the star,  and outside  eclipse we  have very little  hope to
make a  real  breaktrough into  its mystery (observationally,  I  mean;
bright  ideas can come any time).   Unlike the $\zeta$ Aur supergiant eclipsing
systems,  the primary eclipse -- the only one observed -- comes when the
supergiant is eclipsed by -- well, by something.  The  eclipsing object
is the enigma.   It causes  a long  eclipse  about 0.75m deep over a wide
range of wavelengths,  and the eclipse is  reasonably flat,  as if it were
total.   But it cannot be total since the spectrum of the F0Ia supergiant
remains visible without profound changes,  and no other spectrum emerges,
although  judging from  the  depth of  the eclipse,  the other component
should be certainly sufficiently bright to be seen.

     Numerous clever schemes were invented to explain  these paradoxes,
among them the idea that the eclipsing body is essentially a  disk; and,
of  course, as  one alternative for  the central  object of  the disk, a

black hole was suggested. In connection with recent ultraviolet observations, an alternative idea advanced by Hack (1962) becomes very important: The nearly neutral opacity of the disk is explained in terms of electron scattering, and the necessary source for the photons that must ionize hydrogen over a very large volume is sought in a Be star. And indeed, low-dispersion IUE spectra do show a flux excess over that of the F supergiant in the far ultraviolet; the excess flux is detectable at wavelengths shorter than about 150 nm with certainty, and a little beyond this wavelength if the flux of the supergiant can be properly subtracted. After an approximate subtraction, Hack and Selvelli (1979) concluded that the source of the excess flux is most likely a B star, with an effective temperature of about 15,000 K, and with an absolute visual magnitude of about $-1^m$. The supergiant is much brighter in the visual region, $M_v = -6.7^m$ according to van de Kamp (1978), who also finds that the distance to the system is 580 pc from a combination of astrometric and spectrographic observations. A companion of the above temperature and luminosity would be probably an main-sequence star. What puzzles me is the problem how such a modest star of a rather late B spectral type can ionize such a vast volume, whose radius must be about 850 solar radii in order to perform the eclipsing duties properly. I observed the system with the IUE, too, and did find the extra flux in the far UV. From the very short spectral segment observable, it is very hard to conclude anything about the nature of the hotter source; if I fit it by a Kurucz atmosphere model for $T_{eff} = 15,000$ K, I find that the object is a subdwarf rather than a main-sequence star. Its light may be variable; or it may be largely obscured by a disk at whose center it may reside.

Dynamical considerations only augment the puzzle. The radial velocity curve of the F0 supergiant appears to be simple and reliable. It yields a mass function $f(m) = 3.12$. The orbital inclination cannot be too far from 90° because of the long quasi-total eclipse, so adopting sin i = 1 does not introduce a serious error. We also know that the orbit of the supergiant with respect to the center of gravity of the system is $A_F \cong 2.8 \times 10^3$ $R_\odot$. One more assumption then gives us an idea about the masses. We can argue that the evolutionary tracks of massive stars in the H-R diagram are almost horizontal, i.e. their luminosity remains nearly constant. Then the absolute visual magnitude $M_v = -6.7^m$ determined by van de Kamp (1978) suggests $M_F \cong 13.5$ $M_\odot$ and the mass function then gives for the unknown star $M_U \cong 13$ $M_\odot$, and for the separation $A \cong 5.8 \times 10^3$ $R_\odot$. A completely invisible object has the same mass as the luminous F0 supergiant!

This is such an outrageous result that one is tempted to abandon the value of 13.5 $M_\odot$ for the F0 star (although it appears reasonably justified), and to attempt to vary the mass ratio in order to see if anything plausible emerges. It won't! Going to a mass ratio 2:1 in favor of the F0 supergiant quickly increases the masses of both stars above 20 $M_\odot$ and deepens the puzzle of the large secondary mass. If we want to reduce the secondary mass, we must go to an inverse mass ratio, i.e. make the invisible star more massive! For $M_U/M_F = 2$ we get $M_U = 7$

$M_{\Theta}$ , $M_F$ = 3.5 $M_{\odot}$:   now we must explain why 3.5 solar  masses give us a luminous supergiant, while  twice that mass remains invisible.   One may recall the case of β Lyrae, in which a similar situation obtains. But in β Lyrae the more massive component  is not really invisible, we  only do not observe any absorption lines from it; it emits enough continuous radiation to make the secondary eclipse quite perceptible.

Over the range of  mass ratios  considered, the separation  of the components remains of the same order of magnitude, A ≅ 5 × $10^3$ $R_{\Theta}$ ≅ 23 AU.   Thus the F0 supergiant, whose radius we  can estimate from its absolute magnitude and temperature to be $R_F$ ≅ 200 $R_{\odot}$, is far too  small to fill its critical Roche lobe.  If the  secondary is star inside a  disk (an idea which is rather plausible  because of the shape of  the eclipse light curve, see e.g. Wilson, 1971),  why is  it surrounded by  a disk? This can hardly be accretion from the supergiant!

My IUE observations confirm Hack and Selvelli's finding that there is one and just one emission  line visible in the ultraviolet,   namely O I(2) λ 1302 A. I know of only  one other spectrum which shows  just this one emission line, and that is  the symbiotic star CH Cygni,  which consists of a semiregularly variable M6  III giant and a hot  object which, according to Luud (1981) should be white dwarf, while according  to Wing and Carpenter (1981)  most likely is an 0  or early B star close  to the main sequence.   The latter  observations, based on recent IUE  spectra, are probably more reliable, yet in  either case there is most  likely no connection with ε Aurigae  here, only  the similarity of  the underlying physical process (for a discussion, see Hack and Selvelli, 1979).

It   appears that the  number of  puzzles surrounding ε  Aurigae is endless.   Fortunately for us, the next eclipse is  just around the corner.   The partial phase is supposed to start in June/July 1982, the famous "totality" should  last from January/February 1983 through  the end of December  1983 or early  January 1984,  and the partial  phase should then end  in June/August 1984.   The dates  of the contacts are somewhat uncertain and the actual duration of the eclipse appears to be variable, which is not suprising if at least  one of the components is  actually a disk rather than a star.  For the first time, we will be able to observe the eclipse in the  infrared and in the ultraviolet.  Some traditionally accepted concepts, like the greyness of the eclipse, may  disappear just because  of the broad  wavelength range  covered this time.   If nothing else shows up, the least we will get in the ultraviolet is a better look at the mysterious additional light:  if the light of the F0 star is dimmed by about $0.75^m$, then a wider segment of the FUV spectrum of the hotter source should be seen. I will not be surprised, though, if this hotter source  is eclipsed, too!   We have  seen this combination of a ~B8 source with another, F-type continuum in one and the same component in W Serpentis:   I interpreted it as a B star embedded in an optically thick disk.  If the F0 spectrum were due to a flat disk, the flat shape of the eclipse light curve would be easy to understand.   But it is hard to explain the observed high luminosity  and large size of the  eclipsed star by this idea.   What is not hard to explain in  ε Aurigae?   Let's wait, watch and see!

SYMBIOTIC STARS AS BINARIES:   WHAT IS THE DEGREE OF INTERACTION?

The so-called symbiotic objects have long existed at the outskirts of stellar astrophysics as a small group of mysterious objects.   By the classical definition of P. W. Merrill, a symbiotic object displays a combination spectrum:   emission lines indicating a hot source are super-imposed upon a late-type stellar continuum.   In typical cases (if such a thing exists for the  symbiotics), we  observe TiO absorption  bands to-gether with the emission lines of  He II and [O III].   However, the un-derlying continuum  can also be of spectral  type K or G, and  a certain variety in the presence of the emission lines must also be accepted even by purists.

It has long been believed that most if not all symbiotics  are bi-nary systems, but hard evidence was slow to come.  In a few systems, ra-dial velocity  variations suggested Keplerian motion with  long periods, between 1 and 20 years.   Thus large dimensions of the systems are indic-ated, and obviously the nebulosity radiating the emission lines  will be of the same order of size,  otherwise the typical forbidden lines  of [O III], [Ne  III], and occasionally of [Fe  VII] would  not show up.   But there existed harly any direct evidence of the presence of a  hot compo-nent in the system.   In fact,  the veiling of the late-type  absorption lines, often  considered as the  evidence for  a hot blue  continuum, is more likely due to a continuous radiation of circumstellar hydrogen.

The   advent  of   the International Ultraviolet Explorer satellite opened a new epoch in  the investigation of the symbiotics.   We can now directly observe the continuum due to  a hotter object in AG  Pegasi, AG Draconis, Z Andromedae, and other objects.   But it is still not easy to recognize the nature of the  hot components. The slope of  the continuum in several objects resembles that of a B0 star, but the presence  of the emission lines of He II $\lambda$ 164 nm,   C IV $\lambda$ 155 nm,   and N V $\lambda$ 124  nm de-mands a hotter source of ionizing photons:   the Zanstra temperatures are near $10^5$ K.   Thus the FUV continuum we observe with the IUE is probably only the Rayleigh-Jeans tail of the actual stellar continuum. And  it is often contaminated by continuous hydrogen radiation, in particular long-ward  of $\lambda$ 200 nm. In  some symbiotics,  we observe only  an essentially flat, probably circumstellar continuum (AR Pav, CI Cyg, CH Cyg, AX Per). Yet the hot star must be there, since the high-ionization emission lines are strong.   It appears that the hot source must be a small star if  it can be hidden in some sort of a disk or envelope; after all, in spite of its high emissivity,  its contribution  to optical fluxes  is negligible compared to the red giant.

The cool components appear to be normal K-M type giants,  but some are semiregular variables (CH Cygni), others are Miras (R Aquarii). Com-pared to them, the hot components must have very much  smaller effective radiating areas. They appear to be subdwarfs, with radii of the order of 0.1 to 1 $R_0$, and with masses not very different from 1 $M_0$ (but our stat-istics, in particular  of masses, are woefully incomplete!).   A central star of a planetary nebula has just the right temperature, size, and luminosity.  Moreover, spatial distribution of the symbiotics strongly resembles that of the planetary nebulae (Boyarchuk, 1975).   Thus it would be easiest to assume that the hot components of the symbiotics are close relatives of the central stars of the planetaries, and the  red component is present in the system only to  provide (all or most of)  the material for the nebulosity, which is ionized by the photons generated by nuclear burning of the subdwarf.  The cool giant would be losing mass  by stellar wind, as is usual for late-type luminous stars, although we may  have to postulate an "enhanced" wind mass loss on order of $10^{-5}$ to  $10^{-6}$ $M_{\odot}$/year (perhaps enhanced by the  relative proximity  of the photosphere  of the red giant  to its  Roche critical lobe).   This would  be the  simplest, "pure natural" model of a symbiotic object, and I called it a PN symbiotic  or a subdwarf  symbiotic (Plavec,  1982). The difficulty  with this scheme is that the subdwarfs are, according to   theoretical calculations (Paczyński,  1971),  extremely  short-lived objects,  in  particular with masses even a little above 1 $M_{\odot}$. These subdwarfs have degenerate carbon-oxygen cores and  produce energy  in nuclear-burning shells  of hydrogen and helium, located in a fairly thin envelope, which is quickly consumed because of this shell burning.   A slight modification of the same model would  be a helium star as  the hot component,  formed from  a moderately massive Algol subgiant which at the  end of its mass loss  stage ignited helium in its  core.   But  we  encounter another difficulty with  the "natural" model:  it appears  that flares and slow nova-like  eruptions are  typical in  the symbiotics, and  these are  hard to explain  by the above model, which implies  little or no interaction between  the components.  Perhaps the so-called BQ[] stars (Ciatti, D'Odorico and Mammano, 1974) are built on this model.

A very  promising  model was  developed by  Tutukov  and Yungelson (1976) and by Paczyński and Rudak (1980).  Again, the hot component is a subdwarf as described above, but its lifetime is  artificially prolonged by the material which is continually transferred from the red  giant, is accreted in the atmosphere,  and then  consumed in the  nuclear burning shells.   In fact, a degenerate white dwarf can be "rejuvenated" in this way, its nuclear-burning shells ignited, and then maintained by this influx.   The theorists  often speak of this component  as of a degenerate dwarf:  however, because of the formation of the non-degenerate envelope of substantial thickness, it is really a subdwarf by its size, effective temperature, as well as luminosity.  Paczyński and  Rudak (1980), Rudak (1982) and Tutukov and Yungelson (1982) have  shown that this  model is very sensitive to the rate  of mass  transfer, and  can  produce either quasi-periodic  flares  or slow  nova-like eruptions.  Perhaps  the term novalike symbiotics  may be  appropriate for them.   We see  that in the symbiotics built on this model, the red component not only maintains the nebulosity but also stimulates and maintains the production of  the ionizing photons -- at the surface of the other star!   Fairly low rates of accretion are sufficient, in fact needed, of the order of  $10^{-7}$ $M_{\odot}$/year, so again mass loss from the red giant via a stellar wind is all  that is needed.

A third model for the symbiotics postulates accretion not as a stimulant of nuclear burning, but rather as the direct generator of the ionizing photons. Since we need temperatures only of the order of $10^5$ K and the symbiotics are not known to be X-ray emitters (with one or two exceptions), the surfaces of degenerate dwarfs represent too deep potential wells for accretion in this type, and the model postulates accretion on main-sequence stars or on subdwarfs. The required temperature of $10^5$ K is then generated in the innermost parts of an accretion disk surrounding the star, and in particular in the transition zone between the disk and the star itself. This transition zone is thin and therefore has a small effective radiating area, even if the accreting star is fairly large. Thus in this model, the small size of the hot source postulated by its low emission in the optical region, does not necessarily mean that the companion to the red star is a star below the main sequence. Bath (1977, 1981) developed this model as an analogy to his model of optically thick envelopes of novae outbursts (1978). The model requires very high rates of mass transfer between the components of a symbiotic, $10^{-4}$ $M_\odot$/year or higher, and these can be reached only if the red giant fills its critical lobe and loses mass by Roche lobe overflow. The model is again concerned primarily with the eruptive activity observed in many symbiotics, and strongly depends on another theory by Bath (1972), according to which the red giant components of binary stars become temporarily unstable and eject large amounts of gas in spurts. Since the basic mode of mass transfer in this model is the same as in the Algols, and since the gainer is believed to be most likely a main-sequence star as in Algols, I think that the name Algol symbiotics is appropriate.

The cool components of the symbiotics are most likely giants on the second (asymptotic) giant branch of the stellar track through the H-R diagram. This conclusion is less based on a direct determination of the luminosity class of the giant, and more on the fact that the known orbital periods of the symbiotics are of the order of years. The giant should either fill or temporarily fill its critical lobe (as in the case of the Algol symbiotics), or at least it should not be an order of magnitude smaller than the critical lobe (otherwise its wind would probably be too weak). This reasoning suggests that the cool components must be large stars, and therefore lie on the asymptotic branch; the Mira nature of some of them confirms this conclusion. But then, why don't we observe symbiotics with the cool components on the first giant branch? Their orbital periods would be of the order of months. Perhaps the size of the system would not permit the existence of a nebulosity extended enough to display the typical emission lines of the symbiotics. Possibly, the W Serpentis stars (Plavec, 1980) -- or rather some of them, such as RX Cas or SX Cas -- are the relatives of the symbiotics with the cool components on the first giant branch.

At this time, we are unable to decide with certainty which of the above models is the most appropriate for the symbiotics, or if all three apply, each one to different cases. A whole Colloquium, IAU No. 70 has been devoted to them (Viotti and Friedjung, 1982), and the reader will find many answers and even more questions in that publication.

BARIUM STARS: NO SUCH THING OUTSIDE A BINARY SYSTEM?

Remember the many discussions whether abundance anomalies are intrinsic, or due to mass transfer in binary stars? Well, a new twist to the story is here. McClure, Fletcher, and Nemec (1980) found that all stars exhibiting the strong Ba II anomaly vary in radial velocity, and may well all be binaries; in two cases they could go beyond this statement and concluded that the mass functions indicated the presence of a component with a mass between 1 and 2 solar masses. These low masses, low luminosities one must expect for the hypothetical companions, and small radial velocity ranges of the Ba II giants, all suggest that the systems are rather wide and that the companions will probably be degenerate stars. Now Bohm-Vitense (1980) reports that the Ba II class 2 star $\zeta$ Cap, G5 II, indeed has such a component, since the far ultraviolet spectrum shows an increase of the flux shortward of $\lambda$ 150 nm. From the observed flux distribution, the star must have an effective temperature of about 22,000 K, while its mass is near 1 $M_\odot$ : the object is rather similar to Sirius B. These observations strongly suggest that the barium anomaly may be due to mass transfer rather than to an internal mixing process intrinsic to the star.

THE ALGOLS: A BETTER LOOK AT THE COMPONENT STARS IS NOW POSSIBLE

The semidetached binaries of the Algol type, believed to be products of the first phase of mass transfer observed near the end of the mass transfer phase, are easy to detect and study photometrically, but much harder to study spectroscopically. The cooler and fainter subgiant secondary components are as a rule suppressed in the combined spectrum over the spectral range ordinarily explored. As a result, our knowledge of their masses and other characteristics was for years about as crude as were theoretical evolutionary sequences explaining Algols. Recently, however, the situation on the observational front improved substantially with the introduction of red-sensitive image tubes. Popper (1980) lists already 17 reasonably well determined systems; and to them, we should add U Cep (Tomkin, 1981) and U CrB (Batten and Tomkin, 1981).

Although this sample is still insufficient for truly reliable statistical studies, some conclusions can be drawn with more confidence than was possible in the past. I will only mention here an interesting observation about the masses of the subgiants. The masses of the subgiants that accompany the B-type primaries, U Cep, U CrB, and U Sge, are actually not small, not far from 2 $M_\odot$, and reasonably appropriate for G stars above the main sequence. Truly small masses of the subgiants, and hence fairly large overluminosities, are encountered mainly in systems whose primaries are A stars (S Cnc, RY Gem, AS Eri, AW Peg). But there are exceptions among B stars, like RY Per and Algol itself, with rather low-mass secondaries ($\sim$0.8 $M_\odot$).

Studies based on the improved determinations of the masses  of the Algol systems (De Greve and Vanbeveren, 1980; De Greve,  preprint, 1981) confirm  the  suspicion  voiced  earlier  (Kopal,1971; Plavec,  1973) that the present configurations  of the Algols demand considerable  mass loss from these systems at the  earlier stages  of mass transfer.  This would not be so surprising if the principal mode of mass loss from  the losers (the initially more massive components) were an isotropic  stellar wind. But the losers typically have too low luminosities for a  normal stellar wind to be efficient (unless it is tremendously enhanced by  the proximity to the Roche lobe).   Evolutionary calculations postulate Roche lobe overflow with an ensuing gas  stream directed  into the vicinity  of the other component (the gainer).   So why  should the transferred gas leave the system in large  quantities, instead of being accreted?   Why, when, and how  does it happen?   In particular:   can we  identify the systems that are currently in the rapid phase of mass transfer, when this escape from the system must occur?

## THE W SERPENTIS SYSTEMS:   A LINK BETWEEN THE ALGOLS AND THE SYMBIOTICS?

I  believe that we  do observe  interacting systems  in  the rapid phase, and  that they  are  probably  of  the W  Serpentis  type (Plavec, 1980),  and we do observe direct  evidence of mass outflow from  them in the  profiles  of  the  far  ultraviolet emission  lines.   Generally, the presence of any emission lines is a good indicator of the existence of a fairly  large volume around  one component or around the  whole system, filled with fairly dense circumstellar material.  But the emission lines of the Balmer series, observed  in many Algols particularly at  the time of the eclipses, were always believed to come from rings  encircling the gainer; only recently did Crawford (1981) show that this picture  may be oversimplified. The emission lines  discovered in  the FUV by  R.H. Koch and  me (Plavec and Koch, 1978;  Plavec, Weiland and Koch, 1982)  tell a different story.   When they can be observed at high  dispersion (like β Lyrae and KX Andromedae), then all their emission lines display distinct P Cygni profiles.  The lines in question are mostly resonance lines of C II, C IV, N V, Si II, Si III, Si IV,  Al II, Al III, and  some low-level transitions of Fe III.  Thus we observe a stellar wind, and there exists a certain  degree of analogy  with hot  luminous early-type  stars.   In those, too, mass outflow was long  suspected, but only the lines  of the abundant elements observed in the FUV clearly demonstrated the existence of the winds.  But the wind observed in the Serpentids is different from the "classical" wind observed in luminous hot supergiants.  The terminal velocity in  β Lyrae is no more  than 500 km/s, the profile  is asymmetrical with the  emission part  stronger than the  absorption component. Probably collisional excitation  of the upper levels of  the transitions plays a more important role in the Serpentids.   The luminosity and temperature of the central star (the gainer) are too low to provide the necessary driving force.   More likely,  the energy  is ultimately derived from the gravitational potential  energy released in the process  of accretion.

The nature of the components of the W Serpentis stars is not easy to establish because the spectra and photometric light curves are complicated by the circumstellar matter, which, in addition to the emission lines also produces deep shell absorption lines and a hydrogen continuum. In SX Cas, we now believe that the correct spectral types are B7 III + K3 III, plus a fairly strong optically thin hydrogen continuum (Plavec, Weiland and Koch, 1982). In RX Cas, we detected only the late-type component, K1 III, and the hydrogen continuum (Plavec, Weiland, Dobias, and Koch, 1981); the primary component appears to be lost in the hydrogen continuum, and may be either a main-sequence star fainter than A0, or a star below the main sequence. In W Serpentis, we seem to observe only one object, the one that is partially eclipsed at primary eclipse, but two continua appear to be associated with this object: a hotter one, about B8, seen in the FUV, and a cooler one, about F5, dominating in the optical region. Two possible models come to mind: either the primary component is an F5 star, surrounded by an accretion disk, whose innermost part radiates as a smaller B8 object. Or the primary component is actually a B8 star, to a large degree obscured by a thick disk, whose edge radiates as another photosphere simulating a star of spectral type F5. We are now inclined to prefer the latter explanation (Plavec and Sakimoto, 1978; Plavec et al., 1981). But a third explanation, unknown to us at the moment, may be the right one.

It is rather natural to assume that the W Serpentis stars are a natural continuation of Algols toward longer periods. The analysis of SX Cas seems to support this idea, and RX Cas does not contradict it. But it is interesting to realize that their periods are longer than one month, and that the cool components are giants probably on the first giant branch. The flat spectrum of RX Cas obtained when the K1 III giant is subtracted is quite silimar to that of the symbiotic star AR Pavonis. The emission line spectra are not identical: AR Pavonis displays He II emissions and intercombination lines indicative of a moderate electron density ($10^6 - 10^9$ cm$^{-3}$), while RX Cas displays only weak He I emissions, and almost no intercombination lines; the density in its circumstellar envelope must be much higher ($10^{12}$ cm$^{-3}$). But these density differences in the nebulosity may be simply consequences of the different dimensions of the two systems, obvious already from the very different orbital periods (32 days in RX Cas as against 605 days in AR Pav). Otherwise the nature of the objects need not be drastically different. Don't we have here an indication of a possible similarity?

## THE ALGOLS REVISITED: OBJECTS NOT SO DORMANT AS WE THOUGHT

The "classical" Algol systems have periods of only a few days, and have long been considered disappointingly quiescent, "old ladies with an interesting but remote past". Some Algols are probably indeed rather clean of circumstellar matter now (see Fig. 1 for U Sge), but others are more active than we have thought.

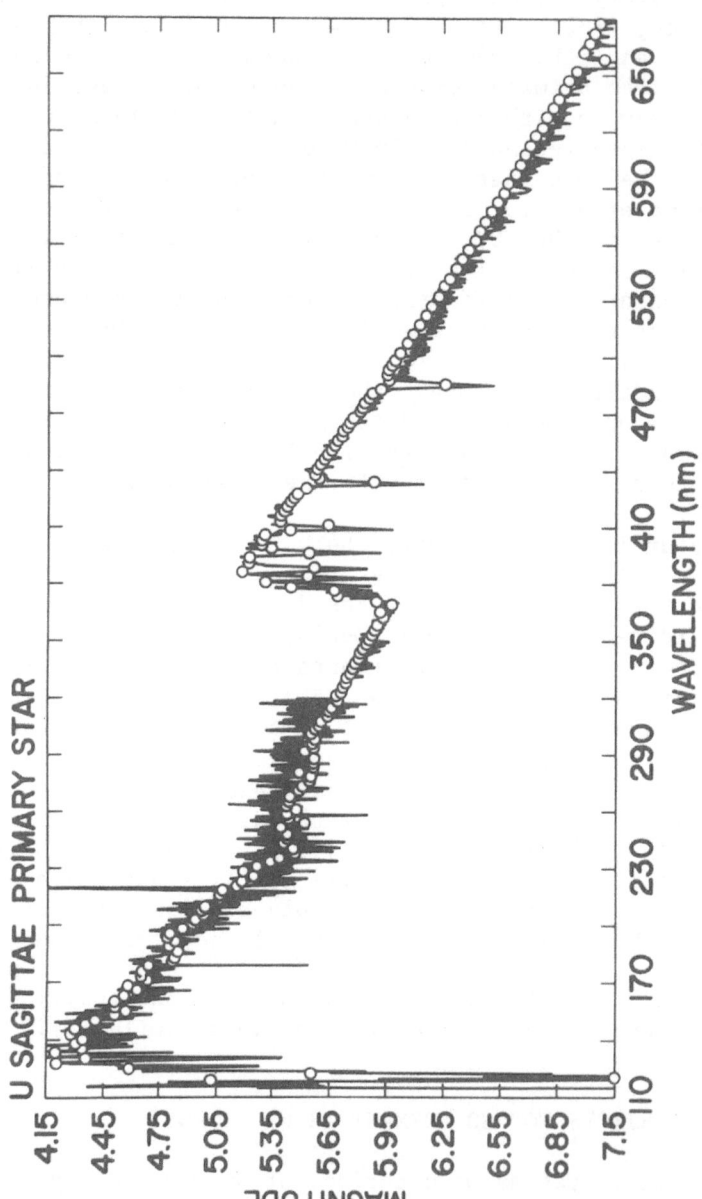

Fig. 1.: The primary component of the Algol system U Sagittae displays an uncomplicated spectrum, which can be well matched by an interpolated Kurucz model atmosphere for $T_{eff} = 12,250$ K, log g = 3.85.

The unexpected flaring up of the Hα emission in U Cephei (Batten et al., 1975; Plavec and Polidan, 1975) called attention to this object, and systematic observations, mainly by Olson (1980) and Crawford (1981), revealed a complex and variable structure of the circumstellar material surrounding the gainer. One could still argue that U Cephei is a uniquely active Algol; but the truth is rather that other Algols have not been studied carefully enough. Olson (1981a) reported a similar phenomena in RW Tauri, and Kaitchuck and Honeycutt (1981) fully confirm his findings. Theirs and Crawford's studies reveal various puzzles. The relative size of the gainer in such short-period systems like U Cep and RW Tau are too large, and the stream from the loser should impact on them directly, rather than form a disk normally expected in longer-period systems (Lubow and Shu, 1975). Yet some sort of transient disks apparently exist in U Cep and RW Tau. Moreover, the emission lines are broadened much more than a Keplerian motion of a simple disk would do. The optical emission lines are not the only evidence of circumstellar activity. Olson (1981b) noticed a near-ultraviolet excess in two Algols of very different period: RS Cephei (P = 12 days) and AI Draconis (P = 1.2 days). In the W Serpentis stars, the near-ultraviolet excess was found to be due to a circumstellar hydrogen continuum with the Balmer jump in emission, but in some systems it can also be the long-wavelength tail of a "hot" far-ultraviolet continuum (originating in a star or in the transition layer between the gainer and the surrounding disk). It would seem that the small system of AI Draconis must be rather similar in its structure.

Another indication that the Algols are far from dormant came with the discovery of the high-ionization emission lines (C IV, N V, Si IV) in the FUV totality spectra of V356 Sagittarii (Plavec and Dobias, 1980) and of U Cephei (Plavec, Dobias and Weiland, 1982). A chromospheric origin of these lines is unlikely: in U Cephei, it would give unusually large surface fluxes, in V356 Sgr we have no star that would be expected to have a chromosphere. Thus, the two stars are probably related to the W Serpentis stars. This means that we must assume the existence of hot circumstellar plasmas even in relatively short-period Algols. Further evidence for the existence of such plasmas comes from the studies of the absorption spectra of the Algols. Kondo, McCluskey, and Harvel (1981) discovered strong absorption lines of Si IV and C IV in U Cephei. Polidan and Peters (private communication) made similar observations in other Algols, such as CX Dra, AU Mon, or U CrB. Our high-dispersion spectra (Plavec, Dobias, and Weiland, 1982) also confirm the presence of absorption lines in a number of Algols of ions of a much higher level of ionization than would be appropriate for the spectral type of the stellar components. Apparently, regions of highly heated plasmas exist in many (if not all) accreting systems, and the transition between the short-period Algols and the W Serpentis stars is only a matter of degree.

BETA LYRAE:  ALWAYS DESERVES A SPECIAL CHAPTER

We have observed β Lyrae at both eclipses and at several intermediate phases, both with the IUE satellite and with the Lick Obsrvatory
ITS scanners. By subtracting the eclipse scans from those taken at full
light, we obtained energy distributions for each component separately
(Plavec, Weiland and Dobias, 1982). No better procedure is available
since the eclipses are not total. But a degree of uncertainty enters
since the light outside the eclipses is not constant. An improvement
will be possible when a better phase coverage is obtained and the observations are tied in with photometric light curve solutions. Nevertheless, even the preliminary results are quite interesting.

The component whose spectral lines are observed at all phases is
usually classified as B8 II. It is surprising to see (Fig. 2) that the
corresponding Kurucz atmosphere providing the best fit ($T_{eff}$ = 11,000 K,
log g = 2) matches the observed flux distribution reasonably well just
only over a part of the optical region (370 –560 nm). There appears to
be a flux deficiency shortward of λ 160 nm; this may be the consequence
of an incomplete inclusion of line blanketing in Kurucz's models of hotter supergiants. Everywhere else, the observed flux exceeds the model
flux. An infrared excess has been known to exist for some time. Now we
see that there exists at least as strong (probably stronger) ultraviolet
excess as well. Both can be probably explained by the same hydrogen
circumstellar cloud. Unfortunately, the important spectral segment in
the vicinity of the Balmer jump has not yet been adequately covered by
our Lick scans.

We have obtained a similar flux distribution for the secondary
component, a truly mysterious object: Although it is more massive than
the primary, it contributes less but still significantly to the continuous radiation, but shows no detectable absorption lines. In the optical region, the secondary's continuum parallel closely that of the B8 II
star, i.e. the two objects have nearly the same color temperature there.
In the far UV, beginning at about λ 160 nm, the secondary component is
brighter, i.e. its color temperature is higher (See Fig. 3). The secondary eclipses are deeper in the FUV than the primary ones. This variation of the color temperature across the spectrum is explained in principle if we assume that the secondary object radiates as a disk. Con-
tamination by circumstellar hydrogen continuum is even stronger than for
the primary component. It is impossible to decide if the secondary star
itself is visible in certain regions of the spectrum. On the whole, the
thick disk model advocated by Wilson (1974) is supported by our observations.

A curious thing happens in the spectral region between λλ 180 –
220 nm: There are practically no eclipses observed in β Lyrae in that
spectral region! Obviously, the circumstellar material surrounding both
stars, or the whole system, extends to such large distances that
eclipses of its parts do not significantly reduce its light; and in the

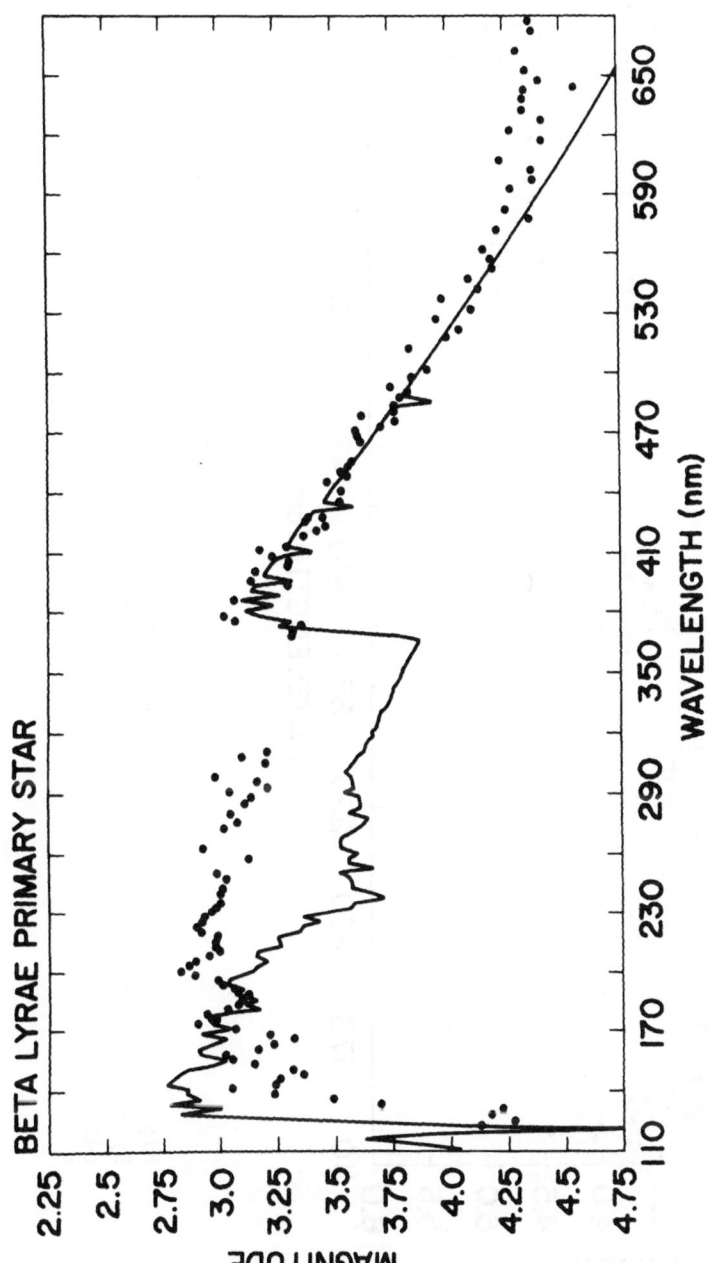

Fig. 2.: The primary (B8 II) component of β Lyrae is well matched by a Kurucz model atmosphere only in the optical spectral region. Over most of the spectrum, the observed fluxes (dots) lie above the model fluxes (full line).

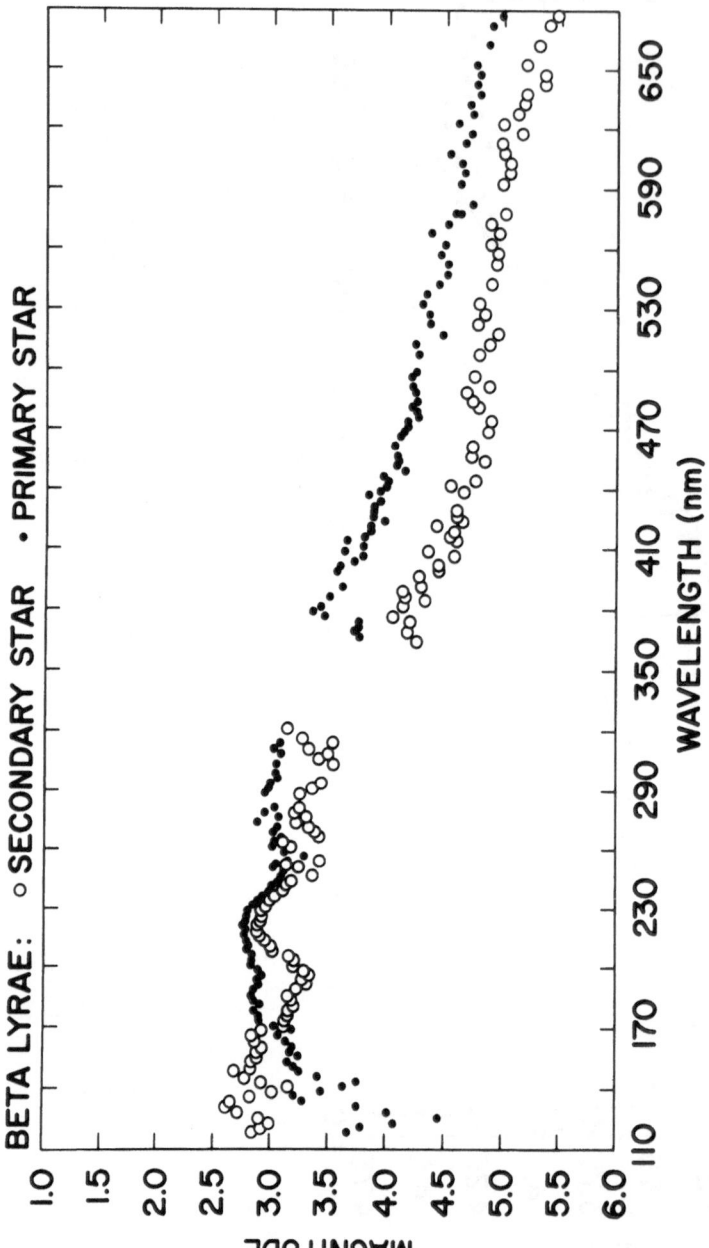

Fig. 3.: The relative energy distributions of the two components of β Lyrae are compared. The secondary star is the brighter one in the far ultraviolet.

spectral region mentioned, the circumstellar material emits morew flux than the two stellar components. This excess flux, or the "$\lambda$ 200 nm bulge" is visible in the combined spectrum at all phases, and was known already from the Copernicus observations. It was explained as a super-position of numerous weak emissiqn lines of Fe III, for example by Viotti (1976). Indeed, a number of prominent isolated Fe III emission lines are visible in the IUE spectrum of $\beta$ Lyrae, and multiplet tables show that very many lines of Fe III cluster just in the above spectral re-gion. Nevertheless, I do not believe that this explanation is complete, in fact it may not even represent the dominant cause of the bulge. Each of the individually observed emission lines of Fe III has a distinct P Cygni profile, and a quasi-continuum consisting of a number of such lines should show traces of this structure, although degraded by super-position. But the continuum is smooth. I would like to suggest that the bulge is due primarily to continuous radiation of hydrogen, with a non-negligible optical thickness at the Balmer limit, and corresponding to an electron temperature near 15,000 K. A comprehensive computer code developed by Drake and Ulrich (1980) at UCLA shows that with a suitable choice of parameters, one can get a local maximum of flux at the ob-served wavelength. Our observations have revealed the presence of sim-ilar bulges in the spectra of all the Serpentids; in some, such as W Crucis, the $\lambda$ 200 nm bulge is very prominent. The Serpentids do not help us to decide between the two above explanations, since Fe III emis-sions are always present in their spectra. But the symbiotics do not show Fe III emissions, yet the bulge is observed in some of them. Moreover, and this is decisive, it is seen displaced to shorter wave-lengths, such as $\lambda$ 160 nm, which is easily possible if we assume an electron temperature of the hydrogen cloud to be closer to 20,000 K, but is impossible to explain by an accumulation of Fe III emissions there.

A FINAL REMARK

I will not attempt to summarize the various topics I mentioned in this paper. There appears to exist a bewildering variety even among the objects most of which we would simply describe as binary systems in the first phase of mass transfer. Yet we also notice surprising links that connect many of them, unexpected similarities: are they Rosetta stones or red herrings?

We could look at the same problem from a different point of view. At times some of the objects: $\varepsilon$ Aurigae, $\beta$ Lyrae, and above all SS 433 appear completely unique. Yet I cannot believe that it is so: I think that much more likely these bizarre systems are just rather extreme cases to which one or more links lead, and for which certain, hopefully simpler, relatives exist. If we manage to identify them, we may be much closer to a better understanding of the greatest puzzles.

ACKNOWLEDGEMENTS

This work has been supported by grants from NSF and NASA. My participation at the meeting was made possible by travel grants by the University of California at Los Angeles, and by the Remeis-Sternwarte in Bamberg. All this support is gratefully acknowledged.

REFERENCES

Proceedings of Symposia, to be referred to by numbers:
1. Chapman, R.D. (ed.): 1981, The Universe at Ultraviolet Wavelengths, NASA Conference Publ. 2171.
2. Chiosi, C. and Stalio, R.: 1981, Effects of Mass Loss on Stellar Evolution, Dordrecht: Reidel.
3. Dommanget, J. (ed.): 1967, On the Evolution of Double Stars. Comm. Obs. Roy. Belgique Serie B, No. 17.
4. Eggleton, P., Mitton, S., and Whelan, J. (ed.): 1976, Structure and Evolution of Close Binary Systems. Dordrecht: Reidel.
5. Plavec, M.J., Popper, D.M. and Ulrich, R.K. (ed.): 1980, Close Binary Stars: Observations and Interpretation. Dordrecht: Reidel
6. Viotti, R. and Friedjung, M: 1982, The Nature of the Symbiotic Stars, Dordrecht: Reidel.
Abt, H.A. and Levy, S.G.: 1978, Astrophys. J. Suppl. 36, 241.
Bath, G.T.: 1972, Astrophys. J. 173, 121.
Bath, G.T.: 1977, Mon. Not. R.A.S. 178, 203.
Bath, G.T.: 1978, Mon. Not. R.A.S. 182, 35.
Bath, G.T.: 1981, in Proc. North Amer. Workshop Symbiotic Stars, (ed. R. Stencel), JILA, 20.
Batten, A.H., Fisher, W.A., Baldwin, B.W., and Scarfe, C.D.: 1975, Nature 253, 174.
Batten, A.H. and Tomkin, J.: 1981, Publ. Dom. Astroph. Obs. Victoria 15, 419.
Boyarchuk A.A.: 1975, in "Variable Stars and Stellar Evolution", (ed. V. E. Sherwood and L. Plaut), Reidel, 377.
Böhm-Vitense, E.: 1980, Astrophys. J. Lett. 239, L79.
Böhm-Vitense, E. and Dettman, T.: 1980, Astrophys. J. 236, 560.
Ciatti, F., D'Odorico, S., and Mammano, A.: 1974, Astr. Astroph. 34, 181
Crawford, R.C.: 1981, PhD thesis, UCLA.
De Grève, J.P. and Vanbeveren, D.: 1980, Astrophys. Space Sci. 68, 433.
Dorren, J.D., Guinan, E.F., and Sion, E.M.: 1981, ref. 1, 381.
Drake, S.A. and Ulrich, R.K.: 1980, Astrophys. J. Suppl. 42, 351.
Hack, M.: 1962, Mem Soc. Astron. Ital. 32, 3.
Hack, M. and Selvelli, P.L.: 1979, Astron. Astrophys. 75, 316.
Kaitchuck, R.H. and Honeycutt, R.K.: 1981, preprint.
Kondo, Y., McCluskey, G.E., Harvel, C.A.: 1981, Astrophys. J. 247, 202.
Kopal, Z.: 1971, Publ. Astron. Soc. Pacif. 83, 521.
Kraft, R.P.: 1963, Adv. Astron. Astrophys. 3, 43.
Lubow, S.H. and Shu, F.H.: 1975, Astrophys. J. 198, 383.
Luud, L.S.: 1981, Astrophysics 16, 262.
Madore, B.F. and Fernie, J.D.: 1980, Publ. Astron. Soc. Pacif. 92, 315.

Mariska, J.T., Doschek, G.A., and Feldman, U.: 1980, Astrophys. J. 238, L87.
McClure, R.D., Fletcher, J.M., and Nemec, J.M.: 1980, Astrophyus. J. 238, L35.
Olson, E.C.: 1980, Astrophys. J. 237, 496.
Olson, E.C.: 1981a,b, preprints.
Paczyński, B.: 1971, Acta Astron. 21, 417.
Paczyński, B. and Rudak, B.: 1980, Astron. Astrophys. 82, 349.
Parsons, S.B.: 1981a, Astrophys. J. 245, 201.
Parsons, S.B.: 1981b, Astrophys. J. 247, 560.
Pel, J.W.: 1978, Astron. Astrophys. 62, 75.
Plavec, M.J.: 1964, Bull. Astron. Inst. Czechosl. 15, 156.
Plavec, M.J.: 1973, in "Extended Atmospheres etc." (ed. A.H. Batten), Reidel, 216.
Plavec, M.J.: 1979, Inf. Bull. Var. Stars (Budapest) No. 1598, May 1979.
Plavec, M.J.: 1980, ref. 5, 251.
Plavec, M.J.: 1981a, ref. 1, 397.
Plavec, M.J.: 1981b, ref. 2, 431.
Plavec, M.J.: 1982, ref. 6.
Plavec, M.J. and Polidan, R.S.: 1975, Nature 253, 173.
Plavec, M.J. and Koch, R.H.: 1978, Inf. Bul;1. Var. Stars No. 1482.
Plavec, M.J. and Sakimoto, P.J.: 1978, Bull. Amer. Astron. Soc. 10, 609.
Plavec, M.J. and Dobias, J.J.: 1980, Bull. Amer. Astron. Soc. 12, 869.
Plavec, M.J. and Weiland, J.L.: 1980, Bull. Amer. Astron. Soc. 12, 869.
Plavec, M.J., Weiland, J.L., and Koch, R.H.: 1982, Astroph. J. in press.
Plavec, M.J., Weiland, J.L., Dobias, J.J., and Koch, R.H.: 1981, Bull. Amer. Astron. Soc. 13, 523.
Plavec, M.J., Dobias, J.J., Weiland, J.L., and Stone, P.R.S.: 1981, in "Be Stars", ed. M. Jaschek and H.-G. Groth, Reidel.
Plavec, M.J., Dobias, J.J., and Weiland, J.L.: 1982, Bull. A.A.S. (Boulder meeting, January 1982)
Plavec, M.J., Weiland, J.L., and Dobias, J.J.: 1982, ibid.
Popper, D.M.: 1980, Ann. Rev. Astron. Astrophys. 18, 115.
Rudak, B.: 1982, ref. 6.
Stencel, R., Kondo, Y., Bernat, A., McCluskey, G.E.: 1980, ref. 5, 555.
Tomkin, J.: 1981, Astrophys. J., in press.
Trimble, V.L.: 1974, Astron. J. 79, 967.
Trimble, V.L. and Thorne, K.S.: 1969, Astrophys. J. 156, 1013.
Tutukov, A.V. and Yungelson, L.R.: 1976, Astrophysics 12, 342.
Tutukov, A.V. and Yungelson, L.R.: 1982, ref. 6.
van de Kamp, P.: 1978, Astron. J. 83, 975.
van Woerden, H. (ed.): 1967, "Radio Astronomy and the Galactic System". London: Academic Press.
Viotti, R.: 1976, Mon. Not. R. A. S. 177, 617.
Wilson, O.C.: 1960, in "Stellar Atmospheres", ed. J.L. Greenstein, Chicago: Chicago University Press, 436.
Wilson, R.E.: 1971, Astrophys. J. 170, 529.
Wilson, R.E.: 1974, Astrophys. J. 189, 319.
Wing, R. F. and Carpenter, K. G.: 1981, ref. 1, 341.
Wright, K.O.: 1970, Vistas in Astron. 12, 147.
Wright, K.O.: 1977, Journ. Roy. Astron. Soc. Canada 71, 152.

# STATISTICAL PROPERTIES OF ALGOL-TYPE SYSTEMS

G. Giuricin, F. Mardirossian, and M. Mezzetti
Astronomical Observatory of Trieste, Trieste, Italy

## ABSTRACT

We have inspected the observational data of 114 Algol-type binaries
in order to clarify some general aspects of their evolutionary scenario.
Through a comparison with the corresponding available statistical analyses
of large sample of eclipsing and spectroscopic binaries, we have found
a larger than normal concentration of low values of masses, binary separations and specific orbital angular momenta for Algol-type stars.
Furthemore, high mass ratios predominate in massive and early-type Algol
systems, whereas high total masses are generally accompanied by less
prominent oversized and overluminous properties of the mass-losing components.

## INTRODUCTION

Extending previous investigations on the observational data of Algol-
type systems, based on a fairly small sample (Ziolkowski, 1969; Popov,
1970; Stothers, 1973; De Grève and Vanbeveren, 1980; De Grève, 1980), we
have attempted a statistical inspection of their properties. To be more
precise, we have examined the distribution of the masses, mass ratios,
radii, orbital periods, spectral types, luminosities, orbital angular
momenta, and binary separations of 114 Algol-type binaries. Although our
sample of Algol-type binaries is seriously biased by observational selection effects, a study of the distribution of their properties, compared
to the corresponding distribution (already discussed in the literature)
for spectroscopic and/or eclipsing binaries as a whole (Kraicheva et al.
1978; Staniucha, 1979; Trimble and Cheung, 1976; Farinella et al. 1979).
furnishes information on how some binary parameters vary as mass transfer
processes occur. We have considered the following binaries: TW And, XZ
And, RY Aql, KO Aql, QY Aql, V337 Aql, V346 Aql, RW Ara, SX Aur, IM Aur,
IU Aur, LY Aur, SU Boo, Y Cam, SZ Cam, S Cnc, RZ Cnc, R CMa, CV Car,

*Z. Kopal and J. Rahe (eds.), Binary and Multiple Stars as Tracers of Stellar Evolution, 183–186.*
*Copyright © 1982 by D. Reidel Publishing Company.*

QZ Car, RX Cas, RZ Cas, SX Cas, TV Cas, TW Cas, AB Cas, U Cep, RS Cep,
XX Cep, XY Cep, GT Cep, TZ CrA, U CrB, RW CrB, SW Cyg, UZ Cyg, VW Cyg,
WW Cyg, ZZ Cyg, KU Cyg, MR Cyg, V448 Cyg, V463 Cyg, V548 Cyg, V729 Cyg,
W Del, Z Dra, TW Dra, AI Dra, S Equ, AS Eri, RW Gem, RX Gem, RY Gem, AL
Gem, X Gru, μ Her, SZ Her, UX Her, AD Her, V338 Her, RX Hya, TT Hya, Y
Leo, T LMi, RS Lep, δ Lib, β Lyr, TT Lyr, RW Mon, TU Mon, AR Mon, AU Mon,
RV Oph, UU Oph, DN Ori, AQ Peg, AT Peg, AW Peg, DI Peg, β Per, RT Per,
RW Per, RY Per, ST Per, DM Per, IZ Per, Y Psc, V Pup, XY Pup, XZ Pup, U
Sge, RS Sgr, XZ Sgr, V356 Sgr, V505 Sgr, $μ^1$ Sco, V453 Sco, RY Sct, RZ
Sct, λ Tau, RW Tau, HU Tau X Tri, TX UMa, VV UMa, W UMi, RT UMi, S Vel,
DL Vir, Z Vul, RS Vul, BE Vul, V78ω Cen.


RESULTS

The values of the masses and radii of the binaries considered have
been taken from the literature. The temperatures of the primary components have been evaluated in accordance with the spectral classifications
found on the literature, on the basis of the temperature scale presented
by Popper (1980). The temperatures of the secondary components have not
been generally taken from the literature, which provides quite inhomogeneous (scale-dependent) evaluations; they are instead our own homogeneous average photometric estimates obtained from the primary' s temperature adopted and the ratio of the surface brightnesses resulting from
lightcurve analyses published in the literature.

The available observational data indicate that several mass-gaining
components tend to be oversized (compared to the main sequence) in the
mass-radius plane and overluminous in the HR diagram; but these components do not exhibit a tendency toward overluminous properties in the mass
luminosity plane. This behaviour is consistent with the view that these
stars are slightly swollen by mass accretion processes, but, at the same
time, have luminosities reduced by fast differential rotation brought
about by mass exchange.

The larger than normal concentration of low values of masses, binary
separations, and specific orbital angular momenta constitutes a strong
argument for an appreciable loss of mass and angular momentum in the
course of Algol evolution.

In spite of the observational selection effects, which act against
the detection of binaries with wide separations and very small mass ratios (and, hence, should lead  to a deficiency of such systems), no significant correlation between binary separation and mass ratio is discern-
ible. This probably means that an intrinsic anti-correlation,  masked
by selection effects, is indeed present in our sample. This is what is
expected as a result of mass transfer processes, provided  that the loss
of angular momentum from Algool binaries is not extremely large.

It is of interest that high mass ratios predominate in massive and early-type Algol systems. Moreover, the larger the total mass of an Algol systems is, the less prominent the oversized and overluminous properties of its mass-losing members, whose spectral types appear to correlate fairly well (especially in the high temperature range) with those of their companions. Large radius and luminosity excesses of the mass-losing members (with respect to main sequence values) have also a tendency to go with long orbital periods. These features are related to the general traits of conservative and quasi-conservative mass transfer theory, since we expect larger final mass ratios and smaller radius and luminosity excesses of the mass-losing components for case A mass exchange remnants (very probably absent in the moderate-mass range) than for case B systems and for binaries less advanced in the final slow phase of mass exchange.

## CONCLUSIONS

To summarize the major conclusions of our study, our survey of the properties of Algol-type binaries provides fresh evidence of the important role played by processes of non-conservative mass transfer. Certainly, the simple available models of non-conservative mass transfer (Yungelson, 1971; Drobyshevski and Reznikov, 1974; Vanbeveren et al., 1979), which yield a qualitative scenario fairly similar to the conservative one, give in several cases an improved agreement with the observations - e.g., final orbital periods, secondary radii and luminosities closer to those observed than in the conservative theory are attained. But, unfortunately, these models are so few in number that a detailed confrontation between theory and observations for a large set of binary parameters is barely meaningful. At this point we cannot yet state whether the relaxation of conservative assumptions may be sufficient to give a fully consistent picture of Algol-type binaries. However, optimistic suppositions in this sense seems to be discouraged by the fact that the available non-conservative evolutionary calculations appear to be somewhat inadequate to account for the observed moderate luminosities of the mass - losing components of Algol-type systems.

## ACKNOWLEDGEMENT

The authors are grateful to Mr. A. Janezich for aid in the preparation of the data.

## REFERENCES

De Grève, J.P.: 1980, Astrophys. Space Sci. 72, 411.
De Grève, J.P. and Vanbeveren, D.: 1980, Astrophys. Space Sci. 68, 433.
Drobyshevski, E.M. and Reznikov, B.I.: 1974, Acta Astron. 24, 29.

Farinella, P., Luzny, F., Mantegazza, L., Paolicchi, P.: 1979,
    Astrophys. J. 234, 973.
Kraicheva, Z.T., Popova, E.I., Tutukov, V., Yungelson, L.: 1978, Astr.
    Zh. 56, 520.
Popov, M.V.: 1970, Perem. Zvezdy 17, 412.
Popper, D.M.: 1980, Ann. Rev. Astron. Astrophys. 18, 115.
Staniucha, M.: 1979, Acta Astron. 29, 587.
Stothers, R.: 1973, Publ. Astron. Soc. Pacific 85, 360.
Trimble, V. and Cheung, C.: 1976, in IAU Symposium 73, Structure and
    Evolution of Close Binary Systems, eds. P. Eggleton, S. Milton,
    J. Whelan (Dordrecht: Reidel), p. 369.
Vanbeveren, D., De Grève, J.P., van Dessel, E.L., de Loore, C.: 1979,
    Astron. Astrophys. 73, 19.
Yungelson, L.: 1971, Sci. Inf. Astron. Council 20, 86 (in Russian)
Ziółkowski, J.:1969, Astroph. Space Sci. 3, 14.

# MASS LOSS IN ALGOL-TYPE STARS: IMPLICATION ON THEIR EVOLUTIONARY STAGE.

F. Mardirossian and G. Giuricin
Astronomical Observatory of Trieste,Trieste, Italy

ABSTRACT

We have examined the observational data of 102 Algols in order to clarify the implications on their evolutionary scenario of various assumptions concerning mass and angular momentum loss during mass transfer. We have found that case B mass exchange is strongly favoured for Algols of relatively low total mass ($\sim M < 7$ M⊙), while case A predominates, though not so widely as expected in Algols of higher total mass.

INTRODUCTION

It is now generally recognized that models of non conservative mass transfer, in which mass and orbital angular momentum loss is taken into account, can improve the agreement between theory and observational data of Algols. In order to investigate the effect of the processes of non-conservative mass transfer on the evolutionary status of Algol-type binaries, we have compiled and discussed the values of the total mass, the orbital period and the mass ratios of 102 Algols, as found in the recent literature. We clarify that in this study we mean by Algols a) semide-tached (sd) systems whose less massive members filling them Roche lobes are apparently more advanced in evolution, (b) the systems with undersize subgiant components (sd-d systems), regarded as post-main sequence mass exchange cooler remnants, and (c) the early-type contact systems having recently undergone or undergoing mass transfer between the components. In our sample we have also included the best known Wolf-Rayet objects GP Cep and V444 Cyg, since Wolf-Rayet objects may be products of mass transfer in very massive binaries. The binaries considered in this study are: TW And, XZ And, RX Aqr, KO Aql, QY Aql, RW Ara, SX Aur, IM Aur, IU Aur, LY Aur, SU Boo, Y Cam, SZ Cam, S Cnc, RZ Cnc, R CMa, CV Car, RZ Cas, SX Cas, TV Cas, TW Cas, U Cep, RS Cep, XX Cep, XY Cep, GP Cep, U CrB, RW CrB, SW Cyg, UZ Cyg, VW Cyg, WW Cyg, ZZ Cyg, KU Cyg, MR Cyg, V444 Cyg,

*Z. Kopal and J. Rahe (eds.), Binary and Multiple Stars as Tracers of Stellar Evolution, 187–189.*
*Copyright © 1982 by D. Reidel Publishing Company.*

V548 Cyg, V729 Cyg, W Del, Z Dra, TW Dra, AI Dra, S Equ, AS Eri, RW Gem,
RX Gem, RY Gem, AL Gem, X Gru, u Her, UX Her, AD Her, V338 Her, RX Hya,
TT Hya, Y Leo, T LMi, $\delta$ Lib, $\beta$ Lyr, TT Lyr, RW Mon, TU Mon, AR Mon, RV
Oph, UU Oph, DN Ori, AQ Peg, AT Peg, AW Peg, DI Peg, $\beta$ Per, RT Per, RW
Per, RY Per, ST Per, DM Per, IZ Per, $\delta$ Pic, Y Psc, V Pup, XZ Pup, U Sge,
RS Sgr, XZ Sgr, V356 Sgr, V505 Sgr, $\mu^1$ Sco, V453 Sco, RY Sct, RZ Sct,
$\lambda$ Tau, RW Tau, X Tri, TX UMa, VV UMa, W UMi, S Vel, DL Vir, Z Vul, RS Vul
BE Vul, V78 $\omega$ Cen.

## RESULTS AND DISCUSSION

Extending previous works generally based on the conservative assumption only (Ziółkowski 1976, Kreiner and Ziółkowski 1978), we have evaluated the initial (i.e., prior-mass-exchange) values of the orbital period
(Po) and total mass (Mo) of each binary, computed according to the assumption of conservative mass transfer and the non-conservative approaches
proposed in the literature (Tutukov,Yungelson, 1971; Plavec et al., 1973;
Drobyshevski and Reznikov, 1974; Djakov and Reznikov, 1979; Vanbeveren
et al., 1979). The positions of the binaries in the bilogarithmic plots
of Po versus Mo-computed according to the above-mentioned approaches –
with respect to the lines representing the initial orbital periods for
a  ZAMS contact system and for the transitions from case A to case B and
from case B to case C mass transfer allow  us to acquire information on
the possible original status and on the evolution of our sample of close
binaries.

Keeping in mind  that the mass ratio of the main sequence progenitor:
of Algols is very lively to be near unity (Svechnikov, 1969; Kraicheva
et al., 1978; Lucy and Ricco, 1979;Garmany et al., 1980; van't Veer,
1981), we have found that conservative calculations of mass exchange
cannot account for the evolutionary scenario of a not negligible fractions  ($\sim$ 20%) of our binaries (especially those  with low total mass and
low mass ratio). Besides, a large amount of angular momentum loss (characterized by a ratio between the specific orbital angular momentum of the
lost matter and that of the original system equal to 2.5-3.0) is required
for bringing the low-mass Algol-type binaries into a consistent picture.

Bearing in mind also the estimates of mass and angular momentum loss
provided by Popov's (1970) surveys of statistical observational data of
main sequence detached binaries and semidetached systems, we can state
that pratically all Algol-type binaries with relatively low total mass
($\lesssim$ 7 M$\odot$) appear to be case B remnants. On the other hand, regarding systems of higher total mass (roughly M $>$ 7 M$\odot$), case A mass transfer predominates, though  not so widely as expected from comparison of theoretical
evolutionary lifetimes.

Our results undoubtedly  indicate a strong deficit of binaries

originating from case A mass exchange. This fact may mean either that
during the early stages of mass transfer a contact configuration,with
the formations of a deep common envelope preventing the discovery of the
binary, occurs. more frequently in case A than in case B candidates or
that progenitors undergoing case A mass transfer are very rare. This latter view is consistent with    Svechnikov's (1969) and Kraicheva et al's
1978) finding that no close binaries with binary separations $A \lesssim 10R_\odot$
and primary to masses in the range $1.5 \lesssim M \lesssim 10\ M\odot$ are observed (perhaps
they cannot form).

## REFERENCES

Djakov, B.B. and Reznikov, B.I.: 1979, Acta Astron. 29, 425.

Drobyshevski, E.M. and Reznikov, B.I.: 1974, Acta Astron. 24, 29.

Garmany, C.D., Conti, P.S., Massey, P.: 1980, Astrophys. J. 242, 1063.

Kraicheva, Z.T., Popova, E.I., Tutukov, A.V., Yungelson, L.R.: 1978,
    Astr. Zh. 55, 1176.

Kreiner, J.M. and Ziółkowski, J.: 1978, Acta Astron. 28, 497

Lucy, L.B. and Ricco, E.: 1979, Astron. J. 84, 401.

Plavec, M., Ulrich, R.K., Polidan, R.S.: 1973, Publ. Astron. Soc. Pacific
    85, 769.

Popov, M.V.: 1970, Prem. Zvezdy 17, 142.

Svechnikov, M.A.: 1969, Katalog orbital'nykh    elementov, mass i sve-
    timostey tesnykh dvoinykh zvezd (Sverdlovsk., izd. UrGU).

Tutukov, A.V. and Yungelson, L.R.: 1971, Scientific Inf. of Astron.
    Council 20, 86 (in Russian)

Vanbeveren, D., De Grève, J.P., van Dessel, E.L., de Loore, C.: 1979,
    Astron.Astrophys. 73, 19.

van't Veer, F.: 1981, Astron. Astrophys. 98, 213.

Ziółkowski, J.: 1976, in IAU Symposium 73, Structure and Evolution of
    Close Binary Systems, eds. P. Eggleton, S. Mitton, and J. Whelan
    (Dordrecht: Reidel), p. 321.

# ULTRAVIOLET LIGHT CURVES OF U GEMINORUM AND VW HYDRI

Chi-Chao Wu, R. J. Panek, A. V. Holm, and F. H. Schiffer, III
Computer Sciences Corporation
Silver Spring, Md., U.S.A.

ABSTRACT.  Ultraviolet light curves have been obtained for the quiescent dwarf novae U Gem and VW Hyi.  The amplitude of the hump associated with the accretion hot spot is much smaller in the UV than in the visible. This implies that the bright spot temperature is roughly 12000 K if it is optically thick.  A hotter spot would have to be optically thin in the near UV.  The flux distribution of U Gem in quiescence cannot be fitted by model spectra of steady state, viscous accretion disks.  The absolute luminosity, the flux distribution, and the far UV spectrum suggest that the primary star is visible in the far UV.  The optical-UV flux distribution of VW Hyi could be matched roughly by our model accretion disks, but the fitting is poorly constrained due to the uncertainty in its distance.

## I.  INTRODUCTION

The optical light curves of dwarf nova systems often show a prominent hump which is attributed to a bright spot at the point where the gas stream from the late type secondary strikes the accretion disk around the white dwarf primary.  We have used the observed amplitudes of the hump in the ultraviolet and optical light curves to derive the temperature of the bright spot of U Gem for the optically thick and thin cases. We have also estimated the temperature and luminosity of the accretion disk of U Gem and VW Hyi by matching the observed flux distribution (UV+ optical) with that predicted by the steady state, viscous disk model computations.  All of the results here refer to the quiescent (non-outburst) state.

## II.  OBSERVATIONS

The UV light curves of U Gem and VW Hyi (Fig. 1 and 2 respectively) were obtained with the 5-channel spectrophotometer on board the Astronomical Netherlands Satellite (ANS) (van Duinen et al 1975).  The channels had almost rectangular response functions with central wavelengths and full widths (given in parentheses) at 1549(149), 1799(149), 2200(200),

*Z. Kopal and J. Rahe (eds.), Binary and Multiple Stars as Tracers of Stellar Evolution, 191–197.*
*Copyright © 1982 by D. Reidel Publishing Company.*

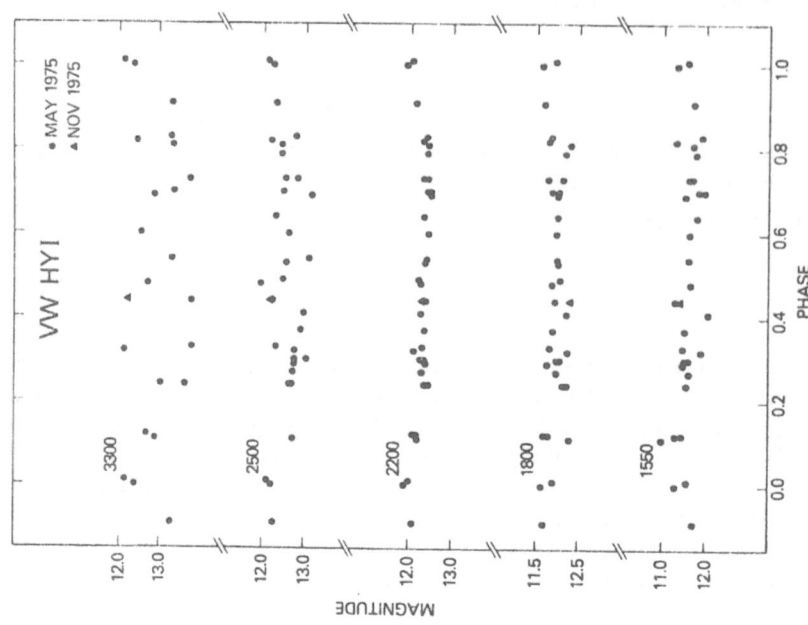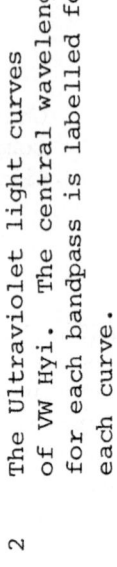

Fig. 2    The Ultraviolet light curves of VW Hyi. The central wavelength for each bandpass is labelled for each curve.

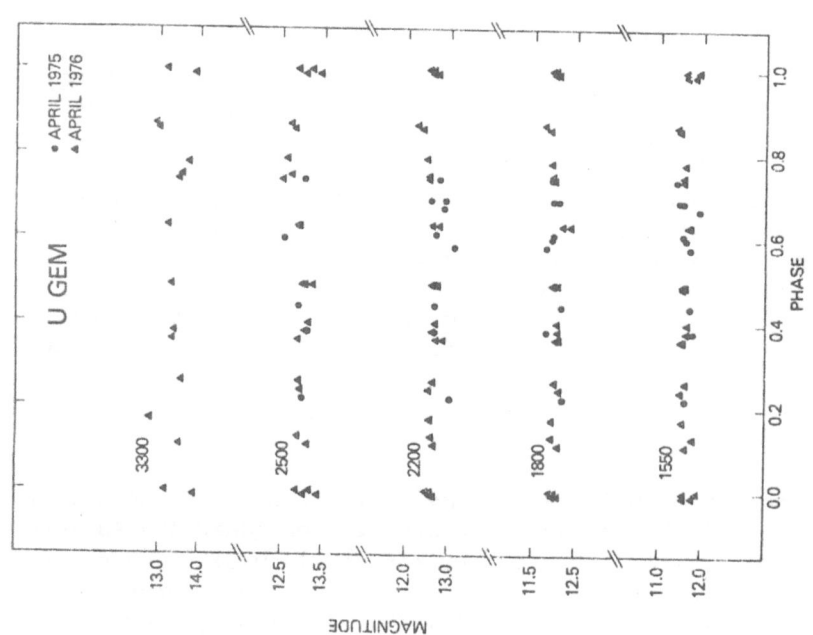

Fig. 1    The Ultraviolet light curves of U Gem. The central wavelength for each bandpass is labelled for each curve.

2493(150) and 3294(101) Å. Most of the observations were made with the offset (sky chopping) mode which had a duration of about 10 minutes. The magnitudes in Figures 1 and 2 are defined such that, for all channels, m = 0.00 for f = 3.64 x $10^{-9}$ ergs $cm^{-2}$ $s^{-1}$ $Å^{-1}$. Spectra of U Gem were obtained with the International Ultraviolet Explorer (IUE) with both the short wavelength prima (SWP) and long wavelength redundant (LWR) cameras. The spectra were taken through the large aperture in the low dispersion mode with a resolution of about 5 and 8 Å respectively for SWP and LWR (Boggess et al. 1978). The exposure time for the images was about 35 minutes. The absolute calibration of Bohlin and Holm (1980) was adopted.

III.  THE BRIGHT SPOT OF U GEM

It is obvious from Fig. 1 and 2 that the bright spot is not prominent in the ultraviolet. From the ANS light curves the hump amplitude is estimated to be 0.18, 0.26, 0.39, 0.35, and 0.69 at 1550, 1800, 2200, 2500, and 3300 A respectively. From optical light curves (Krzeminski 1965), the amplitude is 0.64, 0.87, and 0.66 mag at U,B, and V, respectively. To obtain quantitative estimates for the spot, we have calculated amplitudes for several assumed spot models (Fig. 3). These calculations are normalized to the observed amplitude at V. The assumed spot flux distribution plus the flux distribution of the disk then determines the amplitudes in the other bands. The flux distribution of the disk is taken as the average of the ANS fluxes at orbital phases 0.15-0.55 for which the spot is assumed to be hidden behind the disk. In Fig. 3 we see that the observed low amplitude of the hump in the UV indicates that an optically thick spot must have a temperature near 12000 K. If the spot gas temperature is 30000 K it must be optically thin in the UV. A free-free opacity provides reasonable agreement if the optical depth at V is about 2. Also in Fig. 3, we see that the amplitudes derived from the IUE spectra confirm the ANS results. If the bright spot is a 12000 K blackbody, it has a luminosity of 9 x $10^{-3}$ $L_{\odot}$ and a diameter of 3.2 x $10^{9}$ cm. If it is a 30000 K gas with free-free opacity $\tau$(5500 Å) = 2, it has a luminosity of 8 x $10^{-3}$ $L_{\odot}$ and a diameter of 1.4 x $10^{9}$ cm. These values are consistent with the 2.9 x $10^{9}$ cm estimated by Warner and Nather (1971) from eclipse geometry.

IV.  THE ACCRETION DISK

To calculate model spectra for the disk, we adopt the structure of a steady-state, optically thick, viscous accretion disk as used in several recent detailed comparisons to observations (e.g. see the review by Mayo et al. 1980). The principal novelty of our technique is to represent the energy radiated by each segment of the disk by the empirical flux of a main sequence star of the same effective temperature. The visual brightness is determined from the empirical relation between $T_{eff}$ and bolometric correction (Code et al. 1976), and the flux at the other wavelengths from the relation between $T_{eff}$ and (B-V) and the colors of Wu et al. (1980) and Johnson (1966). The surface gravity of main sequence stars is appropriate for the disks (see the discussion by Mayo

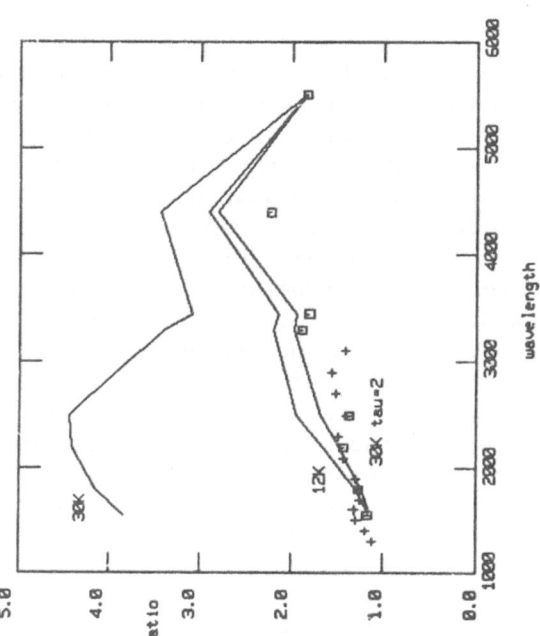

Fig. 3    Ratio of flux distributions for hotspot maximum / hotspot minimum.
The squares are derived from the ANS photometry, the plus symbols
from the IUE spectra.  The solid curves define the expected results
for 3 assumed flux distributions for the hotspot:  a 30000 K
blackbody, a 12000 K blackbody, and a 30000 K hotspot which
has an optical depth of 2 at 5500 A.

et al. ).  Also, the average inclination of a stellar disk is reasonable for the moderately inclined VW Hyi and U Gem.  Our calculations of absolute visual magnitude can reproduce the results of Mayo et al., and we also reproduce the fluxes of the coolest disk model of Herter et al. (1979), except at 1550 A, to within the expected accuracy of 0.1 to 0.2 mag.  Spectra for 3 representative models are shown in Fig. 4.  These demonstrate the manner in which $\dot{M}$, the mass flux, can be traded against disk size to obtain a variety of energy distributions of a given visual luminosity.

a.  The Disk of U Gem

     As discussed above, the UV flux distribution of the disk is assumed to be the average over the phase interval 0.15-0.55.  This is shown in Fig. 5, along with the optical data of Wade (1979) which should also be representative of the quiescent disk.  The UV fluxes were scaled up by 0.3 mag to match the optical data at 3300 A.  No reddening correction should be necessary.

     The distance of U Gem is estimated as 75 pc from the direct observation of the secondary star (Wade 1979).  This provides a valuable constraint in fitting to disk models by fixing the absolute magnitude. Comparison with the model disk spectra of Fig. 4 shows that the models are unable to reproduce the turnup of the flux distribution at the shortest wavelengths.  Furthermore, the disks must be quite small in order not to exceed the relatively low absolute visual magnitude of U Gem. The disk with $\dot{M}$ = 0.25 would provide the closet fit.  This disk has a radius less than 1/10 the radius vector for the spot as derived from the optical eclipse (Smak 1976).

     Our IUE spectra confirm the result of Fabbiano et al. (1981) that the upturn in flux persists down to 1250 A.  However, comparison with stellar spectra (Wu et al. 1981) shows that the far UV energy distribution of U Gem resembles a 30000 K main sequence star.  If we assume that the disk contributes nothing at 1250 A, the corresponding visual magnitude for the central star is 15.6 or absolute 11.2.  This is a reasonable value for a hot white dwarf primary star.  Such a star could be responsible for the observed strong Lyman alpha absorption (see Fig. 7 of Fabbiano et al.).  Thus, we feel that a large contribution to the far UV flux by the primary star is a reasonable alternative for nuclear surface reactions (Fabbiano et al. 1981).

b.  The Disk of VW Hyi

     We assume that the hotspot is hidden behind the disk at orbital phases 0.25-0.75 (the phase convention puts phase zero at the optical spot maximum), so that the flux distribution of the disk is obtained as the average of the ANS fluxes over this interval.  This is shown in Fig. 5, along with the optical fluxes of Panek (1979) which should also be representative of the quiescent disk.  The UV fluxes were scaled up by 0.3 mag to match the optical flux at 3300 A.  Bath et al. (1980) have demonstrated the absence of reddening in VW Hyi.  The fitting to VW Hyi is much less constrained because the distance is very poorly known.

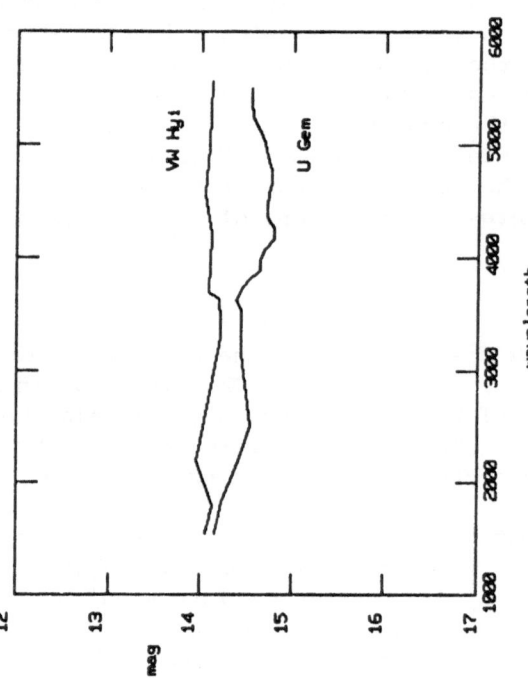

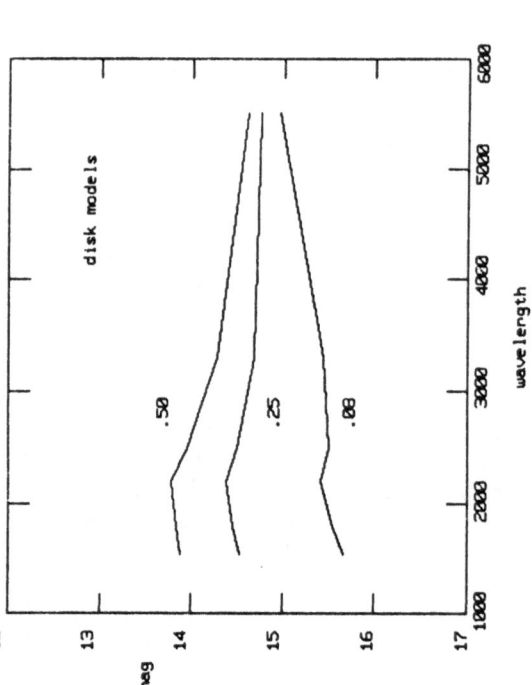

Fig. 4    Flux distributions for 3 representative disk
models, in the units of Fig. 5. The disks are
identified by their mass flux $\dot{M}$ (units of
$10^{16}$ g/s). The other disk parameters are
radius $1.6\ 10^{10}$ cm and luminosity 0.07 L for
the model with $\dot{M} = 0.50$; $1.9\ 10^{9}$ cm and
0.04 L for $\dot{M} = 0.25$; and $2.4\ 10^{9}$ cm and
0.02 L for $\dot{M} = 0.08$. The models have been
placed at the distance and inclination of
U Gem.

Fig. 5    Observed flux distributions for VW Hyi
and U Gem, shown as magnitude per unit
frequency vs wavelength in A. The ultraviolet (ANS) data has been scaled to match
the groundbased fluxes at 3300 A.

The flux distribution shown in Fig. 5 could be obtained with a variety of models if the absolute magnitude is unconstrained. However, we note that the red flux (Panek 1979) and the infrared flux (Sherrington et al. 1980) suggest that the flat spectrum persists to quite long wavelengths. This would require a very large disk to yield low enough temperatures to avoid the Rayleigh Jeans behavior, and would make VW Hyi much more luminous than U Gem.

We wish to acknowledge that this research is supported by NASA research contract NASW 3254 and IUE research contract NAS 5-25774.

## REFERENCES

Bath, G.T., Pringle, J.E., and Whelan, J.A.J.: 1980, Monthly Notices Roy. Astron. Soc. 190, p. 185.

Boggess, A., et al.: 1978, Nature 275, p. 377.

Bohlin, R.C., and Holm, A.V.: 1980, "IUE Newsletter No. 10", p. 37

Code, A.D., Davis, J., Bless, R.C., and Hanbury Brown, R.: 1976, Astrophys. J. 203, p. 417.

van Duinen, R.J., Aalders, J.W.G., Wesselius, P.R., Wildeman, K.J., Wu, C.-C., Luinge, W., and Snel, D.: 1975, Astron. Astrophys. 39, p. 159.

Fabbiano, G., Hartmann, L., Raymond, J., Steiner, J., Branduardi-Raymont, G., and Matilsky, T. : 1981, Astrophys. J. 243, p. 911.

Herter, T., Lacasse, M.G., Wesemael, F., and Winget, D.E. : 1979, Astrophys. J. Suppl. 39, p. 513.

Johnson, H.L. : 1966, Ann. Rev. Astron. Astrophys. 4, p. 193

Krzeminski, W. : 1965, Astrophys. J. 142, p. 1051.

Mayo, S.K., Wickramasinghe, D.T., and Whelan, J.A.J. : 1980, Monthly Notices Roy. Astron. Soc. 193, p. 793.

Panek, R.J. : 1979, Astrophys. J. 234, p. 1016.

Sherrington, M.R., Lawson, P.A., King, A.R., and Jameson, R.F. : 1980, Monthly Notices Roy. Astron. Soc. 191, p. 185.

Smak, J. : 1976, Acta Astronomica 26, p. 277.

Wade, R.A. : 1979, Astron. J. 84, p. 562.

Warner, B., and Nather, R.E. : 1971, Monthly Notices Roy. Astron. Soc. 152, p. 219.

Wu, C.-C., Faber, S.M., Gallagher, J.S., Peck, M., and Tinsley, B.M. : 1980, Astrophys. J. 237, p. 290.

Wu, C.-C., Boggess, A., Holm, A.V., Schiffer, F.H., III, and Turnrose, B.E. : 1981, "IUE Ultraviolet Spectral Atlas, IUE Newsletter No. 14", p. 2.

# STATISTICAL MODELS FOR CLOSE BINARIES

J.L. Halbwachs
Observatoire de Strasbourg, France

ABSTRACT. Statistical models for binaries as a whole were selected; then statistical properties of close binaries -the proportions of spectroscopic and eclipsing binaries and the distributions of the K velocities and of the depths of eclipse- were computed for each model.

## 1. INTRODUCTION

The statistical models describe the distributions of binaries according to their physical parameters. These parameters are: -the semimajor axis of the orbit, "a" ; -the mass ratio of the system , here defined as the ratio $q = M_2/M_1 < 1$; -the inclination of the orbit, "i", and -the eccentricity, "e", which may be neglected in a first approach. The models can be tested on observable features, as will be shown later.

## 2. SELECTION OF THE MODELS

In order to evaluate the observational consequences of the models , some typical distributions were assumed for each of the above mentioned parameters. 3 possible distributions for the logarithm of "a" are illustrated in Figure 1 and 4 distributions of "q" in Figure 2. The inclinations

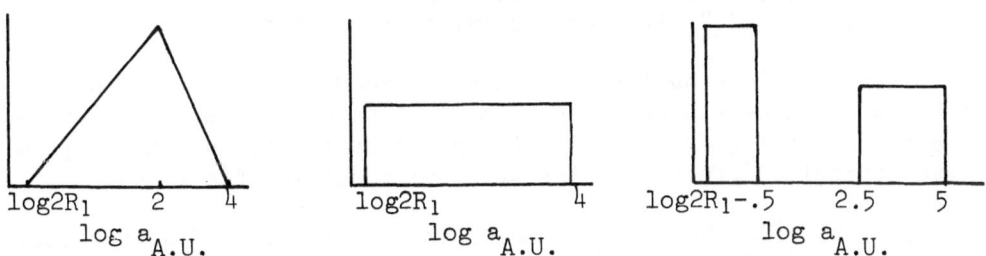

Figure 1. Selected distributions of log a; $R_1$ is the radius of the primary component of the system.

*Z. Kopal and J. Rahe (eds.), Binary and Multiple Stars as Tracers of Stellar Evolution, 199–204.*

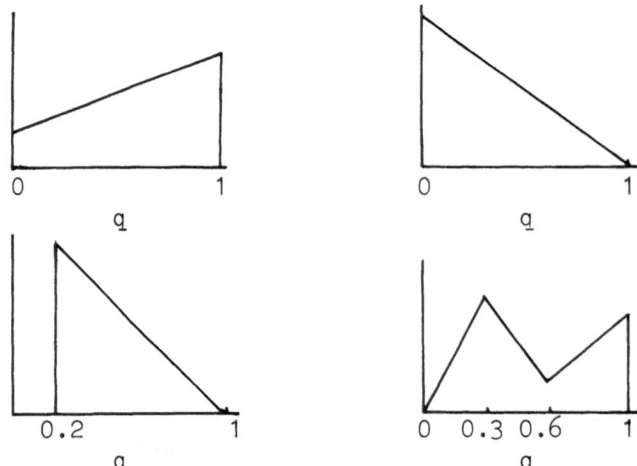

Figure 2. Selected distributions of q.

are assumed to be randomly distributed.

To derive statistical characteristics of the spectroscopic and of the eclipsing binaries , all the 12 combinations of the selected distributions of "a" and "q" were considered.

# 3. APPLICATIONS TO SPECTROSCOPIC BINARIES

## 3.1. Distribution of the semi-amplitudes of radial velocities

The distribution of the semi-amplitudes of the radial velocities of the primary components, "K", was computed by sampling experiments for binaries with solar-type primary components. It was found that when the distribution of log a is unimodal, the resulting distribution of K reflects f(log a) rather than f(q).Figure 3 gives the distributions of K obtained for the constant and for the triangular distributions of log a . The 4 distributions of "q" give results lying inside the hatched areas.

When the distribution of log a is bimodal , the distribution of K may show a gap depending on the distribution of q. When the distribution of q is decreasing from q = 0 , the gap disappears, but the shape of the curve strongly differs from the unimodal cases (cf Figure 4).

When binaries have eccentric orbits , the value of K increases but the curve of radial velocity is distorted and exhibits a narrow peak. Therefore the increase of K can be compensated by the low probability of measuring a velocity near the maximum and it is consequently very convenient to substitute K by half of the range in radial velocity observed if 5 to 10 measures per star are available . As an example, Figure 5 gives the distributions of K obtained from 5 measures per star for 2 synthetic samples which differ only in eccentricities: in the first sample,

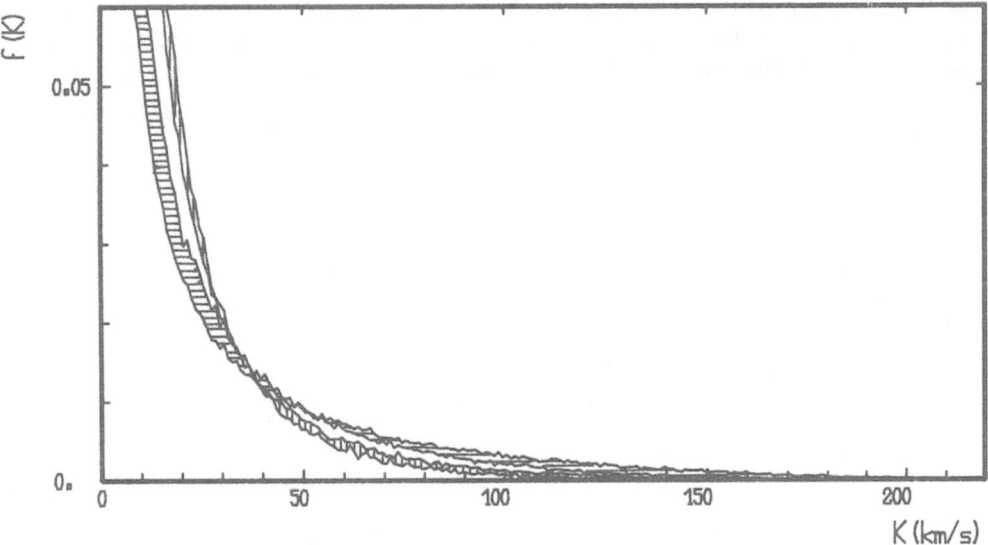

Figure 3. Distributions of K according to the assumed f(loga).
Graphs are normalized for systems with K>15 km/sec . The horizontal hachures denote the constant f(log a)  and the vertical
hachures denote the triangular f(log a).

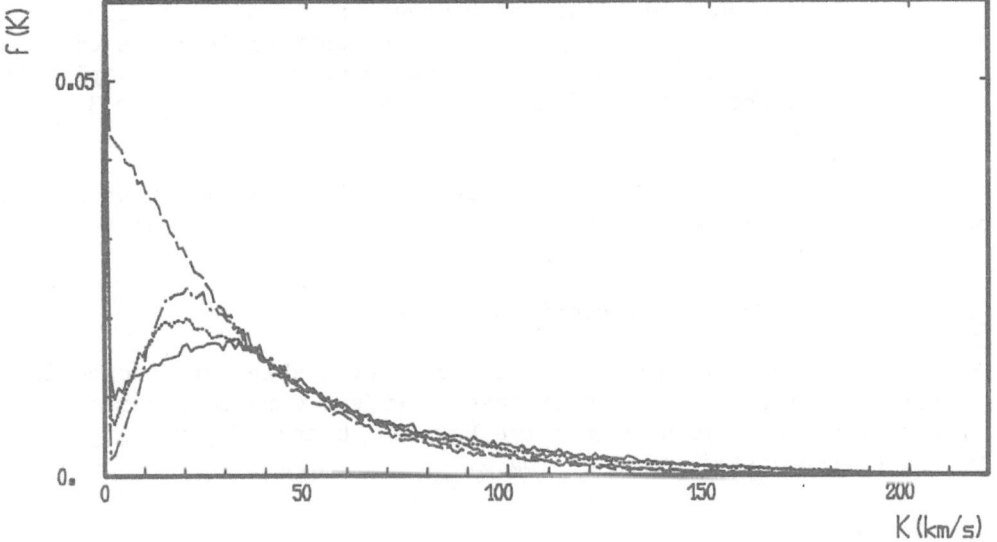

Figure 4. Distributions of K  in the case of a bimodal distribution of log a. The distributions of q are identified as follow : Unbroken line: increasing distribution. Dashed line: decreasing distribution. Dashes and dots : distribution  decreasing from q=0.2. Dotted line: bimodal distribution. Graphs are
normalized as in Figure 3.

binaries have all non-eccentric orbits , while the orbits in the second
sample have all the eccentricity 0.5 . It is easy to see that the agreement between the two resulting distributions of K is very good.

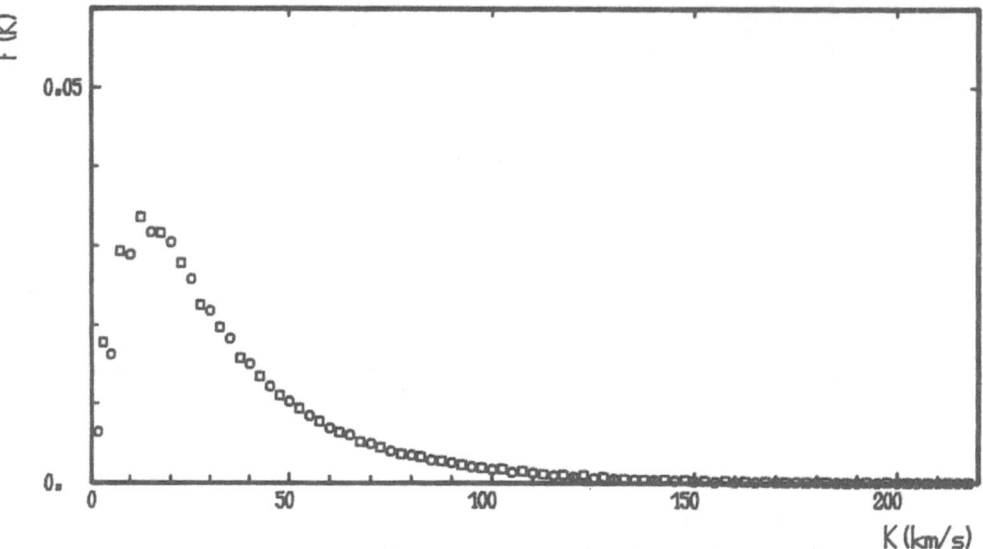

Figure 5. Distributions of K observed in 5 measures per star
in the case of f(log a) bimodal and f(q) decreasing from 0.2
for two samples with different eccentricities: the circles denote a sample with e=0 and the squares denote a sample with e=
0.5. The distributions are normalized as in Figure 3.

Thus it is possible to consider for statistical purposes binaries
with unknown orbit as long as at least 5 measures of radial velocity are
available for each star.

3.2. Proportion of spectroscopic binaries

In order to evaluate the proportion of detectable spectroscopic binaries,the binaries with K higher than 15 km/sec were considered. As shown
in Figure 6, these proportions lye between 6 and 43 per cent and depend
mainly on the distribution of log a.

4. APPLICATION TO ECLIPSING BINARIES

4.1. Distribution of the depths of primary eclipses

The depths of primary eclipses were computed according to the "spherical
model"; this model neglects all proximity effects occurring in close binaries (i.e. stars are considered to be spheres), but includes the limb
darkening. Primary components were assumed to be dwarfs with spectral

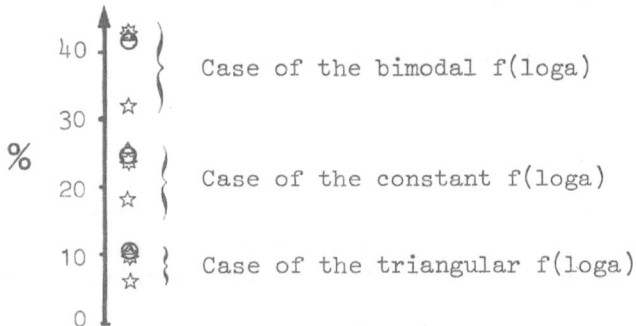

Figure 6. Proportions of binaries for which K > 15 km/sec. The
distributions of q are identified as follow : Δ : increasing
distribution. ☆ : decreasing distribution. ✿ :distribution
decreasing from q=0.2. O : bimodal distribution.

types lying between F and M. As a consequence of the random distribution
of the inclinations , the distribution of  the depths of eclipse depends
only on the distribution of the mass ratios. The distributions   normalized on the systems with eclipse deeper than  0.4  are given in Figure 7.
Notice that the distribution of mass ratios decreasing from q=0.2 produces a gap near  Δm = 0.13 , which is the upper limit for the eclipses of
systems with q=0.2.

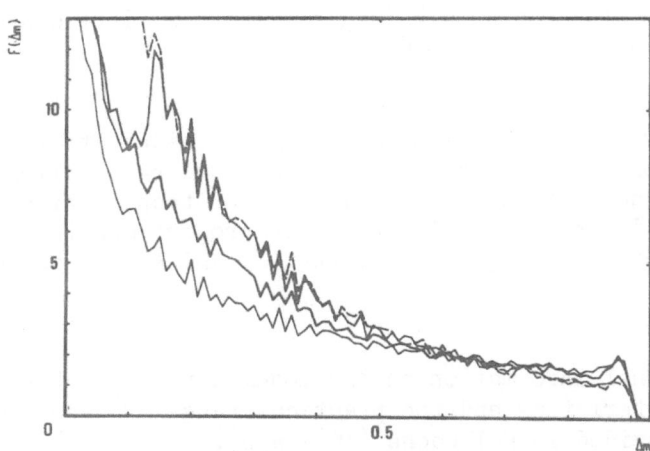

Figure 7. Distributions of the depths of  primary eclipse; the
distributions of q are identified as in Figure 4.

## 4.2. Proportion of eclipsing binaries

The proportion of  binaries that may undergo  an eclipse can be  derived
from :

$$P(\Delta m > 0) = R_1 < 1/a > < 1+q^\alpha > \qquad (1)$$

$\alpha$ being the coefficient of the mass-radius relation :

$$R_1/R_0 = (M_1/M_0)^\alpha \qquad (2)$$

and for dwarfs, $\alpha = 0.75$ . Due to the difficulties of detectability, the proportions of binaries which present eclipse deeper than 0.4 were also computed. Results are shown in Figure 8. As in the case of spectroscopic binaries, the distribution of semimajor axes predominates in the results.

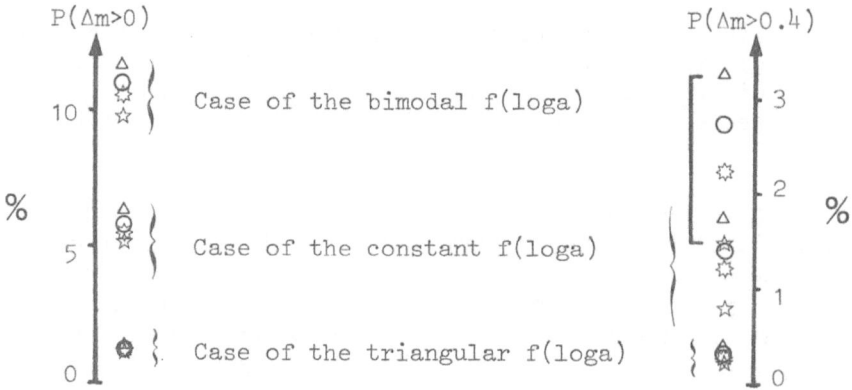

Figure 8. Proportions of binaries with eclipse deeper than 0 (left) or than 0.4 (right). The distributions of q are identified as in Figure 6.

The use of the spherical model being inadequate for very close systems , the proportions of binaries that undergo eclipse were computed again by assuming that all binaries closer than 5 times the primary radius are eclipsing. In this case, the proportions increase by 1.4 when the distribution of log a is triangular and by 1.7 in the two other cases.

Additional information on the models, the computation of the observable characteristics and the practical range of application of the results is provided in Halbwachs (1981 a,b).

It is a pleasure to thank Prof. C. Jaschek for his support and guidance of this project.

REFERENCES

Halbwachs,J.L.:1981a,Astron.Astrophys. in press.
Halbwachs,J.L.:1981b,"modèles statistiques pour l'étude des binaires serrées", thèse de troisième cycle, Observatoire de Strasbourg.

# STRUCTURE AND EVOLUTION OF SV CENTAURI

H. Drechsel, J. Rahe, W. Wargau
Remeis-Sternwarte Bamberg, Astronomical Institute
University Erlangen-Nürnberg / F.R.G.
B. Wolf
Landessternwarte Heidelberg / F.R.G.

ABSTRACT

Optical photometric and spectroscopic observations of
the eclipsing binary SV Cen were used to derive orbital elements and absolute dimensions of the peculiar early-type
contact system. UV spectroscopic observations obtained with
the IUE satellite yielded the detection of mass loss from
the system forming an expanding circumbinary envelope.
SV Cen exhibits a non-uniform rate of period decrease, with
major variations on a time scale of a few years. The period
changes are explained in terms of loss of angular momentum
carried away by the ejected matter. Short-term fluctuations
of the stellar wind on a time scale of a few hours were
found from an analysis of the envelope line profiles of UV
resonance lines. A model of the system is derived, and the
evolutionary stage and interaction processes are discussed.

## A. OBSERVATIONS

Information about the observational material is presented in table 1. We have obtained optical photoelectric
measurements between 1978 and 1981 to serve two purposes:
first, to derive orbital elements from an analysis of the
UBV light curves; and second, to obtain accurate times of
minimum epochs in order to determine the variation and amount
of period decrease.

Further, we took a total of 30 coudé spectrograms between 1977 and 1979, in order to investigate the optical spectral range and to derive the mass ratio.

Finally, we obtained eight UV spectra during two shifts
with the IUE satellite in 1979 and 1981, which cover the
short and long wavelength ranges in both high and low reso-

*Z. Kopal and J. Rahe (eds.), Binary and Multiple Stars as Tracers of Stellar Evolution, 205–216.*
*Copyright © 1982 by D. Reidel Publishing Company.*

lution. The UV spectroscopy yielded the detection of mass
loss from the system, and proved very useful for the investigation of properties of the expanding envelope and for the
analysis of interaction processes of the binary components.

Table 1.   Optical and UV Observations of SV Cen

### 1. Optical Photometry

| Dates | Instrument | Remarks |
|---|---|---|
| 1978, April 2-7 | ESO 50cm | UBV light curves |
| 1979, May 25,28,30 | ESO Bochum 61cm | light curves, minima |
| 1980, Feb.29,March 3 | ESO 50cm | minima |
| 1981, March 21,31 | ESO 50cm | minima |

### 2. Optical Spectroscopy

30 coudé spectrograms, between July 1977 and May 1979;

dispersion: 20 $\overset{\circ}{A}$/mm;   emulsion: Kodak IIaO (B);
instrument: ESO 1.52m telescope

### 3. UV Spectroscopy

| Dates | IUE Image Nos. | Resolution |
|---|---|---|
| 1979, Nov.   3.576 | LWR   6014 | low |
| 1979, Nov.   3.580 | SWP   7078 | low |
| 1979, Nov.   3.631 | LWR   6015 | high |
| 1979, Nov.   3.695 | SWP   7079 | high |
| 1981, Jan.  12.407 | SWP  11036 | low |
| 1981, Jan.  12.435 | LWR   9697 | high |
| 1981, Jan.  12.525 | SWP  11037 | high |
| 1981, Jan.  12.596 | LWR   9698 | low |

B.   ORBITAL ELEMENTS AND ABSOLUTE DIMENSIONS

Orbital elements and absolute dimensions of SV Cen were
derived from the optical photometric and spectroscopic data.
The UBV light curves were analyzed with the light curve and
differential corrections procedure of Wilson and Devinney.
The spectroscopic mass ratio (q=1.25) and the primary effective temperature ($T_1$=23000 K) were used as fixed parameters,
and model assumptions for limb and gravity darkening and for
the color-dependent albedos had to be applied. Satisfactory
convergence was found in the contact configuration mode,
yielding the parameters listed in table 2.

Table 2.  <u>Orbital Elements and Absolute Dimensions of SV Cen</u>

| | | | | | |
|---|---|---|---|---|---|
| $q$ | = | $M_2/M_1$ = 1.25 | $L_1(V)/L_1+L_2$ | = | $0.60\pm0.01$ |
| $i$ | = | $81\overset{o}{.}8 \pm 0\overset{o}{.}1$ | $L_1(B)/L_1+L_2$ | = | $0.64\pm0.01$ |
| $P$ | = | $1\overset{d}{.}6585$ (decreasing) | $L_1(U)/L_1+L_2$ | = | $0.65\pm0.01$ |
| $Sp_1$ | = | B1 V | $L_1$(bolo) | = | $11700\ L_\odot$ |
| $Sp_2$ | = | B6.5 III | $L_2$(bolo) | = | $1900\ L_\odot$ |
| $T_1$ | = | 23000 K | $M_1(V)$ | = | $-3\overset{m}{.}1$ |
| $T_2$ | = | 14000 K | $M_2(V)$ | = | $-2\overset{m}{.}3$ |
| $a_1$ | = | $8.5\ R_\odot$ | $\Omega_1 = \Omega_2$ | = | $3.611\pm0.01$ |
| $a_2$ | = | $6.8\ R_\odot$ | $\Omega_{ic}$ | = | 4.13 |
| $a$ | = | $15.3\ R_\odot$ | $\Omega_{oc}$ | = | 3.57 |
| $M_1$ | = | $7.7\ M_\odot$ | $E(B-V)$ | = | $0\overset{m}{.}27$ |
| $M_2$ | = | $9.6\ M_\odot$ | $A_V$ | = | $0\overset{m}{.}9$ |
| $R_1$ | = | $6.8\ R_\odot$ | $d$ | = | 1800 pc |
| $R_2$ | = | $7.4\ R_\odot$ | | | |

The less massive component of spectral type B1 is the photometric primary. With a mass of about 8 and a radius of about 7 solar units, it is located slightly above the main sequence. The secondary is of spectral type B6.5, and has similar parameters. The system is in overcontact, with both stars overfilling their Roche lobes. The separation of the mass centers amounts to only about 15 solar units.


B.  PRESENCE OF AN EXPANDING ENVELOPE

The period of 1.6 days is rapidly decreasing. SV Cen has previously been regarded as a rare example of a binary being presently observed during the early part of mass exchange, prior to reversal of mass ratio. In order to explain the period decrease, it was assumed that the mass is flowing from the more massive to the less massive component. Since no indication for circumstellar material was found in the optical spectral range, the mass transfer was thought to be fully conservative and to take place within the common contact surface (Wilson and Starr, 1976).

Our UV spectra, however, clearly show the presence of an expanding circumbinary envelope (Drechsel et al., 1980). Apart from photospheric and in some cases interstellar contributions, the most prominent features in the high resolution IUE spectra are highly displaced envelope components of reso-

nance lines of C II, Si IV, C IV, Al II,III, and Mg II, as
well as of the metastable Fe III multiplets 34 and 48. The
central wavelengths of the envelope lines are shifted by -700
to -1200 km s$^{-1}$, and the velocities of the shortward edges
are ranging from -900 to -2000 km s$^{-1}$. A comparison with the
escape velocity of the order of 750 km s$^{-1}$ obviously suggests
the occurrence of mass loss from the system. Radiation pressure provides the driving force for the acceleration of the
expanding envelope. The UV resonance and metastable lines
with envelope components, together with the central wavelengths, $v_c$, and the shortward edges, $v_{max}$, of the displaced
envelope lines are listed in table 3.

Table 3.   UV Resonance and Metastable Lines with Envelope
           Components

| Ion | Multiplet | $\lambda_{lab}$ (Å) | $v_c$ [*] (km s$^{-1}$) | $v_{max}$ [*] (km s$^{-1}$) |
|---|---|---|---|---|
| C II | 1 | 1334.532 | - 710 | -1070 |
|  |  | 1335.705 | - | - |
| Si IV | 1 | 1393.755 | - 895 | -1260 |
|  |  | 1402.770 | - 890 | -1250 |
| Si II | 2 | 1526.708 | - | - 825 |
| C IV | 1 | 1548.188 | -1200 | -2065 |
|  |  | 1550.762 | - | - |
| Al II | 2 | 1670.787 | - 735 | - 970 |
| Al III | 1 | 1854.716 | - 890 | -1230 |
|  |  | 1862.790 | - 830 | - |
| FeIII[+] | 34 | 1895.456 | - 730 | -1055 |
|  |  | 1914.056 | - 790 | -1190 |
|  |  | 1926.304 | - 780 | - |
| FeIII[+] | 48 | 2061.552 | - 770: | -1035: |
|  |  | 2068.243 | - 810 | - |
| Mg II | 1 | 2795.523 | - 645 | - 885 |
|  |  | 2802.698 | - | - |

[*] Radial velocities were measured in the IUE spectra
    SWP 7079 and LWR 6015, obtained on 1979, Nov. 3

[+] Metastable line

## C.  LONG-TERM VARIATIONS OF THE MASS LOSS RATE

In contrast to the previous explanation of the period decrease by conservative mass transfer from the more to the less massive component, which is extremely unlikely to be observed due to the short thermal time scale involved in this process ($P/\dot{P} \approx 50000$ yr), we suggest that the period decrease is due to loss of angular momentum carried away by the ejected matter.

A period-epoch diagram of SV Cen was established for the time between 1894 and 1981, which is based on all published photographic and photoelectric times of minimum, and has been supplemented by measurements of the Bamberg sky patrol plates and by our photoelectric observations during the recent years (Drechsel et al., 1981). First, it is obvious that the period is generally decreasing at the average large rate of $\dot{P}/P = -2.15 \cdot 10^{-5}$ yr$^{-1}$; and second, the rate of decrease is not constant, but highly time-dependent, with intervals of approximate constant rate of decrease, and even those with near constancy of period. Since light time effects can certainly be excluded (see, e.g., Irwin and Landolt, 1972), it is suggested that mass loss and mass exchange of the interacting binary account for the observed unsteady period decrease.

An especially efficient mechanism for accelerating the orbital revolution is ejection of matter through an external Lagrangean point. Under the assumption that the ejected material comes from the Roche lobe overflow of the less massive component through $L_1$, is subsequently transferred to the secondary, and finally lost from the system through $L_3$, we can apply an expression derived by Kruszewski (1966), which gives a correlation between period decrease and mass loss rate:

$$\dot{P}/P = \frac{3}{\mu(1-\mu)} \cdot \left[ x(L_1)^2 + \frac{2}{3}\mu\left(1+\frac{\mu}{2}\right) - d^2 \right] \cdot \frac{\dot{M}}{M_1 + M_2} \quad ,$$

$$\text{where} \quad \mu = \frac{M_2}{M_1 + M_2} \quad ; \quad d = \left| x(L_3) - x(G) \right| \quad .$$

$M_1$ and $M_2$ are the individual masses of primary and secondary; $x$ denotes the rectangular Roche coordinate, i.e. d is the distance of the Lagrangean point $L_3$ from the common center of mass, G.

The time derivative of the period, $\dot{P}$, was determined by differentiating the observed period-epoch correlation in sections of typically a few years. Accordingly, the mass loss rate was computed for the different intervals of approximate constant rate of period decrease; it is plotted in Figure 1 as a function of time between 1894 and 1981.

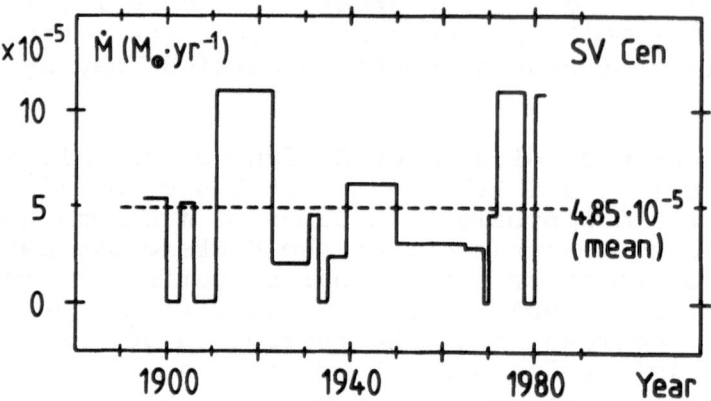

Figure 1.    Mass loss rate of SV Cen for the
time between 1894 and 1981.

Major variations of the mass loss rate with time are obvious, ranging from 0 to $10^{-4}$ $M_{\odot} \cdot yr^{-1}$, with an average value of $4.85 \cdot 10^{-5}$ $M_{\odot} \cdot yr^{-1}$, on a time scale of a few years.

D.    SHORT-TERM FLUCTUATIONS OF THE STELLAR WIND

Besides the long-term variations of the mass loss rate, we found short-term fluctuations of the mass loss process from an analysis of the UV spectra obtained in late 1979 and early 1981. The striking difference between these two epochs is the appearance of several sharp absorption components superimposed on the broad envelope absorption profiles of all strong UV resonance and metastable lines in the IUE spectra taken in January 1981. As an example, the resonance doublets of C II (1334, 1335), Si IV (1393, 1402), and Al III (1854, 1862) are shown in Figure 2. The envelope components contain at least four well-defined narrow absorption dips at nearly identical radial velocities of about −780, −1120, −1350, and −1580 km s$^{-1}$.

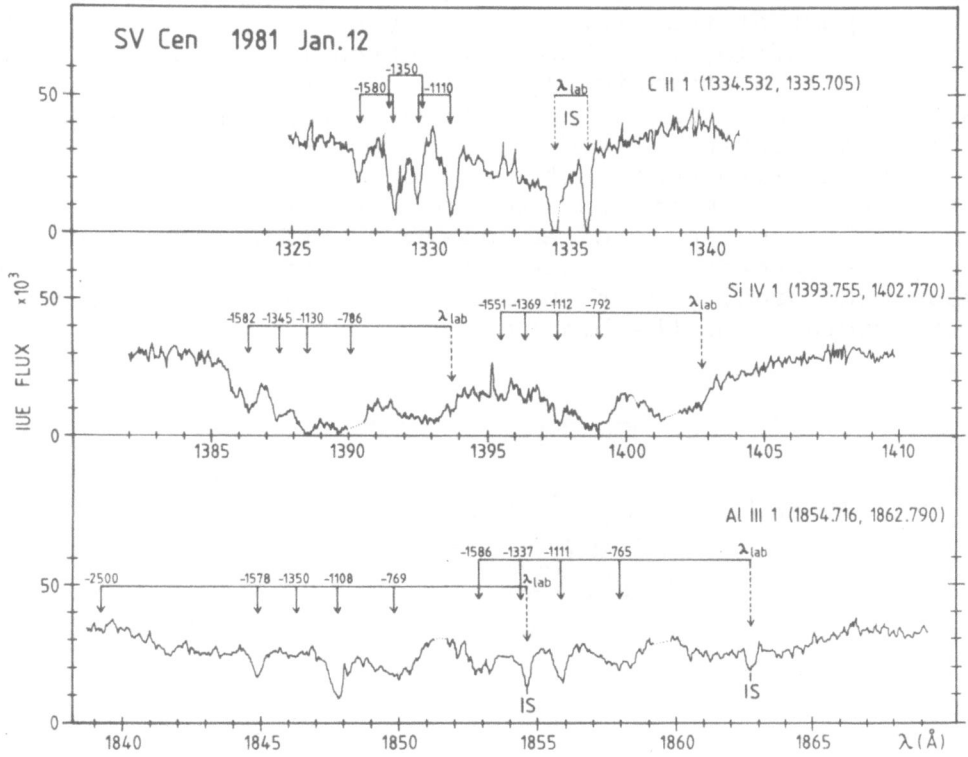

Figure 2.    UV resonance line profiles of C II, Si IV, and
             Al III, as measured with the IUE satellite on
             1981, Jan. 12 in the high resolution mode (SWP
             11037). The envelope components contain at least
             four well-defined narrow absorption dips, which
             are indicated by arrows.


        The uncontaminated envelope absorption of Si IV (1393)
was used for a quantitative fit of its profile. Under the
assumption of isotropic resonance scattering in a spherically
symmetric expanding envelope, a theoretical representation
of the general profile shape and of the narrow components
could be achieved by proper adjustment of the density structure of the envelope, and for an adequate velocity law of
the stellar wind. Four local condensations forming radially
expanding shells can account for the well-defined absorption
components. In the upper part of Figure 3, we have plotted
a quantity proportional to the line optical depth which is
normalized to its maximum observed value as a function of
distance from the mass-losing star. The dashed line gives
the best fitting velocity law in units of the observed ter-

minal velocity of the order of $-2500$ km s$^{-1}$. While the high
velocity part of the flow is in very good agreement with the
predicted profile, the absorption strength at smaller radial
velocities is systematically greater than expected. The reason might be blending with the high velocity part of the
envelope absorption of the Si IV (1402) doublet transition,
and possibly with some photospheric contribution. The appearance of the narrow absorption dips can therefore be attributed to the occurrence of an unsteady outflow of matter
through an external Lagrangean point, giving rise to an inhomogeneous density distribution of the expanding envelope
in the vicinity of the binary.

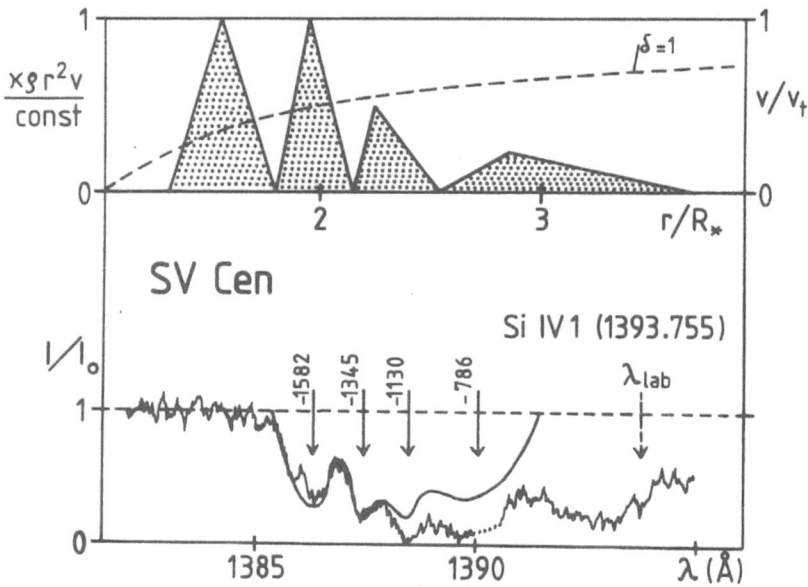

Figure 3.   Measured and predicted profile of the Si IV (1393)
            envelope absorption line. The suggested variation
            of the line optical depth as a function of distance
            from the mass-losing star is plotted in the upper
            part; the dashed line indicates the best fitting
            velocity law.

From an integration of the velocity law

$$v(r) = v_t \cdot (1 - \frac{R_o}{r})^{\delta}$$

($v_t$ = terminal velocity; $R_o$ = radius of mass-losing star),

we can estimate the time scale $\Delta t_{i,i+1}$ for the ejection of
"puffs" of material producing the condensed shells i and i+1
of the expanding envelope:

$$\Delta t_{i,i+1} = \frac{1}{v_t} \cdot \int_{r(v_i)}^{r(v_{i+1})} \left( 1 - \frac{R_o}{r} \right)^{-\delta} dr ,$$

$$\text{with} \quad r(v_i) = R_o \Big/ \left[ 1 - (v_i/v_t)^{1/\delta} \right] .$$

The time scale comes out to be always very short, i.e.
of the order of a few hours, if we assume different values
for the exponent $\delta$ in the velocity law, representing high,
moderate, or even low initial acceleration of the flow;
for $v_i$ we took the radial velocities of the observed narrow
absorption dips. Table 4 gives the times $\Delta t_{i,i+1}$ necessary
to accelerate the stellar wind from the velocity $v_i$ to $v_{i+1}$,
which were obtained by numerical integration of the velocity
law for a value of $\delta$ equal 1, which yielded the best theoretical representation of the Si IV (1393) envelope profile.
Computations for $\delta$ = 0.5, 2, and 3 also gave typical values
of $\Delta t$ equal 0.02 to 0.04 days.

Table 4.  <u>Time scale for the ejection of condensed shells</u>

| $v_i$ (km s$^{-1}$) | $\Delta t_{i,i+1}$ (days) |
|---|---|
| 0 | |
| | 0.$^{d}$096 |
| - 777 | |
| | 0.021 |
| - 1115 | |
| | 0.017 |
| - 1351 | |
| | 0.020 |
| - 1576 | |

This clearly suggests that SV Cen not only shows major
long-term variations of its mass loss rate on a time scale
of a few years, but also exhibits a strong variability of its
stellar wind within only a few hours, due to short-term fluctuations of the mass flow.

E.   PRESENCE OF A HOT SOURCE

     The UV spectroscopic observations yielded the detection
of a hot source in the SV Cen system. The low resolution
short and long wavelength IUE spectra were combined with the
photoelectric B and V measurements to give the absolute flux
distribution in the UV and visual range.

     Figure 4 shows the measured absolute intensity distri-
bution, corrected for interstellar extinction, as filled and
open triangles for the 1979 and 1981 data, respectively. The
observations were obtained around quadrature, i.e. near orbital phase 0.25. The prominent characteristic is a very
strong UV excess, with the maximum of intensity distribution
far beyond the IUE range.

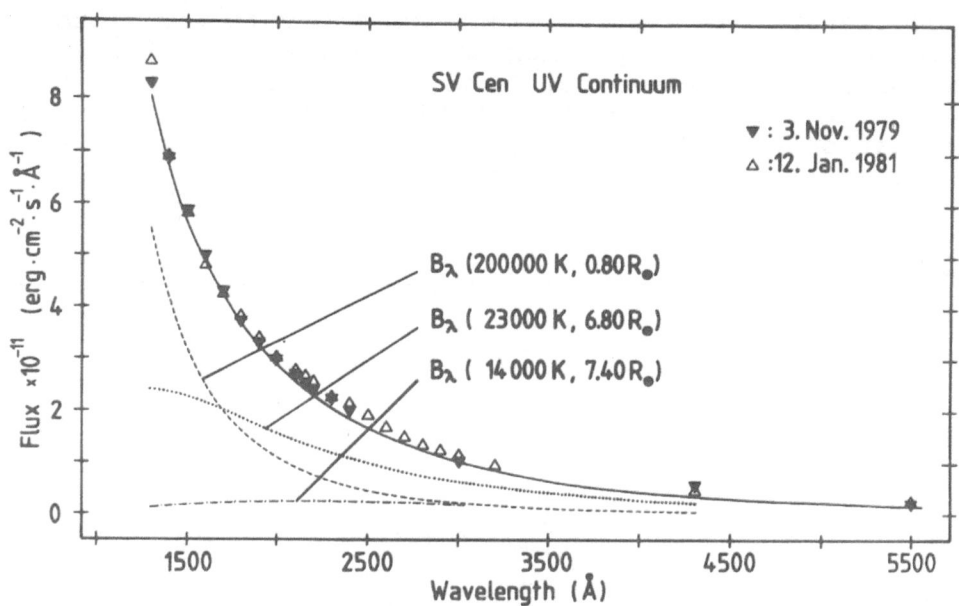

Figure 4.   Absolute flux distribution in the UV and visual
            continuum of SV Cen near quadrature phase 0.25.
            The UV excess is compatible with the presence of
            a hot (200000 K) source, with an extension which
            is small ( <1%) in comparison with the projected
            area of the binary components.


     The measurements were interpreted by means of a three
component black body fit. The dotted and dash-pointed lines
represent the two stellar components, and the dashed line a

hot area of about 200000 K, with an extension of about 1% of
the projected area of the two stars. The integrated absolute
flux of all three components is plotted as solid line, and
represents the measurements quite well.


F.  MODEL OF THE SYSTEM

     From the optical and UV measurements, the following
model of SV Cen can be outlined:

     The hot area can arise from dissipation of kinetic en-
ergy through shock waves connected with a hydrodynamic flow
pattern originating from the Roche lobe overflow of the less
massive but more luminous primary, which is travelling along
the inner critical Roche surface and is approaching the outer
layers of the common envelope in the equatorial region on
the following side of the secondary (see Figure 5). At least
part of the mass flow is leaving the system through the external $L_3$ point, and is forming an expanding envelope which
is driven by radiation pressure. From the UV spectroscopy, we
can certainly exclude the conservative case of mass transfer.
We suggest that the period decrease is caused by loss of angular momentum carried away by the ejected matter.

     From the long-term variation of the rate of period de-
crease, it can be derived that the mass loss rate is variable
on a time scale of a few years, and from an UV envelope line
profile analysis, it is apparent that major short-term fluctuations of the stellar wind occur on a time scale of only
a few hours. The reason for the variable mass loss process
might be twofold:

     First, it can be due to an unsteady Roche lobe overflow
of the primary which is at the beginning of its post main
sequence evolution; and second, it may be caused by a variation of the ratio of ejected to transferred matter. While
the ejection of matter accelerates the orbital revolution,
the transfer of matter from the less massive to the more
massive component tends to increase the period, and both
processes can conceivably combine to yield the observed behavior.

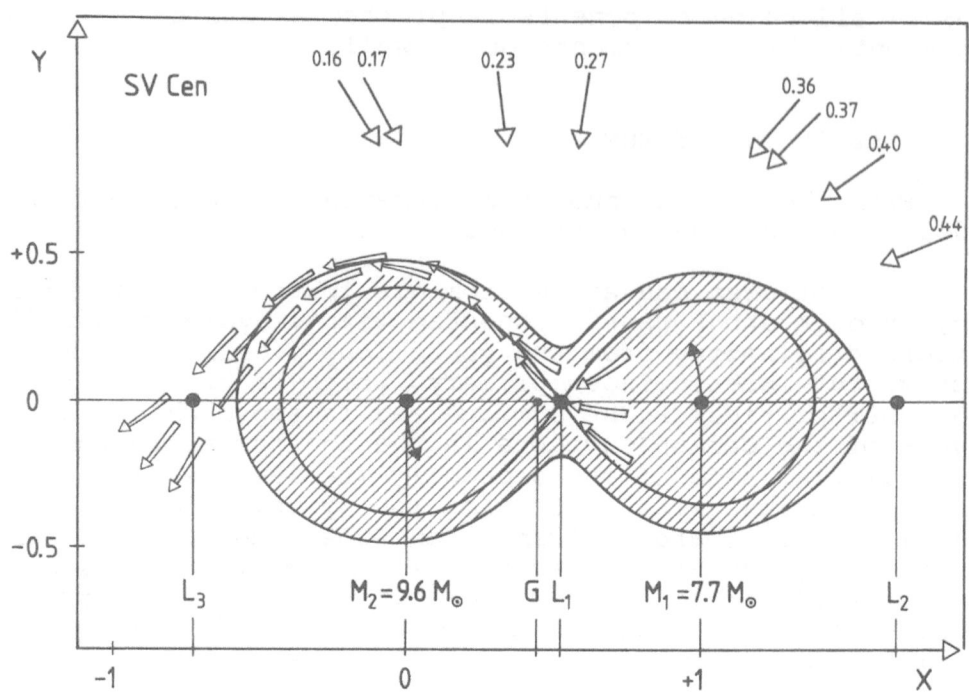

Figure 5.   Model of the SV Cen system. The shaded area is
            the intersection of the Roche equipotential sur-
            faces with the orbital plane. The Roche geometry
            corresponds to the spectroscopic mass ratio
            q = 1.25. Arrows in the upper part indicate the
            orbital phases during which the IUE spectra were
            taken.


REFERENCES

Drechsel, H., Rahe, J., Wolf, B.: 1980, Proc. "Second Euro-
           pean IUE Conference", Tübingen, 26-28 March 1980,
           ESA SP-157, 213.

Drechsel, H., Rahe, J., Wargau, W., Wolf, B.: 1982, in pre-
           paration.

Irwin, J.B., Landolt, A.U.: 1972, Publ. Astron. Soc. Pacific
           84, 686.

Kruszewski, A.: 1966, in "Advances in Astronomy and Astro-
           physics, Vol. 4", ed. Z. Kopal, Academic Press
           New York and London.

Wilson, R.E., Starr, T.C.: 1976, Monthly Not. Roy. Astron.
           Soc. 176, 625.

# AGE DETERMINATION IN CW Eri = BV 1000

H.Mauder

Astronomisches Institut
Universität Tübingen

CW Eri is a double lined spectroscopic binary star. Since the two
components are well detached, it is possible to derive absolute elements  with high accuracy. The photometric elements are based on
observations obtained at Boyden observatory in 1971/72, see Mauder
and Ammann,1976. The spectroscopic data are taken from Popper,1980.
In the following table the system parameters are given.

| | | |
|---|---|---|
| Spectra | F2 + F2 | |
| Period | $2^{d}.728373$ | |
| $M_1 =$ | $1.52\ M_\odot$ | $\pm\ 0.015$ |
| $M_2 =$ | $1.28\ M_\odot$ | $\pm\ 0.010$ |
| $R_1 =$ | $2.10\ R_\odot$ | $\pm\ 0.05$ |
| $R_2 =$ | $1.44\ R_\odot$ | $\pm\ 0.05$ |
| $i =$ | $86°.7$ | $\pm\ 0.1$ |
| $\log T_{e,1} =$ | $3.860$ | (assumed) |
| $\log T_{e,2} =$ | $3.843$ | $\pm\ 0.01$ |

Hejlesen, 1980, has calculated evolutionary tracks for stars of
different chemical composition. In his $\log g$ - $\log T_e$ - graphs
isochronic lines are given which allow for an age determination
of binary stars. In the figure, the two components of CW Eri are
given as crosses, indicating the uncertainty of the elements. It
is evident, that the components are lying on the respective evolutionary tracks according to their masses and that the isochronic line for $1.5 \cdot 10^9$ years fits both stars.
The $\log g$ - $\log T_e$ - diagram used is for a chemical composition of
$X = 0.70$ and $Z = 0.02$. The ratio of mixing length over pressure
scale height $1/H_p = 2$ was used. For $1/H_p = 1.5$ no consistent solution is possible. It is interesting to note, that the respective
values of the sun, drawn into the Hejlesen graphs, yield an age of
about $4 \cdot 10^9$ years for the sun, if $1/H_p = 2$ is used and an age of

*Z. Kopal and J. Rahe (eds.), Binary and Multiple Stars as Tracers of Stellar Evolution, 217–218.*
*Copyright © 1982 by D. Reidel Publishing Company.*

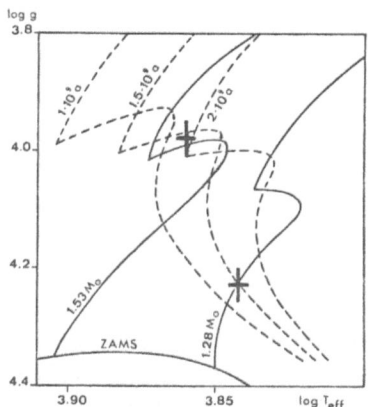

Evolutionary tracks (full) and
isochronic lines (broken) according to Hejlesen (1980).
The two components of CW Eri are
shown as crosses, indicating the
uncertainty of the elements.

about $1 \cdot 10^9$ years if $1/H_p = 1.5$ is adopted.
There are two parameters which might be altered for CW Eri. First,
the chemical composition could be different. Since, however, the
two components must be lying on their respective evolutionary tracks
according to their masses, a consistent solution is possible only
if $X = 0.70 \pm 0.02$ and $Z = 0.020 \pm 0.002$. The second parameter is the
effective temperature of the primary component. The temperature difference $T_{e,1} - T_{e,2}$ is known very accurately from the photometric
solution. However, $T_{e,1}$ was adopted according to the spec -
tral type F2. The consequence of a change in $T_{e,1}$ would be a change
in the chemical composition. For this case, however, it is very difficult to fulfil the condition of both stars lying on a single isochronic line. An independent determination of the chemical composition of CW Eri would be most interesting with respect to the temperature calibration.

References:

P.M.Hejlesen,1980,Astr.&Astrophys.Suppl.39,347
H.Mauder,M.Ammann,1976, Mitt.Astr.Ges.38,231
D.M.Popper,1980,Ann.Rev.Astron.Astrophys.18,115

# NUMERICAL CALCULATIONS FOR ACCRETION DISKS IN CLOSE BINARY SYSTEMS

G. Hensler
Universitäts-Sternwarte
Geismarlandstr. 11
D-3400 Göttingen

## ABSTRACT

A numerical method for 3D magnetohydrodynamical investigations of
accretion disks in close binary systems is presented, which allows
for good spatial resolution of structures (hot spot, accretion column).
The gas is treated as individual gas cells (pseudo-particles) whose
motion is calculated within a grid consisting of one spherical inner
part for 3D MHD and two plane outer parts. Viscous interactions of
the gas cells are taken into account by a special treatment connected
with the grid geometry.
   We present one result of 2D hydrodynamical calculations for a
binary applying the following parameters which are representative for
Cataclysmic Variables: $M_1 = 1\ M_\odot$, $r_1 = 10^{-2}\ R_\odot$, $M_2 = 0.5\ M_\odot$, $p = 0.2$ d,
$\dot{M} = 10^{-9}\ M_\odot\ y^{-1}$.
   Column density and radiative flux distributions over the disk
are shown and briefly discussed by comparison with the theoretical
understanding of these Dwarf Novae drawn from observations.

## I. INTRODUCTION

If in a binary system the one component is filling out its Roche-
volume material is transferred through the inner Lagrangian point
$L_1$ towards its companion, which is called the primary or the gainer.
   According to Kippenhahn and Weigert (1967) one distinguishes
generally between three different cases of mass exchange depending
on the evolutionary stage of the mass giving star, henceforth called
the secondary or loser (as proposed by Plavec (1980)), extended in the
last years to combined cases (e.g. Delgado and Thomas, 1981) due to
a better knowledge of stellar evolution in binaries.
   The matter streaming through $L_1$ does not fall radially onto the
primary's surface but is deflected due to the Coriolis-force in the
corotating frame. Then the increasing gravitational force turns it

*Z. Kopal and J. Rahe (eds.), Binary and Multiple Stars as Tracers of Stellar Evolution, 219–230.*
*Copyright © 1982 by D. Reidel Publishing Company.*

again towards the primary, and if the gas stream is able to pass by
the primary it turns around and strikes itself.

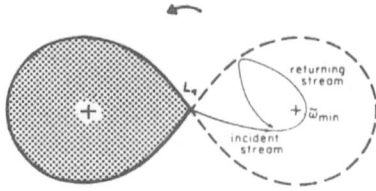

Fig. 1: Stream trajectory for the infalling matter beginning at $L_1$.
We are looking perpendicular onto the equatorial plane with
a counter-clockwise direction of rotation. +: centres of
gravity of the components, $\omega_{min}$: minimum distance of the
stream from the primary's centre (from Lubow and Shu, 1975)

Because of dissipation mechanisms like viscosity and small scale
magnetic fields (Shakura and Sunyaev, 1973; Lynden-Bell and Pringle,
1974) the gas forms a disk where angular momentum is transported outwards so that material flowing inwards can be accreted onto the
primary's surface.

Disks in semidetached binaries can be observed e.g. in Cataclysmic
Variables (Warner, 1976) and most X-ray binaries (van den Heuvel,
1976; Crampton, 1980) where the primary is already a compact object.

To understand the observations it is necessary to investigate
disk structures and the corresponding radiation. Until now only detailed work has been done on these treating special aspects. The
crucial problem for disk calculations is not to know the viscosity
(Lynden-Bell and Pringle, 1974).

Standard accretion disk models (Shakura and Sunyaev, 1973; Novikov
and Thorne, 1973; Lightman, 1974a, 1974b) and investigations concerning
the stability of these disks (Shakura and Sunyaev, 1976; Pringle, 1976),
the time dependent structure of disks in Cataclysmic Variables (Bath
and Pringle, 1981), the vertical disk structure by energy transport
considerations (Tayler, 1980; Meyer and Meyer-Hofmeister, 1981),
spectral flux distributions (Pacharintanakul and Katz, 1980), and continuum and line spectra (Mayo et al., 1980; and references therein)
have made use of the $\alpha$-prescription introduced by Shakura and Sunyaev
(1973) to parametrize the viscosity.

The $\alpha$-disk model, however, is only valid in the inner region of
a disk where the structure is cylindrically symmetric. Asymmetric
phenomena produced by hydrodynamical effects in the outer regions like
a hot spot, or eccentric orbits of the gas and tidal effects due to
the secondary's gravitational influence cannot be considered.

binary system is displayed in Fig. 2. The coordinates are in units of the separation of the two components with the origin at the primary's centre. For simplicity we are calculating in spherical polar coordinates. The following parameters are given:

$Q = \dfrac{M_1}{M_2}$, the mass ratio of the gainer to the loser, PER (= $\tau$, in the text) the orbital period of the binary system, and A (= d) the resulting separation of the components.

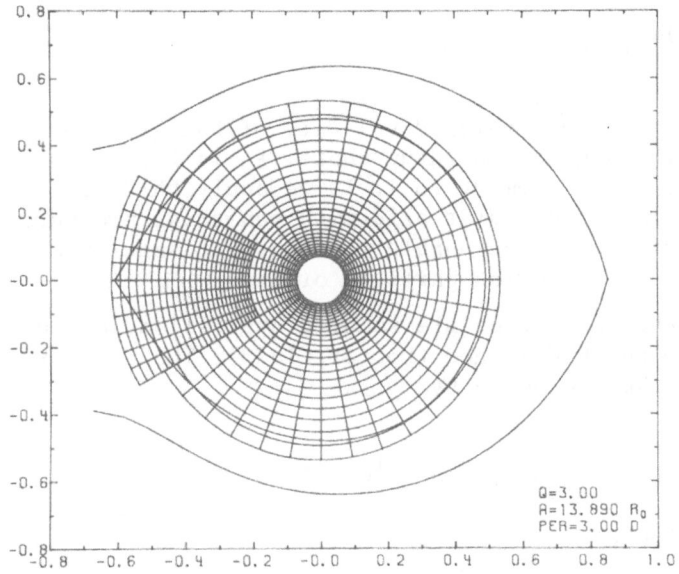

FIG. 2   - THE GRID.

The grid is divided into three parts: a) One spherical part in the inner region which is pinned to the primary because later the magnetic interaction with the matter should be considered within this part, and b) two parts in the outer region, only two-dimensional. The mesh sizes of the grid can be varied separately, but it must be fulfilled that each grid cell is occupied by at least one particle (the PIC method requires an average of 4 - 5 particles per cell) to represent the hydrodynamics. Because the outer grid parts are fixed in the binary system, the inner grid part is able to rotate together with the primary nonsynchronously with respect to the orbital motion.

  With the grid we are calculating the pressure gradient. If we consider, for simplicity, three grid cells in the x-direction only and a particle i in the middle one (index 2) at $x_i$, we express the pressure gradient as

$$\left(\frac{d}{dx}\right)_i P = \frac{(P_3 - P_2)(x_i - x_2)}{(x_3 - x_2)^2} + \frac{(P_2 - P_1)(x_3 - x_i)}{(x_3 - x_2)^2} \; , \tag{2}$$

where $x_2$ and $x_3$ denote the location of the cell boundaries between the cells. Test calculations for a hydrostatic stratification have confirmed this form.

The most important reason for introducing a grid was to consider a viscous interaction of the particles. It consists of two steps:

1) for each grid cell the particle velocities $\underline{u}_i$ are altered by

$$\underline{u}_i^{\,new} = \underline{u}_i^{\,old} (1 - f_\nu) + \underline{\bar{u}}_g \cdot f_\nu \; , \tag{3}$$

where $\underline{\bar{u}}_g$ denotes the averaged velocity of the grid cell in question conserving the angular momentum there. $f_\nu$ is the viscosity parameter in the range from 0. to 1. If we apply this first viscous interaction in our circular grid with vanishing pressure forces, a circular disk would be unable to transport angular momentum radially due to Keplerian particle orbits.

2) Therefore, the velocities are smeared over all adjoining grid cells according to the PIC method (Potter, 1973). The new particle velocity is then computed by

$$\underline{u}_i^{\,new'} = \underline{u}_i^{\,new} (1 - f_\nu) + \underline{\bar{u}}_{ad} \cdot f_\nu \; , \quad 0 \le f_\nu \le 1 \tag{4}$$

The mean velocity (see Fig. 3)

$$\underline{\bar{u}}_{ad} = \frac{\sum\limits_{i,j} a_{i,j} \cdot \rho_{i,j} \cdot \underline{\bar{u}}_{i,j}}{\sum\limits_{i,j} a_{i,j} \cdot \rho_{i,j}} \; , \quad A_g = \sum\limits_{i,j} a_{i,j} \tag{5}$$

also prevents large differences in the fluid velocity of neighbouring grid cells.

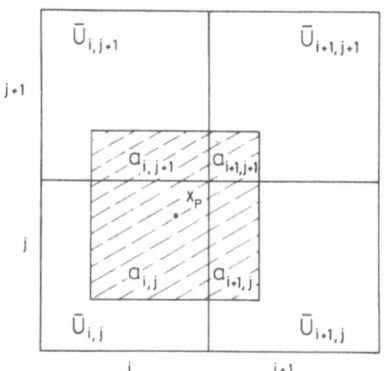

Fig. 3:
2D-interpolation for a 'particle' velocity with the four nearest grid cells. $x_p$: 'particle' location, $a_{i,j}$: fractions of areas around the 'particle' overlapping with the grid cells, $\underline{\bar{u}}_{i,j}$: mean velocity of the grid cells (cf. Potter, 1973, p. 159)

To calculate the gas stream, particle calculations in the restricted three-body problem without any interaction of the particles were conducted by several investigators (for references see the following authors) but cannot represent the hydrodynamics realistically. Also the following hydrodynamical calculations (Prendergast, 1960; Biermann, 1971; Prendergast and Taam, 1974; Lubow and Shu, 1975; Sørensen et al., 1975; Flannery, 1975; Lin and Pringle, 1976) have been carried out either without taking any viscous interaction into account or treating the energy equation only very crudely. The most successful work was done by Lin and Pringle (1976) where the gas is treated as individual gas cells moving in a cartesian grid. This, however, yields an anisotropic viscosity. They did not investigate various disk structures for different viscosities because they only intended to show a method for numerical calculations of disk structures and its feasibility. Standard hydrodynamical finite difference methods as recently also applied by Everson (1981) to time dependent accretion disks are not very suitable because of the limited spatial resolution of the grid due to the rigid grid shape, inability to adjust to the different hydrodynamical structures, and due to the limitation of computer storage.

## II. BASIC EQUATIONS AND NUMERICAL METHOD

Because in many cases the primary has a magnetic field, we are also interested in the interaction of the magnetic field with the matter in the disk. In order to include this effect later, my intention was to develop a numerical method which should allow three-dimensional MHD computations, improve the coupled treatment of hydrodynamics and energy equation, and be applicable to a large range of parameters.

Most promising seemed to me the use of a pseudo-particle method (see e.g. for the "smoothed particle hydrodynamics" (SPH) Lucy (1977), for the "particle-in-cell" method (PIC) Harlow (1964), and for 3D MHD Leboeuf et al. (1979)).

We divide the gas into individual gas cells which carry not only mass but also internal energy with them. Until now we have calculated only two-dimensionally, ignoring for the moment magnetic fields. As we are dealing with pseudo-particles, the momentum equation in the corotating system is expressed in Lagrangian form,

$$\frac{D\underline{u}}{Dt} = -\nabla\phi - \underline{\omega} \times (\underline{\omega} \times \underline{r}) - 2\underline{\omega} \times \underline{u} - \frac{1}{\rho} \nabla P + \nu\Delta\underline{u} \quad , \qquad (1)$$

Here, $\underline{r}$ and $\underline{u}$ are the particle's location and velocity, respectively. $\phi$ consists of the potentials of both components. $\omega$ is the angular velocity $2\pi/\tau$ ($\tau$: the orbital period) with its vector directed perpendicularly to the rotational plane. $\rho$ denotes the density, P the pressure, and $\nu$ the kinematic viscosity.

To calculate e.g. the viscous interaction, besides the particles we introduce a rough grid which fills out the Roche-lobe of the primary. The crosscut of the grid with the rotational plane of the

Test calculations for a viscous isothermal compressible gas flow
between two fixed parallel walls have shown the expected stream profiles. Concerning our input parameters which determine the viscosity,
the test calculations also show that the kinematic viscosity $\nu$ is proportional to the area of the grid cell $A_g$ and nearly proportional to
the square root of our viscosity parameter $f_\nu$.

$$\nu \sim A_g \cdot f_\nu^{1/2} \tag{6}$$

To calculate disk models with an almost constant kinematic viscosity
we accomplish this by taking advantage of this proportionality.

Because the molecular velocity is too inefficient to transport
enough angular momentum to yield a broad disk it is generally accepted
(Shakura and Sunyaev, 1973, 1976; Lynden-Bell and Pringle, 1974) that
the gas flow must be unstable against turbulence.

The energy transport combined with turbulent motion is treated in
the same way as the viscous interaction. The internal energy $e_i$ of a
particle is computed in two steps:

1) Inside each grid cell by

$$e_i^{new} = \frac{\varepsilon_g^{new}}{\rho_g} f_\chi + \frac{E_g^{new}}{E_g^{old}} e_i^{old} (1 - f_\chi) \quad , \tag{7}$$

where $\varepsilon_g$ is the internal energy density and $E_g$ the internal energy
of the grid cell in question. The upper index ("old" and "new")
denotes the internal energy before and after energy dissipation by
viscosity, respectively. $\rho_g$ is the particle density of the grid
cell g and $f_\chi$ the efficiency factor of the energy transport
$(0. \le f_\chi \le 1.)$.

2) Over the adjoining grid cells by

$$e_i = (1 - f_\chi) e_i^{new} + f_\chi \bar{e}_{ad} \tag{8}$$

with

$$\bar{e}_{ad} = \frac{\sum\limits_{i,j} a_{i,j} \cdot \varepsilon_{i,j}}{\sum\limits_{i,j} a_{i,j} \cdot \rho_{i,j}} \tag{9}$$

Assuming the Prandtl number $P_r = \frac{\nu}{\chi}$ to be of order unity for a perfect
gas we perform the calculations with $f_\chi = f_\nu$.

The energy change due to the work done by pressure, $-P\nabla u$, is taken into account similar to the following expression (in dimensionless form), where we again consider only the one-dimensional cartesian case, for simplicity:

$$\frac{dE_2}{dt} = -E_2 \left( \frac{u_3 - u_1}{\Delta x} \right) \tag{10}$$

We obtain the vertical density stratification $\rho$ using the momentum equation (1) in vertical direction $z$ (neglecting the viscosity term) (see Hensler, 1981a) and the pressure $P$ by the perfect gas law. Because we are not interested in the vertical disk structure (for $\alpha$-disks see Meyer and Meyer-Hofmeister, 1981) but only in mean values of $\rho$, $P$, and $T$, we assume the temperature $T$ to be constant in the vertical direction and determine it by the assumption of energy balance. Shakura and Sunyaev (1976) and Pringle (1976) confirmed this to be realized in $\alpha$-disks.

With the further assumption of an optically thick disk the radiation flux from one disk surface

$$Q_{rad} = \frac{Q_{\bar{z}}}{A_g \cdot \Delta t} = \frac{8}{3} \frac{\sigma T^4}{\kappa \Sigma} \tag{12}$$

and Kramer's opacity formula

$$\kappa = 0.4 + 3.2 \times 10^{22} \frac{\rho}{T^{3.5}} \tag{13}$$

determine the equilibrium temperature $T$. Here $\sigma$ is the Stefan-Boltzmann constant and $\Sigma$ the surface or column density.


III. COMPUTATIONS

According to the mass transfer rate $\dot{M}$ through the inner Lagrangian point $L_1$, gas cells are initiated randomly in constant time steps in the vicinity of $L_1$ having an initial velocity of 10 km s$^{-1}$ in the negative radial direction, which is about the sound speed in giant envelopes. The particles are considered in the computation as long as their position is in between the primary's radius $r_a$ and twice the distance of the inner Lagrangian point $r_{L_1}$. They are omitted if they reach the secondary's Roche-lobe.

If the number of the particles in the system remains constant, we have achieved the stationary state of the disk.

The primary is assumed to rotate synchronously. The influence of nonsynchronous rotation will be discussed elsewhere (Hensler, 1981b). The radiation field of the primary is taken into account within the innermost grid part, that of the secondary for the initiated particles around $L_1$.

## IV. A DISK MODEL FOR CATACLYSMIC VARIABLES

To apply the numerical method described above to binary systems in mass exchange, among other models (Hensler, 1981b) we have calculated a disk model for the Cataclysmic Variables (CVs) represented by a system of the following parameters:

primary's mass                          $M_1 = 1\ M_\odot$

primary's radius                        $r_1 = 10^{-2}\ R_\odot$

primary's effective temperature         $T_{eff,1} = 5000\ K$

secondary's mass                        $M_2 = 0.5\ M_\odot$

orbital period                          $\tau = 0.2\ d$

mass transfer rate                      $\dot{M} = 10^{-9}\ M_\odot\ y^{-1}$

From that the mass ratio amounts to q = 2. and the separation of the components to d = 1.65 $R_\odot$. The period denotes a system which is located at larger values of the period gap of the CVs (Warner, 1976).

Though our method would be applicable also to the time evolution of disks (Hensler, 1981b) (as published by Bath and Pringle (1981) for CVs) we only want to calculate the steady state accretion disk. Therefore, here we are not interested to investigate the outburst mechanism for CVs in order to prefer one of the theories discussed (whether there are mass transfer variations (Bath, 1975; Wood, 1977) or disk instabilities (Osaki, 1974)).

In Fig. 4 the column density distribution is displayed. The parameters are those given before (Q = q, A = d, PER = $\tau$, DM = $\dot{M}$), and T is the time in units of the period $\tau$.

The density distribution reveals a (horn-like) ring of maximum density with a shallow decrease inwards but an immediate contact of the disk with the primary's surface. The mass accretion rate remains constant and amounts to more than 95 per cent of the transferred material. Only a small inner disk region is circular and almost Keplerian, while the main disk shows an elliptical form due to hydrodynamical structures. The density is decreasing outwards and the disk is filling the whole Roche area, through this cannot be recognized easily from the picture.

From the outer disk region mass loss, combined with angular momentum loss, takes place. Besides this, angular momentum from the disk is also transferred into orbital motion via tidal friction, in this case as efficiently as estimated by Lin and Pringle (1976) and Papaloizou and Pringle (1978) (results for other systems see also Hensler (1981b)).

The radiation flux distribution (Fig. 5) shows nearly the same structure as the density, but a monotonic increase towards the primary's surface. This is due to an increase in released potential energy and prevails over the density decrease there. No hot spot is obtained where it would be expected.

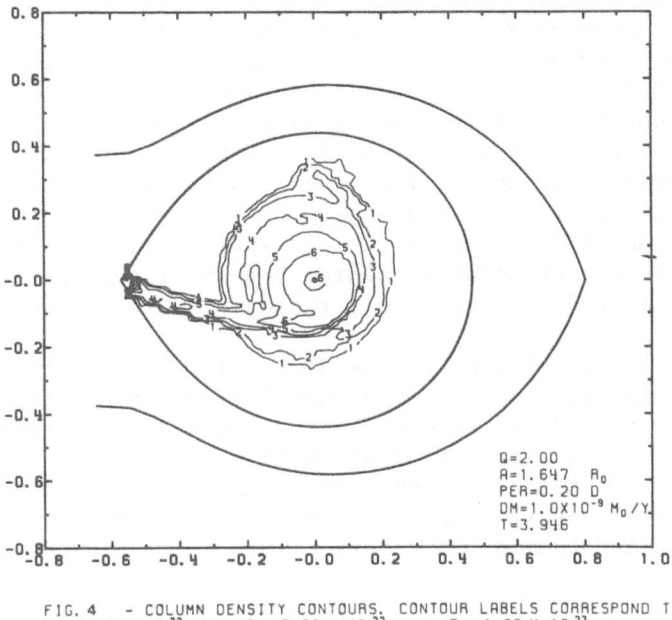

FIG. 4 - COLUMN DENSITY CONTOURS. CONTOUR LABELS CORRESPOND T
1: 3.20 X $10^{22}$     2: 5.60 X $10^{22}$     3: 1.00 X $10^{23}$
4: 1.80 X $10^{23}$     5: 3.20 X $10^{23}$     6: 5.60 X $10^{23}$
7: 1.00 X $10^{24}$     8: 1.80 X $10^{24}$     9: 3.20 X $10^{24}$
PARTICLES / $CM^2$.

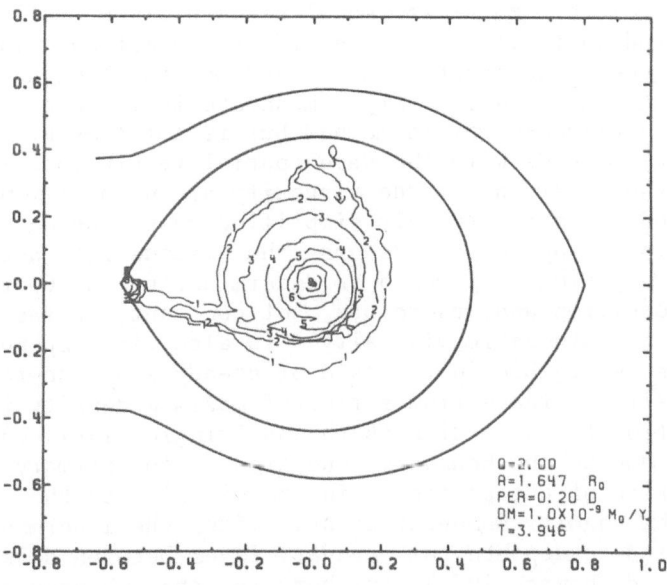

FIG. 5 - RADIATION FLUX. CONTOUR LABELS CORRESPOND TO
1: 1.00 X $10^{11}$     2: 3.20 X $10^{11}$     3: 1.00 X $10^{12}$
4: 3.20 X $10^{12}$     5: 1.00 X $10^{13}$     6: 3.20 X $10^{13}$
7: 1.00 X $10^{14}$     8: 3.20 X $10^{14}$     9: 1.00 X $10^{15}$
ERG / SEC / $CM^2$.

According to the outburst mechanism discussed by Osaki (1974)
due to disk instability this disk model would represent accretion in
outburst. As known from observations (Warner, 1976) the disk in outburst is brighter and larger than in quiescence, and no hot spot is
detectable in the light curve during eclipse (e.g. for OY Car see
Vogt, 1979) due to the bright inner disk region. However, if the viscosity is as large as expected for the outburst phase (Osaki, 1974) the
infalling matter does not strike denser parts of the disk before its
velocity vector nearly coincides with the azimuthal velocity vector
of the disk, and kinetic energy is dissipated only over a large region
and in tongue-like structures towards the infalling gas stream due to
deceleration already in less dense regions. In this case no hot spot
exists.

Because the total mass in the disk (Fig. 4) amounts to $10^{-12} M_\odot$,
in fact the viscosity seems to be somewhat too large with regard to
observed accretion disks in CVs where Lin and Pringle (1977) have
derived a mean total mass of about $10^{-10} M_\odot$.

Opposite the infalling gas stream a warm region is revealed,
which seems from observations to be much more prominent in some CVs
during quiescence (Haefner and Metz, 1981).


V. CONCLUSIONS

As one sees from the results, the numerical method presented here enables us to calculate hydrodynamical effects in accretion disks which
cannot be treated by the $\alpha$-disk model because of asymmetries and cannot be resolved spatially by finite difference methods. This advantage
will be enhanced if we proceed to three-dimensional magnetohydro-
dynamics which would require only a moderate increase of the particle
number in this pseudo-particle method but is not feasible by other
hydrodynamical methods with the same spatial resolution. The main
problems of the treatment of the viscosity are naturally not solved
by this method but are certainly simplified by depending on the mesh
size and viscosity parameter. Because the viscous interaction is
applied in every time step which is determined by the Courant-Fried-
richs-Lewy condition and, therefore, not constant, so far in the
calculations the kinematic viscosity $\nu$ is also time dependent. However, this is negligible in the case of steady state accretion.

The density decrease from a ring of maximum density inwards is
in contradiction to the $\alpha$-disk (Bath and Pringle, 1981) but I suggest
that this is due to the boundary condition of the primary's surface.
The deceleration of the particles in the vicinity of the surface and,
therefore, the angular momentum transfer from the innermost disk to the
star might tend to deplete the boundary layer between disk and surface
(Kippenhahn and Thomas, 1978). Furthermore, the enhanced release of
potential energy combined with a decrease in disk height inwards di-
minishes the density, assuming radial pressure balance there.

During the calculation the spin-up of the angular momentum and
mass gaining primary is not taken into account, because the time intervals are too short.

The model of an accretion disk presented here for the CVs should only
be understood to be one possible disk. Current calculations are carried
out to investigate the dependence of the disk structure on the system
parameters and especially on the viscosity.

More detailed discussions about the numerical method and some
other results can be found in two forthcoming papers (Hensler 1981a, b).

Acknowledgements:

The author is indebted to W. Glatzel, Drs. F. Meyer, H. Ritter, and
M.Schüssler for many helpful discussions. The calculations have been
carried out on the UNIVAC 1100/82 of the Gesellschaft für wissen-
schaftliche Datenverarbeitung, Göttingen.

References:

Bath, G.T.: 1975, Monthly Notices Roy. Astron. Soc. 171, 311
Bath, G.T., Pringle, J.E.: 1981, Monthly Notices Roy. Astron. Soc.
     194, 967
Biermann, P.: 1971, Astron. Astrophys. 10, 205
Crampton, D.: 1980, IAU Symp. No. 88, p. 313
Delgado, A.J., Thomas, H.-C.: 1981, Astron. Astrophys. 96, 142
Everson, B.L.: 1981, preprint
Flannery, B.P.: 1975, Astrophys. J. 201, 661
Haefner, R., Metz, K.: 1981, Astron. Astrophys., in press
Harlow, F.H.: 1964, Methods in Comput. Physics Vol. 3, p. 319
Hensler, G.: 1981, Proceed. V. Göttingen-Jerusalem-Symp., in press
Hensler, G.: 1981, this colloquium, p. 219.
Hensler, G.: 1981a, Astron. Astrophys., submitted
Hensler, G.: 1981b, Astron. Astrophys., submitted
Heuvel, E.J.P. van den: 1976, IAU Symp. No. 73, p. 35
Kippenhahn, R., Weigert, A.: 1967, Z. Astrophys. 65, 251
Kippenhahn, R., Thomas, H.-C.: 1978, Astron. Astrophys. 63, 265
Leboeuf, J.N., Tajima, T., Dawson, J.M.: 1979, J. Comput. Physics 31,
     379
Lightman, A.P.: 1974a, Astrophys. J. 194, 419
Lightman, A.P.: 1974b, Astrophys. J. 194, 429
Lin, D.N.C., Pringle, J.E.: 1976, IAU Symp. No. 73, p. 237
Lin, D.N.C., Pringle, J.E.: 1977, in "Novae and Related Stars",
     Astrophys. Space Sci. Libr. Vol. 65, ed. M. Friedjung, Reidel,
     Dordrecht, p. 35
Lubow, S.H., Shu, F.H.: 1975, Astrophys. J. 198, 383
Lucy, L.B.: 1977, Astron. J. 82, 1013
Lynden-Bell, D., Pringle, J.E.: 1974, Monthly Notices Roy. Astron.
     Soc. 168, 603
Mayo, S.K., Wickramasinghe, D.T., Whelan, J.A.J.: 1980, Monthly
     Notices Roy. Astron. Soc. 193, 793
Meyer, F., Meyer-Hofmeister, E.: 1981, Astron. Astrophys., in press

Novikov, I., Thorne, K.S.: 1973, in "Black Holes", Les Houches 1972,
    eds. C. DeWitt & B.S. DeWitt, Gordon & Breach, New York, p. 343
Osaki, Y.: 1974, Publ. Astron. Soc. Japan 26, 429
Pacharintanakul, P., Katz, J.I.: 1980, Astrophys. J. 238, 985
Papaloizou, J., Pringle, J.E.: 1977, Monthly Notices Roy. Astron.
    181, 441
Plavec, M.J.: 1980, IAU Symp. No. 88, p. 1
Potter, D.: 1973, "Computational Physics", John Wiley & Sons, London
Prendergast, K.H.: 1960, Astrophys. J. 132, 162
Prendergast, K.H., Taam, R.E.: 1974, Astrophys. J. 189, 125
Pringle, J.E.: 1976, Monthly Notices Roy. Astron. Soc. 177, 65
Shakura, N.I., Sunyaev, R.A.: 1973, Astron. Astrophys. 24, 337
Shakura, N.I., Sunyaev, R.A.: 1976, Monthly Notices Roy. Astron. Soc.
    175, 613
Sørensen, S.-A., Matsuda, T., Sakurai, T.: 1975, Astrophys. Space Sci.
    33, 465
Tayler, R.J.: 1980, Monthly Notices Roy. Astron. Soc. 191, 135
Vogt, N.: 1979, ESO Messenger 17, 39
Warner, B.: 1976, IAU Symp. No. 73, p. 85
Wood, P.R.: 1977, Astrophys. J. 217, 530

# QUASI-SIMULTANEOUS PHOTOMETRIC AND POLARIMETRIC
# OBSERVATIONS OF T TAU AND RY TAU

N.P. Red'kina
G.P. Chernova
Institute of Astrophysics
Dushanbe, 734670, U.S.S.R.

## ABSTRACT

Quasi-simultaneous photometric and polarimetric observations of
T Tau and RY Tau show that the proper polarization of these stars
arises in the visual region in their gaseous emitting envelopes.  The
polarization of T Tau shows cyclic variations with a period of $5^h.18$.
That of RY has two different modes of variation at brightnesses
$V > 11^m$ and $V < 11^m$; it is possibly connected with the suspected binary
nature of this star.

## INTRODUCTION

The nature of the polarization of T Tau type stars is subject to
many controversies.  Brager (1974) suggests scattering by dust
particles, while Strom (1977) thinks scattering by electrons to be
more acceptable.  Abuladze et al., (1975) point out the difficulty
of explaining the polarization variations at constant brightness of
T and RY Tau in the frame of the dust cloud hypothesis.

Wavelength dependent linear polarization in the range 3560 to
8410 Å and in the narrow Hα interference filter region was observed
by Bastien and Landstreet (1980) for T and RY Tau and V 866 Sco.  They
showed the polarization to originate in extended circumstellar dust
envelopes.  The considerable change of the wavelength dependent shape
is explained as a result of changing dust particle dimensions.

The observations of linear polarization of RY Tau enabled
Efimov (1980) to conclude that the star has a flattened gas and dust
envelope.  Efimov connects the wavelength dependent variation with the
prevalence of Thomson scattering by electrons at a low brightness of
the star.

Hough et al. (1981) found a considerable linear polarization in
all six T Tau type stars investigated by them in the IR range 0.44 -
2.2 μm.  T Tau (and possibly SU Aur) showed at 2.2 μm, a change of
the polarization plane orientation by 90°.  The polarization spectrum

231

*Z. Kopal and J. Rahe (eds.), Binary and Multiple Stars as Tracers of Stellar Evolution, 231–238.*

of T Tau is flattened, the mean degree of polarization P being about
1%.  P drops to zero at 1.6 $\mu$m and then grows to 0.6% at 2.2 $\mu$m.
To interpret the T Tau polarization in the optical region, Hough et al.
(1981) involve small silicate grains with dimensions $< 0.1 \mu$m; an
admixture of large particles ( $> 1.0 \mu$m) helps to explain the IR
polarization.  The polarization vectors of particles of these two kinds
are orthogonal, and so the inversion of the polarization sign should
occur at some intermediate wavelength.

The peculiar behavior of RY Tau is explained by Nurmanova (1981)
on the basis of her photometric and polarimetric observations as a
consequence of the duplicity of this star.

The discrepancy in the interpretation of T Tau star polarization
is perhaps due to the fact that most authors ignore the interstellar
polarization.

PROPER POLARIZATION OF T TAU

To analyze our T TAU observations in the visual region (Red'kina
et al., 1981) and the results obtained by other authors (Vardanian,
1964 ; Abuladze et al., 1975; Efimov, 1980), we took the effects of
interstellar polarization into account, following Vardanian (1964) and
Efimov (1980).  T Tau shows seasonal variations of its proper polarization from 0.5% to 1.7% at almost constant light in V.  Reciprocal
correlations between long-term (several years) variations of the
polarization and brightness predicted by the dust hypothesis, were
not revealed.

The analysis of our simultaneous photometric and polarimetric
observations of T Tau showed a direct correlation of the proper polarization $P_V$ with the intensity in H$\alpha$ , and a weak correlation with the
brightness in V (Fig. 1a), indicating a connection of the polarization
of the star with its gaseous emission envelope.

Such peculiarities as the direct correlation between $P_V$ and the
intensity in H$\alpha$, the absence of a correlation with the brightness in
V and the essentially flat polarization spectrum of T Tau could hardly be explained by two kinds of scattering particles ( $< 0.1 \mu$m and
$> 1 \mu$m) in a dusty stellar envelope (Hough et al., 1981).  It seems
more realistic to suggest that at $\lambda < 1.6$ $\mu$m, the polarization is
due to scattering in the gaseous emissive shell.  At the envelope
regions where the Thomson scattering cross-section is greater than the
braking radiation cross-section, the polarization spectrum does not
depend on $\lambda$.  The depression in the polarization spectrum at
$\sim 1.6$ $\mu$m is due to the enhanced braking absorption in the gas
envelope.

The peculiarities of the T Tau polarization listed can be explained by radiation scattering in a magnetized optically thin region
of the gaseous envelope of the star.  The proper polarization is

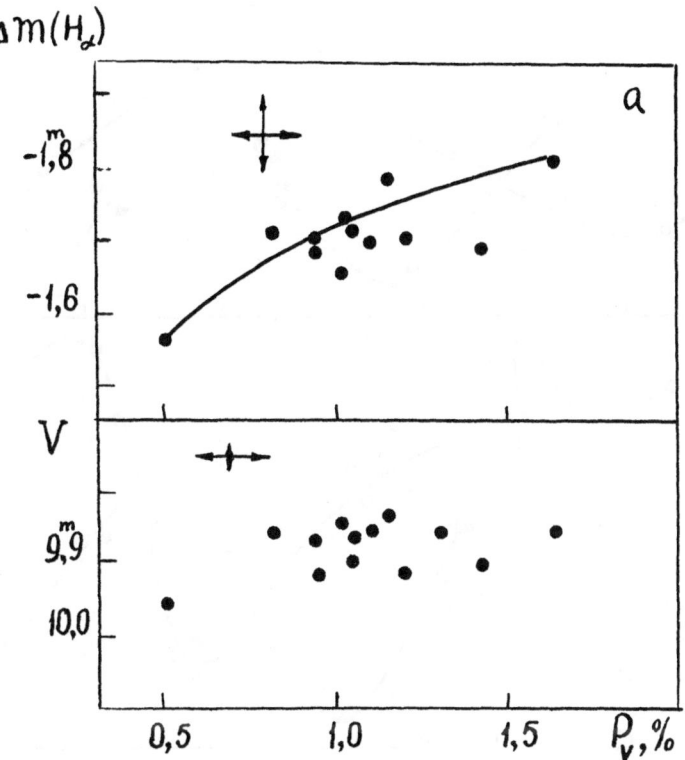

Fig. 1  a)  Interrelation between the proper polarization $P_v^*$, H$\alpha$
intensity (relative to the H$\alpha$ intensity of the comparison
star "c" in Abuladze et al., 1975) and V-brightness for
T Tau.

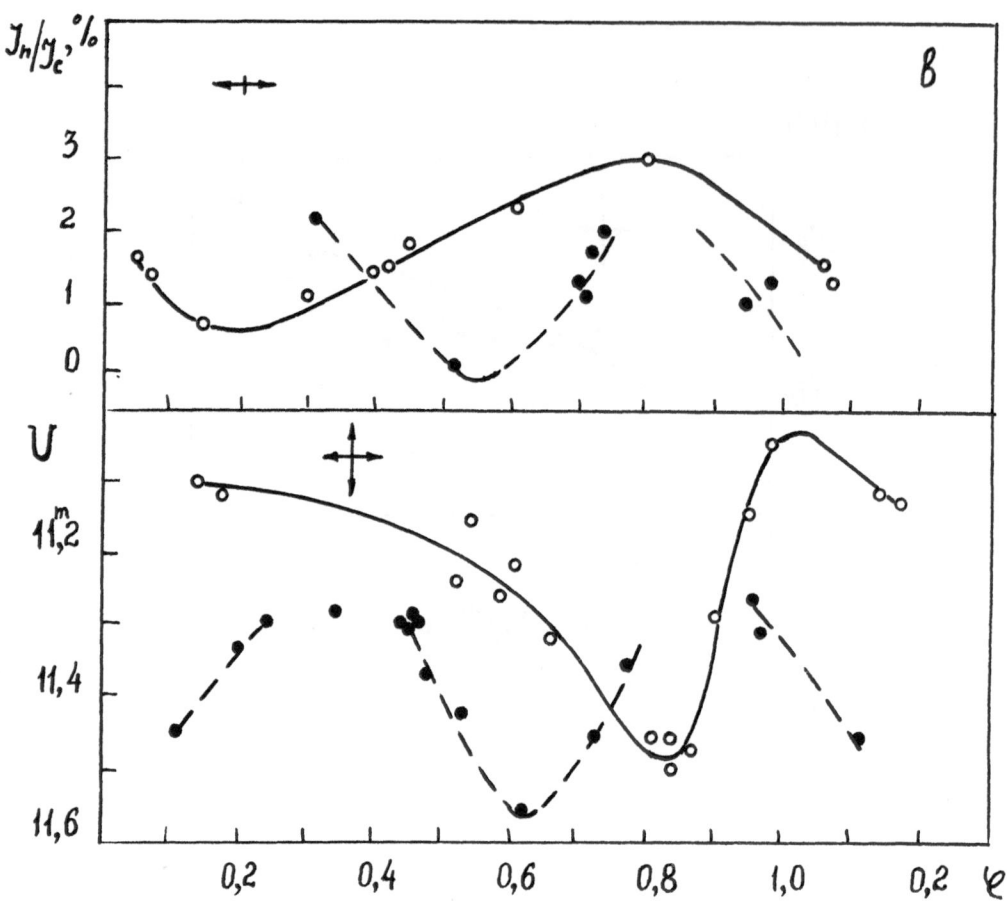

b)  Presentation of T Tau polarized light $I_n$ in the V band
    (relative to the light of the comparison star "c" in
    Abuladze et al., 1975) as a periodic function with
    periods $P=5^h.18$ (open circles) or $P_1 = P/2$ (filled circles)

determined in this case by the depolarization factor $\delta$ which
according to the magnetic active plasma theory is determined by the
formula (Dolginov et al., 1979):

$$\delta = \frac{1}{P^2} \left\{ 9.137 \frac{(1-\mu^2)\,\varphi(\mu)}{H(\eta)} \right\} \cdot \frac{1}{\mu^2}$$ (1)

$$\delta = 7.93.10^7 \, \lambda^2 \cdot B \quad ,$$

where B is the magnetic field strength; $\mu$ = cos $\nu$ with $\nu$ the angle
between the normal to the surface of the medium and the direction of
the magnetic field; H $(\mu)$ and $\varphi$ $(\mu)$ are the generalized
Ambartsumian-Chandrasekhar functions; P is the linear polarization.
The maximal (1.6%) and minimal (0.5%) values of P correspond at
$\mu$ = 0.1 to $\delta$ = 42.52 and 143.81 and B=174 and 520 gauss, respectively. The magnetic fields can also have greater strengths since
$\delta \rightarrow \infty$ when $\mu \rightarrow 0$.

The position angle of the polarization plane $\Theta$ remains
practically unchanged in the visual region. This is shown by our
1979 and 1980 observations as well as by those of other authors
(Efimov, 1980; Abuladze et al., 1975) and by the polarization spectrum
from 0.44 to 1.6 $\mu$m (Hough et al., 1981). The symmetry axis of the
scattering system of the star is supposed to remain fixed in space.

T Tau has possibly a disk-like gas and dust envelope. The jump
of the position angle at 2.2 $\mu$m occurs most probably at the boundary
between the two envelopes. This conclusion is supported by the intensity distribution of T Tau (Rydgren et al., 1976), where the $\Theta$
considered ( $\lambda$ = 2.2 $\mu$m) is the mean of the radiation wavelength of
the gas (free-free transition) and that of the dust envelope. The
geometry of the system is as follows: the electric field vector of
the scattered optical radiation vibrates perpendicular to the symmetry axis of the shell, the electric field vector of the IR radiation
vibrates along the symmetry axis.

Goetz and Wenzel (1970) detected short period changes of the
spectral type of T Tau with a period of $5^h18$. When analyzing our own
observations we found an analogous period for the amount of
polarized light $I_n$. The conversion of the proper polarization $P_v$ to
$I_n$ was made using the formula:

$$I_n (\varphi) = P (\varphi) \cdot I_o (\varphi) \cdot 10^2 ,$$ (2)

where $\varphi$ is the phase, $I_o$ is the light intensity.

The corresponding brightness in U varies in opposite phases
relative to the $I_n (\varphi)$ curve. Fig. 2 shows the possible existence
of vibrations with a period two times shorter than $5^h18$.

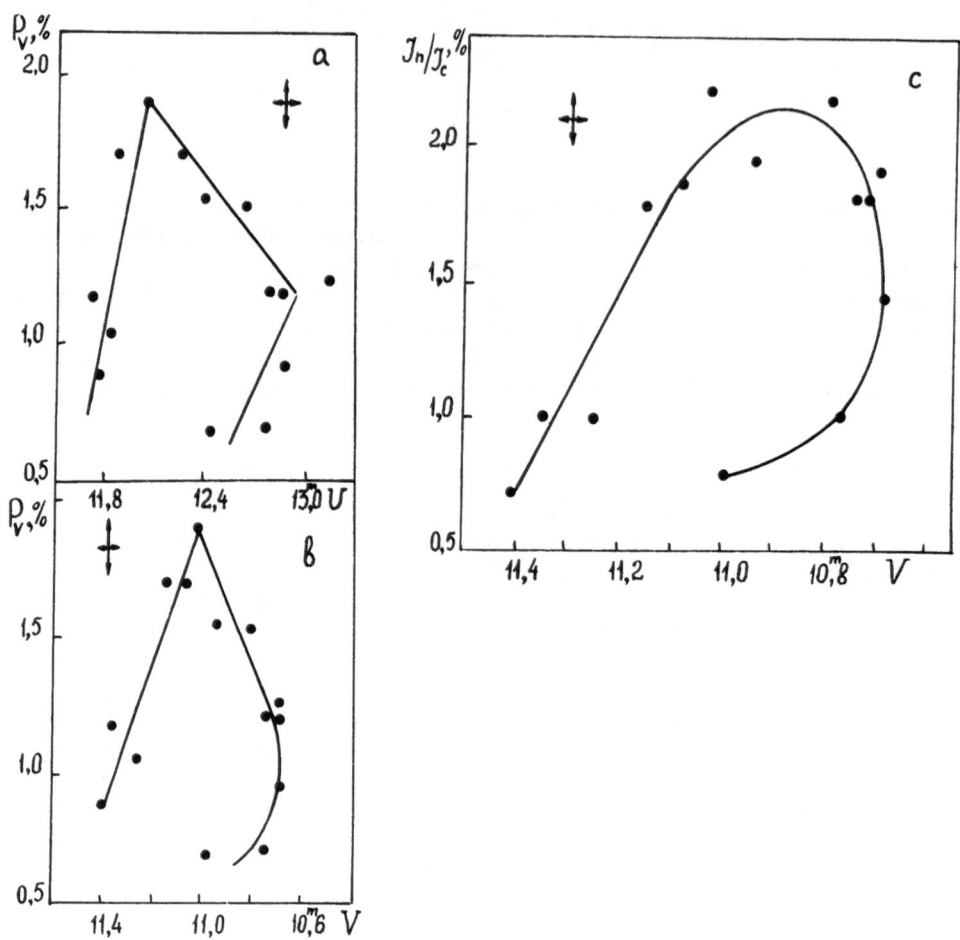

Fig. 2      RY Tau proper polarization as a function of brightness in U (a) and V(b). RY Tau polarized light amount $I_n$ (relative to the light $I_c$ of the comparison star "f" in Zaitseva et al., 1974) as a function of V-brightness (c).

The following interesting peculiarity should be noticed; $I_n$ has its maximum at $\varphi = 0.8$ possibly indicating an inhomogenity of the temperature distribution over the disk of the star. A proof of this through more extensive observations is desirable.

The cyclic variation of physical characteristics of T Tau stars is supported by the recent work by Worden et al., (1981). The analysis of observations in the U band showed the flares of T Tau stars to be a superposition of many solar-like flares. These flares are more frequent and more powerful (in separate cases up to $10^3$ times) than solar flares. The index of the flare activity $\beta$ ($\sim f^{-\beta}$, where f is the frequency) changes considerably, and a variable flare activity resembling the solar cycles is possible.

PROPER POLARIZATION OF RY TAU

RY Tau possesses, in the region studied ($V=10\overset{m}{.}68 \div 11\overset{m}{.}4$), two different types of $P_V$ -V dependency, the critical brightness level being $\sim 11^m$. At $V > 11^m$ (Fig. 2 a,b) $P_V$ increases with increasing V and U. The position angle $\Theta$ lies between $3^o$ and $59^o$. The amplitude of the brightness variation grows towards the red end of the spectrum. At $V < 11^m$, the $P_V$ -V and $P_V$ -U relations become undefined. The position angle varies between $119^o$ and $153^o$. The amplitude of the brightness variation grows towards the blue spectral region. The whole relation considered has a loop-like appearance (Fig. 2 a,b,c). This implies different polarization mechanisms at the minimal and maximal brightness levels, corresponding to different physical states of the star.

Wenzel (1970) suggested RY Tau to have an IR source associated with the dust envelope. Its radiation maximum lies near 1 $\mu$m and its short wavelength wing is seen in the optical region.

Nurmanova (1981) thinks RY Tau to be an eclipsing binary system. This suggestion explains the different properties of RY Tau at $V > 11^m$ and $V < 11^m$, including different dependencies between U and V (Nurmanova, 1981), between the intensity in H$\alpha$ and the color index U-B at $V=10\overset{m}{.}5$ and $V=11^m$ (Zaitseva et al., 1974); different polarization mechanisms at different brightness levels; changes of the wavelength dependency.

SUMMARY

On the basis of our investigations we conclude that the optical radiation polarization of T Tau and RY Tau arises in the gaseous emissive envelopes of these stars due to radiation scattering on free electrons. Further simultaneous photoelectric and polarimetric observations would be of value to confirm the cyclical behavior of the flare activity of T Tau type stars.

The peculiar change of the proper polarization and brightness
of RY Tau suggests the binary nature of this star.

REFERENCES

Abuladze, O., Vardanian, R.A., Kovalenko, V.M., Kumsishvili, Ja.,
     Melikian, J.A., Mironov, A.V., Oshchepkov, V.A., Stepanian, J.A.,
     Totachava, A., Cherepashchuk, A.M., Shanin, G.I., Shpychka, I.V.,
     Shcherbakov, A.G., 1975. Results of Observations of the T Tauri
     Type Stars under 1973 Joint Program. Variable Stars, 20, 47-61.
Bastien, P., and Landstreet, J.D.: 1979, Polarization Observations of
     the T Tauri Stars RY Tauri, T Tauri, and V866 Scorpii. Ap. J.
     Letters, 229, L137-L140.
Brager M.: 1974. Pre-main sequence stars. III. Herbig Be/Ae stars
     and other selected objects. Ap. J., 188, 53-58.
Dolginov, A.Z., Gnedin Ju.N., Silant'ev, N.A.: 1979. Radiation
     Propagation and Polarization in Space. Moscow, Nauka, 131-149.
Efimov, Ju.S.: 1981. Polarization Observations of RY Tau and T Tau.
     Variable Stars, 21, No. 3, 273-284.
Goetz, Von W., Wenzel, W.: 1970. Photoelektrische und Objektiv-
     Prismen-Beobachtungen an T Tauri. NVS, 5, 117-125.
Hough, J.N., Bailey, J., Cunningham, E.C., McCall, A., Axon, D.J.:
     1981. Linear Polarization of T Tauri Stars. Mon. Not. R. Astr.
     Soc., 195, 429-436.
Nurmanova, U.A.: 1981. On the Duplicity of RY Tauri. Variable Stars,
     (in press).
Red'kina, N.P., Zubarev, A.V., Chernova, G.P.: 1981. Photoelectric
     and Polarimetric Observations of T Tauri. Astronom. Zirkuliar,
     No. 1149, 6-7.
Rydgren, A., Strom, S.E., Strom, K.M.: 1976. The Nature of Objects
     of Joy: A Study of the T Tauri Phenomenon. Ap. J., Suppl.,
     30, 307-336.
Strom, S.E.: 1977. Star Formation. Ed. T. de Jong and A. Maeder.
     (Dordrecht: Reidel), p. 179.
Vardanian, R.A.: 1964. The Polarization of T and RY Tau. Soobsch.
     Bjurak. Obs., 35, 3-23.
Wenzel, W.: 1970: Photoelektrische Beobachtungen an RY Tauri. NVS,
     5, 117-125.
Worden, S.P., Schneeberger, T.J., Kyhn, J.R.: 1981. Flare Activity
     on T Tauri Stars, Africana J.L., 244, 520-527.
Zajtseva, G.V., Ljuty, V.M., Cherepaschuk, A.M.: 1974. Photoelectric
     Observations of Intensity Variability of H$\alpha$ Emission Line and
     Continuum in the Spectrum of RY Tau. Astroph. Russ. J.,
     No. 10, 357-364.

S Cnc AS A CASE STUDY OF CLASSICAL ALGOL EVOLUTION

N.S. Awadalla
E. Budding
Department of Astronomy
University of Manchester

ABSTRACT

   Basic information on empirical determinables for the system S Cnc
are reviewed.  Whilst the photometric information can be clearly
analysed along well known lines, the spectrographic information is far
from clear and requires further study, though a semi-detached
configuration, at least, can be confidently deduced.

   The potentially important question of some physical relationship
of the variable to the open cluster Praesepe is considered, but the
result is in the negative.

   The evolutionary status of the binary is examined in a general
way by making use of some summary formulae, some guidelines for which
were taken from previous more general work of Refsdal and Weigert.
S Cnc is in good overall agreement with the low mass Case B
theoretical mode of binary evolution, and even the absence of a
detectable rate of period variation is shown to be not in serious
conflict with this picture.  Interesting close comparisons with the
well known example AS Eri are possible.

INTRODUCTION

   S Cnc (= BD 19° 2090, HD 74307, KW (Praesepe List) 552) has been
a well known example of what is often loosely called an Algol type
variable for the last 135 years.  Its relatively long period ( $\sim$ 9.5
days) has, however, tended to relegate it from the league of intensive-
ly studied close binary systems, particularly in recent years.

   The system may, nevertheless, have some puzzles to offer.  Could
it, for instance, have some connection with the Praesepe cluster
(Kholopov, 1958)?  Does it contain an "undersized" subgiant (Kopal and
Shapley, 1956), or could it be "semi-detached" (Hall, 1974)?  If it
really is semi-detached, doesn't it mean that the secondary is a really
low mass object despite its early K spectral type and $\sim$5.5 $R_0$ radius?

Z. Kopal and J. Rahe (eds.), Binary and Multiple Stars as Tracers of Stellar Evolution, 239–259.
Copyright © 1982 by D. Reidel Publishing Company.

A luminosity excess of some 8 magnitudes over the general mass-luminosity relation is implied. And shouldn't the system be showing a more marked rate of period variation than is apparent (Ziołkowski, 1978; Crawford and Olson, 1980)?

Though the primary is similar to that of AS Eri, a frequently referred to example of classical Algol evolution (e.g., Koch, 1970; Popper, 1973; Plavec, 1973; Refsdal et al., 1974; Plavec and Polidan, 1976; Paczyński and Dearborn, 1980), and the mass ratios may be similar ($\sim 0.1$), the separation between the components in S Cnc is two and a half times greater than that of AS Eri. In the latter case, in fact, the smallness of the present angular momentum seems to require a previous contact or common envelope condition (Paczyński, 1976). Are these two systems simply comparable, or could differences in relation to a possible previous common envelope stage cause any observable consequences?

In what follows we shall explore each of these questions in turn. First we summarise our knowledge of basic parameters characterizing the system as far as we can.

BASIC DETERMINABLES

a. Photometry

S Cnc is an eighth magnitude eclipsing binary system exhibiting deep total eclipses when the early type primary is eclipsed by the cool subgiant companion. Recent out of eclipse magnitudes and colours are given in Table 1.

A particularly good set of photoelectric observations was published by Huffer and Collins in 1962, based on an extensive series of previous observations. This data has been analysed by a number of authors, including ourselves, and some of the various results for the main photometric elements are presented in Table 2. Our optimal curve fit to the $\lambda = 5400\mathring{A}$ (yellow) data is given in Figure 1. Some authors have thought that the data is of sufficient quality that a simultaneous solution for the primary limb darkening can be made, but doubts appear to be cast on this by the work of Linnell and Proctor (1971).

A more recent set of light curves with greater coverage in both wavelengths and phase, including for the first time the very shallow secondary minimum, have been published by Crawford and Olson (1980). These authors refer to some possible slight variability of the secondary star, but it is clear that the light curves are relatively stable for a semi-detached system, and it is unlikely that the more recent photometry permits any clearer definition of the informative primary total minimum over the earlier data.

Table 1

Magnitude and Colour of  S   Cnc

| | GCVS | WOFK | Barnes (1974) | Crawford & Olson (1980) | Awadalla & Budding | Hilditch & Hill (1975) |
|---|---|---|---|---|---|---|
| Mag | 8.45 (pg) | 8.5 (B) | | 8.34 (V) | 8.32 (V) | 8.31 (V) |
| B – V | | | 0.10 | 0.09 | 0.09 | 0.11 |

Values refer to the combined light of the system between eclipses.  This has entailed some small appropriate numerical processing of data from the listed sources in some cases and conversion to the UBV system.

Table 2

Eclipse – Photometric Solution Parameters

a) Geometric elements

| Parameter | Huffer and Collins (1962) | Irwin (1963) | Tabachnik (1969) | | Linnell and Proctor (1971) C FIT | Linnell and Proctor (1971) DIFCORT | Caracatsanis (1977) | Present Work (adopted) |
|---|---|---|---|---|---|---|---|---|
| $r_1$ | .07677 ± .0004 | .0827 ± .0014 | .0934 ± .0013 | .0874 | .076 ± | .07124 ± | .085 | .0838 ± .0005 |
| $r_2$ | .19293  .00382 | .1923  .0012 | .01837  .0007 | .1872 | .196  .011 | .2009  .0034 | .1937 | .1932  .0008 |
| $i$ | 83°.91  .30 | 84°.24  .15 | 85°.342  .0012 | 84°.729 | 83°.55  .37 | 82°.99  .37 | 84°.40 | 84°.05  .05 |

b) Physical elements

— (Adopted Values only)

| | λ = 5400 | λ = 4200 | λ = 3500 |
|---|---|---|---|
| (i) Fractional Luminosities $L_1$ | 0.845 ± 0.001 | 0.923 ± 0.001 | 0.968 ± 0.001 |
| $L_2$ | 0.155  0.001 | 0.077  0.001 | 0.032  0.001 |
| (ii) Limb darkening Coefficient (Linear) $u_1$ | 0.47 | 0.55 | 0.5 |

Remarks: Parameter symbols have their usual designations (c.f. Budding, 1973). Other adopted values come from a simple "spherical model" (8-parameter) fitting to the essentially uncomplicated total eclipse. The solutions are $\chi^2$ consistent ($\chi^2/\nu = 1.1$) for an assumed observational accuracy (individual point) of 0.007. The limb darkening coefficients come from the table of Al Naimiy (1978).

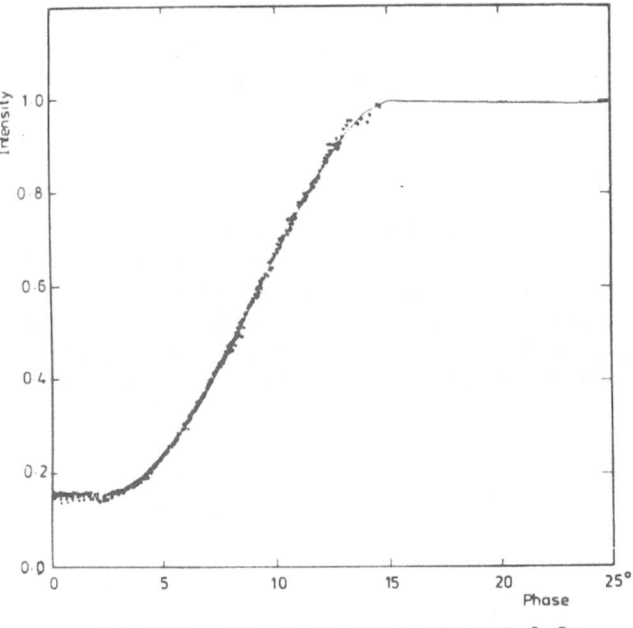

V-Light Curve Fitting for the Primary Minimum of S Cnc.

Figure 1.

b.  Spectroscopy

If the photometry of S Cnc is particularly clear, regrettably the same is not true of spectroscopic data.  Unpublished results from Joy were used by Kopal and Shapley (1956), but when the original compiled values were released by Abt (1970), and further discussed by Weis (1976) and Batten (1976), it became clear that any radial velocity "curve" amounted to little more than a scatter diagram.  In fact, any suggestion of a sinusoidal variation in Joy's data is 180° out of phase with the expected.  Closer examination of the spectra at higher resolution by Batten (1976) still seems to suggest an apparent spurious recession of the primary after its total eclipse, though profile tracings show that there are numerous spectroscopic irregularities suggestive of circumstellar material which are known, from other similar cases, to make for considerable complications to radial velocity specification.  Popper (1980a), on the other hand, appears to have observed S Cnc when such complications were not so troublesome, though he admits uncertainty on his derived secondary mass.  Spectral clas-sification is possibly also subject to uncertainties arising from similar effects.  Older references (e.g., Kopal and Shapley, 1956) seem to prefer A0, while some more recent sources (since, for example,

Azimov, 1963) allow a B9 designation for the primary. A wider uncertainty attaches to the secondary type, which ranges from mid G to
early K - perhaps at least partly due to inherent fluctuations,
though the cooler alternative is supported by Crawford and Olson's
(1980) colours, and Popper's (1980b) recent summary of spectrographic
data.

RELATIONSHIP TO PRAESEPE?

    S Cnc is located close to the normally quoted inner region bound-
ary of the star cluster NGC 2632 (Praesepe), at a separation from the
centroid of some 70' - about the same as member star 35 Cnc or one
and a half times the separation of TX Cnc. In view of the circumstan-
tial consequences which membership of the cluster would give - age
and composition then being determinables - the question deserves close
attention.

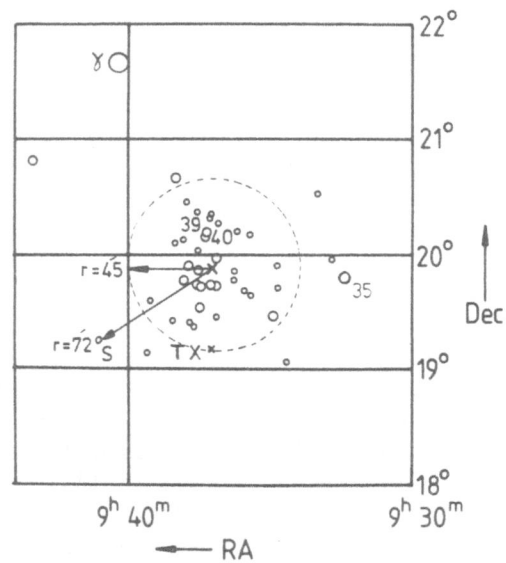

Figure 2.  Coordinate Epoch 1950

    The position of S Cnc in the colour magnitude diagram of
Praesepe though possibly close to the cluster sequence on the strength
of older estimates of its brightness and type, throws doubts on any
possibility of the primary star being like a normal Main Sequence

cluster member when more modern values for these quantities are
adopted. Thus Crawford and Barnes (1969), found, as we do, the star
to be too blue to be on the sequence, and reported it as "probably a
non-member". The metallicity index $m_1$, which from the data of
Hilditch and Hill (1975) would be too little at $m_1 = 0.123$ to be in
agreement with the Crawford and Barnes value, which at the $\beta$ value of
S Cnc (2.781) corresponds to $0.20 \pm 0.02$. Low metallicities appear,
however, to be a characteristic in Algol systems, (see e.g., Plavec and
Polidan's discussion of this subject, 1976), and if the primary has
already accepted a considerable amount of material from the evolved
component the position is not clear. Thus there is theoretical
evidence to suggest that Algol primaries in the latter stages of the
semi-detached phase could lie to the left of the appropriate Main
Sequence, though observational evidence, as far as can be ascertained
from some more well known examples, does not seem to bear this out
(Packet, 1980).

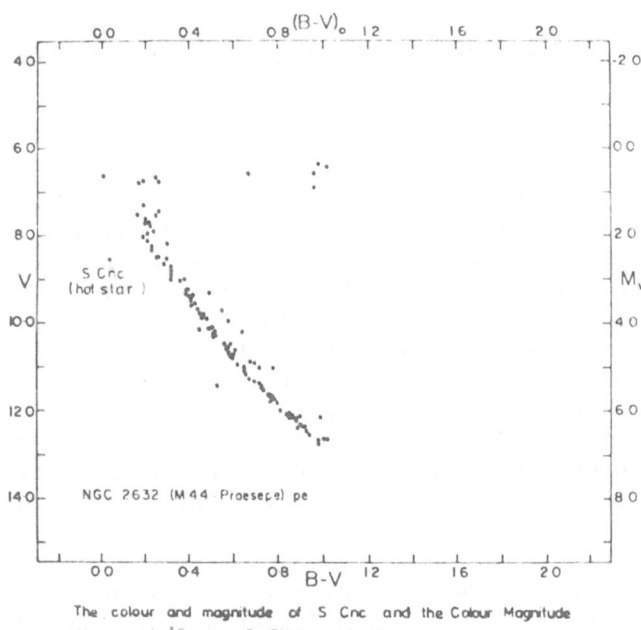

The colour and magnitude of S Cnc and the Colour Magnitude
diagram of 'Praesepe' Cluster (after Hagen, 1970)

Figure 3

Though, in view of spectroscopic complications already mentioned,
arguments based on measured radial velocities can be given little
weight, Dickens et al's (1968) mean radial velocity of NGC 2632 stars

as 31 $\pm$ 2 km/sec could not be ruled out on the basis of Batten's (1976) data on S Cnc.

The most serious doubt on S Cnc's credentials as a Praesepe star, however, comes from the measured proper motions, a point already realized by Heckmann (1937). Though when we first examined the scatter of proper motion values of accepted cluster members in the Smithsonian Catalog (1966) ( $\sim$ 0."02) the membership of S Cnc appeared to be still plausible, the Smithsonian Catalog has compared a number of independent sources of absolute position determinations to produce its proper motions, while a self contained <u>differential</u> system of measurements of a small area on the same machine, such as that of Heckmann (1925) must have a greater inherent precision, probably an order of magnitude higher than the value just mentioned. (The authors are grateful to Professor van de Kamp for pointing this out). On this basis, the separation in proper motion space between S Cnc and the centroid of Praesepe of some 0".019 $(y^{-1})$ makes its present membership "nicht in Frage".

"UNDERSIZED" OR "SEMI-DETACHED"?

A number of authors now appear to support Hall's (1974) arguments in favour of a likely semi-detached configuration to the binary. We briefly review the case.

Kopal and Shapley's (1956) deduction of an undersized subgiant for the secondary appears to have been based on radial velocity data which, as has already been mentioned, is unreliable for the intended purpose. Batten's radial velocity data suggest only that any sinusoidal variation passing through the points in the required sense would have low amplitude. On the other hand, Joy's radial velocities for the secondary, obtained during the totality, give $K_2$ = 140 km s$^{-1}$ a value which is supported by Batten (1976). From this quantity a secondary mass function can be obtained as 2.70. If we use Kopal's (1959) table relating "contact" relative radii to mass ratio the photometric solution yields a mass ratio of 0.09, which then implies a primary mass of 3.3 $M_O$, which is reasonably close to that expected for a Main Sequence primary.

A slightly higher mass than this is found if we combine the Main Sequence Mass-Radius relation (R $\sim$ $M^n$) with Kepler's third law for the binary, i.e.

$$\log M_1 = \frac{1}{3n-1} (2 \log P + 3 \log r_1 + \log (1+q) + 1.872), \qquad (3.1)$$

where P is the period in days, $r_1$ the primary relative radius and q the mass ratio. This results in $M_1$ = 3.6 $M_O$ when the index n is given a value of 0.71 (corresponding to the slope of the ZAMS M:R relation at AO), though with an obvious sensitivity on the choice of n.

An undersized secondary would require the primary mass to be higher and more discordant with the apparent Main Sequence character, while putting up the mass ratio significantly should cause a more noticeable genuine primary velocity variation in the radial velocity curve, and the detected interactive effects should also then become less likely in a non-contact situation.

In so far as a near to Main Sequence character can be assigned to the primary star, therefore, we can see that a semi-detached configuration is supported. The argument used by Hall and Neff (1979), was essentially similar to this, but involving the more reliable mass: luminosity rather than mass:radius relationships.

In Table 3 we give our adopted absolute parameters characterizing the system S Cnc.

Table 3

Adopted Physical Parameters of S Cnc

| | |
|---|---|
| Period | 9.48454 days |
| Primary Mass | 3 $M_o$ |
| Secondary Mass | 0.27 $M_o$ |
| Separation | 28.0 $R_o$ |
| Primary Mean R | 2.35 $R_o$ |
| Secondary Mean R | 5.41 $R_o$ |
| Primary Spectral Type | B9.5 |
| Primary Te | 10300 |
| Secondary Spectral Type | K0 |
| Secondary Te | 4700 |
| $\Lambda M_{bol}$ (luminosity excess of secondary) | $8^m.1$ |
| $P_o$ } (conservative evolution backwards) | $0^d.264$ |
| $A_o$ | 2.57 $R_o$ |

EVOLUTIONARY STATUS

    S Cnc appears to be a natural candidate for rather low mass Case
B type binary evolution, as was indeed proposed by Kreiner and
Ziołkowski (1978). Calculations for this type of evolutionary scheme
are able to demonstrate the observed luminosity excess, and other
physical characteristics of the component stars, which are, in this
case, somewhat more extreme than the observed mode for similar
classical Algol stars. The core of the erstwhile primary has become
like that of a low mass giant branch star. A thin shell source
above this degenerate core is able to maintain the very tenous envelope
with its throughput of relatively excessive luminosity. The basic
reason for the luminosity excess is that core or near core processes
are virtually uncoupled from the removal of the envelope material due
to the Roche lobe overflow (RLOF) mechanism. There is, of course, an
ultimate switching off of the shell source when the envelope has been
sufficiently depleted. Thereafter the evolved star should sink to a
condition like that of a white dwarf (Refsdal and Weigert, 1971) at a
separation of up to about an order of magnitude greater than that which
existed originally.

    One possible discrepancy with the proposed scheme is that there
should perhaps be a larger scale of period variation than is actually
observed (Ziołkowski, 1976; Kreiner and Ziołkowski, 1978). In what
follows we shall try to summarize the situation by reference to some
simplified but general formulation.

    Four or five basic variables can be recognized in the most es-
sential posing of the RLOF problem, together with one or two special
quantities which we may wish to treat as variables, i.e., overall mass
and angular momentum of the binary system, though their proper treatment as variables could be technically awkward; and certain other quantities which relate to the structure and rate of surface expansion
of the mass losing star, which could be regarded as parameters.

    Counting then the mass of either star $m_1$, $m_2$, the radius of the
mass losing star $r_1$, the separation of the two mass centres A and
orbital period P as the basic variables dependent on time t, we have
four fairly clear formulae to interrelate these dependent quantities,
namely: an equation for the overall mass

$$m_1 + m_2 = M, \qquad\qquad (4.1)$$

one for the orbital angular momentum (rotational momentum is usually
neglected)

$$\frac{2\pi A^2}{P} \frac{m_1 m_2}{M} = J, \qquad\qquad (4.2)$$

Kepler's third law

$$(\frac{2}{P} \pi)^2 A^3 = GM , \qquad (4.3)$$

and some equation for the relative size of the mass losing component
(e.g., some "Roche lobe" formula, see e.g., Paczyński, 1971)

$$\frac{r_1}{A} = f (m_1, m_2) , \qquad (4.4)$$

Separate formulae are required for the variation of overall mass and
angular momentum in the binary, but for the present purposes it is
convenient just to regard these as separately specifiable.

A fifth equation for the actual mass transfer is required to
formally close the system in terms of time variation. We shall first
normalize by writing $x = m_1/M$, the fractional mass of the mass losing
star. It can be shown that (using the dot notation for differentiation
with respect to time)

$$\dot{x} = -3\eta \frac{x}{r_1} (s - \dot{r}_1) \qquad (4.5)$$

can express the sought transfer equation. This relates the rate of mass
transfer x to some source function s, due to inherent expansion of the
mass losing star resulting from its own internal structural processes,
minus a retention term $\dot{r}_1$, which comes from the expansion of the available volume for $m_1$. The quantity n expresses the ratio of density
of the surface layers of the mass losing star to the mean density of
the star as a whole; alternatively $\eta$ can be defined as

$$\eta = (\frac{d \log m_1(r)}{d \log r}) \text{ surface .} \qquad (4.6)$$

Let us consider first the "conservative" case, in which M and J
can be taken as constant, at least the former of which may well be
applicable in the slow separation phase of classical Algol evolution.
Manipulation of the first three equations can easily be shown to yield

$$P = \frac{P_o}{64x^3(1-x)^3} , \qquad (4.7)$$

where $P_o$ is the minimum period, obtaining when $x = 1/2$, and given by
$$P_o = \frac{128 \pi J^3}{G^2 M^5} .$$ Differentiating (4.7) with respect to time we obtain

$$\dot{P} = 3g(x) P \frac{\dot{x}}{x} . \qquad (4.8)$$

where we have put $g(x) = (2x-1)/(1-x)$. In the normally encountered
'slow phase' of Algol evolution $x < 1/2$, so $g(x) < 0$, and, of course,
$\dot{x} < 0$ is a necessary condition of mass loss. Under these circumstances,
as x becomes small we might expect that $\dot{P}$ will increase in a rather
sensitive way to the mass transfer. This will evidently depend also
on the behaviour of $\dot{x}$.

The scope of Equation (4.8) can be broadened to include also the
possibly significant case where angular momentum of the overflowing
material is not immediately transferred back into the orbit. At
wider separations the timescale for such angular momentum transfer,
even assuming no matter is actually lost from the system, may well
become appreciable in comparison to Case B evolution timescales. A
phenomenological approach to such effects is to write

$$J/J_{init} = (x/x_{init})^k \; ,$$

where $k = 0$ corresponds to the conservative case, and we expect
$0 \lesssim k \lesssim 1$ in practice. Physically, this expresses the difficulty of
putting the increasing proportion of specific angular momentum in the
RLOF material at low x back into the orbit. $|g(x)|$ then becomes
reduced in size to $(1 - 2x)/(1 - x) - k$.

A combination of (4.5) and the derivative of (4.4) allows us to
write

$$\dot{x} \simeq \frac{- 3\eta s \, x}{r_1 \, (1 - 5.0\eta \, g(x) \,)} \quad , \tag{4.9}$$

where the approximation depends on the form of the Roche lobe approximation (4.4). Hence (4.8) reads on substitution for x

$$\dot{P} = \frac{-9\eta s \, g(x) \, P}{r_1(1 - 5\eta \, g(x))} \simeq \frac{-9\eta s \, g(x)P}{r_1} \quad , \tag{4.10}$$

in view of the expected smallness of $\eta \, (\sim 10^{-2})$.

A suitable combination of the first four equations enables (4.10)
to be rewritten as

$$\dot{P} = - \{ \frac{9}{4} \frac{P_o}{A_o} \frac{g(x)}{(1 - x)} \} \frac{\eta s}{x \, f(x) \, (1 - 5\eta g(x) \,)} \quad , \tag{4.11}$$

where the quantity in curled parentheses tends to a finite constant
as $x \to 0$. For small x, $f(x) \to 0.462x^{1/3}$ (Kopal, 1959) so that

$$\dot{P} \to const. \frac{\eta s}{x^{4/3}} \; . \tag{4.12}$$

In order to ascertain some suitable values for $\eta$ and s some
detailed calculations have been examined.

Considering first $\eta$ ; though a number of calculations of mass
losing stars in approximately similar conditions (low mass, Case B)
have been calculated, a few authors only have published sufficient
numerical details of the subgiant structure to allow appropriate
values of $\eta$ to be directly assessed. Among such calculations, those
of Harmanec (1970) are particularly useful for the present purpose,
and plots of the subgiant structure in the log $m_1$ (r), log r plane,
based on Harmanec's Figure 9 data , are reproduced in Figure 4.

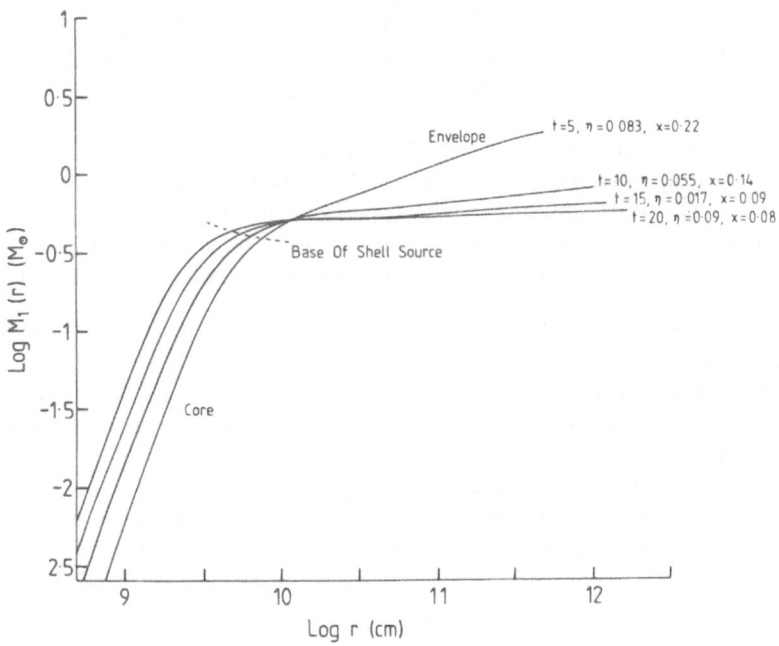

Figure 4.   Structure of Semi-Detached Subgiants.

A relatively simple subdivision into core and envelope regions
can be seen in such plots.  The apparent constancy of $\eta$  through
the envelope region can be justified on the basis of
approximate formulae describing the structure of classical Algol
shell burning subgiants given by Refsdal and Weigert (1970).
Combining together various of the structure equations given by those
authors it is possible to reconcile $\eta$ = constant with a polytrope
n = 3 like behaviour of a constant opacity, tenuous envelope.  If
$\eta$ = constant through the envelope be a reasonable description of the
mean behaviour, it is relatively easy to derive a formula such as

$$\eta = \frac{1}{3} \frac{m_{env}}{m_c} \frac{1}{\ln \frac{R}{R_c}} \quad , \qquad (4.13)$$

so that as a result of envelope depletion and expansion $\eta$ will decrease during the course of the slow phase of mass transfer. Consider, for example, a binary like that of System III of Refsdal and
Weigert (1969), which between points i and k of its calculated
evolution appears to have some resemblance to S Cnc. We have, for
the mid point of this range, (in solar units) $m_c = 0.239$, $m_{env} =$
0.094, R = 9.59, $R_c = 0.028$ (from the $m_c$:$R_c$ relation of Refsdal and
Weigert, 1970) so that $\eta = 0.022$, which compares well with values
obtained from the Harmanec data at a corresponding stage.

In the early stages of the slow phase $\eta \sim 0.1$ and its initial
decrease would appear to be not so rapid, but in the later stages
before eventual envelope collapse the decline to $\eta \sim 0.01$ is more
steep. Thus the last three values obtained from the Harmanec data
show $\eta$ declining with x according to approximately $\eta \sim 3.5 \ 10^3 \ x^5$.

The physical justification for our deductions about the behaviour
of $\eta$ is related to the fact, stressed by a number of authors, of the
controlling influence of the core. The basic envelope expansion s
should also be determined by the behaviour of the core, however it
becomes markedly more rapid just after convection in the outer
envelope becomes established . The role of core mass increase can be
accounted for by differentiating the expression for the mass loss free
overall radius given by Refsdal and Weigert (1970), which we could
write as

$$s = \frac{\beta}{3} \left( 1 - \frac{1}{\alpha} \frac{m_c}{m_{env}} \right) m_c^3 \frac{dm_c}{dt} \qquad (4.14)$$

where $\alpha$ and $\beta$ are numerical parameters; (typically $\alpha \sim 12$ and
$\beta \sim 10^6$ for s in $R_{oy}^{-1}$, however, this formula does not clearly show
the significance of the outer convection zone. Thus in the
evolution sequence of a 2.25 $M_o$ star, as published by Iben (1967), the
surface expansion proceeds at a mean rate of about $3.6 \times 10^{-7} \ R_{oy}^{-1}$
during the phase of shell burning while the envelope is still radiative, but increases to $1.4 \times 10^{-6} \ R_{oy}^{-1}$ during the convective outer
envelope phase. These rates, which show a strong dependence on mass
($\sim m^6$ for first and $\sim m^4$ for second) are somewhat greater than those
expected for S Cnc.

The fact that $\eta$ is a small number in the final slow stages of
mass transfer, however, does allow us to combine the foregoing
equations in an alternative way to the way Equation (4.9) was
derived, to show the "feedback" effect of s on $\dot{r}_1$, to which it
becomes approximately proportional, thus $s \sim \dot{r}_1 / 5\eta g$. The behaviour
of $\dot{r}_1$ from calculated model sequences is relatively easy to study.

While there are some differences of detail the overall pattern appears to be as that shown in Figure 5. After an initial moderate expansion there is a rapid rise during the early establishing of a relatively thick convective envelope. Depending somewhat on initial orbital parameters and mass ratio, published calculations lead us to expect peak Roche lobe expansion rates during this phase to be around $4 \times 10^{-7} \; m_1^4$ init $R_o y^{-1}$.

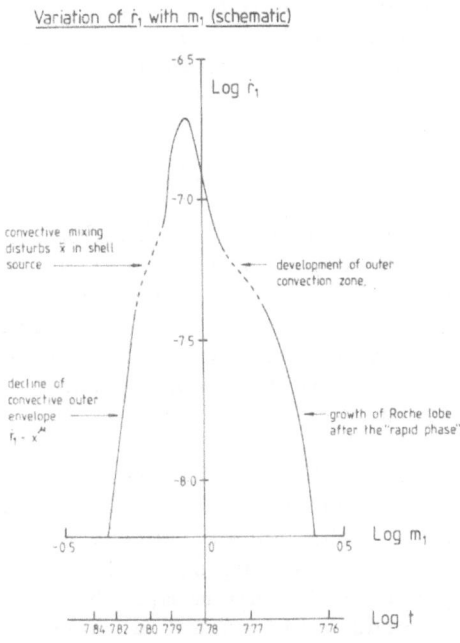

Figure 5. The timescales would be about appropriate for a star of $m_{1 \; init} = 3M_o$

What is of significance to the present discussion is, however, that this maximum rate of expansion during the slow phase is reached relatively early on, whereafter there is a relatively long period during which the expansion rate declines as the convective region of the outer envelope narrows down.

The behaviour of $\dot{r}_1$ with $m_1$ appears to be that of a steep power, i.e., $\dot{r}_1 \sim m_1^{\mu}$ , where $\mu \sim 8$. This phase of slow decline can last several times as long as the phase leading up to the convective maximum. Hence, to return to formula (4.12), what we should expect in the drawn out phase before eventual collapse, is a rate of period variation which declines with a high power of x, i.e., $\dot{P} \sim x^{\mu - 4/3}$. Such a high power dependency means that if we simply take a mean rate

of period variation from a pair of well separated interval points on
a calculated sequence, the derived value is likely to be representative only towards the initial point of the range, while for most of
the interval the actual rate of period variation could be appreciably
less.  Working against such a weighting among actual observational
data would be, of course, the selection effect which favours close
pairs for period determination and study.

Consider again the range i to k of the system already referred
to; the maximum value of s, according to our foregoing estimates,
occurring somewhat before point i, would be $\sim 1.5 \times 10^{-6}$ $R_o^{-1}$.
Substituting appropriate values into the transfer Equation (4.5) we
find,

$$\Delta m_1 \simeq \frac{0.022}{9.595} \left( \frac{1.5 \times 10^{-6}}{\nu} - 7.10^{-8} \right) 7.1 \times 10^7 \, M_o$$

where $\nu$ is the index expressing the power law variation of s with $m_1$
during the slow decline.  A value of $\Delta m_1$ agreeing with that calculated
by Refsdal and Weigert can be found for $\nu \simeq 2$, which gives a tolerable
support to the underlying point in view of other approximations
and uncertainties.  Similar calculations have been performed for other
comparable published examples of evolution towards the end of the slow
phase which show similar effects.

Actually, the period values in the i – k interval of the System
III just referred to are greater than that currently found for S Cnc,
while the model is "conservative" throughout, which, even if possible
in the drawn out phase under consideration, seems less likely in
the range close to minimum period and maximum $m_1$, where extrapolation
back from the present values leads us to expect a stage of contact,
with the possibility of appreciable angular momentum loss from the
system.  Angular momentum loss at this stage of strong interaction
would scale down the periods at subsequent stages, which could otherwise then follow a parallel course to conservative model calculations.

Coming to a closer comparison with actual data on S Cnc, we
find that even the straight mean period variation during the range
i – k is $7.10^{-9}$ when scaled to the S Cnc period, while Kreiner and
Ziołkowski (1978), estimate the accuracy with which such a term can be
evaluated from the available data to be $\sim 2.10^{-9}$.  From what was said
already, such a straight mean would only be representative for a region
of order $3/(3\mu-4)$ of the relevant time interval, where $\eta$ expresses
the full power law dependence of P on x in (4.12); and since $\mu$ is
likely to be quite larger than three, the absence of a detectable
present rate of period variation might not be so worrying, if this
model could indeed be quite appropriate for S Cnc.  The initial mass
for this system is however only $1.4 M_o$, while we know that the original
primary of S Cnc must have been more massive than half the present
total mass (i.e., the adopted $m_1$ init $> 1.6 M_o$).  In view of the high
power dependence of s on $m_1$ init this could mean that the foregoing

period variation should be increased to at least $1.2 \times 10^{-8}$. If we want to still adhere to the conservative model, we are forced to suppose that the subgiant is now relatively close to the end of its period in the semi-detached configuration, but a closer examination of the likely core mass (to be carried out in what follows) will lead us to expect this is not quite so, and that the star should be at an intermediate position along the i - k track.

The likeliest way out of this dilemma appears to be to appeal to the previously mentioned possibility of failure of efficient angular momentum transfer. The scaling constant in Equation (4.12) can thereby be reduced by the factor (1 - k), so that the implied reduction in observable period changes can be achieved without great difficulty. In fact, if we consider Lubow and Shu's (1975) more detailed treatment of the hydrodynamics of the RLOF mechanism in an initially wholly conservative regime it would appear that after the mass fraction x drops below about 0.17 it begins to be possible for stable "disk" like orbits to be performed by material accreting towards the detached component, without impinging directly on its surface. (From Lubow and Shu we observe the disk radius $\pi_d$ satisfies $\pi_d \simeq 0.075x^{-0.43}$, while the detached star radius satisfies $r_2/A \simeq 8x^2(1 - x)^2$ if it does not grow too much during the slow phase and there is no extensive common envelope stage). If thereafter the transferred angular momentum is stored in a disk of radius $\sim\pi_d$ rather than put back into the orbit we find

$$\frac{d \log J}{d \log x} = \frac{0.27y^{1/2}( (0.22-g(x))z+x)}{0.27y^{1/2}z + (1-x)x^{1.22}} \qquad (4.15)$$

where $y = m_2/M$ and $z = m_{disk}/M$. For values of $x \sim 0.1$, $d \log J/d \log x$ is slowly varying around 0.5, so that the scaling factor in (4.12) is halved. The absence of detectable period variation, even with a $1.6M_\odot$ initial primary mass, could then be understood. This does lead us to suppose the existence of some kind of accretion disk around the present primary in S Cnc, however; a point which could be tested by further observational study.

A further clarification of the evolutionary status of S Cnc can be made on the basis of Refsdal et al.'s (1974) detailed considerations of AS Eri. Formulae given by those authors allow values of quantities such as core mass, final core mass and maximum radius to be estimated on the basis of an assumed hydrogen profile parameter X for the subgiant. Details of the two systems evaluated in this way are compared in Table 4.

Table 4

Comparison of Evolved States of S Cnc and AS Eri and Some Models

| | S Cnc | | AS Eri (Aver.Seq. I&II, RRW) |
|---|---|---|---|
| $m_c$ | 0.22 | | 0.18 |
| $m_{env}$ | 0.05 | | 0.03 |
| R | 5.41 | | 2.25 |
| $R_c$ | 0.029 | | 0.030 |
| $\eta$ | 0.014 | | 0.013 |
| $m_{max}$ | 0.26 | (RW, T5) | 0.19 |
| $R_{max}$ | 10.4 | (RRW, E3) | 2.54 |
| $m_{lf}$ | 0.245 | (RRW, E7) | 0.203 |
| $R_{lf}$ | 6.17 | (RRW, E6) | 2.31 |
| $\bar{x}$ (assumed) | 0.5 | | 0.46, 0.55 |

| | System I of GG<br>Middle of range i - k | System III of R Wa<br>Middle of range i - k |
|---|---|---|
| $m_c$ | 0.257 | 0.239 |
| $m_{env}$ | 0.066 | 0.094 |
| R | 12.95 | 9.59 |
| $R_c$ | 0.028 | 0.028 |
| $\eta$ | 0.014 | 0.022 |
| $m_{l(init)}$ | 2 | 1.4 |
| $s_{max}$ | $\sim 6.4.\ 10^{-6}$ | $\sim 1.5.10^{-6}$ |
| $\nu$ | $\sim 2.5$ | $\sim 2$ |
| $\Delta P/P$ | $2.8 \times 10^{-8}$ | $7 \times 10^{-9}$ |

Abbreviations for references

RW  = Refsdal and Weigert (1970)
RWa = Refsdal and Weigert (1969)
RRW = Refsdal, Roth and Weigert (1974)
GG  = Giannone and Giannuzzi (1970)
T   ≡ Table Number;  E ≡ Equation Number

CONCLUSIONS

S Cnc as a well known and relatively bright example of a classical Algol system can provide a good tracer of stellar evolution processes within the circumstance of binarity. The main findings of the present attempt to bring out relevant facts are as follows:

1. A physical connection with NGC 2632 appears very unlikely on purely observational grounds. Moreover, if S Cnc was presently of the same age as the cluster ($\sim 4.10^8$ years) the initial mass of its original primary, on the Case B mass transfer theory, would probably have to be greater than $2M_o$ (c.f., Giannone and Giannuzzi, 1970, 1972), making the currently observed low rate of period variation incompatible with standard Case B theory.

2. The semi-detached configuration following low mass Case B evolution towards the end of the semi-detached phase can provide a good general explanation of the overall system properties.

3. The absence of a noticeable rate of period variation, need not be in such conflict with theory, even on the conservative model, if it is possible to reduce the total mass from the adopted value ($3.27\ M_o$), for example, down to Popper's (1980a) value ($2.8M_o$).

4. The scale of expected period variation could be halved if an "accretion disk" can store angular momentum of the matter currently being lost from the original primary. This situation appears to be more consistent with the expected core mass ($0.22M_o$) required to explain the current overluminosity of the contact component, on the basis of Refsdal and Weigert's (1970) formulae and tables.

5. More observations are required to check on any RW Tau type phenomenon in the system (disk confirmation), or to confirm or otherwise, the apparent disparity between the findings of Batten (1976) and Popper (1980) regarding complications to line profiles.

6. An interesting comparison with the system AS Eri, which was modelled in great detail by Refsdal, Roth and Weigert (1974), is possible. AS Eri probably started with initial separation or mass ratio ($m_2/m_1$) less than that of S Cnc, bringing it into deep contact during the rapid phase, when non-conservative effects would have been enhanced. Higher proportions of processed material may therefore be present in the outer regions of the stars in AS Eri compared with S Cnc.

REFERENCES

Abt, H.A.: 1970, Astrophys. J. Suppl., 19, 387.
Al Naimiy, H.M.K.: 1978, Astrophys. Space Sci., 53, 1981.
Azimov, S.M.: 1963, Izv. Pulkova, 23, 76.

Barnes, R.C.: 1974, Publ. A.S.P., <u>86</u>, 195.
Batten, A.M.: 1976, in IAU Symp. No. 83, <u>Structure and Evolution of</u>
    <u>Close Binary Systems</u> (ed. P. Eggleton, S. Mitton and J. Whelan),
    D. Reidel, Dordrecth, Holland, p. 303.
Budding, E.: 1973, Astrophys. Space Sci., <u>22</u>, 87.
Caracatsanis, V.A.: 1977, Astrophys. Space Sci., <u>72</u>, 369.
Crawford, D.L. and Barnes, J.V.: 1969, Astron. J. <u>74</u>, 818.
Crawford, R.C. and Olson, E.C.: 1980, Pub. A.S.P., <u>92</u>, 833.
Dickens, R.J., Draft, R.P. and Krezeminski, W.: 1968, A.J., <u>74</u>, 818.
Giannone, P. and Giannuzzi, M.A.: 1970, Astron. and Astrophys., <u>6</u>,
    309.
Giannone, P. and Giannuzzi, M.A.: 1972, Astron. and Astrophys., <u>19</u>,
    298.
Hagen, G.L.: 1970, An Atlas of Open Cluster Colour-Magnitude
    Diagrams, <u>4</u>, Publ. David Dunlap Obs. Univ. of Toronto.
Hall, D.S.: 1974, Acta Astron. <u>24</u>, 7.
Hall, D.S. and Neff, S.G.: 1979, Acta. Astron., <u>29</u>, 641.
Harmanec, P.: 1970, Bull. Astron. Inst. Czech., <u>21</u>, 113.
Heckmann, O.: 1925, Astron. Nach., <u>225</u>, 49.
Heckmann, O.: 1937, Astron. Nach., <u>264</u>, 25.
Hilditch, R.W. and Hill, G.: 1975, Mem. R. Astr. Soc., <u>79</u>, 101.
Huffer, C.M. and Collins, G.W.: 1962, Astrophys. J. Suppl., <u>7</u>, 351.
Iben, I., Jr.,: 1967, Ann. Rev. Astron. and Astrophys., <u>5</u>, 571.
Irwin, J.B.: 1963, Astrophys. J., <u>138</u>, 1104.
Kholopov, P.N.: 1958, Peremenniye Zvezdyi <u>11</u>, 325.
Koch, R.H.: 1970, in IAU Coll. No. 6, <u>Mass Loss and Evolution in</u>
    <u>Close Binaries</u> (ed. K. Gyldenkerne and R.M. West), Copenhagen
    University Publication, p. <u>65</u>.
Kopal, Z.: 1959, <u>Close Binary Systems</u>,Chapman and Hall Ltd., London.
Kopal, Z. and Shapley, M.B.: 1956, Jodrell Bank Ann. <u>1</u>, 141.
Kreiner, J.M. and Ziołkowski, J.: 1978, Acta. Astron., <u>28</u>, No. 4,
    497.
Kukarkin, B.V., Kholopov, P.N., Efremov, Yu. N., Kukarkina, N.P.,
    Kurochkin, N.E., Medvedeva, G.I., Perova, N.B., Pskovskii, Yu.
    P., Fedorovich, V.P. and Frolov, M.S.: 1976, <u>Third Supplement</u>
    <u>to the Third Edition of the General Catalogue of Variable Stars,</u>
    Publishing House "Nauka", Moscow.
Linnell, A.P. and Proctor, D.D.: 1971, Astron. J., <u>164</u>, 131.
Lubow, S.H. and Shu, F.H.: 1975, Astrophys. J., <u>128</u>, 190.
Packet, W.: 1980, in IAU Symp. No. 88, <u>Close Binary Stars:  Observa-</u>
    <u>tions and Interpretation</u> (ed. M.J. Plavec, D.M. Popper and
    R.K. Ulrich), D. Reidel, Dordrecht, Holland., p. 211.
Paczyński, B.: 1971, Ann. Rev. Astron. and Astrophys., <u>9</u>, 183.
Paczyński, B.: 1976, in IAU Symp. No. 73, <u>Structure and Evolution</u>
    <u>of Close Binary Systems</u>, (ed. P. Eggleton, S. Mitton and J.
    Whelan), D. Reidel, Dordrecht, Holland, p. 75.
Paczyński, B. and Dearborn, D.S.: 1980, Mon. Not. R. Astron. Soc.,
    <u>190</u>, 395.
Plavec, M.: 1973, in IAU Symp. No. 51, <u>Extended Atmospheres and</u>
    <u>Circumstellar Matter in Spectroscopic Binary Systems</u> (ed. A.H.
    Batten), D. Reidel, Dordrecht, Holland, p. 216.

Plavec, M. and Polidan, R.S.:  1976, in IAU Symp. No. 73, Structure and Evolution of Close Binary Systems, (ed. P. Eggleton, S. Mitton and J. Whelan), D. Reidel, Dordrecht, Holland, p. 289.

Popper, D.M.:  1973, Astrophys. J. 185, 265.

Popper, D.M.:  1980a, in IAU Symp. No. 88, Close Binary Stars: Observation and Interpretation, ed. M.J. Plavec, D.M. Popper and R.K. Ulrich), D. Reidel, Dordrecht, Holland, p. 203.

Popper, D.M.:  1980b, Ann. Rev. Astron. and Astrophys. 18, 115.

Refsdal, S. and Weigert, A.:  1969, Astron. and Astrophys. 1, 167.

Refsdal, S. and Weigert, A.:  1970, Astron. and Astrophys. 6, 426.

Refsdal, S. and Weigert, A.:  1971, Astron. and Astrophys. 13, 367.

Refsdal, S. Roth, M.L. and Weigert, A.:  1974, Astron. and Astrophys. 36, 113.

Smithsonian Astrophysical Observatory - Star Catalog, (prepared by the Staff, Smithsonian Astrophys. Obs.), Cambridge, Mass., Smithsonian Institution, Washington, D.C., 1966.

Tabachnik, V.M.:  1969, Soviet Astronomy - A.J., 12, 380.

Weis, E.W.:  1976, The Observatory, 96, 9.

Wood, F.B., Oliver, J.P., Florkowski, D.R. and Koch, R.M.:  1980, A Finding List for Observers of Interacting Binary Stars, Univ. of Pennsylvania Press.

Ziołkowski, J.:  1976, in IAU Symp. No. 73, Structure and Evolution of Close Binary Systems (ed. P. Eggleton, S. Mitton and J. Whelan), D. Reidel, Dordrecht, Holland, p. 289.

# STRUCTURAL MODELS FOR BETA LYRAE-TYPE DISKS

R. E. Wilson
Department of Astronomy, University of Florida

ABSTRACT

  Equilibrium structural models are computed for a thick, self-gravitating disk in a binary system. Accretion onto the star is limited by the star's rapid rotation (the system is a double-contact binary). The potential formulation is taken from a previous paper, and represents the gravitational potential as that of a massive wire. Corrections to the stellar structure differential equations for the distorted geometry are applied, and the equations are integrated and solved by the fitting point method. The energy is supplied by viscosity. Energy transfer is by convection, and is appreciably superadiabatic throughout the disk. A mass of 0.5 $M_\odot$ is assumed. Representative results are: "central" temperature, 67000 K; "central" pressure, $5 \times 10^{11}$ dynes/cm$^2$; "equal volume" radius, 17 $R_\odot$; luminosity, $5 \times 10^3$ $L_\odot$. The model "radius" is in excellent agreement with the observational value for β Lyrae. The model luminosity is slightly higher than the available rate of expenditure of gravitational energy, indicating that a lower disk mass (perhaps 0.25 $M_\odot$) should be tried.

## I. INTRODUCTION

  Over the past several decades, many questions have arisen concerning various unusual features of the β Lyrae system. Underlying all of these is the central question: Why is β Lyrae so different from other semi-detached, mass transferring binaries? We would like to know how β Lyrae is able to maintain a geometrically thick, opaque disk (viz. Huang, 1963; Wilson, 1974) about its more massive component, while a system such as U Cep cannot. Of course, one can point to the rather large rate of period change, which places β Lyrae in the rapid phase of mass transfer. However there are grounds to suspect that the explanation runs deeper, as the predicted time scale for collapse of the disk onto its central star is quite short compared to the historical interval of accurate β Lyr observations ($\approx$ 65 years). There is now fairly uniform agreement regarding the existence of a massive ($\approx$ 10 to 15 $M_\odot$)

261

*Z. Kopal and J. Rahe (eds.), Binary and Multiple Stars as Tracers of Stellar Evolution, 261–273.*
*Copyright © 1982 by D. Reidel Publishing Company.*

main sequence star at the center of the β Lyr disk. Had the disk ceased to exist or had it even become substantially thinner at any time during the star's observational history, the system would have brightened considerably and become optically prominent in the constellation Lyra. Since there is no record of such an occurrence, one can conclude that the disk is quite stable in its large-scale aspects. In this paper, the suggestion (Wilson, 1979, 1981) is adopted that centrifugally limited rotation of the central star is responsible for the disk's persistence, so that β Lyr is a double-contact binary (Wilson, 1979). Note that Packet (1981) has shown that relatively little mass transfer may be needed to spin an accreting star up to its centrifugal limit.

In the past two years, Plavec (1980) has called attention to a class of binaries which have very extensive circumstellar disks and are similar in many ways to β Lyr. These he calls the W Serpentis stars. Members of the class include SX Cas, RX Cas, AR Pav, W Ser, V367 Cyg, W Cru, V356 Sgr, and β Lyr, all of which show spectacular spectroscopic phenomena, including evidence for high velocity ejection of matter from the entire system. Our fundamental question may now be widened, and we ask: Why are the W Serpentis stars so different from other semi-detached, mass transferring binaries? A plausible reason is that they all are in the double-contact phase.

It seems appropriate to place β Lyr, the most thoroughly observed W Ser star, at the focus of our attention, so a brief review of its overall features follows. The optically dominant member of β Lyr is a mass-losing giant star, apparently rotating synchronously with the orbit, and filling its Roche lobe. The spectral type is approximately B9 II and well-defined radial velocity curves exist (Sahade, et al. 1959), which give a mass function of 8.5 $M_\odot$. Primary eclipse in optical passbands is the eclipse of this star. The other component has long been regarded as enigmatic. In the optical region it shows no line spectrum (probably because of rotational line broadening), but we know that it does emit significant light - comparable in amount to that of the B9 star - because there is a well defined secondary eclipse. Also its luminosity has been estimated by fairly detailed modeling of the light variation (Wilson, 1974; Wilson and Lapasset, 1981). The mass of this object has been difficult to determine, although there now seems to be unanimous agreement that it is more massive than the B9 star by a factor of at least 2.5, and probably 6 or more (see Wilson, 1974 for a summary of mass ratio determinations). Here we assume a mass for this component of about 12 $M_\odot$.

Our working model for the dim but massive component will be that of a main sequence star surrounded by a thick disk which entirely obscures it from direct view. Perhaps "ring" might be a better word than "disk", since the structure is certainly thick, while the term "disk" usually describes a highly flattened object. However, we retain the name "disk" here because real progress in understanding began with Huang's (1963) paper, which made it clear for the first time that we are dealing with a relatively flattened circumstellar "disk". While Huang's disk was actually much too thin for its predictions to be in quantitative agreement with the light curves, the "Huang disk model" is well known and should continue to be remembered for its start in the right direction.

Our lack of physical models for β Lyr-type disks is a reflection of the fact that we do not understand the accretion process in the rapid phase of mass transfer in sufficient detail. This is a very unsatisfactory state of affairs, for it is now well accepted that mass and angular momentum are not conserved in mass exchange episodes. If these quantities are not conserved, computations of binary star evolution must eventually follow mass transfer events in detail. We must understand what is happening within β Lyrae type disks because spectroscopic observations show that mass is ejected from the disks and from the systems, and all later evolutionary stages are thereby affected.

This paper is intended as a start toward full modelling of β Lyr type disks. Certainly one should be able to answer simple questions concerning the order of magnitude of physical variables within the disks. What are reasonable masses, internal pressures, and temperatures? What is the relative importance of one source of luminosity compared to another? Is energy transferred primarily by radiation or by convection? Can we say anything about the rotation law? What are the approximate distributions of mass and other physical variables? Toward these ends we adopt the following strategy. The disk is to be approximated by an equilibrium structure. If we can establish the form of the potential field, and define a reasonable mode of luminosity generation, it should be possible to compute static structural models in the same manner as for single stars, provided that we apply corrections for the non-spherical geometry. A logical basis for the potential has been given in an earlier paper (Wilson, 1981, hereafter Paper 1). Essentially, the idea is to represent each binary component as a mass point, with the disk mass concentrated into a circular wire, centered on one of the stars. Rotation velocity is to be constant on cylinders so as to give a conservative potential. Numerical experiments have shown that the wire model gives a satisfactory approximation to the disk-mass potential. Note the analogy of the wire model with the Roche point mass approximation for stars. The centrifugal potential is computed from a rotation law in which angular velocity varies as $u^n$, with u the distance from an axis of rotation and n a parameter which turns out to be constrained in several ways by various features of the problem (Paper 1). Actually, two regimes with different n are required if the solutions are to be fully consistent. Figure 1 shows a computed disk equilibrium figure (cf. Paper 1).

While in a formal sense the present computation of disk models assumes static equilibrium, this is only a simplification, needed to keep the first efforts computationally managable. In reality the situation would be one of dynamical equilibrium, with matter continually accreting from disk to star, but with that flow balanced by a return injection from star to disk. Probably the latter occurs at zero latitude, while the former occurs slightly above and below zero latitude.

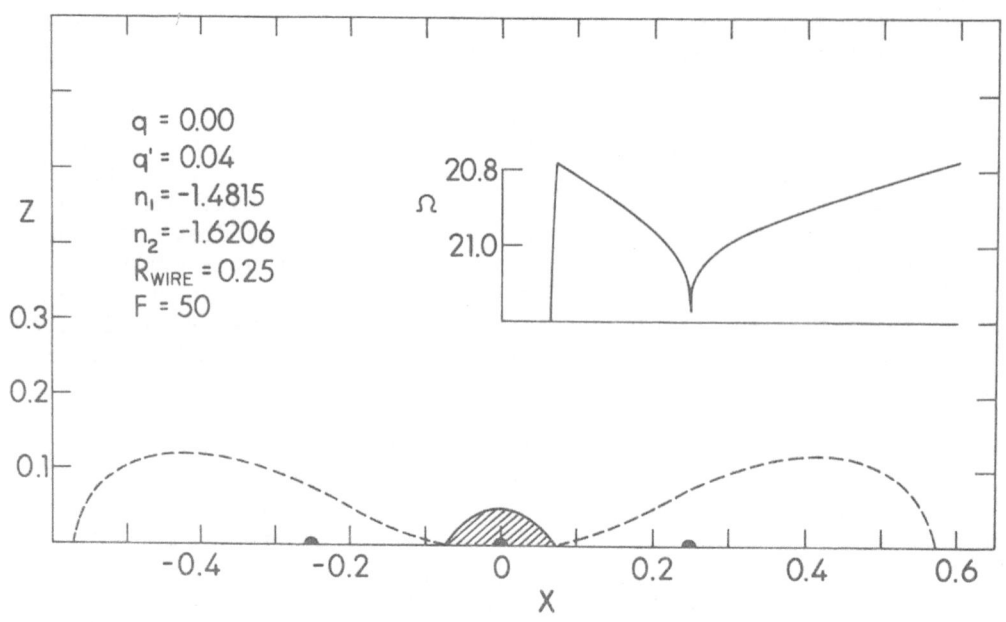

Figure 1.  A cross-section of the surface equipotential for the
case used in this paper.  The hatched area is the critically rotating central star and the three half-dots along the x-axis
depict the locations of mass concentrations used in computing
the potential.  The inset shows the potential as a function of x.

## II.  INPUT TO THE PROBLEM - GROSS CHARACTERISTICS OF THE DISK

In order to be sure that our disk dimensions are reasonable, we
adopt those of the β Lyrae disk, as estimated in earlier papers (Wilson,
1974; Wilson and Lapasset, 1981).  According to the disk potential
model (Paper 1), the full potential is determined when the star masses,
disk mass, disk radius, and disk thickness have been specified.  The
numbers for our β Lyr calculation are $R_{disk}$ = 0.57a, $Z_{max}$ (half-
thickness) = 0.12a, $M_{disk}$ = 0.04 $M_{star}$ and $M_{disk}$ = 0.5 $M_{\odot}$.  We take the
mass of the other star as zero so as to make these first computations
axisymmetric.  It is shown in Paper 1 that the disk mass should be non-
neglible, although it is not clear that 0.5 $M_{\odot}$ is a particularly good
estimate.  Hopefully, it is of the correct order of magnitude (cf.
arguments in Paper 1).  The relative chemical abundances by mass are
taken to be X = 0.70, Y = 0.27, and Z = 0.03.  More detailed abundances
needed when solving for the ionization equilibrium were taken from
Allen, 1973.  The same source was used for ionization potentials and
partition functions.  We cannot expect the rotation law to be Keplerian
because thermal gas pressure support is important in the "z" dimension
(the disk is quite thick), so it certainly would be important also in
the radial dimension.  Only part of the radial support would therefore

come from rotation.

Is the disk radiative or convective?  Since it has no nuclear
luminosity, one might think that radiation could easily carry the energy
flux.  However on two grounds we know that it must have a luminosity
which is enormous for a 0.5 $M_\odot$ (or so) object.  First, we have the
directly observable luminosity, which is of the same order as that of
the B9 star (Wilson, 1974).  Second, the considerable thickness of the
disk requires a large luminosity to provide adequate thermal pressure
support.  If we substitute approximate values for the physical variables
into the Schwarzschild criterion for convective instability, we find
that the disk will be convective by a large margin, and (drawing on the
results of the computed models) full structural computations show that
it will be convective throughout.  Energy transfer by convection is an
advantage if the disk is to be an equilibrium structure because (non-
nuclear) energy will not be generated at constant rates on equipotentials.
Therefore convection, as a mechanism for lateral redistribution of
energy, will improve the correspondence between the real disk and the
model.  On the other hand, there are difficulties in understanding how
the necessary differential rotation can be maintained in the presence of
convection.  However there is no choice - order of magnitude calculations
show that the disk will certainly be convective.  Therefore we compute
our first models in the spirit of making a beginning.  Hopefully something will be learned from them, even if it is only a rough indication
of where to go from here.

Although we expect the energy transport to be convective, we cannot
expect adiabatic convection because temperatures and densities are too
low.  The outer convection zone of a normal star such as the sun is
superadiabatic only in its very outermost part - perhaps half of one
percent of the stellar radius.  However our disk will have internal temperatures of the order of a hundred times smaller and densities of the
order of $10^3$ to $10^4$ times smaller than those of a star (viz. estimates
in Paper 1), so that we expect it to be superadiabatic throughout, or
almost throughout.  Thus the temperature gradient should be computed by
means of a full theory of convective transport.  The computations here,
including solutions for the ionization equilibrium, are based on the
method outlined by Baker and Kippenhahn, 1962 (BT), which is a variation
on that by Bohm-Vitense, 1958.  The convection routine was checked
(satisfactorily) against the "Tables of Convective Stellar Envelope
Models" by Baker and Temesvary, 1966.

III.  COMPUTATIONAL METHOD

Viewing our problem from the standpoint of stellar structure theory,
we can write the potential (Paper 1, eqns. 1, 3, 12) in a form which
satisfies all the exterior conditions, such as rotational continuity
with the central star, which seem logically to apply for β Lyr and the
other W Ser stars.  The potential so formulated shows sufficient topo-
logical similarity with the standard stellar structure problem that a

one-dimensional formulation becomes natural. Let us make a meridional
slice through the disk. There will be a point of maximum density,
which should lie on the symmetry axis of the section. This point is the
analog of the center of a normal star. We are to integrate the differential equations of stellar structure outward from this point. A similar inward integration from the surface of the disk is to meet the outward integration smoothly, as in the usual fitting point method
(Haselgrove and Hoyle, 1956).

Corrections for the distorted geometry can be made very conveniently
by the method given by Kippenhahn and Thomas, 1970 (herafter KT). While
the KT method was intended for distorted stars, there is no reason not
to use it for our disk model. The radial coordinate for the integration
is the radius of a sphere having the same volume as the toroidal equipotential. We label this coordinate $r_s$ (for $r_{sphere}$). We need to compute the volume, surface area, mean acceleration due to gravity, and
mean spacing of the level surfaces for equipotentials as they are encountered while integrating the differential equations. Combinations
of these functions (viz. KT) are lumped into two correction factors -
one, $f_t$, for the $dT/dM_r$ equation and one, $f_p$, for the $dP/dM_r$ equation.
The continuity equations, $dr/dM_r$ and $dL_r/dM_r$ require no correction factors. The volume, surface area, mean gravity, and mean inverse gravity
(proportional to level spacing) were computed, respectively, by

$$V = 4\Pi \int_{u_A}^{u_B} zu\, du \tag{1}$$

$$S = 4\Pi \int_{u_A}^{u_B} u\sqrt{1 + \left(\frac{dz}{du}\right)^2}\, du, \tag{2}$$

$$\bar{g} = \frac{4\Pi}{S} \int_{u_A}^{u_B} gu\sqrt{1 + \left(\frac{dz}{du}\right)^2}\, du,\ and \tag{3}$$

$$\overline{g^{-1}} = \frac{4\Pi}{S} \int_{u_A}^{u_B} g^{-1} u\sqrt{1 + \left(\frac{dz}{du}\right)^2}\, du. \tag{4}$$

Here u is the distance from the rotation axis, which passes through
the center of the star and is normal to the disk-orbit plane. The
quantities $u_A$ and $u_B$ are the inner and outer limits of a given (toroidal)
level surface, and z is the coordinate, measured vertically to the disk-
orbit plane, of a given point on the level surface. Eqn. (1) [supple-
mented by eqns. 3 and 12] of Paper 1 and its u derivatives provide
$dz/du$, $g$, and $g^{-1}$. The integrals in eqns. (1, 2, 3, 4) were evaluated

by a seventh order gaussian quadrature. V, S, $\bar{g}$ and $\overline{g^{-1}}$ were then combined into correction factors, as prescribed by KT.

We must also find a means for computing the energy deposition. From the several likely contributors, let us adopt viscous dissipation for the present. In a simple treatment we may consider this proportional to the local density and to the velocity shear, dv/du. The rate of viscous energy generation per gram will be independent of the density. Averaged over a given toroidal shell between two adjacent level surfaces, it can be shown to be given by

$$\epsilon_v = \frac{2\Pi |n_2 + 1| KF}{u_1 \; n1 \; u_2}(n2 - n1) \frac{\displaystyle\int_{u_A}^{u_B} \frac{u^{n_2 + 1}}{\left(\frac{d\Omega}{dz}\right)} du}{\displaystyle\int_{u_A}^{u_B} \frac{u}{\left(\frac{d\Omega}{dz}\right)} du} , \qquad (5)$$

where $u_1$, $u_2$ are the values of u at the disk-star contact point and the disk density maximum, respectively. F is the ratio of the rotation rate to the synchronous rate at $u_1$, P is the binary orbital period, and $\Omega$ is the potential according to eqn. 1 of Paper 1. We leave K, the effective efficiency of viscous energy conversion, as a free parameter, and we shall compute models for several K values.

As in normal stellar structure integrations, it is necessary to make short analytic integrations at the "center" and surface, to avoid singularities. These may be done in the usual way (viz. Schwarzschild, 1958, pp. 114-6) except that we need values of $f_p$ and $f_t$ at the "center" and surface. At the surface these functions vary slowly with $r_s$ and may be computed in the same way as elsewhere within the disk. However, near the density maximum (our "center") they vary rapidly. Furthermore our quadrature schemes for finding $f_p$ and $f_t$ may be expected to lose accuracy near the center. Therefore we need analytic forms for $f_t$ and $f_p$ in the limit of small $r_s$. The $f_t$ factor for a toroidal equipotential of very small $r_s$, surrounding the ring-like density maximum, becomes

$$f_t (r_s \to 0) = \frac{r_s}{3\Pi u_2} . \qquad (6)$$

The factor $f_p$ is not so simple, and goes to

$$f_p (r_s \to 0) = \sqrt{\frac{3}{2\Pi u_2}} \frac{r_s^{5/2}}{g^{-1}} \qquad (7)$$

near the center. For very small $r_s$, the wire potential alone determines the force field and $\overline{g^{-1}}$ depends only on $r_s$, $u_2$, and q'. The behavior of $f_p$ (eqn. 7) depends on the ratio $r_s^{5/2}/\overline{g^{-1}}$, and for small $r_s$

one can show that $\overline{g^{-1}}$ is proportional to $r_s$. Thus $f_p$ is proportional to $r_s^{3/2}$ and approaches zero for small $r_s$. Unfortunately, without resorting to series approximations, $\overline{g^{-1}}$ cannot be represented by a simple expression which will improve the usefullness of eqn. 7. In future work it may be worthwhile to develop such series, but for now we adopt $f_p = 0$ for the short analytic integration from the center. As the analytic integration covers only a small range of $r_s$ (from 0 to 0.001a) and $f_p$ is indeed small, only a minute error is involved.

We also need an expression for the energy generation rate for the "center" analytic integration. Eqn. 5 simplifies at the center and becomes

$$\varepsilon_{v_c} = \frac{2\Pi|(n_2 + 1)| KF}{P} \left(\frac{u_2}{u_1}\right)^{n1}. \qquad (8)$$

The independent variable for the integrations through the disk will be $M_r$ (mass interior to a given $r_s$). However $M_r$ is not a good choice for the independent variable in the outer parts of the disk because the derivatives $d(\log P)/d(\log M_r)$, $d(\log T)/d(\log M_r)$, etc. become extremely large and non-linear. Therefore, a switchover is made to $r_s$ as the independent variable in the difficult region, beginning where the surface analytic integration terminates and ending where the density is found to be reasonably high. For the present computations the surface analytic integration extends from $r_s/R = 1.00$ to $0.996$, and the Eulerian numerical integration from there to $r_s/R = 0.550$. Here R is the surface value of $r_s$ (not the value of u at the density maximum as in Paper 1). The fitting point was placed at $M_r/M = 0.400$.

Unless some special steps are taken, a program of the type we need will use an enormous amount of computing time, even on a rather fast machine. There are two reasons for this. First and more important, the geometrical correction factors must be computed many times for each integration step. In fact the situation is far worse than one might first think, because the independent variable is not the potential $(\Omega)$, but rather $r_s$, so an additional nested step is required to find $\Omega(r_s)$ by (for example Newton-Raphson) iteration. Naturally, even after $\Omega$ is known, many inversions of the potential equation are needed to evaluate eqns. 1, 2, 3, 4, and 5. The second problem concerns the need for doing non-adiabatic convection calculations at every integration step.

A great reduction in computer time is potentially available if the first problem can be circumvented. Note that the entire disk geometry is permanently specified as soon as the potential has been formulated. This point suggests that we establish the functions $f_t(r_s)$, $f_p(r_s)$ and $\varepsilon_v(r_s)$ at the beginning and represent them by simple approximation polynomials for later use. Adequate accuracy was obtained here with cubic polynomials, as the three special functions are all smooth and generally well behaved. Actually, all three special functions are nearly linear over long ranges, so a combination of a straight line and a cubic was used for each. The improvement in running speed was more than a factor of ten.  Unfortunately the program is still slow because of the convection calculations and the relatively slow convergence of the fitting point scheme for disks, compared to that for normal star models.


IV.  RESULTS AND DISCUSSION

Table 1 lists the main results of the first experiments, with masses and dimensions as given in section II.  Figure 2 shows the results in graphical form.  The precision (i.e. repeatability, given the assumptions of the procedure) of the numbers is set by tolerances within the fitting point iteration scheme.  For all four physical variables (radius, luminosity, "central" temperature and "central" pressure) the error tolerances were 0.3 percent.  Four values of K, the coefficient of viscous energy generation, were tried.  Convergence was much faster for the middle two K values than for the high and low value, which could not quite satisfy the 0.3 percent error tolerances.  In fact, in all cases convergence was much slower than for a stellar model, and was only reliable within regions of parameter space fairly close (within perhaps 20 percent) of the final answers.  Therefore to a certain extent it is necessary to "discover" solutions, although once a solution has been found, it is easy to predict neighboring solutions for other values of K, disk mass, etc.  The radius and luminosity are in excellent agreement with the observational values for β Lyr (Wilson, 1974, 1981).  The central temperature and pressure agree well with the order of magnitude predictions in Paper 1.  It is important to have found solutions over a range of K values.  If only eigensolutions were possible, the entire idea would be rendered implausible, as the real disk could not be expected to know the correct physical value of K.


Table 1

Disk Structural Models

| K(erg/g) | $\overline{R}_s/R_\odot$ | $L/10^3\ L_\odot$ | Tc(K) | Pc $\left(10^{11}\ \dfrac{dynes}{cm^2}\right)$ |
|---|---|---|---|---|
| $5.0 \times 10^8$ | 15.7 | 3.7 | 69,800 | 7.2 |
| $7.0 \times 10^8$ | 17.0 | 5.1 | 67,100 | 4.9 |
| $8.0 \times 10^8$ | 17.5 | 5.9 | 66,500 | 4.3 |
| $11.0 \times 10^8$ | 18.8 | 8.1 | 63,000 | 1.6 |

The run of physical variables through the disk is shown in Figures 3 and 4.  Application of the Schwarzschild criterion shows the disk to be convective everywhere.  Notice the high central condensation, which shows that the wire approximation for computing the potential is a good one, or at least that the overall scheme is self-consistent in this respect.  Taken at face value, the coincidence between the observed and theoretical radii and luminosities is encouraging.  To put this in perspective, we are finding a radius of the order of 20 times the normal radius for a 0.5 $M_\odot$, chemically uniform object.  The corresponding luminosity factor is of the order of $10^5$.  The apparent agreement with

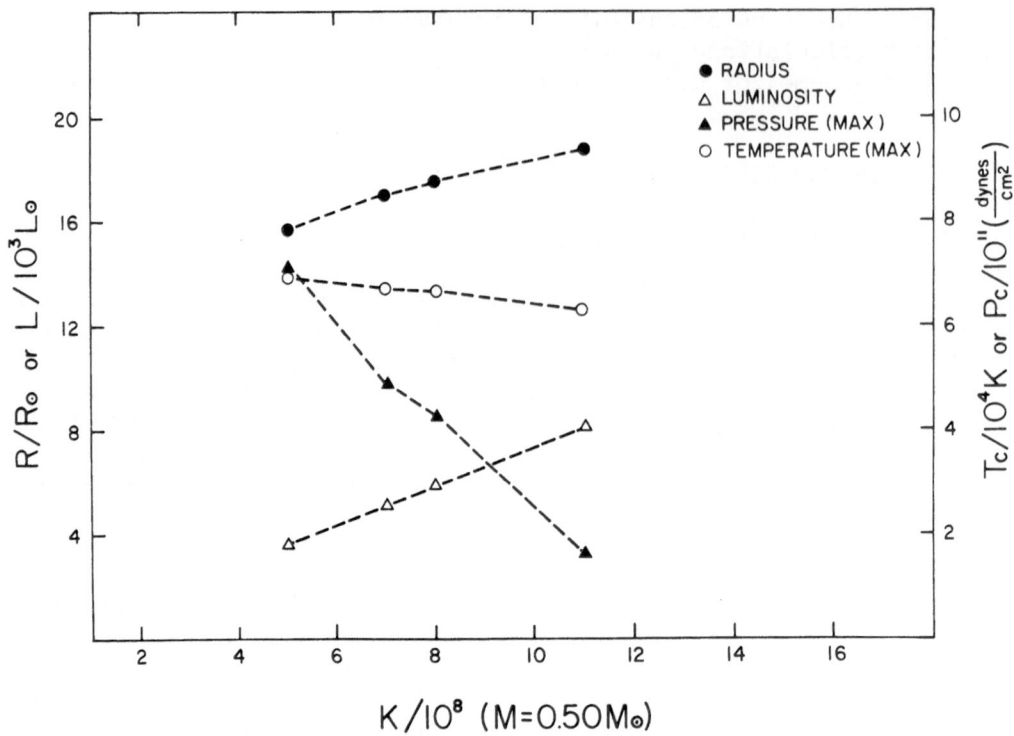

Figure 2. Variation of radius, luminosity, "central" temperature and "central" pressure with K, the factor governing the rate of viscous energy deposition. The solutions for the high and low K-values did not converge as well as for the middle values.

observation does seem remarkable. In fact, as we shall presently see, there is still another coincidence in the numbers which suggests that we are on the right track.

The released viscous energy ultimately comes at the expense of the gravitational energy of the accreting matter, which may be calculated approximately from the relation,

$$\frac{dE}{dt} = \frac{GM}{R} \frac{dm}{dt} , \qquad (9)$$

where M, R are the mass and radius of the embedded (supposedly main sequence) star, and dm/dt is the mass accretion rate. Reasonable estimates would be M = 2.4 x $10^{34}$ g, R = 2.9 x $10^{11}$ cm, and dm/dt = 3 x $10^{-5}$ $M_\odot$/yr (2 x $10^{21}$ g/sec). The last number comes from the rate of period change and can be found in Wilson, 1974. Eqn. 9 then gives dE/dt ≈ 1.1 x $10^{37}$ ergs/sec (2.8 x $10^3$ $L_\odot$) for the accretion luminosity, which differs from our middle theoretical estimates by only a factor of 2. For a slightly less massive disk we should find complete agreement.

One can imagine many discordances between theoretical and observa-

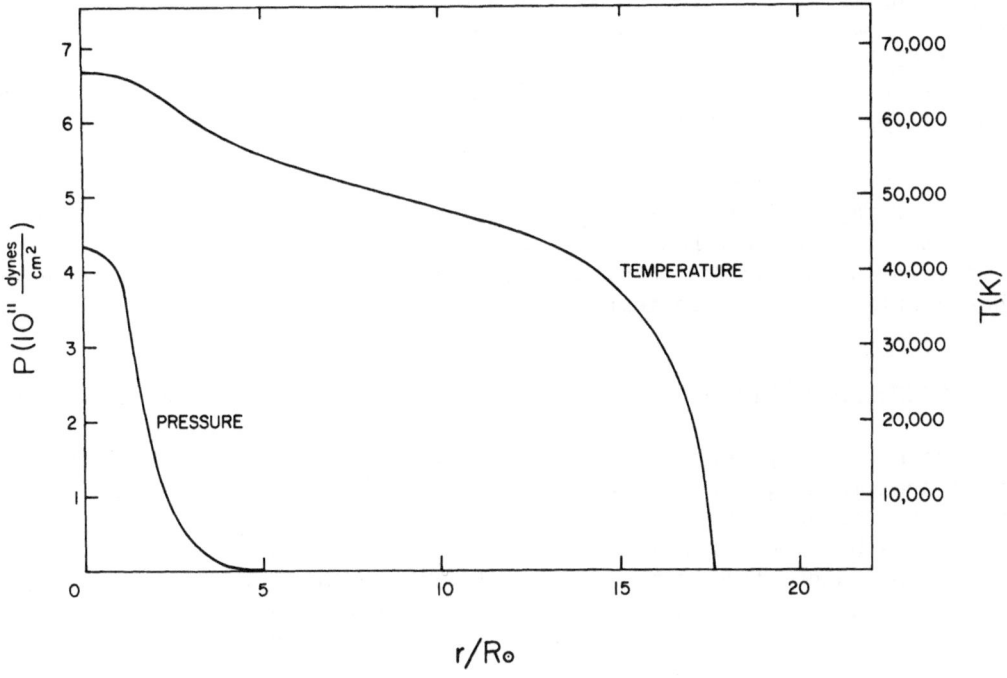

Figure 3.  The run of pressure and temperature with "equal volume" radius within the disk.

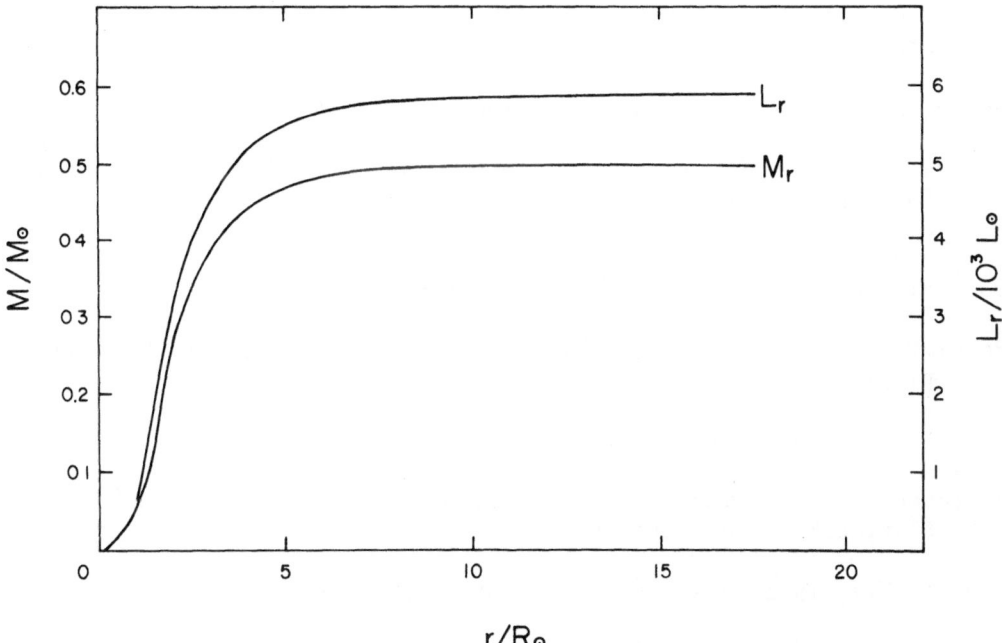

Figure 4.  The run of $M_r$ and $L_r$ with "equal volume" radius.  Note the rather high "central" condensation, which shows that the wire approximation for computing the potential is a good one.

tional numbers which might have appeared, and such difficulties were expected. Somehow they have not materialized. Despite the simple nature of the disk structural model, it seems to be - as much to my astonishment as anyone else's - virtually in complete agreement with the observational properties of β Lyr.

The original purpose of this paper was to make estimates of the orders of magnitude of the physical variables within β Lyrae type disks. It seems that the models may now have made that possible for β Lyr itself. In further work it may be possible to predict and account for the characteristics of other members of the W Serpentis class.

## V. ACKNOWLEDGEMENTS

Work on the potential of massive disks, which led the work on structural models, was done during a year's stay at the Max Planck Institute for Astrophysics (Garching, F.R.G.). That visit was made possible by a U.S. Senior Scientist Award of the Alexander von Humboldt Foundation, under the sponsorship of Professor R. Kippenhahn. Attendance at the meeting was made possible by travel grants from the American Astronomical Society and the host Institute.

## REFERENCES

Allen, C. W.: 1973, Astrophysical Quantities, 3rd ed. (London: Athlone Press).

Baker, N. H. and Kippenhahn, R.: 1962, Zt. f. Astrophys. 54, p. 114.

Baker, N. H. and Temesvary, S.: 1966, "Tables of Convective Stellar Envelope Models" (New York: NASA Goddard Institute of Space Studies) (BT).

Bohm-Vitense, E.: 1958, Zt. f. Astrophys. 46, p. 108.

Haselgrove, C. B. and Hoyle, F.: 1956, Mon. Not. R. Astr. Soc. 116, p. 515.

Huang, S.: 1963, Astrophys. J. 138, p. 342.

Kippenhahn, R. and Thomas, H.-C.: 1970, in "Stellar Rotation", ed. A. Slettebak (Dordrecht: Reidel), p. 20 (KT).

Packet, W.: 1981, Astr. and Astrophys. (preprint).

Plavec, M.: 1980, U.C.L.A. Astr. and Astrophys. Preprint No. 86.

Sahade, J., Huang, S., Struve, O., and Zebergs, V.: 1959, Trans. Amer. Phil. Soc., 49, Part 1.

Schwarzschild, M.:  1958, "Structure and Evolution of the Stars" (Princeton:  Princeton Univ. Press).

Wilson, R. E.:  1974, Astrophys. J. 189, p. 319.

Wilson, R. E.:  1979, Astrophys. J. 234, p. 1054.

Wilson, R. E. and Lapasset, E.:  1981, Astr. and Astrophys. 95, p. 328.

Wilson, R. E.:  1981, Astrophys. J. 251, (in press for issue of Dec. 1) (Paper 1).

# EVOLUTIONARY EFFECTS IN CONTACT BINARIES[+]

H.Mauder

Astronomisches Institut
Universität Tübingen

The structure and evolution of contact binaries is an extremely difficult problem in theoretical astrophysics.Several attempts have
been made to set up a satisfactory model, see e.g. Lucy,1976, Flannery,1976, Shu, Lubow and Anderson, 1976, Robertson and Eggleton,
1977, Rahunen and Vilhu, 1977 and others. There are, roughly, two
lines of sight: Shu, Lubow and Anderson postulate the existence
of a contact discontinuity in one of the two components of a contact
binary (DSC Theory); in this case age zero models in thermal equilibrium are possible. The other authors find that unevolved contact
binaries cannot achieve thermal equilibrium; therefore, relaxation
oscillations on a thermal time scale  are expected (TRO Theory).
For a detailed discussion of the different effects see Lucy and
Wilson,1979.

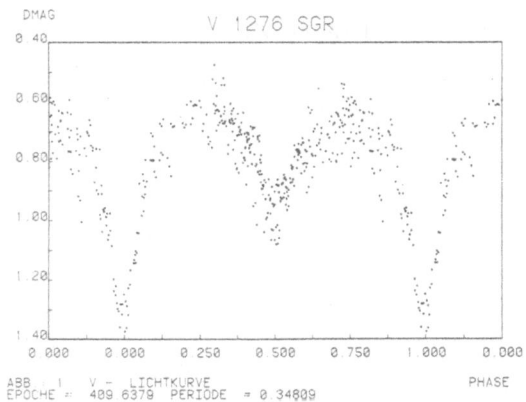

Figure 1:

V light curve of
V1276 Sgr

[+] Based on observations obtained at ESO/La Silla

*Z. Kopal and J. Rahe (eds.), Binary and Multiple Stars as Tracers of Stellar Evolution, 275–278.*

Though W UMa systems are very numerous, only a limited number has
been studied in detail. It is necessary, therefore,to look for systems which might eventually show effects that are predicted or excluded by one of the theories,at least in their present form. We
have started an observational campaign concentrating on stars with
periods below 0.5 days and minima of unequal depths.
One example is RW Dor with a period of $0^d.285$. This is an A-type
system with a deep common envelope, a mass ratio of 0.38 and spectral type K0. In secondary minimum the system is 0.16 mag brighter
than in primary minimum. Systems like RW Dor are difficult to understand in DSC theory. Another example is V 1276 Sgr with
P = $0^d.348$, spectral type K3 and mass ratio about 0.6. V 1276 Sgr
is remarkably active with large intrinsic scatter in the light
curve, see fig.1. For the moment it is not possible, therefore, to
decide, whether the two components are in contact or not. The difference in the depths of the minima is 0.45 mag in this case. No
emission lines are seen in the spectra, the masses are about 0.7
and 0.4 solar masses. Additional observations are planned for a
better definition of the light curve.
A very striking system we found is FT Lup = BV 851. The period is
P = $0^d.470$, spectral type F1 and mass ratio about 0.4. The differe-
rence in the minima is 0.6 mag, primary minimum is due to an annu-
lar eclipse. The light curve is well defined, see fig.2, and a
preliminary solution shows that the two components are just in
contact. The masses of the two components are about 2 $M_\odot$ and 0.8 $M_\odot$.

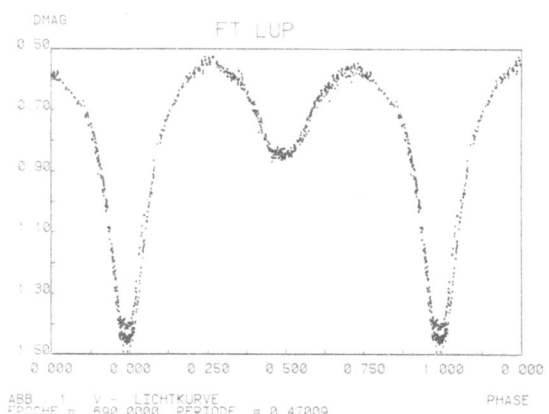

Figure 2:

V light curve
of FT Lup

No peculiarities are seen in the spectra. With radii of about
1.65 $R_\odot$ and 1.06 $R_\odot$ and log $T_{e,1}$ = 3.87, log $T_{e,2}$ = 3.71, the two
components are clearly not normal zero age main sequence objects.
The two stars show the typical properties of the components of
W UMa systems, namely too small radius and too low temperature of
the primary star, too large radius and too high temperature for

the secondary, as compared with zero age main sequence objects of
the same mass. The only difference against W UMa systems is the
high difference in the surface temperatures between the two components.
The period of FT Lup has shortened remarkably within the last 80
years. In figure 3 the O-C diagram is shown , according to the results of Bauernfeind, 1968 and Strohmeier,1967. In addition, the
O-C values of the six photoelectric minima from March and June
1981 are indicated as a cross. From the 1981 p.e. minima a period
of $0\overset{d}{.}4700730 \pm 0.0000015$ is derived. The period decrease can be
interpreted as mass transfer from the primary to the secondary at
a rate of more than $3 \cdot 10^{-7} M_{\odot}$ per year.

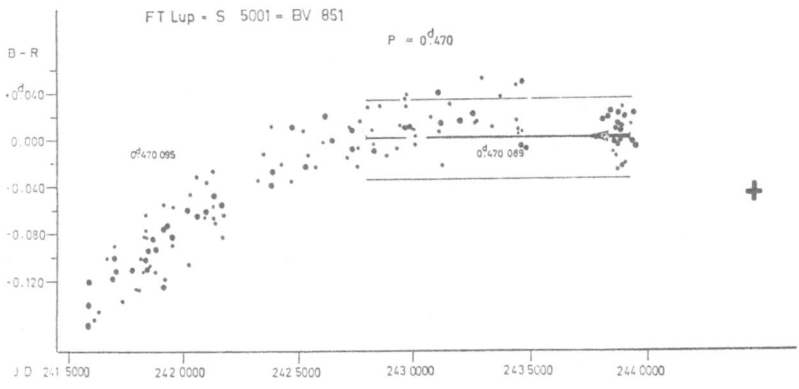

Figure 3:    O-C diagram of pg minima of FT Lup. The
             cross indicates the O-C of the 1981 p.e.
             minima

It seems to be obvious that FT Lup is in a phase where the evolution
towards the normal W UMa stage is at its very beginning. The degree of contact is not yet sufficient to build up a common envelope
with constant specific entropy, but already strong enough to produce  the deviations from the zero age main sequence. A comparison
of the position of the components  of FT Lup in the log g - log $T_e$
diagrams of Hejlesen, 1980, with the possible evolutionary tracks
of stars of equal mass,taking into account the possible alterations
of R and $T_e$, indicates a remarkably high content of heavy elements,
Z about 0.04, and an age of probably less than $5 \cdot 10^8$ years for
FT Lup. The properties of this system, therefore, seem to be of
special importance for the future development of DSC and TRO theories.

References:
H.Bauernfeind,1968,Veröff.Bamberg $\underline{8}$,Nr.81
B.P.Flannery,1976,Ap.J.$\underline{205}$,217
P.M.Hejlesen,1980,Astr.&Astrophys.Suppl.$\underline{39}$,347
L.B.Lucy,1976,Ap.J.$\underline{205}$,208
L.B.Lucy and R.E.Wilson,1979,Ap.J.$\underline{231}$,502
T.Rahunen and Q.Vilhu,1977,Astr.&Astrophys.$\underline{56}$,99
J.A.Robertson and P.P.Eggleton,1977,MNRAS $\underline{175}$,279
F.M.Shu,S.H.Lubow and L.Anderson,1976,Ap.J.$\underline{209}$,536
W.Strohmeier,1967,IBVS 184

EVOLUTIONARY POSSIBILITIES AND IMPOSSIBILITIES FOR SOLAR TYPE CONTACT
BINARIES IN NGC 188.

Frans VAN 'T VEER
Institut d'Astrophysique de Paris
98 bis, boulevard Arago - FRANCE.

## Abstract

We give a great number of arguments for the hypothesis that several e-
pochs of star formation have taken place at different times in the old
galactic cluster NGC 188.
From the last burst of star formation in this cluster, not more than a
few times $10^8$ years ago, 4 contact binaries are still now visible as W
UMa stars. With the aid of a simplified probability calculation we argue
that these 4 contact binaries are physically related and that the alignment of the orbital axes is not accidental.

## Introduction

We consider as W UMa binaries all late type (later than F5) contact binaries which show light variations due to tidal distortion and/or eclipses. Supposing a,(not generally admitted) random orientation of the orbital axes they represent about 25 to 30% of the total number of late
type contact binaries. It has been recognized for a long time that they
outnumber all others types of close and wide binaries together, and
their importance for the evolution of small mass stars was rapidly understood (Shapley, 1948; Struve, 1950; Strömgren, 1952). Also their possible relationship with planetary systems was sometimes suggested (Struve, 1950; Van 't Veer, 1975.
One can also find in the literature on this subject a certain number of
suggestions concerning their alleged progenitors and descendants. Most
people believe indeed that their appearance is a stage of limited duration in a cycle of stellar phenomena. The problem however is to identify
the other stages of the cycle and the lifetime of the W UMa stage. The
estimation of their lifetime varies considerably from one author to the
other (see for example the discussion in Van 't Veer, 1976) but we now
possess modern observations in UV and X-ray wavelengths. These are in

*Z. Kopal and J. Rahe (eds.), Binary and Multiple Stars as Tracers of Stellar Evolution, 279–287.*
*Copyright © 1982 by D. Reidel Publishing Company.*

favour of a model with high angular momentum loss (Van 't Veer, 1979,
Vilhu and Rahunen, 1980) and a resulting short lifetime not exceeding
$10^8$ years as originally advocated by Van 't Veer (1975). Nevertheless
the file concerning the question of age is certainly not closed. We
think however that the conclusion, now and then found in the literature, that contact binaries should be evolved, does not seem tenable.
The simple reason for this was formulated recently (Van t Veer, 1980) by
the simple question : if all contact binaries are evolved, what do they
look like when they are young or simply on the main sequence? We therefore believe that all studies reporting higher than main sequence radii
for contact binaries do not demonstrate that these objects are evolved,
but that on the contrary most of them are still contracting towards the
main sequence.
All these facts make the W UMa stars one of the most fascinating groups
of close double stars. We now shall discuss the puzzling presence of 4
of them in the old galactic cluster NGC 188.

## The problem

The whole question may be formulated as follows : if these 4 W UMa stars
discovered by Hoffmeister (1964) in NGC 188, are of the same age as the
cluster ($5 \times 10^9$ years deduced from its colour - magnitude diagram by
Demarque, 1979) we are confronted with the problem of how to keep them
alive after such a long time.
In the following pages we shall try to analyze what is possible  and
what is not possible in this domain. Since however we do not pretend to
detect the true possibilities of nature we shall limit ourselves to such
notions as probable, less probable, improbable and highly improbable.
These man made probabilities will be the basis of our conclusions formulated at the end of this paper.

## The definitions of age and life time

This question was recently treated by us (Van 't Veer, 1980) in response
to a note on young clusters (Ruciński, 1980) and it is perhaps good to
give some clear and succint definitions concerning the ages and lifetimes of contact binaries :

Contact age = t = time elapsed since contact, or first contact if we
                  suppose that the contact may be broken temporarily one or
                  more times.
Contact lifetime = $t_{max}$= lifetime in the contact stage of evolution.
Pre-contact lifetime = $\tau$ = total lifetime of all evolutionary stages
                  preceding contact.
Stellar age = $t + \tau$
Stellar lifetime = $t_{max} + \tau$

Every body will agree with the following statement :
The age of a contact binary cannot exceed its maximum possible stellar
lifetime.

NGC 188 -

We now return to NGC 188. It is the only old galactic cluster which
seems to possess more than one contact binary. This fact is important
and it makes the difference with clusters, like M 67, which only have
one contact binary. In such cases its presence may be more easily explained by the coîncidence of a fore - or background object. For NGC 188
this explanation is highly improbable : the presence of 4 W UMa stars
sitting at not more than $0.^{m}5$ distance from the main sequence would suppose a local space density of these objects in the neighbourhood of the
cluster of more than 20 times the mean space density.
This hypothesis would be a highly improbable solution and as far as we
know everybody believes that the 4 W UMa stars of NGC 188 are members of
the cluster.

How many contact binaries - The first conflict.

We will continue with the hypothesis of membership and try to find out
first how many contact binaries must be present in the cluster. This
problem may also be treated statistically. We observe 4 W UMa stars, the
amplitudes of their light curves lie in the interval $\Delta m = 0.^{m}45 - 0.^{m}8$,
the B-V colour varies from $0.^{m}8 - 0.^{m}9$. If we reasonably admit that they
all have a mass ratio q = 0.3 - 0.6 and a surface temperature of 5000 ±
500°K we may calculate the range of inclinations ι which is necessary to
produce the observed range of amplitude. Table 1 reproduces the result
of our calculation. We see that the probability p for a contact binary to
be a W UMa star with a light curve exhibiting an amplitude between $0.^{m}45$
and $0.^{m}80$ (group I) is p = $\Delta i/90$ = 0.17 in the case of a random orientation of the axes. The probability is p = 0.28 for an amplitude between
0.15 and 0.45 (group II) which should also be detectable in a well studied cluster like NGC 188. The probability becomes p = 0.55 when we consider amplitudes smaller than $\Delta m$ = 0.15 (group III) that means difficult
to detect.
So we immediately see the surprising observational fact that 4 W UMa
stars are present in group I and none in group II. Group III will contain a number of contact binaries which is unknown because of their un-
detectability. With these observational facts and probability results in
mind we may now derive some interesting statistical conclusions concerning the answer to the following question : What is the most probable
total number of contact binaries present in NGC 188, so that 4 of them
have orbital inclinations between 75 and 90° as observed? The answer
requires some statistical preparation. We imagine first the cluster without contact binaries. We then throw a certain number (n) of contact
binaries of the observed colour range into the cluster in a random way
and finally examine the probabiliy to find the really observed situation.
From binomial theory we can derive that the probability b to find k
contact binaries in I, l in II and m in III is :

Table 1 -

| Amplitude of life curve $\Delta M_v$ | Corresponding mean inclination i | Range of inclination $\Delta i$ | Group | Probability of being member of group I, II or III $\Delta i/90$ |
|---|---|---|---|---|
| 0.80 | 90 | | | |
| | | 15 | I | $P_I = 0.17$ |
| 0.45 | 75 | | | |
| | | 25 | II | $P_{II} = 0.28$ |
| 0.15 | 50 | | | |
| | | 50 | III | $P_{III} = 0.55$ |
| 0 | 0 | | | |

Table 2 -

| n | $q_n$ |
|---|---|
| 4 | 0.00084 (0.17) |
| 15 | 0.138 |
| 20 | 0.203 |
| 23 | 0.212 |
| 24 | 0.213 |
| 25 | 0.212 |
| 30 | 0.183 |
| 35 | 0.142 |

$$b(k,l,m) = \frac{(k+l+m)!}{k!\,l!\,m!}\ p_I{}^k\ p_{II}{}^l\ p_{III}{}^m$$

the first factor at the right-hand side being the total number of possibilities to find k, l, and m contact binaries in I, II and III respectiveley. Evidently

$$\sum_n b(k,l,m) = 1$$

when counted over all possible combinations k+l+m = n.

In the case of NGC 188 we know that k = 4 and it is now easy to calculate the possibilities for different values of n. Some of the results can be seen in table 2 which gives the probability $q_n$ that 4 W UMa stars may be found in the group I when we randomly throw n contact binaries in the cluster. It is clear that we reasonably need between 20 and 30 contact binaries. About 1/3 of them would have, at least statistically, light curves with amplitudes between $0.^m15$ and $0.^m45$ (group II) and hence would be also detectable as W UMa stars.

If we now admit the most probable solution of table 2 we are confronted with 2 unsolvable problems which we will call the first conflict :

1. There are about 20 unidentified contact binaries (thus seen as normal main sequence stars) in the range $0.^m8 < B-V < 0.^m9$ that means more than half the number of observed main sequence stars in that range (Sandage, 1961; Mc Clure and Twarog, 1977);

2. among these 20 stars there should be 6 or 7 W UMa stars with an amplitude $0.^m15 < \Delta m < 0.^m45$. These detectable contact binaries have never been seen.

These results make us believe that NGC 188 does not possess 20 or 30 contact binaries. There are not more than 4 contact binaries which, perhaps thanks to a statistical accident, have about the same orientation in space. One can also think that the alignment of the axes is the result of some physical cause. In that case, the probability to find 4 W UMa stars with only 4 contact binaries is much greater (0.17). In our conclusions we will return to this question.

How old are they? - The second conflict.

The next problem is at least as difficult to solve as the preceding one. From our definitions of ages and the results of laborious work, cited here above, on the colour-magnitude diagram of NGC 188, we may infer that we need stellar ages for our contact binaries which are not far from $5 \times 10^9$ years. In that case we evidently suppose coeval formation of all stars of the group. We further know with a high probability that the contact lifetime $t_{max} < 10^8$ years so it follows that a pre-contact lifetime of $4.9 \times 10^9$ years is necessary. Two different pre-contact stages may be envisaged :

1 - Single star progenitor (before splitting),
2 - Binary progenitor (detached or semi-detached).

Splitting of a single star may only be conceived during evolutionary stages of rapid rotation accompanied by sufficient contraction. These are only found during pre-main sequence evolution when the star is very young. The second process may have better chances. Detached and semi-detached binaries, which have been formed by the fragmentation process may evolve into contact binaries. For that it is necessary, as was pointed out by Huang (1966) and Mestel (1968) that the combined action of angular momentum loss and tidal coupling between spin and orbital momentum brings about a slow approach during the time $\tau$ estimated above. In principle one can imagine a loose interaction so that it is possible to bring together two late type stars in about $5 \times 10^9$ years. In most cases however the interaction is much stronger and will only take some $10^8$ years to achieve the approach (Van 't Veer, 1979). Needless to say that it is highly improbable that this mechanism will produce 24 or even 4 contact binaries $5 \times 10^9$ years after the birth of a cluster, and no other sorts of late type close binaries.

Summarizing we can say that it is highly improbable to explain the number or the age, let alone both, in the classical way by coeval formation of non aligned contact binaries.

Hence our conclusion is that these 4 contact binaries were formed not only much later than the main part of the cluster but also with a non-random distribution of the orbital axes. This somewhat crude statement is in agreement with the observational fact that no binaries are observed in the old globular clusters (Liller, 1979). The real or observed absence of close binaries in globular clusters may be due to a lower rate of formation (angular momentum, viscosity) but also to an increased disappearance due to:

1 - Angular momentum loss by magnetic activity,
2 - disruption by close encounters,
3 - disruption by explosion of one of the components,
4 - invisibility due to death of one of the components.

Coeval or non-coeval star formation.

We give the following formulation to the problem: is it reasonable to suppose that star formation in a cluster may take place at different times, and even until rather recently in the case of old galactic clusters? The idea is not completely new. For example for the young Pleiades we know that age determination from massive (turn off) and light stars (turn on) differ from $7 \times 10^7$ to $2.2 \times 10^8$y (Stauffer, 1980) indicating a non-coeval formation of these stars. In a spectroscopic study of the galactic cluster M67($4 \times 10^9$y.) Barry et al.(1981) concluded that several bursts of star formation should have taken place at different epochs. They used the H and K line emission as a criterion of age. There are still other examples but we shall return to NGC 188 and have a look to the abundant literature concerning this cluster.

We first see that the determination of the age raises some problems. From the diagrams with isochrones for M67 and NGC 188 published by Sandage and Eggen (1969) and Twarog (1978) we see that it is difficult to give an unambiguous age to these clusters even when we limit ourselves to stars in the neigbourhood of the turn off. The reason for this is that the giant branch intersects different isochrones. Furthermore the absolute magnitude range for the NGC 188 giants later than G8 is about $3^m$.

This corresponds to a factor of 10 in luminosity and indicates that late type giants up to $3m_\odot$ are present in the cluster. These G8 giants cannot be older than a few times $10^8$ years. For statistical reasons it is improbable that these stars may be eliminated as foreground stars. We shall not say more about this question. These and other problems were studied by Eggen and Sandage (1969) who already suggested that two epochs of star formation could have existed.

The next point is the presence of blue stragglers in the cluster. They are characterized by a position in the colour-magnitude diagram located on the blue side extrapolation of the main sequence.

Different explanations for the origin of blue stragglers in NGC 188 and many other old clusters have been put forward.

Among these explanations we find :

1 - binaries with special mass exchange conditions, (McCrea, 1964; Van den Heuvel, 1968).

2 - single stars with quasi-homogeneous evolution and extensive internal mixing, (Wheeler, 1979).

3 - or simply normal main sequence stars which are formed later (Hintzen et al.,1974).

Without discrediting the often ingeneous processes invented in order to support the first and second explanations, we believe that the third hypothesis is the more natural. It supposes that a burst of star formation has taken place at an epoch less than $10^9$ years ago. This epoch explains the presence of stars with $m = 2m_\odot$ on the main sequence.

Another effect may be considered as important in this context. There is a considerable scatter in the main sequence for stars with masses $m <$ $m_\odot$. This scatter is much greater than the one observed for younger clusters like the Hyades, Praesepe and Coma. McLure and Twarog (1977) conclude that this scatter is intrinsic and probably attribuable to abundance differences. Without denying this effect we should like to remark that this diffuse appearance is exactly what we will see in an old cluster where the stars were formed at different epochs. We must never forget that the main sequence locus varies with time as may be seen from the diagrams given by Iben (1965, 1967). Other explanations for the diffuse character of the main sequence using the presence of binaries or rotational effects should play a much less important role for both these influences decrease with increasing age.

The last intriguing problem concerning NGC 188 (and other old clusters) may be called the abundance problem. We find intrinsic vatiations of CN strength correlated with UV excesses (McLure and Twarog, 1977) for the main sequence stars and different metallicity parameters for the later typc giants (Gottlieb and Bell, 1972) which suggest the presence of normal , metal-rich and super-metal-rich stars in the cluster. The current explanations propose the mixing of processed material towards the surface or intrinsic variations in the initial cluster cloud. It is also possible that processed material from supernova explosions, planetary nebulae, stellar winds, etc...participates in the repeated formation of new stars.

A possible solution.

Our main conclusion from this incomplete survey of difficulties with the age calibration from isochrones, the magnitudes of late type giants, the presence of blue stragglers, the dispersion of the main sequence stars, the understanding of the chemical composition without forgetting the existence of 4 W UMa stars in NGC 188 is the following :
All these difficulties may be solved if it is true that different bursts of star formation have taken place at different epochs in the cluster. The periodicity of these bursts is difficult to establish, but a succession of 3 bursts with intervals of about $1 - 2 \times 10^9$ years would be sufficient to explain the different "anomalies". The last burst should have been at the origin of the 4 W UMa stars and have occurred not longer than some times $10^8$ years ago.

Random or non-random axes?

We now come to the second point raised at the beginning, which concerns the hypothesis of a random or non-random distribution of the orbital axes. If the first hypothesis is true we may expect the presence of 4 contact binaries in the cluster. If the seond hypothesis is true a total number of 20 or 30 contact binaries has to be expected, with the statistically difficult problem why no W UMa stars are observed in the amplitude range $\Delta m < 0.45$. We already discussed the problem of aligned orbital axes some years ago (Van 't Veer, 1975a). We then concluded from a study of the bibliography on the subject that young galactic clusters may show a preferred orientation of the spin of their members. Little work seems to have been done on this subject since that time. So if we still admit this conclusion we find no reasons to suppose that there is a preferred orientation of the axes in a galactic cluster as old as NGC 188. However the facts seem to contradict this view and perhaps we may reverse this question by saying that the observation of 4 W UMa stars with inclinations between 90° and 75° in NGC 188 is not a simple coïncidence but a convincing piece of evidence for the hypothesis of preferred orbital orientation of binary stars in certain groups. It may perhaps be considered as a supplementary demonstration that these 4 contact binaries were formed only recently and still remember the original vortex.

References.

Barry,D.C., Cromwell,R.H., Hege K., Schoolman,S.A : 1981, Astrophys. J. 247, 210.
Demarque,P.: 1979, I.A.U. Symp. N° 85, ed. Hessen,J.E., p. 281.
Eggen,O.J., Sandage,A. : 1969, Astrophys. J. 158, 669.
Gottlieb,D., Bell,R. : 1972, Astron. Astrophys. 19, 434.
Heuvel,E.P.J. van den : 1968, Bull. Astron. Inst. Netherlands 19, 326.
Hintzen,P., Scott,J., Whelan,J. : 1974, Astrophys. J. 194, 657.
Hoffmeister,C. : 1964, Inf. Bull. Variable Stars No. 67.
Huang,S.S. : 1966, Ann. Astrophys. 29, 331.

Iben,I. : 1965, Astrophys. J. 141, 993.
Iben,I. : 1967, Astrophys. J. 147, 624.
Liller,M. : 1979, IAU Symp. n° 85, ed. Hessen,J.E.p.357.
McLure,R.D., Twarog,B.A. : 1977, Astrophys. J. 214, 111.
McCrea,W.H. : 1964, M.N.R.A.S. 128, 147.
Mestel,L. : 1968, M.N.R.A.S. 138, 359.
Rucinski,S. : 1980, Acta Astron. 30, 373.
Sandage,A. : 1961, Astrophys. J. 135, 333.
Sandage,A., Eggen,O.J. : 1969, Astrophys. J 158, 685.
Shapley,H.: 1948, Harvard Obs. Monographs 7, 249.
Stauffer,J.R. : 1980, Astron. J. 85, 1341.
Strömgren,B. : 1952, Astron. J. 57, 65.
Struve,O. : 1950, Stellar Evolution, Princeton University Press.
Twarog, B.A. : 1978, Astrophys. J. 220, 890.
Van 't Veer,F.: 1975, Astron. Astrophys. 40, 167.
Van 't Veer,F. : 1975a, Astron. Astrophys. 44, 437.
Van 't Veer,F. : 1976, TAU Symp. N°73, eds. Eggleton et al., p.343.
Van 't Veer,F. : 1979, Astron. Astrophys. 80, 287.
Van 't Veer,F. : 1980, Acta Astron. 30, 381.
Vilhu,O., Rahunen,T. : 1980, IAU Symp. N°88, eds. PLavec,M.J. et al.
          p.491.
Wheeler,J.C. : 1979, Astrophys. J. 234, 569.

# ORIGIN AND EVOLUTION OF CONTACT BINARIES OF W UMa TYPE

Timo Rahunen and Osmi Vilhu
Observatory and Astrophysics Laboratory
University of Helsinki
SF-00130 Helsinki 13, Finland

ABSTRACT

Using angular momentum loss estimates from single star studies, it is shown that detached binaries are good candidates as progenitors of contact binaries. Three theories constructed for contact binary evolution (DSC, TRO and AML, see Fig. 1) are discussed. The DSC and TRO theories require a contact binary formation mechanism which produces unequal components (fission) while the AML theory can start from equal components (initially detached binaries). In all theories the end-product is a single star.

The stability of unequal entropy models was studied using a formula (2) which couples the energy transfer with the depth of contact and with the entropy difference. The models experience cyclic behaviour on a time scale of $10^6$-$10^7$ years (Fig. 2) and the contact never breaks even on the nuclear time scale. This is the important consequence of a formula of type (2). Similar behaviour (with similar formula) is expected also for TRO models and even for DSC models if the discontinuity can be preserved during one cycle period.

The DSC and TRO theories, which at first sight look quite different, are in fact complementary. The most probable contact binary theory is perhaps a suitable combination of all three. In this theory angular momentum loss is the new and important factor which may manifest itself in the UV- and X-ray activity observed in W UMa stars.

## I. PROGENITORS OF W UMa STARS

Two different origins of W UMa stars have been proposed:

1. The fission of a rapidly rotating star at the end of the pre-main sequence contraction phase. This process produces roughly the correct amount of angular momentum (Roxburgh, 1966), and small mass ratios (Lucy, 1977) as are in fact observed.

*Z. Kopal and J. Rahe (eds.), Binary and Multiple Stars as Tracers of Stellar Evolution, 289–299.*
*Copyright © 1982 by D. Reidel Publishing Company.*

2. Evolution from a detached or semidetached binary by angular
momentum loss. Huang (1966) suggested that magnetic torques
could bring together the separate components of a detached binary.
This is Schatzman's (1962) and Mestel's (1968) mechanism of magnetic
braking.

We have studied this second formation mechanism in a more
quantitative way and tried to estimate its consequences for close
binary statistics (the details will be published elsewhere by
Vilhu, 1981b). In order to do this we need three rather poorly
known factors: a) the initial distribution of physical parameters
such as the mass ratio and the period, b) the angular momentum loss
rate and c) the life time of the contact phase.

Lucy and Ricco (1979) conclude that short period binaries of small
mass are formed by a mechanism that would create binaries with equal
components and identify this with the hierarchical fragmentation scheme
during the final dynamical collapse of a rotating protostar (Bodenheimer,
1978). A similar conclusion has also been reached by Kraicheva et al.
(1979) and they also claim that most probably the initial period distribution is flat (in the sense $dN/d \log P$ = const) with a short period cut-
off somewhere around 2 days and a long period cut-off around $10^6$ days
(the exact size of this last number is not essential).

An important piece of evidence for the braking of rotation in
solar type stars comes from the observations of rotational velocities
in clusters. Using the sun, solar type dwarfs in Hyades, Pleiades
and in some other clusters, Skumanich (1972) and Smith (1979) find
that the surface rotation (as well as $Ca^+$-emission) decays as the inverse square root of the age: $v_{rot} \propto t^{-1/2}$. Assuming (as seems reasonable) that this behaviour is connected with the rotation itself (by
e.g. a rotationally driven dynamo in the outer convective zone and
subsequent braking by a magnetic co-rotating wind), we can easily
derive a formula for the decline of the spin angular momentum
(assuming solid body rotation).

Chromospheric $Ca^+$-emission correlates with the rotational velocity
and during the last few years similar activity – rotation relationships
have been extended to much higher rotational velocities. This gives us
an important argument in favour of an increasing angular momentum loss
rate when the rotational period decreases, but the exact form of this
behaviour is unknown. For this reason we have parametrized it by $\alpha$
($\alpha$ = 3 for periods longer than 3 days which is the revolution period
of a typical Pleiades solar type dwarf):

$$dJ_{spin}/dt \approx 2 \cdot 10^{41} (P/3)^{-\alpha} \, g \, cm^2 \, s^{-1} \, year^{-1} \tag{1}$$

where P should be expressed in days.

Next we couple this angular momentum loss rate with the orbital
angular momentum assuming that (due to tidal effects) the spin and

orbital motions are synchronized.  In this way the orbital angular
momentum serves as a reservoir for the spin losses.

We mention some general conclusions derived from this
detached → contact treatment (for details see Vilhu, 1981b):

1.  The process produces roughly the correct mean field density
(0.5 % - 2 % of all stars of the same spectral type are contact
binaries) if the contact life time $\tau_{cont}$ is about $5 \cdot 10^8$ years.

2.  Contact binaries can be produced by this process in old
($\gtrsim 5 \cdot 10^9$ y) as well as in intermediate age ($\sim 5 \cdot 10^8$ y) clusters.

3.  A smooth period cut-off around 2 days in the initial period
distribution produces roughly equal numbers of evolved and non-
evolved contact systems.

4.  The ratio of low mass binaries (excluding W UMa's) with
periods less than two days to those with periods greater than
two days is about 1/60.

5.  It is difficult to choose the "best" values for the parameters
involved, but we make the choice (which may not be unique):
$\alpha = 1.5$, $\tau_{cont} = 5 \cdot 10^8$ years and dN/d log P in the initial period
distribution goes from P = 2 days linearly to zero at P = 1 day.
If the contact life time turns out to be e.g. 10 times longer,
then practically the same conclusions hold if we had 10 times
fewer initial systems in the period interval 1 - 10 days.

These results demonstrate that the values of the parameters involved need not be extremely peculiar for the process to work.  It is
perhaps just a coincidence that the angular momentum loss rate, as computed from our basic formula (1) with $\alpha = 1.5$, is roughly the same
($10^{43}$ g cm$^2$ s$^{-1}$ year$^{-1}$) as the AML theory predicts (see Vilhu and Rahunen,
1980 and 1981; Rahunen 1981a).

## II. THREE THEORIES FOR CONTACT BINARY EVOLUTION

So far three different theories for contact binaries have been
proposed, on the basis of which discussion on evolutionary paths is
possible.  We consider these theories and especially their predictions
in the period-colour diagram shown in Fig. 1 where Eggen's (1961, 1967)
period-colour relation for W UMa stars together with theoretical boundaries for zero-age and evolved systems are plotted.  It is worth noting
that the blue zero-age systems with small mass ratios fall outside the
observed region (see Vilhu, 1981a).  The evolutionary possibilities as
predicted by the different theories are also shown.

The contact discontinuity theory (DSC) traces back to Biermann
and Thomas (1972 and 1973) and to Vilhu (1973) who were able to

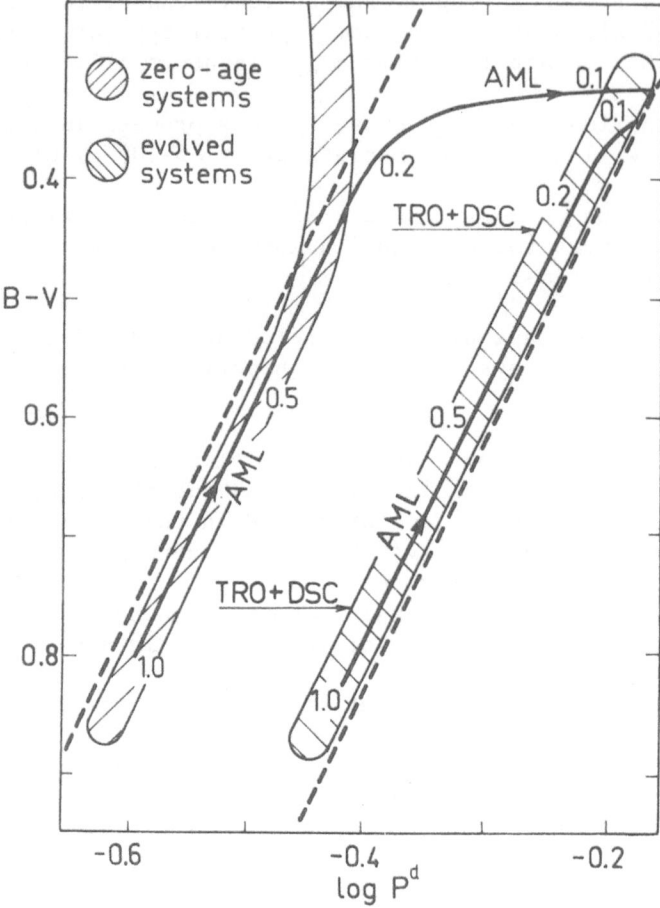

Figure 1.  The period-colour diagram for contact binaries of W UMa
type.  The dashed lines show the observed boundaries.  The shaded
regions show the positions of thermal equilibrium models with different mass ratios (shown by the numbers 1.0 – 0.1).  Evolutionary
tracks of the three main theories DSC (contact discontinuity), TRO
(thermal relaxation oscillations) and AML (angular momentum loss)
are shown by the heavy lines.

construct zero-age models by allowing unequal entropies for the component stars. Their thermal equilibrium models evolve towards more extreme mass ratios on the primary's nuclear time scale. These models
were thought to represent the interiors of the component stars. However, it was clear that the models needed an exterior common envelope
to produce the observational property of W UMa stars: both components
have nearly the same surface temperatures. However, an ad hoc-assump-
tion was made that this common envelope does not have any crucial
influence on the internal structure of the stars.

Later Shu, Lubow and Anderson (1976 and 1979) and Lubow and Shu
(1977) gave a physical (but heavily criticized) argument for this
common envelope concept and published models which are now known as
contact discontinuity (DSC) models. Although no evolutionary sequences
have been published for DSC models, it seems clear that they (if stable)
evolve in the same way as the unequal entropy models by Biermann and
Thomas and Vilhu. In the period-colour diagram (Fig. 1) these models
move from left to right with slightly decreasing mass ratios on the
nuclear time scale of the primary.

In the thermal relaxation oscillation (TRO) theory (Lucy, 1976,
Flannery, 1976, Rahunen and Vilhu, 1977, Robertson and Eggleton, 1977,
Rahunen, 1981a) the component stars are unable to find thermal equilibrium either in or out of contact. Therefore the system performs oscillations with alternating contact and semi-detached phases. During the
contact phase, the thermal contact is good and the mass ratio q decreases (q < 1 by definition). During the semi-detached phase thermal
contact is poor and q increases. The attempt of the system to reach
the nonexistent thermal equilibrium with equal entropies, coupled with
the Roche geometry, is the driver for the cycling behaviour.

All constant angular momentum models (i.e. DSC or TRO) need a
formation mechanism which can produce unequal components, because with
constant angular momentum it is not possible to cover a large range of
mass ratios (see e.g. Vilhu, 1981a, Rahunen, 1981a). We identify this
mechanism to be the fission mechanism.

In the period-colour diagram (Fig. 1) DSC and TRO models evolve
in the same way from the left boundary to the right one on the primary's
nuclear time scale. At low mass ratios (where most W UMa stars are)
only a relatively small range of mass ratios are reached in one evolutionary sequence.

The angular momentum loss (AML) theory is simply a link between
different TRO models (in good thermal contact) at a fixed nuclear
evolutionary state of the primary (Vilhu and Rahunen, 1980 and 1981;
Rahunen, 1981a). In this theory the cycles are avoided by continuous
angular momentum loss. The most suitable contact binary production
mechanism for the AML theory is the detached → contact picture as
discussed in Ch. I. Whatever the production mechanism is in reality,
one AML sequence can cover all the mass ratios observed, which is not

the case with the DSC and TRO theories.

AML models evolve parallel to the boundaries of the period-colour
diagram with decreasing mass ratio on the secondary's thermal time
scale.  The nuclear evolutionary state of the primary at the time when
contact was established determines the horizontal distance from the
observed boundaries, so that zero-age systems reproduce the left boundary whereas evolved ones populate the right boundary.  But in fact also
the initially zero-age AML models turn to the right in the period-colour
diagram when the mass ratio has decreased sufficiently.  This occurs
because finally the nuclear time scale of the primary becomes shorter
than the thermal time scale of the secondary.  Using simple expressions
for the time scales $[\tau_{nuc} \sim 10^{10} (M/M_\odot)^{-3.5}$ y and $\tau_{th} \sim 3 \cdot 10^7 (M/M_\odot)^{-3.5}$ y]
we find that these are equal when the mass ratio is $q \sim 0.2$.  After this
point the evolution is essentially determined by the nuclear evolution
of the primary.

In reality the angular momentum loss may not be so large as the
ideal AML models require.  In this case also AML models might have
cycling behaviour which, however, would not be so violent as with constant angular momentum.  But, provided that the time scale of angular
momentum loss is shorter than the nuclear time scale of the primary,
all the essential features of the AML theory remain.  So a mixture of
AML and TRO theories is perhaps the best choice for contact binary evolution.

We can summarize some of the features of the three theories as
follows:

DSC and TRO theories (constant angular momentum) require a formation mechanism which produces unequal components.  We identify this as
fission.  Otherwise systems with very small mass ratios cannot be obtained.  In these theories the systems move across the period-colour
diagram (from left to right, see Fig. 1) with only slightly decreasing
mass ratio and without much change in colour.  The evolution takes
place on the primary's nuclear time scale.

For the AML theory we find the detached → contact picture as the
most natural production mechanism (see Ch. I).  The theory can start
from a detached system having $q = 1$ which evolves into contact at zero-
age or at some later hydrogen burning stage.  The contact system evolves
towards $q \sim 0.2$ on the secondary's thermal time scale and continues towards $q = 0$ on the primary's nuclear time scale.  Zero-age systems move
along the left boundary while the evolved ones (if near the end of the
main sequence phase) move along the right boundary of the period-colour
diagram (see Fig. 1).

In all three theories zero-age models with very small mass ratios
($q < 0.2$) would be located outside the observed region of the period-
colour diagram.  This suggests that all W UMa stars with rather small
mass ratios are partially evolved.

In all theories the most probable end-product of contact evolution
is a single star, and the final coalescence takes place presumably
after the main sequence. We identify FK Com stars (rapidly rotating
giants) as the most probable candidates for these end-products (Webbink,
1976, Bopp and Rucinski, 1981).

## III. ON THE STABILITY OF UNEQUAL ENTROPY MODELS

In the previous chapter (Ch. II) we considered the unequal entropy
models of Biermann and Thomas (1972 and 1973) and of Vilhu (1973), as
well as the discontinuity models of Shu et al. (1976 and 1979) and of
Lubow and Shu (1977). We shall call these models the "DSC theory".
Although the physical basis of these models may seem different, they
have one common feature: the internal structure of the models inside
the Roche lobes is similar. The essential assumption is that the
components, which have unequal entropies in their convective envelopes,
can maintain thermal equilibrium for a nuclear time scale. Therefore
the question of the thermal stability of unequal entropy models is
crucial for the DSC theory.

TRO theory at its early stage was criticized for postulating
oscillations around a nonexistent equilibrium configuration. However,
later Lucy and Wilson (1979) stressed that the oscillations take place
around unequal entropy configurations. Therefore the question of the
thermal stability of unequal entropy models is crucial for the TRO
theory as well.

Hazlehurst and Refsdal (1980) studied the stability problem using
"stellar response functions" introduced by Hazlehurst et al. (1977).
They found the system of two zero-age stars ($1\,M_{\odot} + 0.6\,M_{\odot}$) to be unstable with an e-folding time of $3 \cdot 10^4$ years. Unfortunately they were
not able to trace the further evolution with their linear stability
analysis. They used a formula which couples the energy transfer ($\Delta L$)
with the depth of contact (d) and with the entropy difference ($\Delta S$):

$$\Delta L = k\, d^m\, \Delta S^n , \tag{2}$$

where k, m and n are constants. In the lack of a detailed physical
theory for the energy transfer this formula may give some physical
insight into the problem, if different choices for the parameters k,
m and n are considered. Moreover, without such a formula it is not
possible to study the stability problem at all.

We have studied the thermal stability and evolution of a
$1.0\,M_{\odot} + 0.6\,M_{\odot}$ system using a Henyey-type stellar evolution code
(the details will be published by Rahunen, 1981b). We applied as
boundary conditions the usual equipotential condition and the formula
(2) for the energy transfer. The stability was studied by introducing
a small perturbation in the form of an energy pulse from the primary
to the secondary. We assumed several values for the transport coeffi-

cients (k, m and n), and studied the stability in each case separately.

As an example consider the following case: With luminosity transfer $\Delta L = 0.144\ L_\odot$ two ZAMS-stars of $1.0\ M_\odot$ and $0.6\ M_\odot$ can be brought into contact while still remaining in equilibrium. In this case there is an entropy difference $\Delta S = 0.813$ leading to different surface temperatures $\Delta \log T_{eff} = 0.076$. Here $\Delta S$ is expressed as $\Delta S = -\Delta \ln K$ where K is the adiabatic constant $= P/T^{2.5}$. We may now ask how stable this kind a configuration is against small perturbations if the luminosity transfer is computed from the formula (2). Without detailed model computations it is not at all obvious what will be the result.

Results of computations (with specific values for the parameters k, m and n) are shown in Fig. 2. As can be seen the system is thermally unstable and behaves much like the TRO models. After $10^7$ years steady cycles with a period of $6 \cdot 10^6$ years seem to be established. The important consequence of the energy transfer formula (2) is that the contact never breaks. On the other hand, the entropy difference is fairly large, the minimum corresponding to $\Delta \log T_{eff} = 0.04$. This minimum is also very sensitive to the choice of the transport coefficients.

To see how the system behaves in different stages of nuclear evolution similar unequal entropy models were constructed with various central hydrogen contents of the primary. It turned out that at first the most prominent effect of the nuclear evolution was to shorten the cycle period, but finally at a certain hydrogen content the instability was removed. A stable configuration in thermal equilibrium was found after a reduction of the initial central hydrogen by about 15 % corresponding to an age of about $10^9$ years. This age is achieved after about 200 cycles. Then such a system evolved on the primary nuclear time scale with decreasing mass ratio and with decreasing entropy difference resembling more and more the A-type W UMa stars.

Although these ZAMS cycling unequal entropy models cannot explain the equal temperatures of W UMa stars, they help us to understand some of their problems. Our computations show that an unequal entropy model is unstable and its subsequent evolution closely resembles the evolution of TRO models. We have a good reason to believe that a slight modification of the energy transfer formula (2) might result in cycles which are consistent with the TRO models published so far. Further, it seems probable that in a similar manner TRO models can be constructed in better agreement with observations (components are in contact permanently).

Our results speak against the assumption of thermal equilibrium in unequal entropy models. However, our results also show that the contact does not necessarily break even on the nuclear time scale. Further, if we do not insist on thermal equilibrium and take the DSC concept more generally, our unequal entropy models can be made much

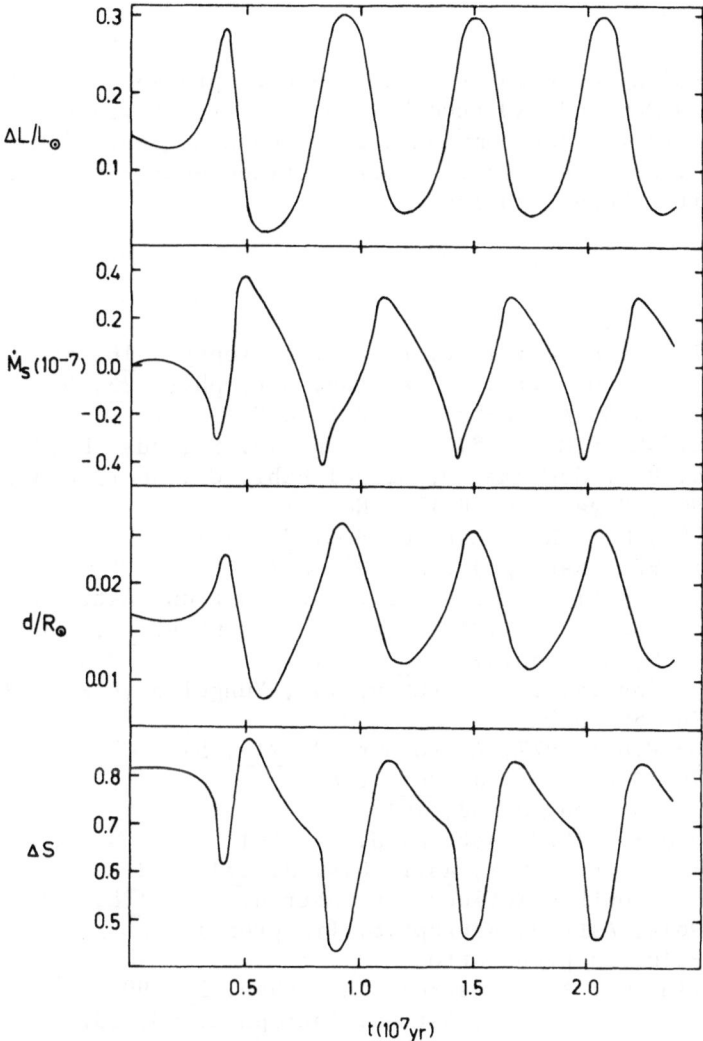

Figure 2. The cycling behaviour of an unequal entropy model with component masses 1.0 $M_\odot$ and 0.6 $M_\odot$. The luminosity transfer $\Delta L$ was computed from the formula $\Delta L = k\, d^m\, \Delta S^n$ with the energy transport coefficients $m = 3$ and $n = 1$. $\dot{M}_S$ is the rate of mass transfer $(M_\odot/y)$ from the primary to the secondary. d is the depth of contact and $\Delta S$ the entropy difference between the component envelopes.

closer to DSC models assuming a common envelope around the component stars. As far as we know nobody has proved that the contact discontinuity cannot be preserved on the short time scale of one cycle ($6 \cdot 10^6$ years, see Fig. 2). In fact our models show a tendency to develop a temperature inversion (a contact discontinuity) at a certain stage of the cycle.

In conclusion, we think that the results presented can be applied to improve both DSC and TRO theories which at first sight look quite different but are in fact complementary (see also Shu, 1980). Perhaps it is possible, with the aid of a more refined energy transfer formulation, to unite these theories.

# REFERENCES

Biermann,P., Thomas,H.-C.: 1972, Astron. Astrophys. 16, 60
Biermann,P., Thomas,H.-C.: 1973, Astron. Astrophys. 23, 55
Bodenheimer,P.: 1978, Astrophys. J. 224, 488
Bopp,B.W., Ruciński,S.M.: 1981, IAU Symp. No. 93, eds. D. Sugimoto, D.Q. Lamb, D.N. Schramm, D. Reidel Publ. Co., Dordrecht, p. 181
Eggen,O.J.: 1961, Royal Obs. Bull., No. 31
Eggen,O.J.: 1967, Mem. Roy. Astron. Soc. 70, 111
Flannery,B.P.: 1976, Astrophys. J. 205, 217
Hazlehurst,J., Refsdal,S., Stobbe,C.: 1977, Astron. Astrophys. 58, 47
Hazlehurst,J., Refsdal,S.: 1980, Astron. Astrophys. 84, 200
Huang,S.-S.: 1966, Astrophys. J. 29, 331
Kraicheva,Z.T., Popova,E.I., Tutukov,A.V., Yungelson,L.R.: 1979, Astron. Zh. 56, 520
Lubow,S.H., Shu,F.H.: 1977, Astrophys. J. 216, 517
Lucy,L.B.: 1976, Astrophys. J. 205, 208
Lucy,L.B.: 1977, Astron. J. 82, 1013
Lucy,L.B., Ricco,E.: 1979, Astron. J. 84, 401
Lucy,L.B., Wilson,R.E.: 1979, Astrophys. J. 231, 502
Mestel,L.: 1968, Monthly Notices Roy. Astron. Soc. 138, 359
Rahunen,T.: 1981a, Astron. Astrophys. (in press)
Rahunen,T.: 1981b, in preparation
Rahunen,T., Vilhu,O.: 1977, Astron. Astrophys. 56, 99
Refsdal,S., Stabell,R.: 1981, Astron. Astrophys. 93, 297
Robertson,J.A., Eggleton,P.P.: 1977, Monthly Notices Roy. Astron. Soc. 179, 359
Roxburgh,I.W.: 1966, Astrophys. J. 143, 111
Schatzman,E.: 1962, Ann. Astrophys. 25, 18
Shu,F.H.: 1980, IAU Symp. No. 88, eds. M.J. Plavec, D.M. Popper, R.K. Ulrich, D. Reidel Publ. Co., Dordrecht, p. 477
Shu,F.H., Lubow,S.H., Anderson,L.: 1976, Astrophys. J. 209, 536
Shu,F.H., Lubow,S.H., Anderson,L.: 1979, Astrophys. J. 229, 223
Skumanich,A.: 1972, Astrophys. J. 171, 565
Smith,M.A.: 1979, Publ. Astr. Soc. Pacific 91, 737
Vilhu,O.: 1973, Astron. Astrophys. 26, 267
Vilhu,O.: 1981a, Astrophys. Space Sci. 78, 401

Vilhu,O.: 1981b, in preparation
Vilhu,O., Rahunen,T.: 1980, IAU Symp. No. 88, eds. M.J. Plavec,
    D.M. Popper, R.K. Ulrich, D. Reidel Publ. Co., Dordrecht, p. 491
Vilhu,O., Rahunen,T.: 1981, IAU Symp. No. 93, eds. D. Sugimoto,
    D.Q. Lamb, D.N. Schramm, D. Reidel Publ. Co., Dordrecht, p. 181
Webbink,R.F.: 1976, Astrophys. J. 209, 829

# SOME ASPECTS OF THE EVOLUTIONARY STATUS OF W URSAE MAJORIS BINARIES DEDUCED FROM OBSERVATIONAL DATA

Walter Van Hamme[1]
Sterrenkundig Observatorium, Rijksuniversiteit Gent, Belgium

ABSTRACT

We have developed a test for the evolutionary state of W Ursae Majoris binaries by comparing the observed spectral type of 31 of these systems (14 of type W and 17 of type A) with the expected one when their primary component is an unevolved main sequence star. It appears that both the W- and A-type systems have a primary with a mass and radius too large to be compatible with the observed spectral type, so there is no indication that each type should mark a different evolutionary stage.

## 1. INTRODUCTION

In the paper of Wilson (1978) a method is presented to examine whether the radius of the components of 8 A-type W Uma binaries, selected for having accurately known orbital parameters, was a ZAMS radius corresponding to a reasonable value of the mass (say 1-2 solar masses). Let us have a brief look at this method. The actual orbital distance $a_k$ (in solar radii) between the components of a binary system is given by

$$a_k = 4.2060 \, P^{2/3} (1+q)^{1/3} m_1^{1/3} \tag{1}$$

with P the orbital period in days, q the mass ratio $m_2/m_1$ and $m_1$ the mass (in solar masses) of the primary, i.e. the more massive component. The orbital separation $a_z$ on the assumption that both components are ZAMS objects, fitting the mass-radius relation

$$R = m^{0.6} \, , \tag{2}$$

can be written as

---

[1] Aangesteld navorser van het Belgisch Nationaal Fonds voor Wetenschappelijk Onderzoek

*Z. Kopal and J. Rahe (eds.), Binary and Multiple Stars as Tracers of Stellar Evolution, 301–304.*

$$a_z = F(1+q^{0.6})m_1^{0.6} \qquad (3)$$

if we denote  by F the ratio of $a_z$ and the sum of the mean stellar radii :

$$F = \frac{a_z}{\overline{R}_1+\overline{R}_2} = \frac{1}{\overline{r}_1+\overline{r}_2} \qquad (4)$$

This ratio is completely specified by the Roche geometry for a given value of the mass ratio and the degree of overcontact. For all 8 A-type binaries the $a_z$ value was smaller than the $a_k$ value in the $m_1$ mass range of 1–3 solar masses. So Wilson concluded that the A-type systems have larger than ZAMS radii and are evolved from the main sequence.

Now a slight modification of Wilson's method will allow us to extend it to a larger group of W Uma binaries and especially to the group of W-type systems.

## 2. METHOD

Since we know from Kuiper's paradox (Kuiper, 1941) that it is impossible for the components of a contact binary both to fit a main sequence mass-radius relation of type (2) and to have sizes according to the Roche geometry, we introduce the orbital separation $a_z$ on the assumption that only the primary component is a main sequence star obeying the mass-radius relation

$$R = m^{0.8} \qquad (5)$$

Note the exponent value 0.8 instead of Wilson's lower value 0.6. The quantity $a_z$ can easily be calculated and we get

$$a_z = m_1^{0.8}/\overline{r}_1 \qquad (6)$$

We then can solve the equation $a_z = a_k$ in which we consider $m_1$ as the unknown. This gives us the value $m_1^{\star}$ we expect the mass of the primary to have when it is a main sequence object fitting the condition (5). Finally we derive the corresponding expected spectral type $Sp^{\star}$, which can be compared with the really observed one $Sp'$, from the mass-spectral type relation appropriate for the main sequence. We have applied this method on a sample of 17 A-type and 14 W-type W Uma contact binaries for which a modern lightcurve solution and an estimation of the spectral type could be found in the literature.

## 3. RESULTS

Fig. 1 shows us the diagram of the expected spectral type $Sp^{\star}$ versus the observed one $Sp'$. The horizontal bars indicate the spectral

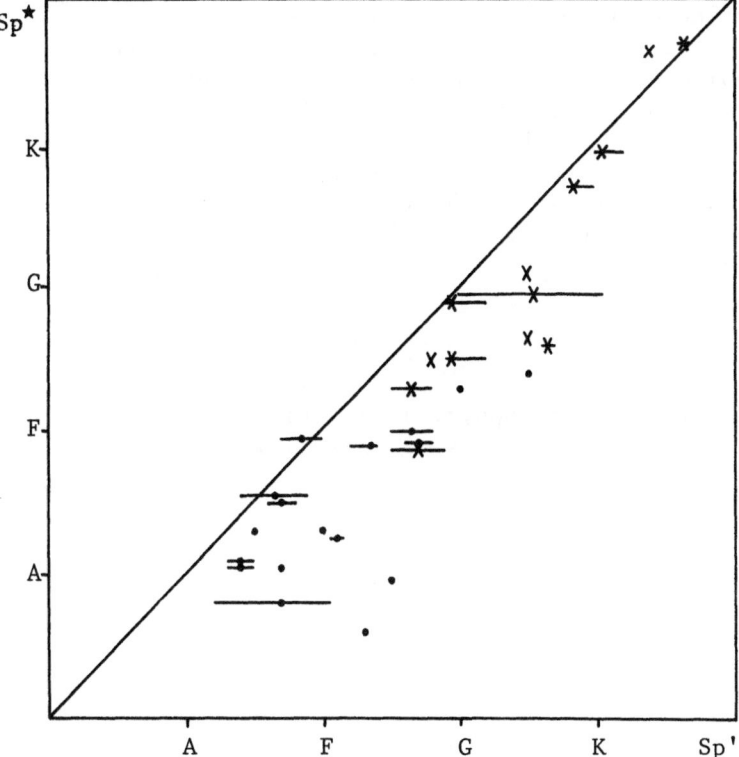

Figure 1. The expected spectral type Sp[*] versus the really observed one Sp' for 17 A-type (.) and 14 W-type (X) W Uma
binaries.

range in which the spectral type estimation of different authors are
lying. Nearly all systems, the A-types as well as the W-types, appear
to be shifted in the direction of the later types, i.e. they are showing a later spectral type than we expect if their primary would be a
main sequence star. In other words the mass and radius of the primary
component of W UMa binaries are too large to correspond to a main sequence
star with a spectral type the same as the one we observe, and this is true
for both the W- and A-type systems. Again using the main sequence mass-
spectral type relation we can derive the value $m_1'$ of the primary's mass appropriate to its observed spectral type Sp' and the corresponding main sequence radius $R_1' = a_z' \, \bar{r}_1$. On the other hand the actual radius will be
larger and given by $R_1' + \Delta R_1' = a_k' \, \bar{r}_1$ so the fraction by which the radius
of the primary is too large to be a main sequence object with the ob-
served spectral type is given by

$$\frac{\Delta R_1'}{R_1'} = \frac{a_k'}{a_z'} - 1 \tag{7}$$

The mean value of this fraction amounts to $0.21 \pm 0.05$ for the A-type
and $0.07 \pm 0.02$ for the W-type group. One could argue that the hypothesis of evolved A-type systems remains valid. However, we think that
such a conclusion is not as obvious as it looks at first sight. Both
the A- *and* W-type W Uma binaries behave essentially in the same way,
the figure 0.07 is indeed significant. We would conclude that A-type
binaries can be found in a wider range of possible evolutionary stages,
evolved and unevolved ones. Not all A-type W Uma binaries are evolved
from the main sequence.

REFERENCES

Kuiper, G.P. : 1941, Astrophys. J. 93, 133
Wilson, R.E. : 1978, Astrophys. J. 224, 885

A POSSIBLE EXPLANATION OF DISTORTIONS IN THE LIGHT CURVES OF VW CEP
BY STARSPOTS

A. Yamasaki
Dept. Earth Science and Astronomy,
University of Tokyo, Tokyo 153, Japan

ABSTRACT

An attempt is made to explain the variations in the light curve of
VW Cep in terms of a starspot model. Based upon Kwee's (1966) observations, an unperturbed light curve without spots is constructed and analysed for the geometrical elements of VW Cep. Then, by assuming that
starspots exist on the surface of the primary component, Kwee's individual light curves are further analysed for locations and intensities of
the starspots.

VW Cep (P = $0.^{d}278$) is a well-known W UMa type system showing variations in the light curve and in the orbital period. Several important
papers (e.g., Walter, 1979) have been published so far to interpret these
variations. In the present paper, it is shown that variations in the
light curve will be explained by a starspot hypothesis.

Starspot models for late type variables such as BY Dra stars and
RS CVn stars have given successful explanations of distortions in the
light and flare-like activities sometimes observed in these stars.
Because of the similarities between VW Cep (∿KOV) and RS CVn stars in the
spectral types and in the nature of light curve distortions, it may be
worth trying to apply a starspot model to VW Cep.

To determine distributions and intensities of starspots, an unspotted light curve which is free from starspots should be necessary as the
reference. However, each observed light curve seems to suffer from some
amounts of distortions. Therefore, an unspotted light curve has to be
constructed from observed light curves by taking the maximum light among
observed light curves for every corresponding phase (no hot spots, no en-
hancement of the continuum radiation are assumed).

This process has been applied to Kwee's (1966) observations which
were made between 1957 and 1959 at λ5600, to obtain the unspotted light
curve. Then, the unspotted light curve has been analysed by the light-

Z. Kopal and J. Rahe (eds.), Binary and Multiple Stars as Tracers of Stellar Evolution, 305–308.

curve synthesis method for the geometrical elements.

For the mass ratio  q = 0.41 (Binnendijk, 1967), the elements

i = 64° ± 1°
f = 1.10 ± 0.05
j = 1.82 ± 0.05

have been obtained, where  f  and  j  denote the fill-out-ratio and the
ratio of the surface brightness (Yamasaki, 1981), respectively, while
for the mass ratio  q = 0.33 (Popper, 1948), the slightly different element

i = 65° ± 1°
f = 1.10 ± 0.05
j = 1.77 ± 0.05

have been obtained.  Note that adopting the effective temperature for the
primary component  $T_{eff1}$ = 5300 K  for K0V, we obtain that the temperature
for the secondary component can be estimated to be around  $T_{eff2}$ = 6100 K,
hence the system reveals a rather large temperature difference.

Then, according to Mullan (1975), starspots on the surface of the
primary component have been considered.  Four light curves, on JD2436124
(epoch I), JD2436285(epoch II), JD2436448(epoch III) and JD2436679(epoch
IV), among Kwee's (1966) observations have been used to find out positions
and intensities of starspots.

The surface of the primary component has been divided into 66 regions
where S, B, P and P' represent $L_1$ point ($\lambda$ = 0°, $\phi$ = 0°), the back point
($\lambda$ = 180°, $\phi$ = 0°), the pole visible from the Earth ($\phi$ = 90°) and the
pole invisible from the Earth ($\phi$ = -90°), respectively.  Phases 0°, 90°,
180° and 270° correspond to the transit minimum (the photometric secondary
minimum), the second maximum (points $\lambda$ = 90° face the Earth), the occultation minimum (the photometric primary minimum) and the first maximum
(points $\lambda$ = 270° face the Earth), respectively.

Figure 1 shows Kwee's observations, the starspot model solutions
(solid line) and the distribution of starspots (the intensities of starspots are represented by the darkening of the shades).

The movement of starspots can be clearly seen in Figure 1, where
starspots are gathered around $\lambda$ = 0°, 270°, 180° and 90° for epochs I,
II, III and IV, respectively.  This may indicate that starspots rotate on
the surface of the primary component a little faster (by 0.04%) than the
synchronized rotation.

Recent discovery of X-ray emission from VW Cep (Carroll et al., 1980)
and the intense UV emissions (Dupree et al., 1979) would suggest activities which are associated with starspots on the surface of VW Cep.

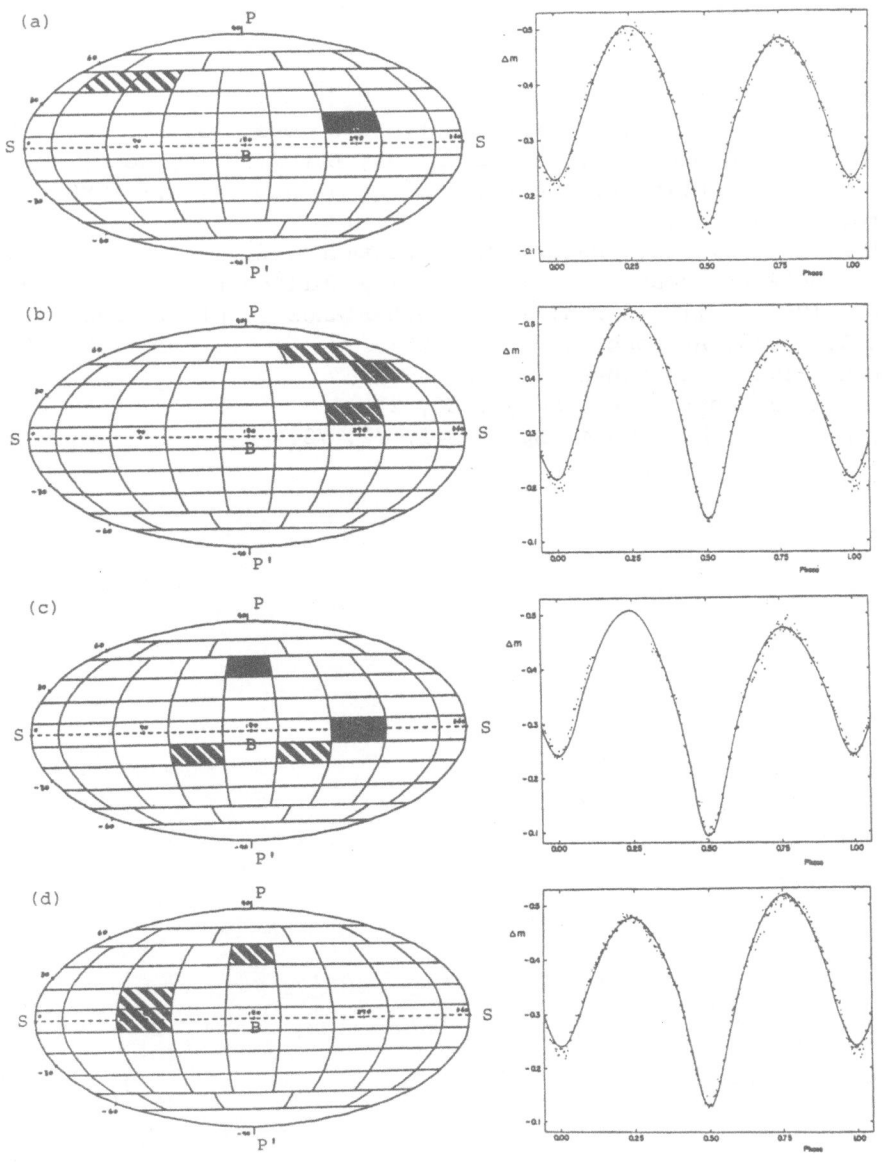

Figure 1.  Individual observations (Kwee, 1966), the starspot model solutions (solid line) and the distribution of starspots on the surface of the primary component of VW Cep for (a) epoch I (JD2436124), (b) epoch II (JD 2436285), (c) epoch III (JD2436448) and (d) epoch IV (JD2436679), respectively.

This work was supported in part by the Scientific Research Fund of the Ministry of Education, Science and Culture (56540137).

REFERENCES

Binnendijk, L.: 1967, Publ. Dominion Astrophys. Obs. 13, 27.
Carroll, R.W., Cruddace, R.G., Friedman, H., Byram, E.T., Wood, K.,
    Meekins, J., Yentis, D., Share, G.H., and Chubb, T.A.: 1980,
    Astrophys. J. 235, L77.
Dupree, A.K., Black, J.H., Davis, R.J., Hartmann, L., and Raymond, J.C.:
    1979, The First Year of IUE, University College London.
Kwee, K.K.: 1966, Bull. Astron. Inst. Netherlands Suppl. 1, 245.
Mullan, D.J.: 1975, Astrophys. J. 198, 563.
Popper, D.M.: 1948, Astrophys. J. 108, 490.
Walter, K.: 1979, Astron. Astrophys. 80, 27.
Yamasaki, A.: 1981, Astrophys. Space Sci. 77, 75.

# THE IUE OBSERVATIONS OF W UMA[†]

S.M. Rucinski[*]
Max Planck Institute for Astrophysics, Munich

J.E. Pringle and J.A.J. Whelan[+]
Institute of Astronomy, Cambridge

Abstract. The low resolution, phase averaged SWP spectra reveal the trend of increased transition-region line intensities for higher temperatures of ion formation which was previously observed for other active systems. Two series of low resolution LWR spectra covering two full orbital periods have been used to check the photometric results from the ANS satellite and a good agreement is found for the 2200 Å band. Wavelength shifts and changes in the intensity of the blended Mg II emission feature are not inconsistent with the ANS result that the more massive component is the more active one in the W UMa system.

## 1. The SWP Spectra Region

Two low resolution, large aperture SWP spectra (Fig. 1) were obtained on April 4, 1980 by integrating the rather weak stellar signal during a few of the LWR camera read-out intervals (cf. next section). The exposures lasted effectively 140 and 85 minutes and covered the phase intervals 0.66-0.11 (SWP 8653) and 0.17-0.43 (SWP 8654). Therefore, the fluxes in Fig. 1 are phase averaged over considerable portions of the light curve. The phasing of our observations has been based on the light curve obtained with the FES detector and described by Rucinski et al. (1980). The tentative line identifications are marked on the figure. To convert the observed fluxes into the surface fluxes, we used the formulae given in Linsky and Ayres (1978). When compared with the solar data (Fig. 2) the emission line fluxes for lines originating in the transition region reveal the same trend as observed for other active stars and for other W UMa-type systems in particular (Dupree and Preston 1980), i.e. the trend of increased line intensity for higher temperatures of ion formation. This increase ranges between a factor of about 30 for the Si II lines up to about 250 for the Si IV

---

[†] Based on observations collected at the Villafranca Satellite Tracking Station of the European Space Agency.
[*] On leave from the Warsaw University Observatory.

[+] While this paper was in the press, Dr. Whelan passed away on 21 December 1981, aged 35 years.

*Z. Kopal and J. Rahe (eds.), Binary and Multiple Stars as Tracers of Stellar Evolution, 309–315.*

lines and can be interpreted by the relatively high proportion of
closed magnetic tubes confining the high temperature plasma.

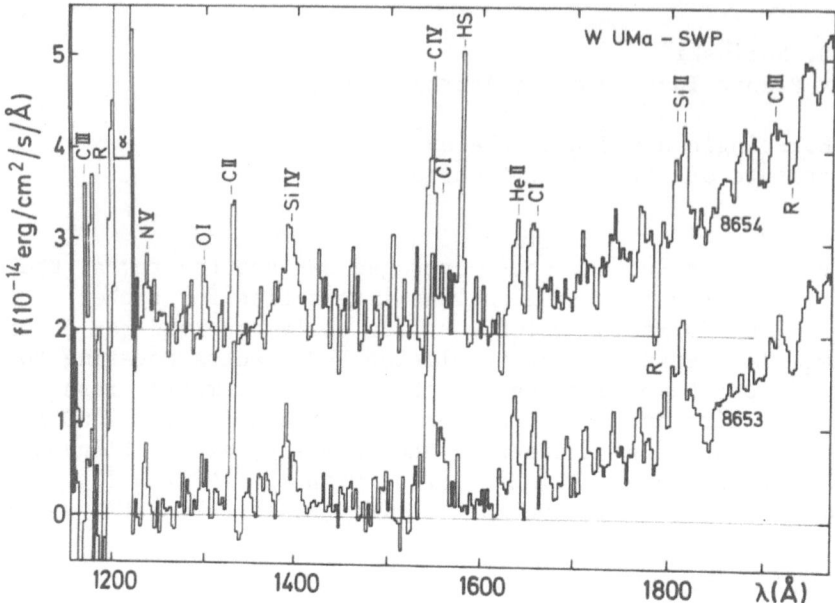

Figure 1.  The SWP spectra of W UMa.

Figure 2.  The surface line fluxes relative to the solar values versus
           the temperature of ion formation.

## 2. The LWR Spectral Region

Two series of low resolution, large aperture LWR spectra were obtained on April 3 and 4, 1980 during two consecutive European IUE shifts, i.e. shifts separated by 16 hours or, almost exactly, by two orbital periods of W UMa. Because of the modest dynamic range of the IUE cameras and the rapid fall-off of the continuum towards shorter wavelengths it was necessary to observe W UMa with different exposure times. Seven strongly exposed (25 minutes) images obtained during the first shift permitted analysis of the spectrum shortward of 2650 Å; ten images with exposures ranging between 5 to 10 minutes obtained during the second shift were used for the longward portion of the spectrum. Two representative spectra taken at orbital quadratures of W UMa are shown in Fig. 3; spectra at other phases are very similar in the general shape and intensities of spectral features.

The LWR spectra are dominated by broad blends of many Fe I and Fe II lines with especially prominent blends at 2400-2630 Å (Fe I, Fe II, Si I, Mg I) and close to the 2800 Å line of Mg II (Fe II, Cr II, Mg I, Mg II).

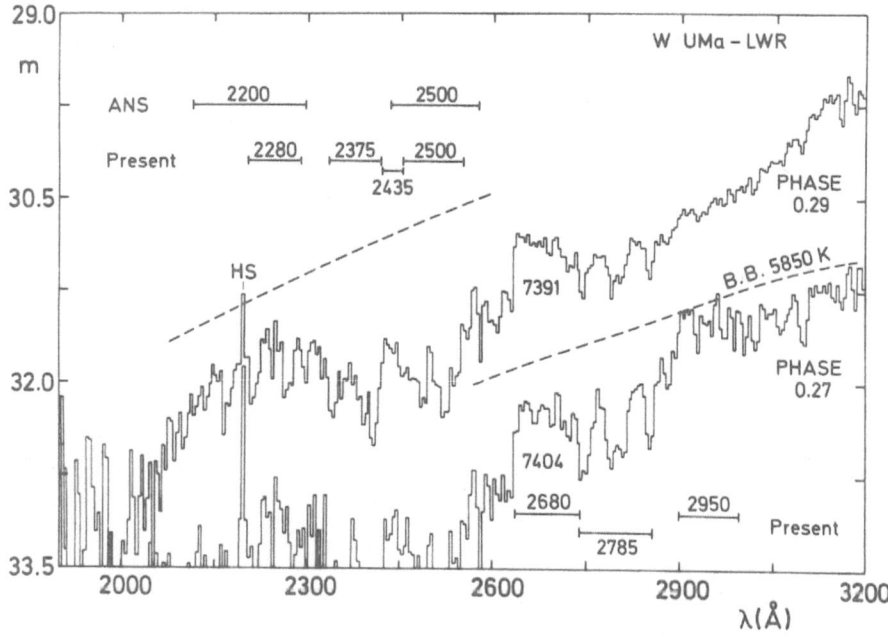

Figure 3.  Two low resolution LWR spectra taken at phases close to orbital quadratures. The scale of stellar magnitudes (the ANS system) is appropriate for LWR 7391 (upper spectrum); the spectrum LWR 7404 is shifted down by 1.5 mag. Notice the overexposure of LWR 7391 extending from about 2650 to 3100 Å. A blackbody curve corresponding to the mean effective temperature of W UMa has been arbitrarily located in the vertical coordinate.

The behaviour of these blends could have some influence on the results
obtained previously with the ANS satellite photometry by Eaton et al.
(1980). They found that the gravity darkening must be rather small and
that the dark spots on the more massive component are the only reasonable explanation for the W-type light curve. Their basic assumption
was that the ANS spectral bands at 2200 and 2500 Å measured the photospheric continuum radiation and the ultraviolet-optical colours are
good indicators of averaged effective temperatures. Locations of the
almost rectangular ANS bandpasses relative to the spectrum (Fig. 3)
suggest that the 2200 Å band was indeed a useful continuum indicator for
W UMa whereas the 2500 Å falls into the choppy region what might explain,
in addition to reasons listed by Eaton et al., the much worse definition
of the ANS light curve at 2500 Å. Unfortunately the 2200 Å band contains
a "hot-spot" at 2190 Å so that the simulated ANS fluxes (the IUE fluxes
convolved with the ANS bandpasses) contain at least a 3% contribution
(if one pixel is affected) from this blemish. The simulated ANS magnitudes at 2200 Å are presented in Fig. 4 together with the original ANS
results.

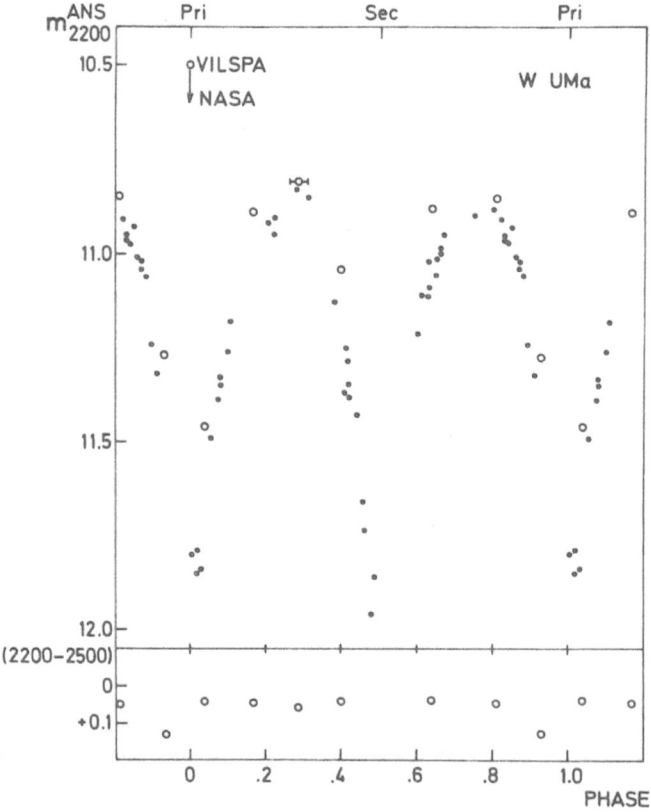

Figure 4. Comparison of the simulated ANS magnitudes for the 2200 Å band-
          pass (circles) with the original ANS data (dots).

The uncertainty of the calibration is shown by the arrow giving the change when the original Vilspa calibration used here (Bohlin et al., 1980a) is replaced by the new one adopted for the first time by NASA in May 1980 (Bohlin and Holm 1980b). Taking into account these instrumental uncertainties which together might reach 12%, the agreement is satisfactory.

In addition to the ANS bandpasses, we analysed also the phase dependence of the LWR spectrum in bands selected to represent the major low resolution features, as marked in Fig. 3. The obtained integral changes are very small, never exceeding 5-7%, i.e. rather close to errors of measurements; however, they do seem to be systematic, e.g. the absorption features at 2375 and 2500 Å become somewhat stronger at orbital quadratures (double cosine behaviour), whereas the triple blend contained in the 2785 Å band shows a weak single cosine dependence. This suggests that those spectral regions should be avoided in future continuum studies.

## 3. The Mg II Emission Line

The low resolution spectra reveal only a weak emission feature at the center of a broad absorption around 2800 Å. The emission peak shows small wavelength shifts resulting in variable blending with the steep branches of the surrounding absorption. The blending is phase dependent with indications of opposite in direction and largest shifts at quadratures (Fig. 5). The sense of shifts suggests that the more massive com-

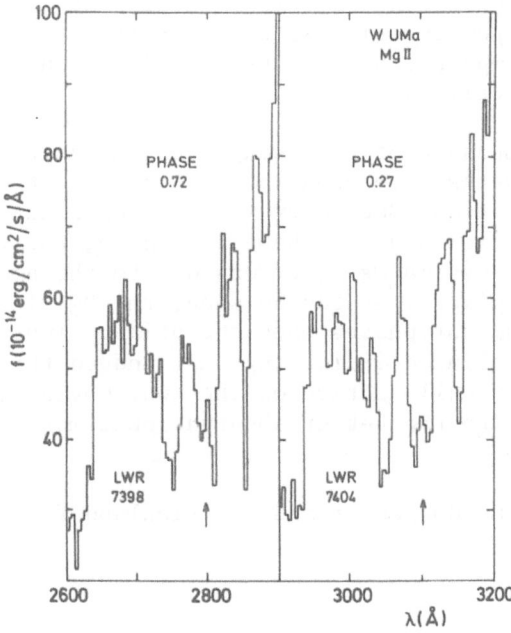

Figure 5. The region of LWR spectrum close to the Mg II line at two orbital quadratures of W UMa.

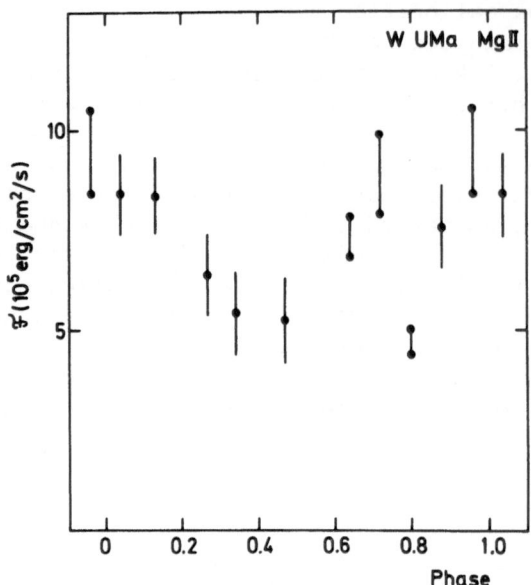

Figure 6. Approximate measure of the Mg II emission surface flux versus
     phase.


ponent contributes most of the Mg II emission but with the present resolution it remains an open question if this merely reflects the larger
emitting area of this component or the stronger line flux indicating its
elevated chromospheric activity.

     To obtain a crude measure of the changes in intensity of the Mg II
emission, the emission peaks have been integrated above the lowest points
in the surrounding absorption. Such approximate fluxes have been converted to the surface fluxes and are shown in Fig. 6; the vertical bars
in that figure represent estimates of errors due to the ambiguity in
inclusion of individual pixels. As can be seen, the Mg II feature seems
to reveal a broad minimum for phases when the primary (more massive)
component is partially eclipsed indicating that indeed this component
might be more chromospherically active of the two. Obviously, this
result must be checked with the better observational material.


We thank L. Bianchi, P. Gondhalekhar and D. Stickland for help.


References

Bohlin, R.C., Holm, A.V., Savage, B.D., Snijders, M.A.J., Sparks, W.M.:
     1980a, Astron. Astrophys. 85, 1.
Bohlin, R.C., Holm, A.V.: 1980b, IUE-NASA Newsletter, No. 11, June 1980.

Dupree, A.K., Preston, S.: 1980, in "The Universe at Ultraviolet Wave-
    lengths - The First Two Years of IUE", NASA Conference Publ.
    No. 2171, p. 333.
Eaton, J.A., Wu, C.-C., Rucinski, S.M.: 1980, Astroph. Journ., 239, 919.
Linsky, J.L., Ayres, T.R.: 1978, Astroph. Journ. 220, 619.
Rucinski, S.M., Gondhalekhar, P., Pringle, J.E., Whelan, J.A.J.: 1980,
    Inf. Bull. Var. Stars, No. 1844.

# MASS FLOW DUE TO HEATING IN A BINARY SYSTEM: APPLICATION TO U CEPHEI

Yoji Kondo
Laboratory for Astronomy and Solar Physics, NASA/GSFC
Greenbelt, MD  20771, U.S.A.

Jerry L. Modisette
Modisette, Inc.
4223 Richmond Avenue, Houston, TX  77027, U.S.A.

Abstract

We have investigated the possibility of mass flow due to the heating
of the cooler component in a close binary system.  The heating may
be caused by irradiation from the hotter companion or by other
mechanisms such as the spacial coincidence of non-linear "g-mode"
oscillations in the cooler star.  The 2.4-day period binary U Cep,
in which gas streaming has been observed, has been chosen for model
calculations.  Preliminary results show that such a heating of the
lower atmosphere of the cooler star could lead to mass flow at an
average rate of $10^{-9}$ to $10^{-7}$ solar mass per year without the
star's necessarily filling its critical Roche surface.

## 1. Introduction

We recently reported our initial results (Modisette and Kondo 1980a,
b) on the problem of mass flow due to the heating of the surface of
the cooler component in a close binary; in those previous papers a
binary system analogous to Her X-1 = HZ Her was used as a model.
The source of the heating was not specified in that study but the
beat of harmonics in the non-linear "g-mode" oscillations might be
responsible for periodic heating in HZ Her (Wolff and Kondo 1978).
We wish to report on our follow-on work since these prior
publications.

## 2. Model

We have used U Cephei for model calculations.  The reasons are
partially that gas streaming observed in U Cep (Kondo, McCluskey and
Stencel 1980; Kondo, McCluskey and Harvel 1981) requires a source of
energy in addition to that available from the simple conversion of
gravitational potential into kinetic energy in what is commonly
known as Roche lobe overflow.  Also, irradiation by the hotter B8

317

star in that binary could heat the facing hemisphere of the cooler G
star up to a few thousand degrees, which might provide the energy
necessary for causing mass surge.

Batten's (1974) values have been adopted as the physical parameters
for U Cep.  That is, the mass of the B8 star is 4.2 $M_\odot$, the mass
of the G5 III companion, 2.8 $M_\odot$, the radius of the B component,
2.9 $R_\odot$, the radius of the G component, 4.7 $R_\odot$, the separation of
the centers of the two stars, 14.7 $R_\odot$ and the orbital period,
about 2.493 days.  Thus, the G star does not fill its critical Roche
lobe; its radius is 70 percent of the distance to the first
Lagrangian point.

If the surface temperature of the B star is 13500K and that of the G
star 4700K, for the dilution factor appropriate for the above
geometry, the amount of heating at the photosphere of the G star
should be in the range of 1000 to 2000K, provided that the
absorption of the irradiation in the photosphere of the G star is at
the efficiency of 30 to 50 percent.  However, our current knowledge
of the ultraviolet absorptivity of the atmospheres of various type
stars is rather incomplete.

The previously used theoretical program (Modisette and Kondo 1980b)
was modified to enable calculations of continuous heating as well as
sinusoidal heating.

3.  Discussions

The results show that the continuous heating of the lower
chromosphere could lead to mass surge lasting a few hundred thousand
seconds.  Total mass flow would depend on the density of the layer
where the effective absorption of the irradiation occurs as well as
on a number of other boundary conditions in the model.  The rate
could correspond to $10^{-9}$ to $10^{-7}$ solar mass per year.  The
outflow does not continue for more than a few days if the heating at
the original layer remains constant.  However, the opacity of the
atmosphere at higher level would increase as the mass surge
progresses higher; this would in turn decrease the irradiation at
the original layer.  If hot spots should develop on the B star as
the result of the accretion of the ejected matter from the G star,
this could also change the amount of irradiation from the B star.
At any rate, the heating at the base of the chromosphere is likely
to fluctuate.

We have thus carried out computations for the case in which the
heating at the base varies in a sinusoidal fashion over the periods
of 100 to a few hundred seconds.  Note that non-linear oscillations
could also give rise to periodic heating of the lower
chromospheres.  Calculations show that although the mass surge
oscillates wildly at lower levels, pulses of surges move more or

less continuously through the layers of large distances, say, of 1 solar radius from the photosphere. The average rate of mass flow is in the range of $10^{-9}$ to $10^{-7}$ solar mass per year. Unlike the continuous heating case, the mass surge due to the pulsed heating could continue for an apparently indefinite period. The calculations have been carried out thus far for the duration of up to $10^6$ seconds.

We have had to make a number of gross simplifications in performing these calculations. What our preliminary results show is merely what could be happening rather than what must be happening in a close binary like U Cep. Nevertheless, the disturbances at the lower atmosphere in the form of heating by its hotter companion or by non-linear oscillations could lead to significantly higher degrees of disturbances at higher levels. It has been well known that a disturbance at a higher density medium may propagate into a lower density medium increasing its amplitude. As the smaller quantity of matter tries to carry the same amount of energy, the amplitude of the disturbance must be increased. Examples of such phenomena include the breaking of waves on the beach or the cracking of a whip. Solar wind too is harnessed by propagation of the disturbances at lower atmospheres through chromosphere and corona.

The physical parameters due to Batten (1974) indicate that the G star in U Cep does not fill its Roche lobe; its radius reaches only 70 percent of the critical radius. The gas stream observed in this binary would, therefore, seem to require that there be a mechanism, other than the conversion of gravitational potential, to harness the mass flow.

According to the present results, mass flow, caused by heating, could start off long before the star fills its critical Roche lobe. When the star fills the critical equipotential surface, mass flow could be further assisted and accelerated by this process. This also means that the presence of a gas stream emanating from the cooler component in a binary system would not necessarily indicate that the cooler star has filled its critical Roche equipotential surface.

We wish to thank Jim Sewell for a very capable assistance in carrying out computer programming and calculations.

References

Batten, A.: 1974, Pub. Dom. Astrophys. Obs. 11, 191.
Kondo, Y., McCluskey, G. E., and Stencel, R. E.: 1980, Astrophys. J. 906.
Kondo, Y., McCluskey, G. E., and Harvel, C. A.: 1981, Astrophys. J. 202.
Modisette, J. L. and Kondo, Y.: 1980a, Proc. IAU Symp. No. 88, 195.
Modisette, J. L. and Kondo, Y.: 1980b, Astrophys. J. 240, 180.
Wolff, C. L. and Kondo, Y.: 1978, Astrophys. J. 219, 605.

# THE VERY UNUSUAL ULTRAVIOLET SPECTRA OF R ARAE

Yoji Kondo
Laboratory for Astronomy and Solar Physics, NASA/GSFC
Greenbelt, MD  20771, U.S.A.

George E. McCluskey
Department of Mathematics, Lehigh University
Bethlehem, PA  18015, U.S.A.

Abstract

The high resolution ultraviolet spectra of the 4.4-day period binary
R Arae, observed in 1980 with the International Ultraviolet Explorer
(IUE), show that its continuum flux level varied outside of eclipse
by more than a factor of two in ten days, and by over 50 percent
within a same orbital cycle.  The flux level varied
non-monotonically at different wavelengths.  The resonance lines of
Mg II and Si IV exhibited shortward-shifted absorption components
near phase 0.4, indicating the presence of a gas stream toward the
observer at a velocity of some -450 to -500 km s$^{-1}$.  Our
observations of R Arae with the Einstein satellite show it to be an
X-ray source.

## 1.  Introduction

R Arae is a relatively bright ($m_v$ at maximum = 6.5) eclipsing
binary with a period of about 4.4 days.  Its primary component is
classified as B9p, but its secondary component is spectroscopically
unidentified (Sahade 1952).  He derived a mass function of 0.10 $M_\odot$
but cautioned about large uncertainties involved.  Taking the mass
of the primary to be 4.0 $M_\odot$, Sahade derived an estimated mass for
the secondary of 1.4 $M_\odot$.  If it is a main-sequence star, it is an
F-type object.  He also noted a number of peculiarities in the
ground-based spectra of this binary.

We have included this binary in our study of the mass flow and
evolution in close binary system, which is being conducted with the
IUE.  The IUE has been described in detail by Boggess et al. (1978).

## 2.  Observations

In all, four exposures with the far ultraviolet SWP camera covering
the 1150-1900Å range and six exposures with the mid-ultraviolet LWR

*Z. Kopal and J. Rahe (eds.), Binary and Multiple Stars as Tracers of Stellar Evolution, 321–326.*
*Copyright © 1982 by D. Reidel Publishing Company.*

camera covering the 1900–3200Å range, were obtained in the high-resolution mode ($\lambda/\Delta\lambda = 10^4$) using the large spectrograph entrance aperture. The pertinent information is given in Table 1. The spectra were obtained during a 12 day period in May through June 1980. All except those for phase 0.13 were obtained during the same orbital cycle.

## 3. Discussion

### 3.1 Spectra

The resonance lines of C II (1335Å), Si IV (1393, 1402Å), Si II (1526Å), Fe II (2599Å) and Mg II (2795, 2802Å) were observed as P-Cygni features. The emission in those P Cygni features varied significantly but no clear-cut changes were detected in the absorption equivalent widths. The radial velocities in the absorption lines showed small (10 to 50 km s$^{-1}$) but real deviations from what is expected for the orbital velocities of the B star; the deviation was in the direction of short wavelengths. Near phase 0.41, the Mg II and Si IV resonance doublets showed edge velocities of –440 and –500 km s$^{-1}$, respectively, indicating that a gas stream was approaching toward the observer at the time of observation. Unless the gas was located very close to the B star, the observed velocity was probably in excess of the escape velocity from the star and possibly even out of the binary system. A somewhat analogous gas streaming was observed in a 2.4-day binary U Cep and reported by Kondo, McCluskey and Stencel (1980) and Kondo, McCluskey and Harvel (1981).

The presence of the Si IV resonance doublet indicates a temperature higher that that appropriate for a B9 star. The Si IV lines probably originated in the gas stream and possibly also in a hot spot on the B star resulting from the accretion of infalling matter; if accretion on to the B star is occurring, then it is the secondary component that is losing the matter. The resonance doublets of Si IV and C IV (1548, 1550Å) were also observed in U Cep, whose hotter component is classified as a B8 star; these Si IV and C IV lines were interpreted in the above-mentioned work as evidence of mass accretion onto the B8 star in U Cep. However, no emission feature was detected in U Cep.

### 3.2 Flux Variations

It became apparent as the exposures were being taken, that the continuum flux was varying significantly outside eclipse. In order to study the variation quantitatively, we selected the regions of flat continuum that were free of any emission or absorption features for several angstroms. After normalization for different exposure times for each spectrum, the ratios of flux levels were taken with respect to those at phase 0.28, which we took arbitrarily as standard; thus, the flux ratio = $F_\lambda$ (phase X)/$F_\lambda$ (phase 0.28). The uncertainty in these ratios is estimated to be 10 to 15 percent.

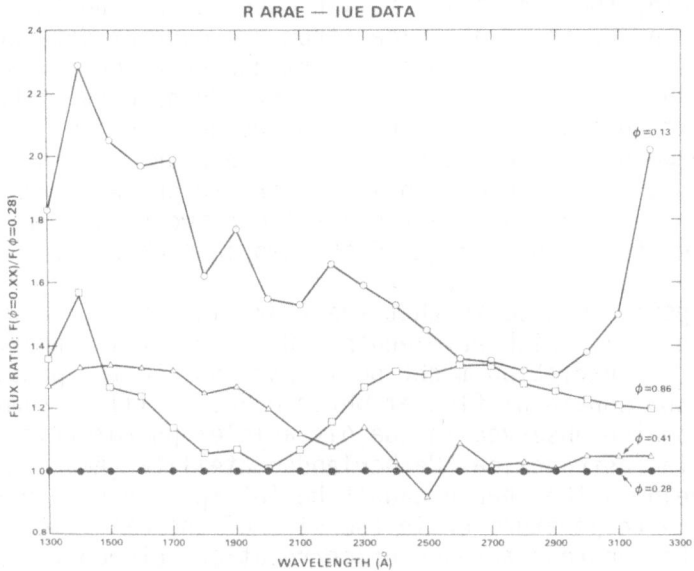

Figure 1.   IUE Flux Ratios of R Arae

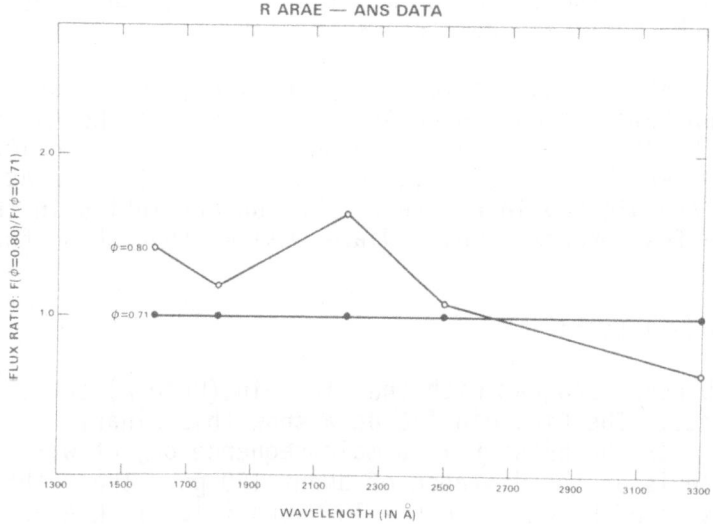

Figure 2. ANS Flux Ratios of R Arae

The flux level at phase 0.13, the only data taken at another orbital cycle 10 days apart, was higher by a factor of up to two. During one orbital cycle, the highest variation observed was up to 50 percent. However, it is not clear from the current data whether the variation was a phase dependent phenomenon or a secular one. We note that the changes recorded for a pair of LWR spectra, taken practically next to each other with an interspacing of only one SWP exposure in between, were quite small and within the errors of determination. Such a pair of exposures were taken twice at phases 0.13 and 0.86; the flux levels given for these two phases were, therefore, taken to be an average of the two in each case.

The important point to note is that the flux did not increase or decrease monotonically with wavelength. Of course this apparently peculiar manner of variation might be in part due to the peculiarity of the comparison continuum flux at phase 0.28. Still, a similar peculiarity was also observed in the ultraviolet photometric data obtained with the Astronomical Netherlands Satellite (ANS), as will be discussed below. The change cannot be interpreted simply as variation in the temperature or in the size of the emitting area in the source. The apparent absence of correlation between the absorption equivalent widths and the observed flux levels present an additional challenge in interpreting the data. The correct interpretation might lie in a combination of several factors, such as an uneven variation in the temperature of the emitting plasma and in the source size. The plasma causing the variation could pertain to the B star, in which case hot or cool spots as well as an unstable expanding or contracting optically thick outer atmosphere might be involved. The variation might in part be caused by optically thick, variable gas streams or disks.

After the current IUE observations were obtained, the 5-band ultraviolet photometric data of R Arae obtained with the ANS became available. The data, obtained during one orbital cycle in 1974, are reported elsewhere by Kondo, McCluskey and Wu (1981). The ANS observations are plotted in Figure 2. The uncertainties in the ANS data are at a few percent level and are smaller than those for the IUE data.

## 3.3 X-ray Observations

On a guest observer program with the Einstein (HEAO-2) satellite, we observed R Arae. The Einstein IPC data show this binary to be an X-ray source. If the B9 star is a main-sequence object with $M_V = 0$, R Arae is at the distance of about 160 pc. Then, the X-ray luminosity of R Arae is $6.5 \times 10^{30}$ erg s$^{-1}$. If R Arae is a luminous supergiant with $M_V = -7$, the binary is at the distance of 4000 pc. The X-ray luminosity, then, is $3.5 \times 10^{33}$ erg s$^{-1}$.

If the foregoing considerations are correct, R Arae is not a strong X-ray source.

## 4. Concluding Remarks

The close binary R Arae exhibits very unusual ultraviolet spectra. So far as we are aware, no other close binary system, observed with any of the orbiting satellites, show outside-eclipse ultraviolet continuum flux variations of this nature. The apparent "seesawing" of the continuum level with respect to the reference spectrum is puzzling. The shortward shifted absorption component in the Si IV and Mg II resonance doublet with a Doppler velocity of some $-500$ kms s$^{-1}$ indicates probable mass loss from the system. In addition, the possible existence of hot spots is hinted by the presence of the Si IV doublet.

Clearly, this binary is passing through a very unusual stage in its evolutionary path. The current results suggest the presence of an optically thick plasma that vary as a function of the phase, or secularly, or both. The plasma causing these variations could involve gas streams, circumstellar or circumbinary disks or an unstable outer atmosphere of the B component.

From the currently available data, we are unable to infer the exact nature of this binary. However, several possibilities, including the following, may be considered. (A) R Arae is in that much talked about but never observationally verified state in which the originally more massive component is currently losing mass. (B) The binary is just entering or coming out of the short-lived dynamic phase of mass transfer, which ß Lyr is purportedly undergoing. (C) The spectroscopically undetected companion is a compact object; this would place the binary in an advanced stage of evolution. If so, the weakness of its X-ray emission might be explained as the result of blanketing of the inner X-ray emission by feeding of the matter at the super-critical rate of $10^{-6}$ solar mass per year or more.

Clearly, much more observations are needed in all spectral ranges, including hitherto unobserved infrared and radio regions. In particular, we need continuous ultraviolet observations during one orbital cycle. Since the IUE telescope time is quite valuable, one practical approach may be to obtain a set of SWP and LWR spectra at high-resolution twice daily for 4.4 days, thus providing a coverage approximately at an interval of one-eighth the period. In addition, a few additional exposures several orbital cycles later will be helpful in investigating possible secular variations.

New spectroscopic and photometric observations from Southern Hemisphere observatories will also be of much value. Of course, except from an Antarctic observatory, it will not be possible to cover this binary continuously during one orbital cycle. Infrared and radio observations would likely add to our information base. For instance, infrared observations might help us pin down the nature of the secondary component; it might at least impose more

Table 1
IUE Observations of R Arae

| Exposures (min.) | JD2444300+ | Phase |
|---|---|---|
| LWR 7781 ( 6) | 77.3338 | 0.127 |
| SWP 9026 (18) | 77.3542 | 0.132 |
| LWR 7782 (11) | 77.3698 | 0.135 |
| LWR 7869 (20) | 86.8696 | 0.282 |
| SWP 9138 (30) | 86.8911 | 0.287 |
| LWR 7877 (30) | 87.4329 | 0.409 |
| SWP 9144 (50) | 87.4703 | 0.418 |
| LWR 7888 (34) | 89.4161 | 0.857 |
| SWP 9151 (55) | 89.4503 | 0.865 |
| LWR 7889 (25) | 89.4815 | 0.872 |

strict conditions on its nature. Radio observations would augment our knowledge of the non-thermal emissions from this binary. Finally, a continuous monitoring of this object in the X-ray region, when a satellite observatory for such purposes becomes available again, would certainly be of much interest.

We wish to thank the U.S. IUE project team, headed by Dr. A. Boggess, for their competent support in obtaining the observations used in this work.

References

Boggess, A. et al.: 1978, Nature 275, 377.
Kondo, Y., McCluskey, G. E. and Stencel, R. E.: 1980, Astrophys. J. 233, 906.
Kondo, Y., McCluskey, G. E. and Harvel, C. H.: 1981, Astrophys. J. 247, 202.
Kondo, Y., McCluskey, G. E. and Wu, C-C.: 1981, Astrophys. J. Suppl. in press (Dec. 1).
Sahade, J.: 1952, Astrophys. J. 116, 27.

ON A POSSIBLE LINKAGE BETWEEN W-TYPE WUMa SYSTEMS AND THE SHORT PERIOD
RSCVn-LIKE BINARIES.

MILANO L., RUSSO G. and MANCUSO S.
Capodimonte Astronomical Observatory, Naples, Italy

ABSTRACT: We analize the general properties of a group of WUMa-type
binaries, which show RSCVn-like activity. The position of these stars
in the colour-density and mass-orbital momentum diagrams is studied,
but no definite answer can be given on the linkage between short period
RSCVn systems (SPG) and WUMa systems with RSCVn-like activity (WWG).

I. INTRODUCTION

In a review paper on RSCVn-like stars, Hall (1976) introduced the
possibility that some systems, which shared with the RSCVn group some
of the observed properties, like the H and K emission, the spectral
type and the presence of a migration wave (more or less pronounced),
could be subdivided into new groups, and introduced the definition of
a 'Short Period Group' (hereafter SPG) and a 'WUMa Group' (WWG). The
general properties of SPG have already been analyzed by one of us
(Milano, 1981). In this paper we want  study the general properties of
the WWG in a similar way, and then to discuss a possible evolutionary
linkage between SPG and WWG. We also introduce an empirical classification criterion based on the type of primary eclipse (transit/occultation).

II. DATA ON SPG AND WWG STARS

a) SPG stars
The lightcurve analysis of the stars belonging to SPG and some speculative considerations showed that these stars are detached systems (see
fig. 1). Up to date, unfortunately, there is no homogeneous study of
the lightcurves of these systems with modern lightcurve synthesis
programs, so the data we present in Table 1 are collected from different
sources and we think are affected by a certain degree of bias. As one
can see from Table 1 the SPG has orbital periods that range from $0.5^d$
to $0.9^d$, the spectral classes range from F8V to G9V for the primaries

327

*Z. Kopal and J. Rahe (eds.), Binary and Multiple Stars as Tracers of Stellar Evolution, 327–335.*
*Copyright © 1982 by D. Reidel Publishing Company.*

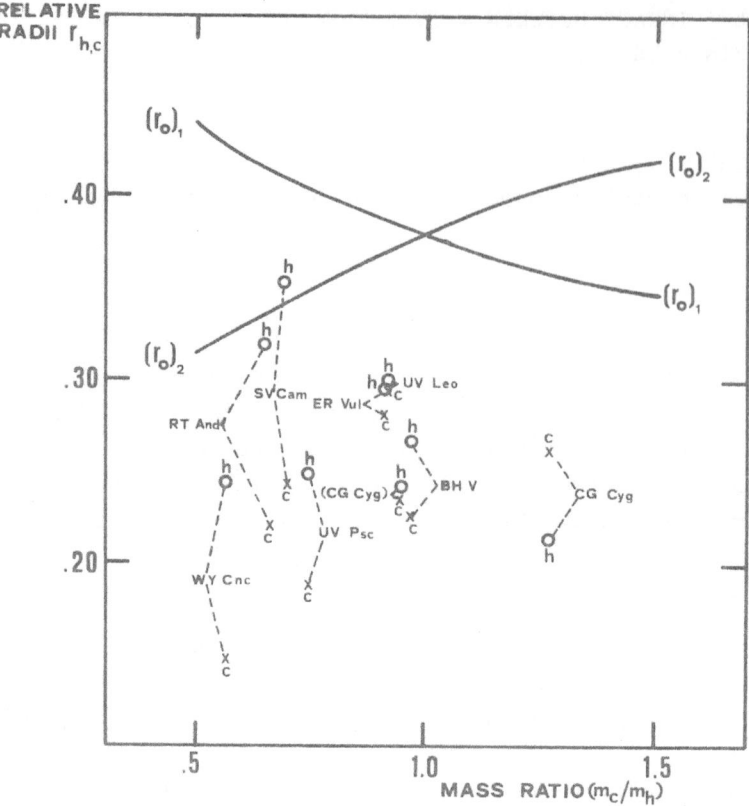

Figure 1: Relative radii for the SPG. The solid lines correspond to the
          contact configuration.

and practically from K0V to K5V, apart from a few exceptions for the
secondaries. H and K emission lines of Ca II have been detected for
almost all the components of the SPG.
There is no information on this subject for SV Cam.
From IUE observations by Budding et al. (1981) it has been ascertained
h and k emission lines of Mg II on almost all the systems of the group
(RT And, XY UMa, WY Cnc, ER Vul, SV Cam, UV Leo doubtful, CG Cyg,BH Vir).
This is a clear index of active coronae and chromospheres. An interesting
problem is the determination of the source of H and K emission, in other
words which component is responsible for the emission. Practically we
have information on this subject only for WY Cnc, UV Psc and for XY UMa.
An interesting fact that came out from a previous analysis (Milano,1981)
concerns the type of primary eclipse for these systems. It was ascertained that almost all the systems have a transit at the primary eclipse.
This fact might be a clue to distinguish different type of RS CVn stars.

| | Period(days) | $m_c+m_h$ | Spectral type | $q=m_c/m_h$ | B-V | T |
|---|---|---|---|---|---|---|
| RT And | 0.629 | 0.98 + 1.50 | F8V+G5-K0V | 0.65 | 0.50 | 6210 |
| SV Cam | 0.593 | 0.7 + 1.0 | G0-G5V+K4V | 0.70 | 0.72 | 5750 |
| WY Cnc | 0.829 | 0.53 + 0.93 | G5V+M2V | 0.57 | 0.61 | 5520 |
| CG Cyg | 0.631 | 1.04 + 0.82 (0.78)+(0.82) | G1V+G9V (G9V+K0V) | 0.79 (0.95) | 0.78 | 5970 |
| UV Leo | 0.600 | 1.25 + 1.36 | G9V+K05 | 0.92 | 0.61 | 5980 |
| UV Psc | 0.861 | 0.9 + 1.2 | G2V+K0IV | 0.75 | 0.81 | 5740 |
| XY UMa | 0.479 | 0.7 + 0.95 | G2-G5V+(K5V) | 0.74 | 0.85 | 5660 |
| BH Vir | 0.817 | 0.86 + 0.87 | F8V+G2V | 0.97 | 0.57 | 6250 |
| ER Vul | 0.698 | 1.13 + 1.23 | G0V+G5V | 0.92 | 0.60 | 5980 |

Table 1: Fundamental data for the SPG. Masses are in solar units and temperatures in Kelvins.

In other words we think it is possible to classify from an observational point of view the RS CVn-like stars in two subcategories, that is, O Type (occultation) and T Type (transit), whatever the period may be. We think, in this way, the class of this stars can be enlarged and it will be possible to understand the behavior of many other eclipsing binaries of the solar type.

b) WWG stars

The data we present in Table 2 on WWG are collected mainly from Mochnacki's paper (1981) on W UMa's and as it is possible to see there are presented the nine candidates displayng RS CVn like activity. Apart from the periods that range between 0.23 to about 0.4 days and the mass ratios it is possible to see that the other parameters are quite similar to the ones of SPG. In fig. 2 is shown the period-colour diagram and the period-mass ratio diagram with the position of SPG and WWG components. As it is possible to see both SPG and WWG are in the same interval of colour D(B-V) = .27 and, obviously there is a shift in the periods of WWG and SPG with a gap practically between 0.4 and 0.5 - 0.6. If we consider the relations

$$C_1 = -.50 - 2.26 \log P$$

$$C_2 = -.50 - 2.26 (\log P - .18)$$

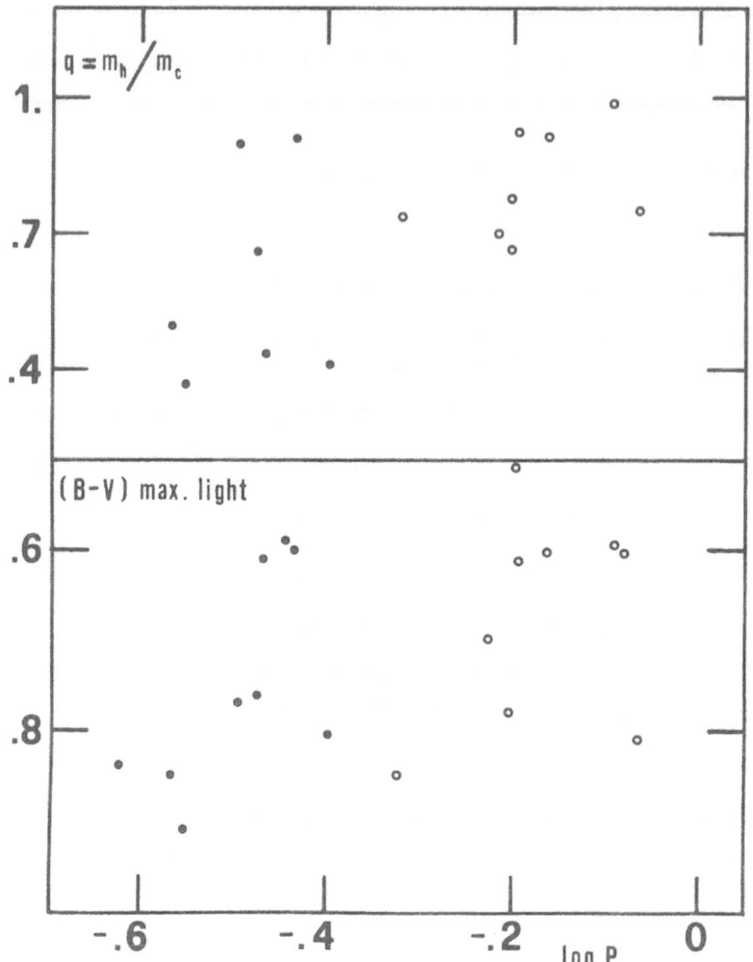

Figure 2: The period-colour and period mass ratio diagram for the SPG
(open circles) and WWG (filled circles).

for WWG, it is also possible to consider a third relation for SPG

$$C_3 = -.50 - 2.26 (\log P - .54)$$

Apart this considerations we also know that WWG components are in a
state of thin or marginal contact whilst SPG components are detached
systems. To try to understand the evolutionary status of these systems
without the knowledge of their absolute dimensions, a good way is the one
suggested by Mochnacki (1981), that is the use of a colour-density
diagram. The procedure we adopted here to derive this graph for SPG and
WWG is derived from the paper already quoted above and the results are
shown in fig. 3.

| | Period(days) | $m_c+m_h$ | Spectral type | $q=m_c/m_h$ | B-V | T |
|---|---|---|---|---|---|---|
| RW Com | 0.2373 | — | G2e+G2e | — | 0.841 | — |
| 44i Boo | 0.2678 | 0.97+0.47 | dG1+dG2 | 2.01 | 0.85 | 5150 |
| VW Cep | 0.2783 | 1.1 +0.4 | G8:n+KOV | 2.75 | 0.86 | 5100 |
| SW Lac | 0.3207 | 1.2 +1.1 | G3p+G3p | 1.11 | | 5400 |
| W UMa | 0.3336 | 1.3 +0.9 | dF8p+dF8p | 1.52 | 0.76 | 5670 |
| RZ Com | 0.3385 | 0.9 +2.1 | F7+GO | 2.3 | 0.61 | 7250 |
| AM Leo | 0.3658 | — | F8V | — | 0.59 | 6350 |
| U Peg | 0.3748 | 1.3 +1.1 | F3+F3 | 1.1 | — | 5830 |
| AH Vir | 0.4075 | 1.4 +0.6 | KOV+(K1) | 2.38 | 0.81 | 5380 |

Table 2: Fundamental data for the WWG. Masses are in the solar units
and temperatures in Kelvins.

In the figure the full lines are the ZAMS and TAMS for Z = 0.002 while
the broken line represents the ZAMS for Z = 0.003, to show the effect
of the variation of metal abundances.  This graph contains the same
information as the period-colour diagram, but with the effects of
different mass ratios and fill-outs removed.
It is to be noted the peculiar position of the primary of CG Cyg, either
for occultation or transit solution (point 1 and 2) and the anomalous
position of AH Vir. This diagram will be considered again during the
final discussion on the evolutionary status of SPG and WWG after having
considered the angular momenta of these systems.
In Tab. 3 we give $H_{orb}$ and $H_{rot}$ and the ratio $H_{orb}/H_{rot}$ for the
components of SPG and WWG. Using the relations between orbital angular
momenta and total mass estabilished by Chaubey (1979) and modified by us
we got the graph of fig. 4. In this figure are shown the WWG, SPG, CG
and LG. It is easy to see that, if there is an evolutionary linkage between
the groups, it must be in the sense SPG → WWG. This might be considered
a working hypothesis to be verified. An interesting fact, which can be
seen, is the position of SPG on the empirical straight
lines deduced for semidetached systems and, taking into  account that

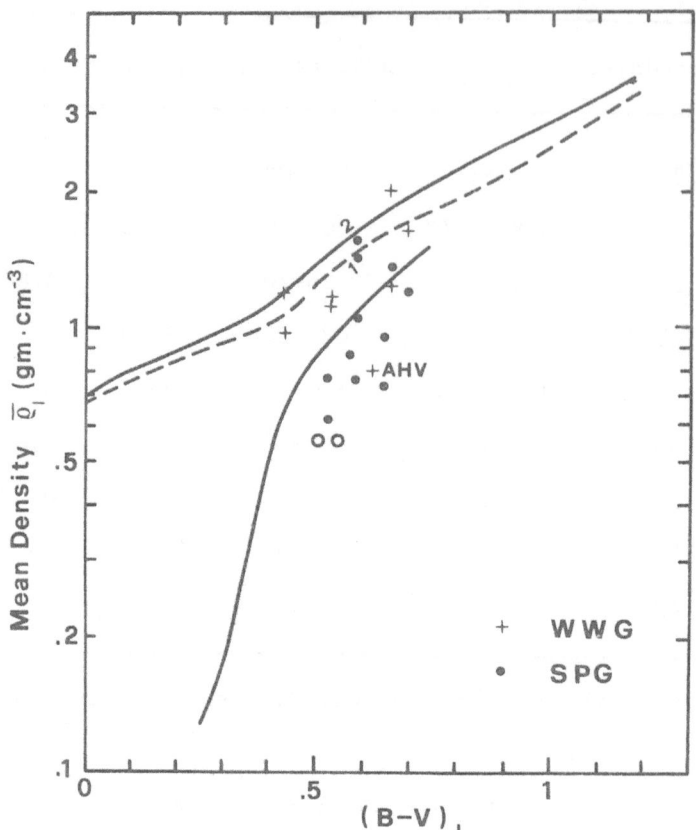

Figure 3: Mean primary density versus corrected primary color $(B-V)_I$.

The ZAMS and TAMS are shown, together with a dashed line to show the effect of varying the metal abundance.

SPG are detached systems, this might be strong indication of orbital angular momentum loss. In other words these systems would be detached systems of low masses that are losing orbital angular momentum.
This fact might be an observational evidence of a linkage between SPG and WWG.
Following Vilhu (1981), taking into account the mechanism of orbital angular momentum loss by magnetic braking, we could again hypothize an evolution from SPG to WWG.
We tried to verify the time of braking of some systems applying the relation by Mochnacki

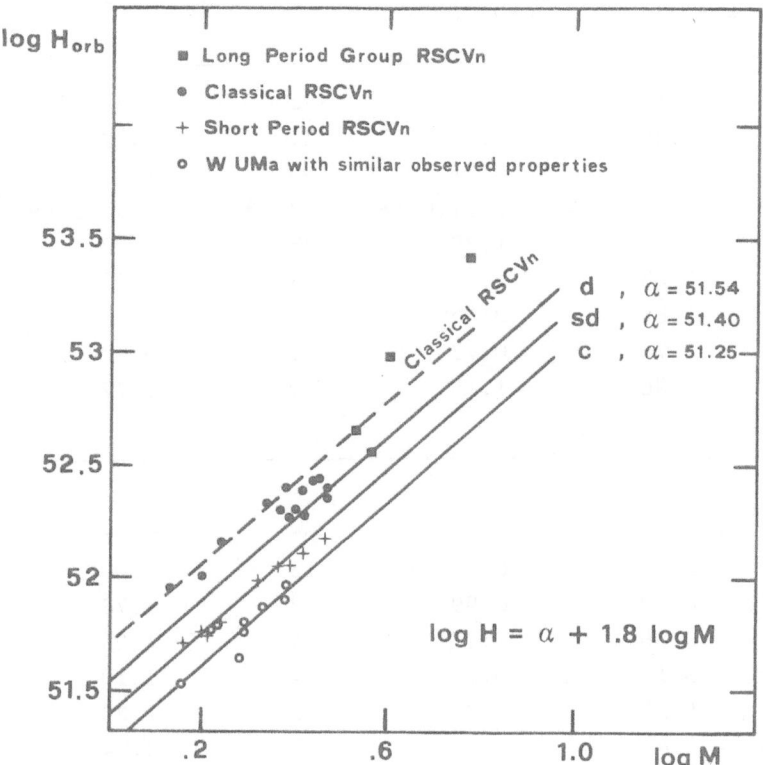

Figure 4: The angular momentum as a function of the total mass of the
system. The lines are taken by Chaubey (1979).

$$\tau_B = \left(\frac{A}{R_1}\right)^3 \left(\frac{R_1\,M^2}{G\,\dot{M}\,\alpha_p^4}\right)^{1/3} \quad f(F, q)$$

The time scale of SPG come out of the order of $\sim 6. \times 10^9 \div 2 \times 10^{10}$ years,
with $\dot{M}$ typically $\sim 10^{-12} \div 10^{-13}$.
Considering now the density-colour diagram for the primaries, we note
an inconsistency in the scenario we exposed above because we have that
the densities of SPG are smaller than those of WWG. This fact might
be an indication of a more advanced evolutionary state of SPG than WWG.

CONCLUSIONS

Two conclusions are possible: a) there is not evolutionary connection
with WWG and SPG.

| | $H_{rot}$ (x$10^{50}$) | $H_{orb}$ (x$10^{52}$) | ratio | $\varrho_1$ (gr cm$^{-3}$) | $\varrho_2$ (gr cm$^{-3}$) |
|---|---|---|---|---|---|
| RT And | 4.12 | 1.15 | 28 | 0.87 | 1.38 |
| SV Cam | 2.46 | 0.61 | 25 | 0.72 | 1.52 |
| WY Cnc | 1.02 | 0.51 | 50 | 1.17 | 3.07 |
| CG Cyg | 3.59(1.34) | 1.51(0.58) | 42(43) | 1.47(1.69) | 2.13(1.65) |
| UV Leo | 4.59 | 1.29 | 28 | 1.02 | 0.96 |
| UV Psc | 2.08 | 0.99 | 48 | 0.93 | 1.59 |
| XY UMa | 1.86 | 0.55 | 30 | 1.44 | 2.53 |
| BH Vir | 1.76 | 0.72 | 41 | 0.75 | 1.25 |
| ER Vul | 3.86 | 1.15 | 30 | 0.77 | 0.83 |
| RW Com | ——— | ——— | —— | ——— | ——— |
| 44i Boo | 2.45 | 0.33 | 13 | 2.00 | 2.60 |
| VW Cep | 8.50 | 0.99 | 12 | 1.78 | 2.51 |
| SW Lac | 8.86 | 1.27 | 14 | 1.26 | 1.30 |
| W UMa | 5.17 | 0.70 | 14 | 1.10 | 1.18 |
| RZ Com | 6.55 | 0.79 | 12 | 1.22 | 1.69 |
| AM Leo | 4.45 | 0.49 | 11 | 0.94 | 1.25 |
| U Peg | 5.87 | 0.95 | 16 | 1.18 | 1.10 |
| AH Vir | 5.14 | 0.59 | 11 | 0.79 | 1.07 |

Table 3: Data for angular and rotational momentum, and for the densities
of the SPG and WWG.

b) there is somewhat of wrong either in Mochnacki's density-colour
diagram or in our computations.
However, our work is still in progress and this very preliminary report
on the hypothesis of a possible evolutionary linkage between SPG and
WWG cannot permit us to derive definitive conclusions. In other words
the problem is open and we are studying the way to overcome the
difficulty we exposed above.

Acknowledgements

This work has been supported by Consiglio Nazionale delle Ricerche (C.N.R.)

## References

Budding, E. et al.: 1982, Astrophys. Space Sci., submitted.
Chaubey, U.S.: 1979, Astrophys. Space Sci., 74, 177.
Hall, D.S.: 1976, Multiple Periodic Variable Stars, IAU Colloquium no. 29
          (Budapest), part 1, p. 297, D. Reidel Publ. Co., Dordrecht,
          Holland.
Milano, L.: 1981, in NATO ASI "Photometric and Spectroscopic Binary
          Systems, E.B. Carling and Z. Kopal (eds.), D. Reidel Publ. Co.,
          Dordrecht, Holland, p. 331.

RECENT OBSERVATIONAL EVIDENCE ON W URSAE MAJORIS STARS

H. Rovithis-Livaniou
P. Rovithis
National Observatory
Athens, Greece

During the four last years (1978-1981), some close binaries of
W Ursae Majoris type have been observed from the National Observatory
of Athens.

In all cases a two-beam, multi-mode, nebular-stellar photometer
has been used, attached to the 48-inch Cassegrain reflector at the
Kryonerion Station.

The observed systems are:  44i Boo, VW Cep, AB And, BV Dra,
BW Dra, U Peg, V508 Oph.  Especially VW Cep and 44i Boo, are very
interesting objects; in addition to the light curves that have already been published (Rovithis and Rovithis-Livaniou, 1980-1981),
other observations were obtained during the last year and are now at
computational stage.  The observations of these two systems will be
continued during the next years.

In figures 1-4, the light curves of 44i Boo in U,B,V,R, are
represented (observations of 1978).

From them, it is obvious that the two minima are of unequal
depth and Max I (the one following primary minimum) is higher than
Max II.  Moreover, in the B, V and R light curves of figures 3 and 4
there is a very good fit between the observations on 5/6 of April and
3/4 of May.  In the U curve of figure 3, there is not much dispersion
in the ascending branch towards Max II while it is large in the one
descending towards the primary minimum.  In the other curves of
figures 3 and 4, the dispersion is quite large between phases 0.570
and 1.0.

Figure 5 represents the B and V light curves of VW Cep (observations of 1979).

In all light curves of VW Cep we have obtained until now, the
two minima are of unequal depth.  This was the case in Kwee's (1966)
observations, but was not so in 1969 when Scarfe and Brimacombe (1971)

337

*Z. Kopal and J. Rahe (eds.), Binary and Multiple Stars as Tracers of Stellar Evolution, 337–342.*
*Copyright © 1982 by D. Reidel Publishing Company.*

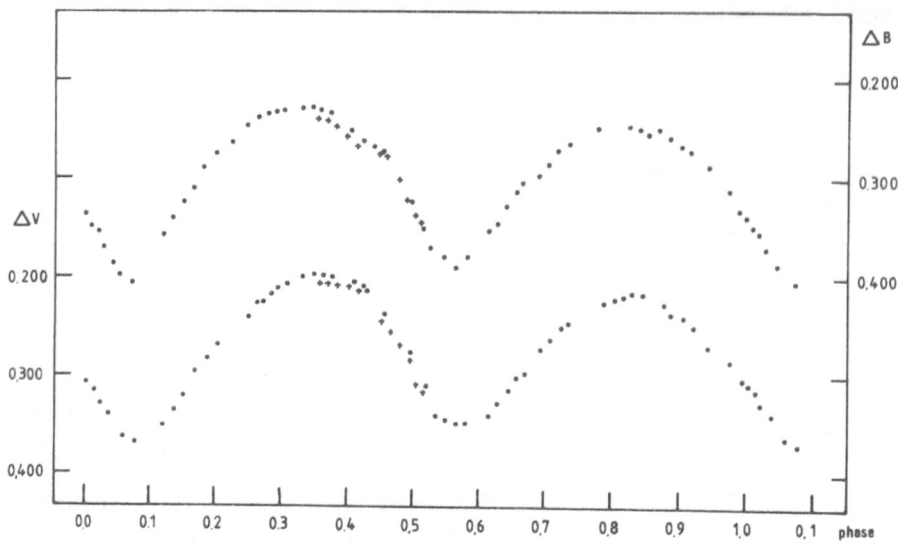

Fig. 1.    B and V light curves of   44i Boo on 15/4/1978. The phases
have been computed using Kukarkin's et al. ephemeris formula.

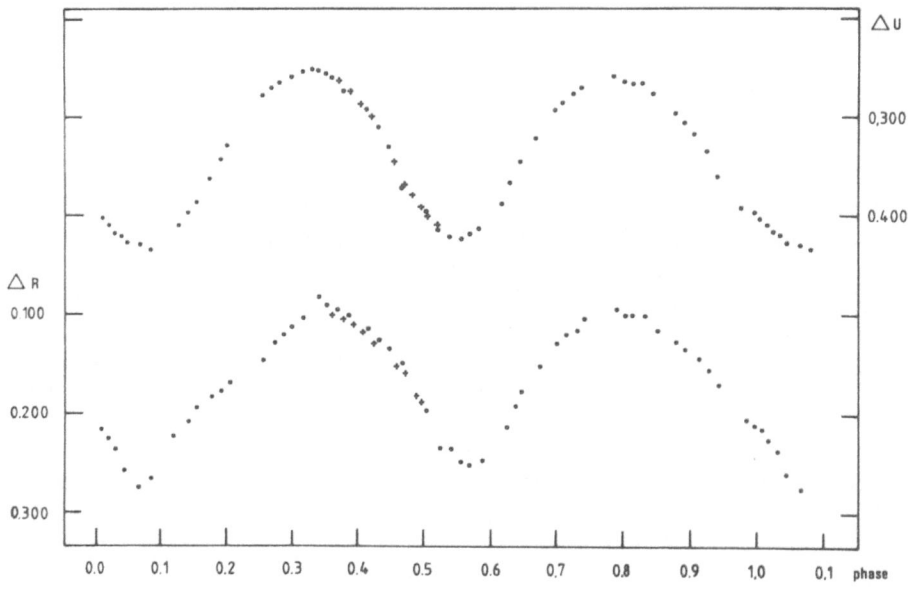

Fig. 2.    U and R light curves of 44i Boo on 15/4/1978.

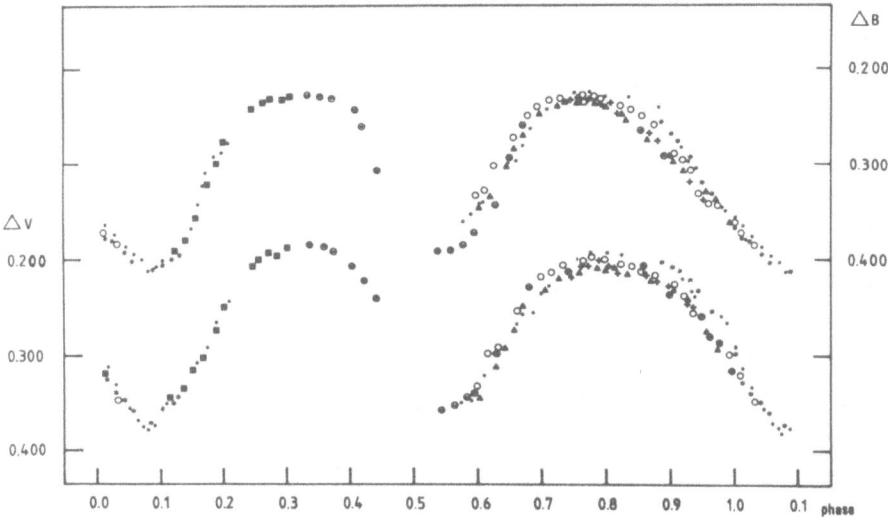

Fig. 3.   B and V light curves of 44i Boo. The individual measurements
          were obtained on 1978 April 5 (⊖), 11 (〇), 12 (▲), 16 (•),
          17 (✻), and 1978 May 3 (■), 4 (+).

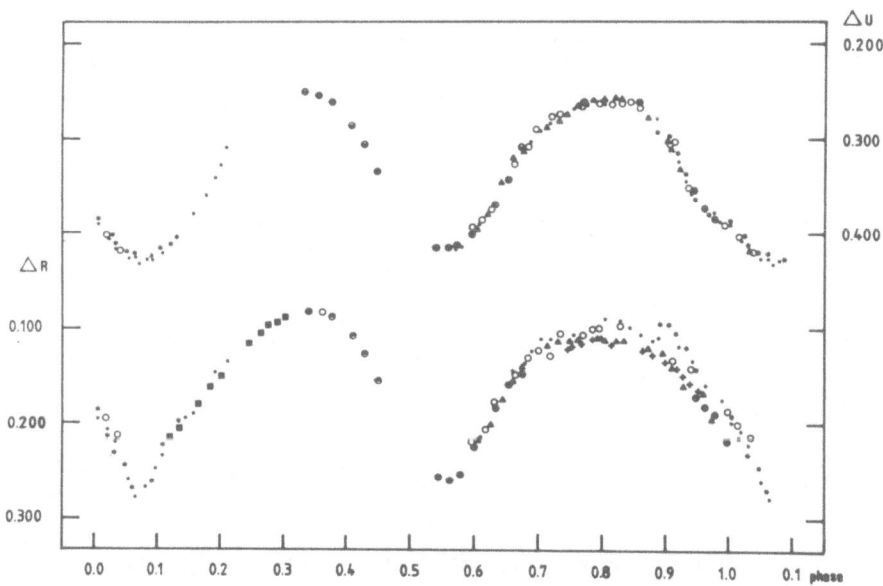

Fig. 4.   U and R light curves of 44i Boo.

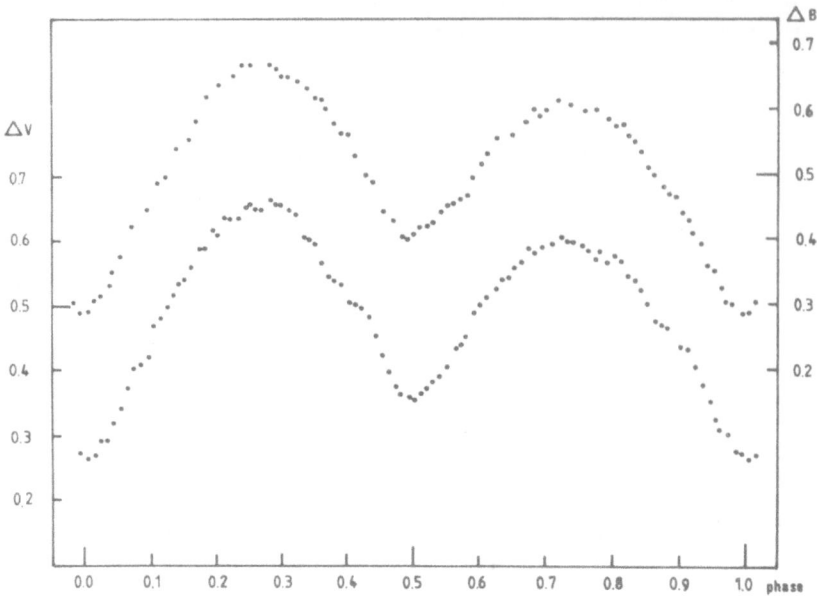

Fig. 5.   B and V light curves of VW Cep during 1979, relative to
HD 192889. (Normal points; unpublished light curves).

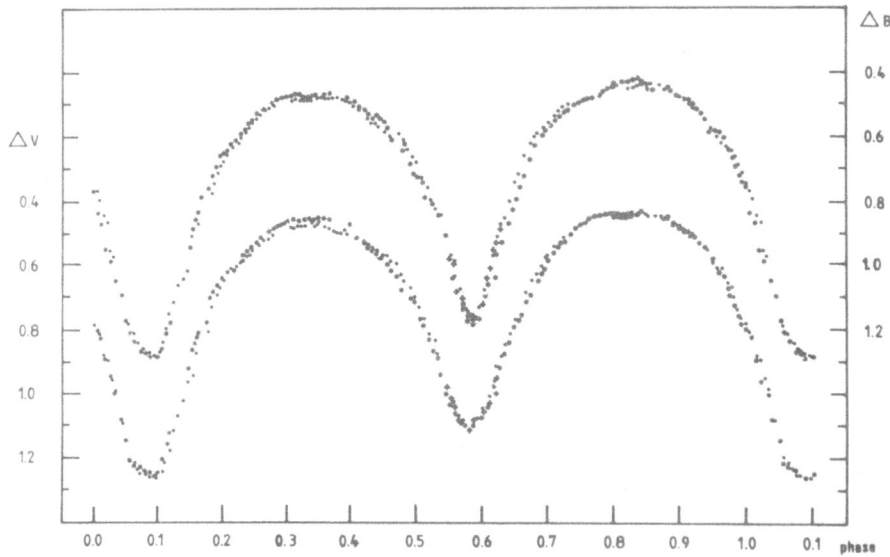

Fig. 6.   B and V light curves of AB And relative to HD 219372, obtained
on 1979 Sept. 19/20 ( • ), 20/21 ( ✱ ), 21/22 ( + ). The phases
have been computed using Kukarkin's et al. formula.

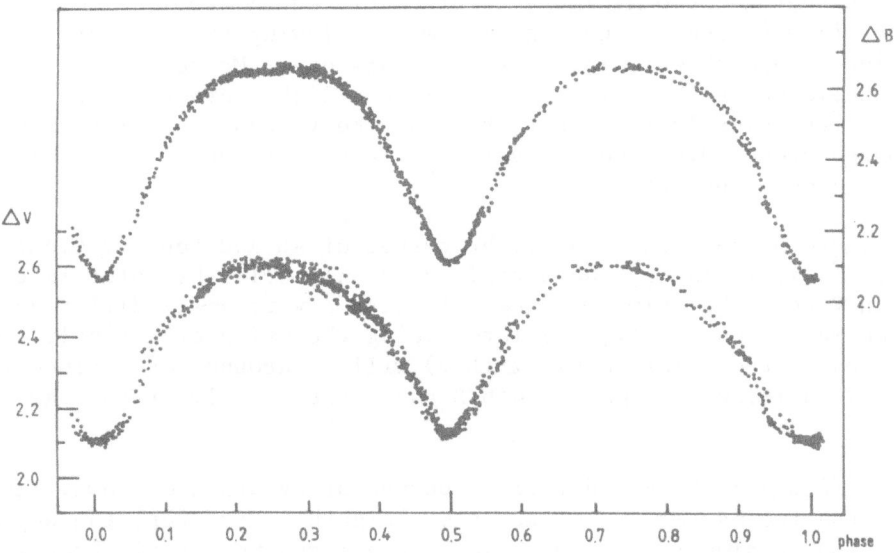

Fig. 7.  B and V light curves of BV Dra obtained in 1980, relative to
SAO 166029 (DM +62°1390).  The phases have been computed
using the formula Min I (hel.) = 2443244.1904 + 0ᵈ35006732 E
(in press).

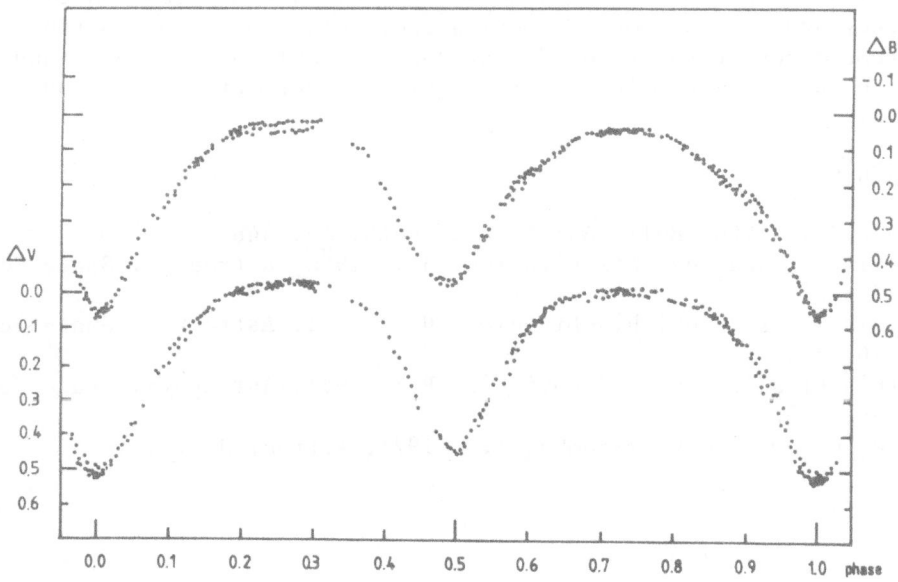

Fig. 8.  B and V light curves of U Peg, obtained in 1980. (Unpublished
light curves).

reported that the two minima were of nearly equal depth.  It is also obvious from figure 5, that the maximum following the secondary minimum is fainter than that following the primary one.  Moreover, there are two "shoulders" visible in almost all four light curves, appearing more or less clearly in either the B or the V band; the first appears at phases around 0.4, and the second at phases around 0.9 (the last one being more obvious).

Figure 6 represents the light curves of AB And (observations of 1979).  As it is shown, the secondary is only $0^{m}115$ brighter in B and $0^{m}16$ in V than the primary.  There is also a very small difference between Max I and Max II, the later being the brighter.  Finally, the variation of the color index  $\Delta(B-V)$ will be around zero, since the magnitude differences $\Delta m$ in both B and V vary by almost the same amount.

In figure 7 the B and V light curves of BV Dra are shown.  In both curves the two maxima are of the same heighth.  Moreover, the depth of the minima is almost the same.  We have a large number of observational points and the dispersion is very small in the B curve.  Unfortunately, this is not the case in the V curve in which the dispersion is quite large.

Figure 8 represents the B and V light curves of U Peg.  Both minima are relatively deep and Maximum I is brighter than Max II (more clear in B curve).

Finally, photoelectric observations of BW Dra have been completed and they are now at computational stage, while for V508 Oph there are some observations taken during June and July of this year and we hope to have them completed either this October, or sometime next Spring.

REFERENCES

Kwee, K.K.:  1966, Bull. Astr. Inst. Neth. 18, 448.
Rovithis, P. and Rovithis-Livaniou, H.:  1980, Astrophys. Space Sci. 73, 27.
Rovithis, P. and Rovithis-Livaniou, H.:  1981, Astrophys. Space Sci. 76, 351.
Rovithis-Livaniou, H. and Rovithis, P.:  1981, Astrophys. Space Sci. 76, 465.
Scarfe, C., D. and Brimacombe, I.:  1971, Astron. J. 76, 50.

# THE TIME CHANGE OF THE AW UMa SYSTEM

M. Kurpińska
K. Otmianowska
Obserwatorium Astronomiczne
Uniwersytet Jagielloński
Kraków, Poland

ABSTRACT

In connection with the period change of the AW UMa system, an A-type contact binary, as it was indicated by Woodward (1980) and elaborated by Kurpińska (1980), we try to analyse with the Wilson-Devinney code all accessible light curves of this star before and after the observed shortening of the period.

REFERENCES

Kurpińska, M.: 1980. IBVS No. 1843.
Woodward, W.: 1980. Astron. Journ. 85, 85.

*Z. Kopal and J. Rahe (eds.), Binary and Multiple Stars as Tracers of Stellar Evolution, 343.*
*Copyright © 1982 by D. Reidel Publishing Company.*

THE [illegible] CHANGE OF THE AI Dra SYSTEM

M. [illegible]
K. [illegible]
Obserwatorium Astronomiczne
Uniwersytetu Jagiellońskiego
Kraków, Poland

ABSTRACT

In this paper the period change of the AI Dra system [illegible] [illegible] indicated by [illegible] (1980) and [illegible] by Kurpińska (1980), we try to analyse with the Wilson-Devinney code all accessible light curves of this star before and after the observed [illegible] of the [illegible].

REFERENCES

[illegible]
[illegible]

[illegible]

# THE SHORT-PERIOD NON-CONTACT BINARY SYSTEMS UU LYN AND GR TAU

A. Yamasaki
Dept. Earth Science and Astronomy,
University of Tokyo, Tokyo 153, Japan
A. Okazaki and M. Kitamura
Tokyo Astronomical Observatory,
University of Tokyo, Tokyo 181, Japan

ABSTRACT

Two binary systems UU Lyn (P = $0\overset{d}{.}468$) and GR Tau (P = $0\overset{d}{.}430$) are discussed on the basis of BV photoelectric observations and spectroscopic observations which were made between 1980 and 1981 at the Tokyo Astronomical Observatory. The component stars in both systems are found to be still inside their Roche lobes but very close to them.

## UU LYN

The eclipsing variable UU Lyn was discovered by Geyer, Kippenhahn and Strohmeier (1955) in Bamberg. Strohmeier, Knigge and Ott (1963a) gave the orbital period (P = $0\overset{d}{.}468461$) and presented the photographic light curve for this system. The spectral type was given to be A4 by Götz and Wenzel (1962b). We are interested in this system because of its rather flat maxima and of the early spectral type for such a short orbital period.

We have made photoelectric and spectroscopic observations for this system. BV photoelectric observations were done with the 0.9-m reflector at the Dodaira Station of Tokyo Astronomical Observatory on 6 nights between February 1980 and March 1981. The V magnitude and B - V colour outside eclipse have been determined to be

$$V = 11.60 \quad \text{and} \quad B - V = +0.41.$$

We found an (O - C) in times of minima calculated by the ephemeris of SAC (1981), and it amounts to -0.0037 days or $-0\overset{p}{.}0079$. Therefore, our observations together with many other visual and photographic observations were used to improve the ephemeris to be

$$\text{Min I} = \text{HJD } 2444674.0507 + 0.46846023 \text{ E.}$$
$$\pm 19 \qquad \pm 6 \text{ (p.e.)}$$

*Z. Kopal and J. Rahe (eds.), Binary and Multiple Stars as Tracers of Stellar Evolution, 345–349.*
*Copyright © 1982 by D. Reidel Publishing Company.*

Our B and V light curves (Fig. 1) show continuous variations, even outside eclipse, which reflect strong proximity effects.

These light curves have been analysed by the light-curve synthesis
method assuming that the surface of either component is represented by
the Roche equipotential surface.  The results are

$$q = 0.40 \pm 0.03\,(p.e.)$$
$$i = 88\overset{\circ}{.}5 \pm 1\overset{\circ}{.}5$$
$$r_1 = 0.45 \pm 0.01$$
$$r_2 = 0.28 \pm 0.02$$

(V)  $u_1 = 0.4 \pm 0.1,$   $u_2 = 0.4 \pm 0.1$
     $\ell_1 = 0.98 \pm 0.01,$   $\ell_2 = 0.02 \pm 0.01$   (at maximum light)

(B)  $u_1 = 0.4 \pm 0.1,$   $u_2 = 0.4 \pm 0.1$
     $\ell_1 = 0.98 \pm 0.01,$   $\ell_2 = 0.02 \pm 0.01$   (at maximum light)

for assumed albedos $A_1 = 1.0$  and  $A_2 = 0.5$, and assumed gravity-dark-
ening $\beta_1 = 0.25$  and  $\beta_2 = 0.08$, where subscripts 1 and 2 represent the
primary and the secondary stars, respectively.

From the analysis both components have been found to be still inside,
but very close to the respective Roche lobes.

Spectroscopic observations have also been done for UU Lyn with the
1.9-m reflector at the Okayama Astrophysical Observatory between December 1980 and April 1981.  Spectra (with the dispersion 37 Amm$^{-1}$) outside
eclipse show the spectral type of F3V.  No significant changes in the
spectral features were found within the orbital cycle although we need
spectra of much higher dispersion to conclude this point definitely.

Fifteen spectrograms have been measured for the radial velocities
of the primary component giving

$$K_1 = 100 \pm 6\,(p.e.) \text{ km s}^{-1}$$
$$V_0 = -43 \pm 7 \qquad \text{km s}^{-1} \quad (e = 0 \text{ assumed})$$

which leads to the mass function

$$f(M) = 0.049 \text{ M}_\odot.$$

Adopting $M_1 = 1.5$ M$_\odot$ for the mass of the primary component for an
F3V star, the mass of the secondary component is estimated to be $M_2 =$
0.6 M$_\odot$.  The resulting mass ratio is  $q = 0.4$, being in good agreement
with the results of the light curve analysis.  The absolute dimensions
of the system have been estimated as

$$A = 3.2 \text{ R}_\odot \quad \text{(separation)}$$
$$R_1 = 1.5 \text{ R}_\odot$$
$$R_2 = 0.9 \text{ R}_\odot$$

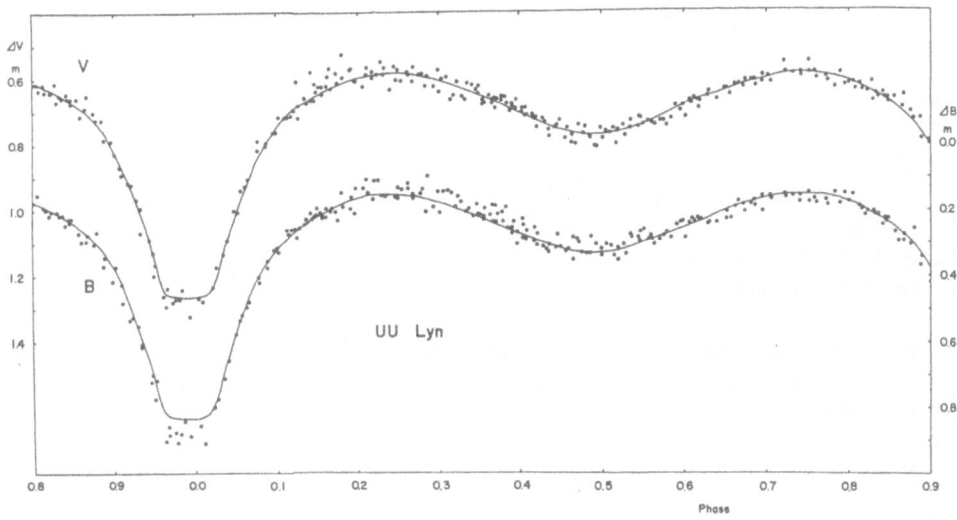

Figure 1.  Photoelectric BV light curves of UU Lyn and the theoretical
solutions (solid line).

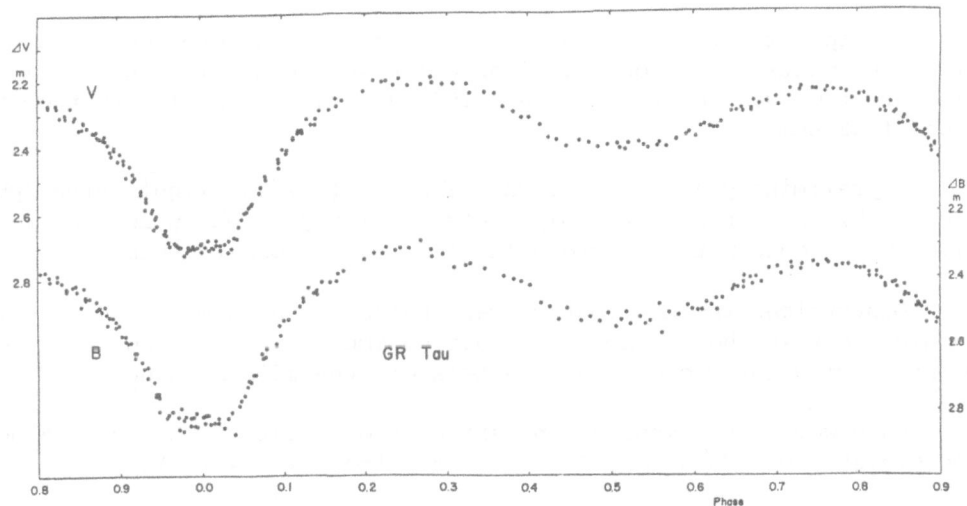

Figure 2.  Photoelectric BV light curves of GR Tau.

(it is difficult to assign the errors to these numbers since they depend on the assumption of the mass for the primary component of UU Lyn). Note that the radius of the primary component is a little larger than that of a main sequence star of F3V.


## GR TAU

This eclipsing binary system was also discovered in Bamberg (Strohmeier, Kippenhahn and Geyer, 1957). Strohmeier, Knigge and Ott (1963b) gave the photographic light curve and the orbital period (P = $0.^d474012$). Götz and Wenzel (1962a) determined the spectral type to be A9. This system looks very much like UU Lyn.

We have made UBV photoelectric observations with the 0.9-m reflectors at the Dodaira and Okayama Observatories on 7 nights between October and December 1980. Magnitude and colours have been determined as

$$V = 10.26, \quad B - V = +0.32 \quad \text{and} \quad U - B = +0.24.$$

We have found that the previous period is incorrect, and the ephemeris is given by

Min I = HJD 2444573.1070 + 0.42985 E.
                              ±9              ±2(p.e.)

B and V light curves show distortions and asymmetries (Fig. 2). The second maximum was fainter than the first maximum by 0.03 mag in V and 0.05 mag in B, respectively.

The spectral type outside eclipses has been determined to be A3V from the spectroscopic observations which were made with the 1.9-m reflector at Okayama between January 1980 and January 1981 with the dispersion of 32 Amm$^{-1}$.

A preliminary analysis of the first half of the light curve (phases 0.0 - 0.5) shows that both components are found to be inside the respective Roche lobes but they are definitely very close to them.

Distortions in the light curves indicate that some kind of activity should occur in the system which must be important from the evolutionary point of view for close binary systems off the main sequence.

This work was supported in part by the Scientific Research Fund of the Ministry of Education, Science and Culture (56540137).

REFERENCES

Geyer, E., Kippenhahn, R. and Strohmeier, W.: 1955, Kleine Veröffentl.
    Bamberg, No. 11.
Götz, W. and Wenzel, W.: 1962a, Mitt. Veränderliche Sterne, No. 628.
Götz, W. and Wenzel, W.: 1962b, Mitt. Veränderliche Sterne, No. 702.
SAC: 1981, Supplemento ad Annuario Cracoviense, No. 52.
Strohmeier, W., Kippenhahn, R. and Geyer, E.: 1957, Kleine Veröffentl.
    Bamberg, No. 18.
Strohmeier, W., Knigge, R. and Ott, H.: 1963a, Veröffentl. Bamberg,
    5, No. 16.
Strohmeier, W., Knigge, R. and Ott, H.: 1963b, Veröffentl. Bamberg,
    5, No. 17.

ON THE SPATIAL DENSITY OF W UMa TYPE STARS

E. Budding
Department of Astronomy
University of Manchester
England

ABSTRACT

   Attention is directed to the anomalous incidence of W UMa stars, which can be regarded as coming from not only a disproportionately large accumulation among close binary systems with primaries later than around mid-F spectral type, but also as a deficit at early types.

   Doubt is placed on the necessity of a straightforward identification of W UMa type light curves with contact binaries; and this allows some reduction in the estimated spatial incidence of contact binaries, from the figure of Van't Veer (1975), to $8 \times 10^{-4}$ of all stars.

   The incidence is considered, with the aid of some simplifying assumptions, as an example of the general evolution of the distribution of binary systems in the primary spectral type - orbital period plane, subject to some known mechanisms of binary evolution.

1. INTRODUCTION

   There are three main empirical points concerning W UMa type systems which have caused them to be at the focus of considerable attention. These relate to (i) their incidence, which is generally taken to be spatially relatively high (Shapley, 1948), (ii) the form of their light curves, which has been closely associated with a proximity so close as to actually imply physical contact or "over-contact" of the two photospheres (Kopal, 1955; Lucy, 1968), and (iii) the existence of a correlation between their periods and colours (Eggen, 1967).

   There have been numerous attempts to reconcile the information summarized by such empirical points within the framework of established physical theory, especially, perhaps, with regard to the contact condition - its mechanism and stability - though, according to Mochnacki's (1981) recent survey, still with only partial success.

351

*Z. Kopal and J. Rahe (eds.), Binary and Multiple Stars as Tracers of Stellar Evolution, 351–370.*

The purpose of this paper is to look more closely again at the underlying empirical points in the hope of eliciting fresh guidelines for
theory.


2.  IS  THERE  A  PECULIARITY  ASSOCIATED  WITH  W  UMa  TYPE  LIGHT
    CURVE  INCIDENCE?

In Figure 1 is presented some information on the incidence of Main
Sequence stars - brighter and fainter single stars, as well as the
primaries of some commonly observed kinds of close binary system.  This
is taken from a recent collection of data (Budding, 1981) on close binary
systems of short period.  For comparison some modified Salpeter function
curves, showing theoretical distributions of dwarf stars brighter than
magnitude 10(1) and 15(2), are shown, where the transformation from mass
to spectral type was made on the basis of curves representing unevolved
Main Sequence stars drawn from the empirically based compilation of
Popper (1980).

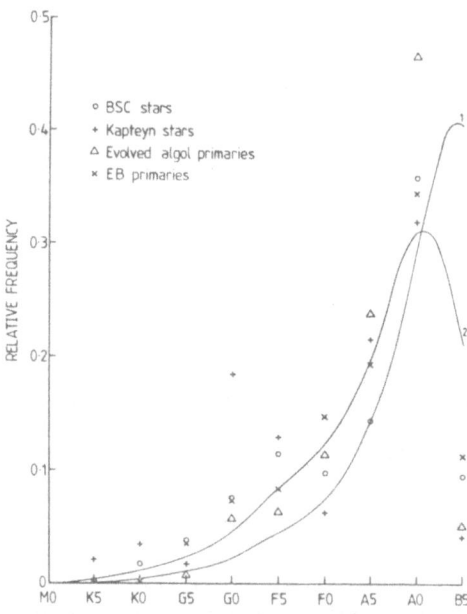

Figure 1.   The relative frequency of dwarf stars brighter than
            some given magnitude in relation to spectral type.

The general trend of fewer stars at types later than AO can be discerned, though there are complications associated with such things as aging effects, extinction and homogeneity of source material.

In any case, there is a sharp contrast between the distribution of normal (brighter) Main Sequence single stars and primaries in relatively unevolved close binary systems, and the primaries of W UMa systems. This can be seen from Figure 2 where, in addition to the variation of frequency of incidence with primary spectral type, the distribution of the systems in the period - spectral type plane is shown. Whereas from Figure 1 we might expect to observe three or four times the number of early A to early G systems if the primaries were distributed like the other Main Sequence stars, we actually find ten times as many early G as early A W UMa type systems.

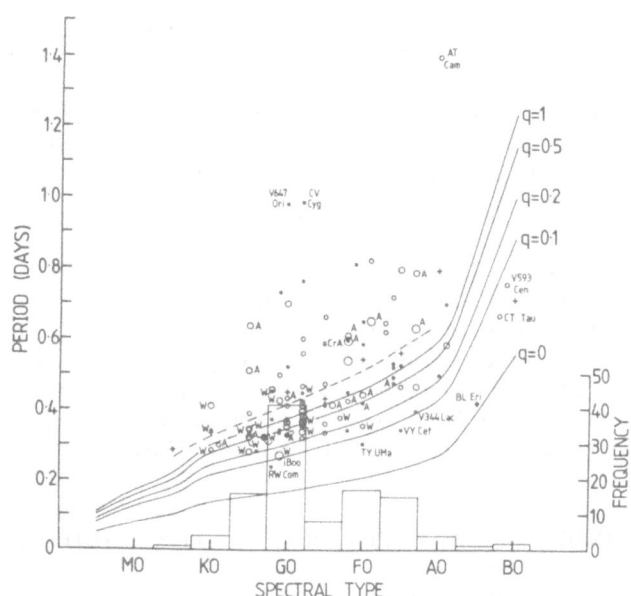

Figure 2.   The distribution of W UMa type stars in the orbital period - primary spectral type plane. The dashed line follows the trend of mean periods. The continuous lines give periods at which ZAMS primaries of given spectral type accompanied by stars of relative mass q are in contact with their Roche lobes.

Before leaving Figure 2 some other relevant points may be noted as follows:

1.    The correlation between period and colour of these systems, noted by
Eggen, is broadly confirmed in the period spectral-type diagram, though
there is an appreciable scatter.  Though spectral type has been quoted
for more systems than those studied by Eggen and is free of reddening
complications, it may reflect some qualitative factors in the matter of
type assignation, and the sample shown, which comes from various recent
catalogues and papers, is certainly inhomogeneous.  Comparison of such
sources reveals, however, that the type assignation is rarely likely to
be in error by more than five type class subdivisions.

2.    There is an overlap in the spectral type-period plane between systems
of W UMa type and other types of close binary system, i.e. a system like
V753 Cyg of primary type F8 and period $0.476^{d}$ is actually classified as
having an Algol type light-curve (Kukarkin et al., 1969) though it lies
close to the centroid of the distribution of W UMa-type systems in the
spectral type-period plane.  Another similar case might be the variable
BD And, also of type F8 and period 0.463 but with light curve classified
as β Lyr type.

3.    Though there is a pronounced peak in the distribution around spectral
type GO, light curves which have been classified as of W UMa type do
occur at all spectral subgroups from BO to K5.

4.    There are a number of exceptional systems, with either anomalously
long or short periods.  These include:  AT Cam, CV Cyg, V647 Ori, AZ Gem
and DV Peg (all with periods around 1 day or greater), and V593 Cen,
CT Tau, BL Eri, V344 Lac, VY Cet, TY UMa and RW Com (all with periods so
short as to suggest "deep contact" if the primaries are at all comparable
to Main Sequence stars of the same type).  It seems quite possible that
some, if not all, of the long period anomalies are not really W UMa stars
at all, but RRc type variables (c.f. Eggen, 1967).


3.  TOO  MANY  COOL  OR  TOO  FEW  HOT  W  UMa  SYSTEMS?

        Having confirmed the existence of a remarkable overabundance of
W  UMa systems with spectral type around GO, we may next wish to consider
whether this anomaly arises from a surfeit among the later type stars,
a deficit among the early types, or perhaps both.

        In order to examine this we consider first the distributions, with
period, of close binary systems of unevolved type of different spectral
type groups.  In Figures 3 and 4 the frequencies of unevolved close
binaries of spectral types A, F and G are plotted normalized against the
quoted sample sizes.  For comparison we have plotted a curve which could
be approximately proportional to the frequency of detected eclipsing
systems on the basis of some simplifying assumptions.

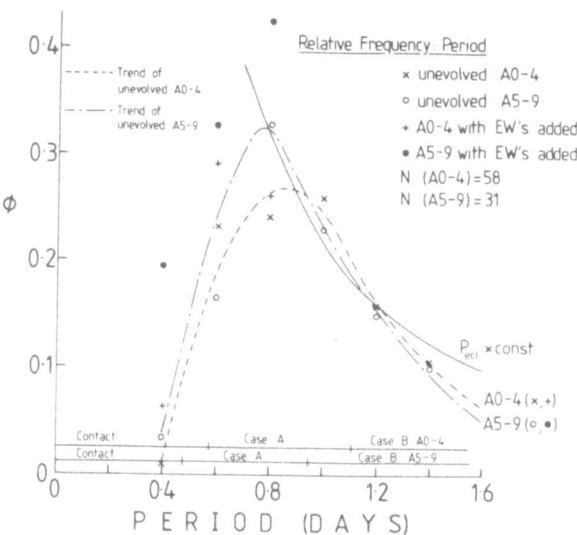

Figure 3.   Short period binaries of primary spectray type A.


Such a curve has the form

$$\phi = \text{const.} \times \left(1 + \frac{\Delta P}{P_*}\right)^n \qquad\qquad (3.1)$$

where the constant factor has been chosen so that the curve lies close to
the observed points in some selected range.   $P_*$ corresponds to the period
at which a pair of unevolved Main Sequence stars of equal mass, representative of the type range in question would be in "contact" and $\Delta P$ is the
difference between any particular period P and $P_*$.   The spatial incidence
of such binaries in a given interval around P would be proportional to
$P^{n+2/3}$, i.e. if spatial incidence was constant, observed incidence would
fall off with a $-\frac{2}{3}$ power dependence on period.   The form which has been
chosen in Figures 3 and 4 sets n = -5/3, which would correspond to a
power law distribution of semi-major axes A of the form $g(A) \propto A^{-1}$.   Such
a power law form has been considered appropriate by, for instance, Tutukov
and Yungelson (1980), though they had in mind somewhat wider systems than
those considered here.

   What can be judged from such diagrams are the following:
1.   Though there is some indication of more of a bunching together at
shorter periods for the later type stars, the distributions are feasibly
of the same form in dependence on period for different mass binaries, i.e.
there could exist some function which when suitably scaled for masses or

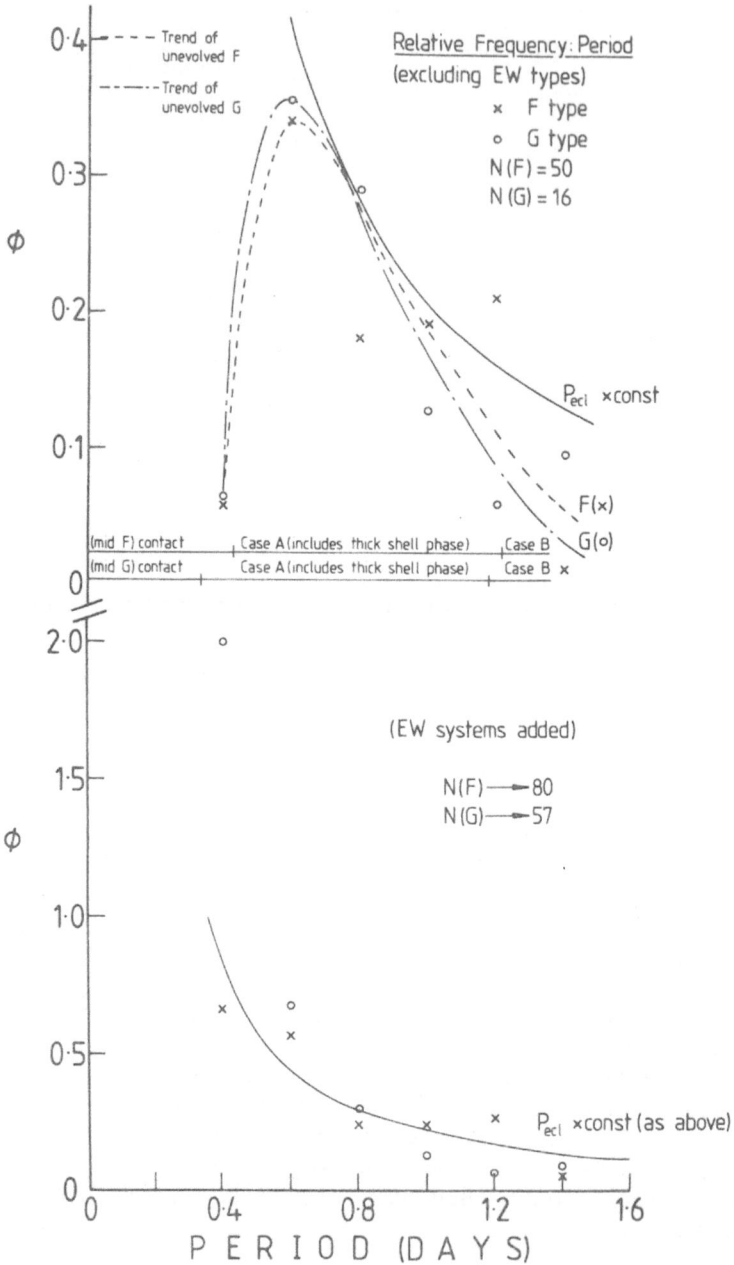

Figure 4.  Short period binaries of spectral types F and G.

sizes of the constituent stars could describe the distribution of
observed eclipsing unevolved systems of all spectral types with period
(c.f. Farinella et al., 1979).

2.    The maxima of these distributions occur at period values rather in
excess of the contact values, i.e. there is some falling away of the
incidence of unevolved close binaries which are not W  UMa type systems,
but close to contact.  This suggests the possibility that some systems,
classified as W  UMa type on the basis of their light curves, are actually
not in contact but just very close unevolved pairs.  This is perhaps more
noticeable at earlier type.

3.    There appears to be some fall-off at longer periods in the number of
observed systems relative to the comparison curves.  This could be due
to some failure of any of the foregoing assumptions, which at least in
the spherical star respect are oversimplified, while with respect to the
power law distribution have no necessary physical basis.  Farinella and
Paolicchi (1978) for instance, consider an underlying distribution of
truncated Gaussian form in the mass per unit volume of binaries.

4.    When the W  UMa type stars are added into the distributions, in the
case of the A-type systems there is, of course, some rise at the short
period end; but the early A systems still lie appreciably below the
comparison curve at periods just greater than the contact value.  The
later A systems exhibit also a deficit at periods just greater than the
contact value, though with suitable averaging to include the incidence at
slightly longer period, the discrepancy is not as great.  Also it is
clear that there is a build up at periods less than[*]the shortest period
for unevolved (equal mass) contact for these stars.

5.    For F type systems, inclusion of the W  UMa stars allows the distribution to match the comparison curve tolerably well right up to the
contact period - though, of course, this need not mean that we are dealing
with a uniform class of object in which proximity only is varying.

6.    However, it is among the G systems, where the previously noted
difference in the distribution from that of normal Main Sequence stars
was strongest, that we observe a conspicuous overabundance over the
comparison curve.

      Hence the overall answer to the question posed at the beginning of
this section is that the discrepancy in the distribution of W UMa type
primaries compared with normal unevolved Main Sequence primaries in
binary systems (which distribute like Main Sequence single stars) clearly
involves a surplus of late spectral type stars, but could also imply a
deficit of earlier spectral type systems if there should be a uniform
distribution of close binaries.

*
      Systems of still earlier type exhibiting a similar trend were considered
by Wilson and Rafert (1981).

## 4.  DO  W  UMa  TYPE  LIGHT  CURVES  NECESSARILY  IMPLY  CONTACT?

In considering possible interpretations of the foregoing information on W UMa system incidence, and the relationship of these systems to the more conventional unevolved detached kind, the question of whether such stars are necessarily in contact arises (e.g. point (2) in Section (3)). In this connection let us note in Figure 5 a W UMa type light curve which has been generated from a pair of stars not actually in contact.

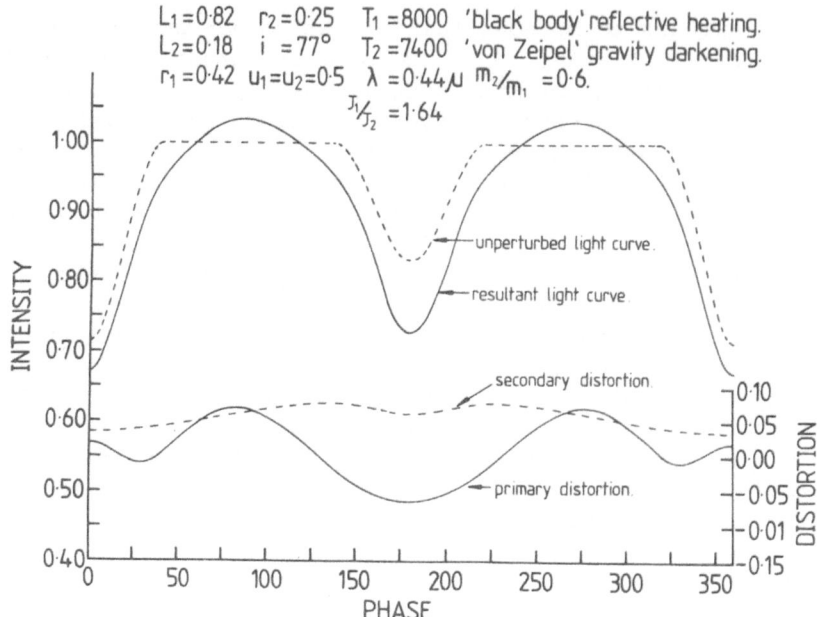

Figure 5.   A light curve produced by a pair of close, but not energy transferring "over-contact" stars, which looks like that of a W UMa system.

The difference in luminosities for the given mass ratio (0.6) would be appropriate for a Main Sequence primary, originally of early A type, but having evolved somewhat from ZAMS, for which the ratio of radii has appropriately decreased (from an initial 0.72 down to 0.6). The otherwise generally standard modelling parameters are indicated on Figure 5. The main cause of the W UMa like light curve comes from the large scale of the "ellipticity" distortion of the primary star, which adds to the depth of the secondary minimum, though it is largely eclipsed out at primary minimum. Though the depths of both minima are approximately equal the mean surface flux of the primary can then be more than 1.6 times that of the secondary at the wavelength of observation.

The possibility of this kind of light curve was presented in somewhat general and approximate terms in Budding's (1981) article, where it
was argued that close but non-contact pairs for which the primary was of
type earlier than, say, mid F could give rise to an EW type photometric
variation, though no actual case was cited. In Figure 6, however, we
have, with RR Cen, a possible real example of a system of similar type.
In Table 1, we present our optimal parameter set for this light curve -
values obtained by procedures discussed by Budding and Najim (1980).

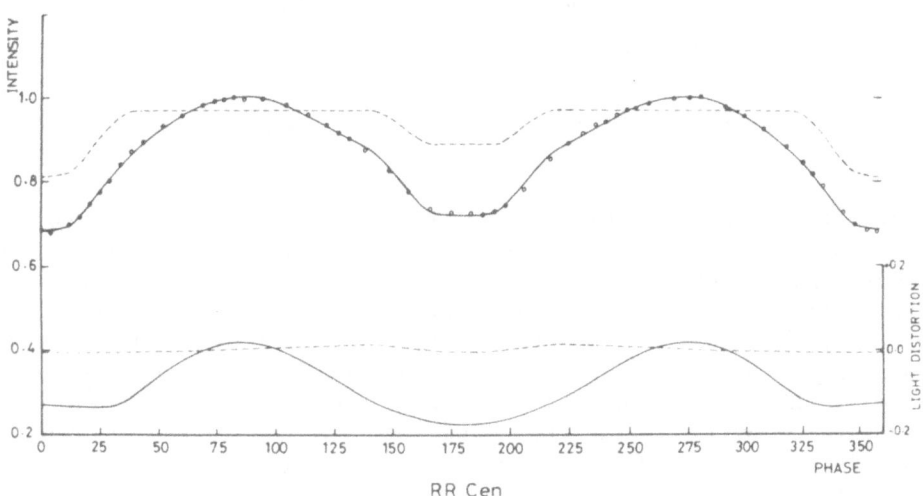

RR Cen

Figure 6.  Knipe's $\lambda 5280$ light curve of RR Cen as modelled by the
parameters given in Table 1. (The reference light level is
first normalized to unity to provide the quantities
$L_1$, $L_2$.).

The luminosities $L_1$, $L_2$ are constrained so that their sum is unity.
Radii $r_1$ and $r_2$ correspond to the volumetrically defined quantities $(r^*)$,
in Kopal's (1959) terms, whose sum, being approximately constant at 0.75
in the condition of "contact" of the two stars, can be used as a dis-
criminant about this condition. The totality at the second minimum acts
as good constraint on geometrical elements, serving, for instance, to show
that the inclination i could not be much less than the derived 80°. The
close to unity value of reduced chi-squared ($\nu$ represents the number of
degrees of freedom, which amounts to 50 for Knipe's (1965) normal points),
indicates that the observations are in excellent accord with the fitting
function on the basis of Knipe's accuracy assessment $\Delta \ell$. The mass ratio
$m_2/m_1$, limb darkening (u), gravity darkening ($\tau$) and reflection coefficients (E) are adopted quantities. Some other details of notation or
method may be found in Budding and Najim (1980). A slight, probably in-

| | | | |
|---|---|---|---|
| $L_1$ = 0.913 | ± 0.02 | $u_1$ = 0.58 | |
| $L_2$ = 0.087 | | $u_2$ = 0.5 | |
| $r_1$ = 0.43 | ± 0.01 | $\tau_1$ = 1.0 | |
| $r_2$ = 0.19 | ± 0.03 | $\tau_2$ = 1.1 | |
| $i$ = 80° | ± 1° | $E_1$ = 1.0 | |
| $\Delta\theta$ = 0.2 | ± 0.3 | $E_2$ = 1.1 | |
| $m_2/m_1$ = 0.6 | | $\Delta\ell_{(s.d.)}$ = 0.0055 | (Knipe's value) |

$$\chi^2/\nu = 0.93$$

Table 1. Optimal parameter set for Knipe's (1965) light curve of RR Cen ($\lambda_{eff}$ = 5280 Å).

significant, correction to the zero point of the listed phases, $\Delta\theta_o$, was also found.

At first sight it seems difficult to reconcile the observed ratio of radii ($\sim$ 0.45) with the assumed mass ratio (0.6) on the basis of the "standard" Main Sequence mass radius relation. However, if we compare a pair of stars like, for example, the 1.5 and 1 $M_\odot$ models whose evolution tracks were computed by Iben (1967), it can be observed that the ratio of radii has dropped from an initial 0.77 to 0.37 by the time of the end of the (Main Sequence) thick shell burning phase of the more massive star. By this time the bolometric luminosity ratio would have reached 14, which, allowing for a slight bolometric excess to the more massive star, which would appear as of early F type, still surpasses somewhat the derived value for RR Cen of about 10.5. The observed parameters could, however, be feasibly matched by a pair of stars still evolving in the Main Sequence band, the primary, of mass about 1.5 $M_\odot$ towards the end of this stage, with a secondary not far off 1 $M_\odot$ and relatively little evolved.

A possible difficulty rests with the fractional radius of the primary which, for the adopted mass ratio 0.6, already seems too big for its Roche lobe mean radius ($r_1 \sim y_5$ - in Kopal's 1959 notation - = 0.42). There are various remarks one might make about this; concerning, for example, the appropriateness of underlying approximations, the Roche model formulae, effects of truncation of series of terms, or the sizes of probable errors. However, the main point of the present section is not to show that one or other model is the definitive one, but that the scale of uncertainty when dealing with light curves such as this is such as to allow models of inherently quite different kinds to be able to reproduce the observations plausibly. In a word, the contact model need not be unique.

In this way it could be argued that a good many, perhaps most, of the W UMa systems with primary spectral type earlier than about F5 and periods greater than half a day need not be in contact at all. A much stronger case for contact comes with the W UMa systems of later type and low periods, as was argued by Budding (1981). Such stars may well represent the bulk of the classical contact W UMa systems, which form the basis of numerous special studies.

## 5. AN ASPECT OF THE PARAMETER DETERMINACY QUESTION FOR W UMa LIGHT CURVES

Though the results of the foregoing section imply that light curves of W UMa type do not necessarily imply contact, there is clearly an ambiguity since numerous authors have generated light curves of the same general form from models of stars which are in over-contact. The general problem of determinacy and uniqueness in curve fitting is rather broad and cannot be fully dealt with here, but there is one particular aspect of W UMa light curve generation which, as more data becomes available, might be capable of receiving further empirical testing. This refers to the differing possibilities with regard to gravity darkening (or brightening), about which different authors have presented different ideas.

The main point of present relevance about this is that the generation of light curves requires some description of the extent of gravity darkening (usually by means of a single pair of parameters), but the adopted position on this will influence the resulting values of other quantities treated as unknowns. This affects the degree of observational support for the contact hypothesis (Anderson et al., 1980; Kopal, 1968). Anderson et al. (op cit.) urge detailed consideration of spectrographic evidence to help resolve this question. Alternative suggestions may be offered, as follows.

Firstly, the main geometrical elements determined at different observation wavelengths should be sensibly the same. Any systematic variation with wavelength may reflect model inadequacies such as an imposed incorrect gravity darkening parameter. Secondly, the promising new method of differential polarimetry might be applied to advantage to a few of the brighter W UMa systems. It could, in this way, be possible to provide some independent check on orbital inclinations, whose values can be seen to correlate with assumed gravity darkening parameters in published lists of parameter values.

In connection with this latter point a further test is, in principle, possible. In Figure 7 we compare a distribution of the quantity $x^2$, where $x = (\sin i - \sin i_{min}) \div (1 - \sin i_{min})$, $\sin i_{min}$ being given by $\{1 - (r_1 + r_2)^2\}^{\frac{1}{2}}$, for a sample of 49 essentially uncomplicated and well determinable detached pairs, derived originally from Svechnikov's (1969) compilation, but with some additions and modifications based on more recent analyses of 13 of these systems, with that coming from the tabulation of geometric elements of 35 W UMa systems published by Mochnacki (1981).

The range of inclinations available to a system composed of spherical stars of relative radii (in terms of the mean separation) $r_1$ and $r_2$ in

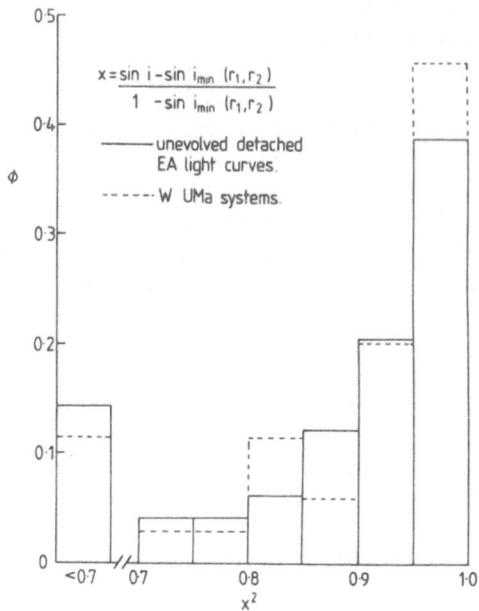

Figure 7.   Relative frequency $\phi$ of systems of different inclination.

which they can be seen to eclipse is from $90^{\circ}$ to $\cos^{-1} (r_1 + r_2)$.   The
probability of finding a system at inclination i within this range would,
in principle, be proportional to sin i, assuming an arbitrary distribution of orbital revolution axes over the sphere and if there were no
selection effects related to the amplitude of the light variation associated with the eclipses.   If we are comparing distributions whose range of
possible inclinations is different, due to different proximity, we can
standardize by use of the aforementioned variable x (Note $0 \leqslant x \leqslant 1$).   The
more realistic approach to the probability of a certain value of x is
therefore to write $P(x) = x\, S(x)$, where $S(x)$ expresses the selection
effect referred to.   It does not seem plausible to suppose that $S(x)$ can
be precisely specifiable for the general context we are considering, in
which the spherical star geometry is, in any case, an oversimplification.
However, in the combined interests of simplicity and clarity we have, in
Figure 7, scaled the abscissae to values of $x^2$.

Though the sample sizes are rather small to allow any decisive interpretation at this stage, it could be regarded as odd that the distribution
of derived inclination values for W UMa systems is relatively somewhat
more compressed towards $90^{\circ}$ than that of the detached systems.   In fact,
since the underlying arguments have nowhere referred to proximity effects,
which should enhance the chance of discovery over a purely spherical case
at lower inclination, one would expect the biassing to be in the opposite
sense to that observed.   A possible explanation for such a discrepancy
could be due to a systematic error of procedure in relation to the assigned

gravity darkening parameter used in the sources quoted by Mochnacki (1981).
Other things being equal, an assigned low value of gravity darkening
coefficient would require a higher inclination to produce the same
"photometric ellipticity".


## 6.  THE  SPATIAL  INCIDENCE  OF  CONTACT  BINARIES

It is clear, in a general way, from the high relative incidence of
W  UMa stars among cooler stars, and the known high spatial density of
low mass stars that the number of W  UMa systems as a whole, in the
galactic field, must be comparatively large.  About 400 EW variables
($\equiv$ W  UMa type) are listed among 4062 eclipsing variables of all types
in the "General Catalogue" of Kukarkin et al. (1969) (Van't Veer, 1975;
Yamasaki, 1975).  Lucy's (1976) data suggests that he counted about 120
such systems brighter than magnitude 12, a sample which is probably close
to the.118 systems of known spectral type used to form Figure 2.

Comparing the distribution of such stars having photographic magnitudes brighter than a given value, $N_{EW}$, with the average numbers of
stars $N_*$ given by Allen (1973), we find ratios as given in Table 2.

| $m_{pg}$ | $(N_{EW}/N_*) \times 10^{-4}$ | $(N_{EAU}/N_*) \times 10^{-4}$ |
|---|---|---|
| 7 | 4.7 | 0.6 |
| 8 | 5.6 | 1.5 |
| 9 | 3.8 | 1.3 |
| 10 | 2.4 | 1.9 |
| 11 | 1.7 | 1.3 |
| 12 | 1.0 | 0.6 |

Table 2.   Incidence of W  UMa (EW) and unevolved, detached binaries
           (EAU).


A selection effect operating against the discovery of fainter
variables can be assumed to become significant, at least by the ninth
magnitude, and a reasonable estimate for the discovered incidence
frequency $\alpha_d$ would appear to be about $5 \times 10^{-4}$ (c.f. Van't Veer, 1975).

What this means in terms of the actual spatial incidence $\alpha_s$ of
genuine common envelope W  UMa systems depends on how light curves are
interpreted.  If we assume that this configuration really only refers to
the aforementioned considerable excess at around spectral type G0, a conservative estimate of $f_1$ = 50% may be put for systems classified as
having EW type light curves to be of over-contact type.  Figure 7 suggests
that the selection effect in inclination $f_2$ is so severe that perhaps only
the range $1 \gtrsim x \gtrsim 0.98$ of the eclipsing sample is actually complete.  It
will be assumed, however, that this range, which should account for 12%

of the entire group if their orbital axes are distributed randomly to the
line of sight, actually corresponds to 39% of the observed set. This
percentage really refers to the more populated comparison group of detached
binaries - the percentage of EW systems in this range is somewhat more
than 39% according to Figure 7, but the point of that comparison was to
suggest possible systematic error in the photometrically derived inclinations of W  UMa systems.

We finally derive for the spatial incidence of contact W  UMa type
binaries a proportion

$$\alpha_s = f_1 f_2 \alpha_d = 5 \times 0.5 \times (0.39/0.12) \times 10^{-4} = 8 \times 10^{-4} \tag{6.1}$$

of all stars. This figure is rather less than that proposed by Van't Veer
(though subject to essentially the same uncertainty, i.e. $\sim$ 50%), chiefly
because of the more conservative estimate of what is likely to be a contact system ($f_1$), and also because of some difference in the estimated
proportion of W  UMa stars actually seen ($f_2$).

## 7.  POSSIBLE  LINES  OF  EXPLANATION

In considering the origin and high incidence of contact W  UMa
systems two major lines of approach have been followed:
  (i)   some stars may originate in the contact condition (incomplete
        fission) and remain in, or indistinguishably close to, such a
        state for nuclear timescales; or
  (ii)  stars may become like this from originally detached binaries
        through some process involving a loss of angular momentum. The
        high incidence should then be related to relevant properties
        of the supposed antecedents of the contact systems.

Of the two approaches the second would appear to be more pragmatic
than the first, in the sense that the first requires us to explain why
just binaries should be formed in this way, and why such binaries should
be confined to a particular spectral range. The stability of the configuration through the various stages of its formation should also be
examined in order to establish its required duration. It may be possible
to do this; but in the second approach less presumption is possible, since
both binaries and angular momentum loss are known to exist independently
of the existence of W  UMa stars. Moreover, since angular momentum loss
would normally imply also mass loss, some path may be open to account for
the excess of low mass systems, perhaps together with the deficit of comparable high mass systems. Of course, many more possible contact W  UMa
stars exist than could directly be accounted for by the observed deficit
of very close spectral type A systems. The situation might be interpreted in terms of a relatively rapid degradation of close and more massive
binaries, through some of the considered kinds of interactive evolution,
to a slower accumulation of the remnants of such evolution at lower mass

and period values.

It has been argued that the numbers of unevolved binaries at lower mass are insufficient to account for the large incidence of W UMa systems (e.g. Kraft, 1969). Some relevant quantities will be considered shortly. Let us first note possible difficulties in comparing the expected numbers of "protomorphs" of contact systems with the observed population of un-evolved binaries in the presence of light curve ambiguities, selection effects relating to discoveries, and general processes associated with mass and angular momentum loss and aging. Thus, for example, magnetic braking (Huang, 1966; Mestel, 1968) might be of key significance in ex-plaining why a close low mass pair could spend only a fraction of the primary core hydrogen burning lifetime as a detached system (Van't Veer, 1976; Vilhu, 1981; Mochnacki, 1981).

Before introducing such an extra degree of freedom into the problem, however, certain points can be made from considering the situation in which only the normal processes of binary evolution are involved, but with the well known possibility that such processes can lead to a common envelope phase, around which time significant mass and angular momentum loss may occur. Such a mechanism could certainly enhance the persistence of the contact or close to contact condition; without it, i.e. in a purely conservative regime, comparable numbers of pre and post-mass transfer binaries, in which the close to contact phase appears as a relatively short episode, should be expected. The comparisons of Kraft (1969), at least if we assume that the separations of low mass binaries should in-itially distribute like those of higher mass, appear sufficient to dis-allow such a line of explanation.

The line of explanation that we now seek to investigate is that the number of contact binaries $N_w$, brighter than a given apparent magnitude $m_o$, which, in view of the Eggen correlation can be essentially associated with unit variation of a single independent variable, which we shall choose to be primary absolute magnitude M, can be related to a correspond-ing number of protomorphs $N_p$, via some relation of the type

$$N_w(M) = \int_{\Delta'M} N_p(M - \delta M) \, \nu(M, \delta M) \, \tau(M, \delta M) \, dM \quad , \qquad (7.1)$$

where the protomorphs come from a range $\Delta'M$ of, by implication, somewhat more massive close binaries, with compensating factors $\nu$ and $\tau$ accounting for the greater volume occupied by the brighter protomorphs, and the relative timescales which they spend in detached and contact conditions, respectively. If the required distribution $N_p(M)$ could be obtained from such a relation it might be compared with the observed incidence of un-evolved systems.

To do this in a more complete way involves a number of possibly com-plicated factors, about which consideration is deferred. We proceed, at this stage, by making a linearized trial solution to the foregoing integral equation. In order to make comparisons we retain the notion of a uniform

distribution of unevolved binaries, of the form $\phi = \text{const.} \times f(M) P^n$ and
also suppose a more or less constant rate of binary formation of all kinds.
Keeping in mind the range of total mass loss (0 - 40%) considered plausible
by Refsdel et al. (1974) for the much discussed example AS Eri, we consider the possibility that the protomorphs of the anomalous accumulation
of contact binaries of spectral type G0-6 may be essentially found among
low period systems with F3-8 type primaries, estimating that $\sim$ 20% of the
original mass of the system may be lost when the separation of centres
is small.  If such a matching can be successful it might be applied in a
parallel way to more massive systems.  Let us assume that faintness affects
the detection of both kinds of system to the same extent, (in fact, the
detection of EW systems falls away somewhat more rapidly with magnitude
than that of unevolved EA stars as may be seen from Table 2, though the
effect does not have proportionately serious consequences).  Equation (7.1)
can now be approximated by

$$N_p(M - \Delta M) \; \Delta'M = N_w(M) \; \frac{V(M - \Delta M)}{V(M)} \; \frac{T_p(M - \Delta M)}{T_w(M)} \quad , \qquad (7.2)$$

where $V(M)$ is the volume of space out to which a star at M appears brighter
than $m_o$ and $T_{p/w}$ is the expected lifetime in the protomorph/contact condition.  Now, in parallel with (6.1),

$$N_w = f_1 f_2 N_d \qquad . \qquad (7.3)$$

From the data indicated in Figure 2, with $m_o$ taken to be 12, we
have $N_d = 40$, while, since genuine contact was considered more likely
among the cooler type EW light curves, we can set $f_1 = 1$.  As before,
$f_2 = 3.3$ so that $N_w = 132$.  If $N_p$ refers, in the present example, to the
foregoing type range, the increment $\Delta'M$ can be taken as effectively unity,
from which it probably differs little in any case.  Taking M (blue) and
$\Delta M$ to be 5.0 and 1.0 respectively (remembering the generally increased
proportion of secondary light in W  UMa systems), we find
$V(M - \Delta M)/V(M)$ to be about 4, though with a strong dependence on the
assumed mass loss.  $T_p/T_w$ could be taken to be $\sim$ 1, at least if Case A
type mass transfer operates.  $N_p$ then turns out to be 530.

The period $P_{max}$ up to which close binaries in the type range considered (F3-8) would need to be drawn from in order to provide such
protomorphs can then be expressed by

$$N_p = \Sigma_{P_\star}^{P_{max}} \; n_d(P) \; f_1 \; f_2(P) \; \Delta P \qquad , \qquad (7.4)$$

where $n_d(P)$ is the number of protomorphs in a given period interval about
P (in days).  From the sample of candidate stars referred to in connec-

tion with Figure 4, and matching the observed numbers to the theoretical comparison form we have, when $\Delta P = 0.2$ days, $n_d(P)\,\Delta P \simeq 5.8\,P^{-5/3}$. The selection factor $f_2(P)$ works out at $f_2 = 5.8\,P^d$ which reduces to the same value as that used for the W UMa stars at the equal mass unevolved contact period (0.43 days). $f_1$ is again taken to be unity and though this might overestimate detached protomorph numbers close to $P_*$, $N_w$ may also have been slightly overestimated as a result of setting $f_1 = 1$ in (7.3).

On these assumptions, it can be found that it would be necessary to look for supposed protomorphs of the G0-6 type contact systems among middle-late F-type primary binaries with periods up to about 10 days. The upper limit period for Case A mass transfer among such systems is only about 1.2 days, however. Only about a third of the considered group of contact binaries could therefore be accounted for in this way, based on the adopted statistics.

Turning to the possibilities of Case B, it can be noted that a number of low mass evolved Algols exist, which if "evolved" backwards (conservatively) must have passed through a common envelope stage. Such systems include R CMa, RW CrB, RZ Dra, AS Eri, DN Ori, RT Per, VV UMa S Vel; all with total mass around 2 $M_\odot$, and which, in a common envelope configuration, must have looked like W UMa type systems. However, a problem now is that, even allowing that angular momentum loss during the common envelope stage might allow that contact persist into the "slow phase" of mass transfer, the entire semi-detached stage of low mass Case B evolution is still only $\sim 10^{-1}$ of the Main Sequence lifetime. The factor $T_p/T_w$ in (7.2) is thus increased appreciably. There are, moreover, other difficulties: if the mass losing star is able to swing in sufficiently close for the final product to look like a contact-system the initial mass ratio seems required to be small. Such systems are believed to represent rather a minority among close binaries (Lucy and Ricco, 1979; Plavec, 1982) and, in any case, do not correspond to the observed candidates on which the comparison statistics are based. Then further ad hoc discussion is required to explain how the supposed relatively small total mass loss fraction carries away the larger angular momentum loss required.

All in all, it seems difficult to account for the large incidence of W UMa type binaries on the basis of a simple uniform distribution of separations for unevolved binaries of all masses, allowing only for the well known mass transfer modes in binary evolution, accompanied by systemic mass and angular momentum loss when transfer commences. When such processes are included, though, the disparity between possible protomorphs and observed W UMa systems need not be as great as previously estimated ($\sim$ factor 10 according to Kraft, 1969). Also it seems likely that certain cases of low-mass semi-detached systems evolving in Case B should have once looked like contact binaries. It is unsatisfactory that no distinguishing mark of such binaries currently in the contact state has been pointed out. As well it should be noted that the failure of our simple trial solution has not proved the impossibility of some appropriate choice of factors in Equations (7.1) and (7.4) from

allowing some explanation along these lines. In particular, the assumed
constancy in the rate of binary formation may be a weak point in the
foregoing comparisons. Also, since the foregoing treatment implies that
some of the protomorphs may be drawn from systems evolving relatively
rapidly, e.g. by already being in a common envelope phase, the factor
$T_p(M - \Delta M)$ would have been overestimated by simply setting it equal to
$T_w^p(M)$. In such ways the "slower accumulation" mentioned at the outset
could be effected.

Of course, the introduction of magnetic braking in the evolution of
cool close binaries, for which there appears to be accumulating evidence
(Ruciński et al. 1982, Budding et al. 1982), may help to clarify and
remove many problems connected with their incidence. In the simple
terms of (7.2) for example, magnetic braking relieves the requirement for
a very large $N_R$, by reducing $\Delta M$, and therefore the volume ratio, as well
as possibly reducing the time ratio $T_p/T_w$, since estimated rates of
angular momentum loss can produce coalescence, in some cases, in much less
than a nuclear timescale. On the other hand, the same mechanism will
entail a breakdown of the uniform distribution idea, producing changes in
the factors $f_2$ and possibly also $f_1$ in (7.4) which are not obvious. The
introduction of the extra degree of freedom associated with angular
momentum loss ab initio, due to magnetically driven processes, would
then detract from the effectiveness of observational evidence, of the kind
considered in this paper, in providing unambiguous tests of theory, unless such processes could be separately quantified.


8. SUMMARY

This paper has been aimed at bringing out the peculiar incidence of
W UMa type binaries, the essence of which was shown in Section 2.
Section 3 pointed out the possibility that this peculiarity could be
regarded as a deficit among early type systems of this kind, as well as
a surplus at spectral types later than mid F.

Sections 4 and 5 were intended to emphasize ambiguities associated
with the photometric evidence alone. The fact that a light curve is
classified as of W UMa type does not force us to assume contact, and
an analysis of Knipe's (1968) data on RR Cen was used to illustrate the
point. Also a comparison was made between the distribution of determined
inclination values from analysis of W UMa type light curves and that of
the better determinable detached systems. Taking such ambiguities into
account, the spatial incidence of contact binaries, though clearly high,
need not be as high as that considered by Van't Veer (1975) (Section 6).

The problem posed by the incidence of W UMa systems was found to
be open to analysis, on the basis of a number of simplifying assumptions,
in the light of what was judged in Section 7 to be the more pragmatic
approach to explaining their origin. It may be worthwhile to conclude by
indicating such simplifications or limitations, and suggesting possible
areas in which the problem could be developed.

Firstly, though the numbers of stars involved in the statistics are moderate, they still could not be regarded as large - large enough, for example, to permit smaller increments than a few spectral type subdivisions or one magnitude range in dealing with the representative stars considered in relation to Equations such as (7.1) or (7.4). In a similar way the latter equation, relating back to Section 3, utilizes the notion of an underlying uniform distribution, which while feasible, cannot be regarded as definitely established by the statistics given in this paper. The trends shown in Figures 3 and 4 could, in fact, be better represented by the truncated Gaussian considered by Farinella and Paolicchi (1978). The form actually used in (7.4) was chosen for reasons of simplicity - but the major results of the discussion are not seriously affected by the particular form chosen.

Throughout Section 7 there was, apart from a qualification added to the discussion of Case B, a concentration on the properties of one star only, which implies fairly constant behaviour or properties of the protomorph's secondary, or that primaries and secondaries in the considered systems tend to have a fixed relationship to each other, such as via a preferred mass ratio, for example. This, like the effects of evolution within the Main Sequence band in relation to the possibility of characterising stable primaries by a single independent variable, has been tacitly associated with small scatter effects, such as that found within the Eggen correlation. (This point does, however, raise an issue which could merit further investigation, namely, the possibility of another "compensating factor" in (7.1) associated with a difference between the range of mass ratios of protomorphs with that of observed W UMa systems. If, for instance, a wider range of mass ratios among the protomorphs is implied, the requirement for a high $N_D$ in (7.4) could be eased, since, as with the small initial mass ratio Case B possibility, some such protomorphs would not easily be observed as binaries.)

Then, of course, details of the supposed scale or mechanism of mass and angular momentum loss that underlie the approach of Section 7 were also dealt with in a purely summary way. The circumstances may differ so much in individual cases as to cast doubts on the reliability of the straightforward linearization of (7.1) into (7.2).

By way of a positive response to such doubts, the main purpose of Section 7 has been not only to offer one approach to observational testing of theories of the origin of contact binaries, but, more generally, to suggest a future possible area of work, in connection with binary evolution. In a parallel way to the use of two dimensional diagrams relating single star evolution to observational data, the distribution of binaries in the plane of, for example, primary type and orbital period could be studied in its dependence on time and in relation to proposed paths of binary evolution starting from a given distribution of initial conditions.

REFERENCES

Allen, C. W.:   1973, *Astrophysical Quantities*, Athlone Press, London.
Anderson, L., Raff, M. and Shu, F.:   1980, in *Close Binary Stars: Observations and Interpretation* (eds. M. Plavec et al.) Reidel, p. 48
Budding, E.:   1981, in *Investigating the Universe* (ed. F. D. Kahn) Reidel, p. 271.
Budding, E., Kadouri, T. H. and Gimenez, A.:   1982, Mon. Not. R. astr. Soc (in press).
Budding, E. and Najim, N. N.:   1980, Astrophys. Space Sci., 72, 369.
Eggen, O. J.:   1967, Mem. Roy. Astron. Soc., 90, 54.
Farinella, P. and Paolicchi, P.:   1978, Astrophys. Space Sci., 54, 389.
Farinella, P., Luzny, F., Mantagazza, L. and Paolicchi, P.:   1979, Astrophys. J., 234, 973.
Huang, S. S.:   1966, Ann. d'Astrophys., 29, 331.
Iben, I.:   1967, Ann. Rev. Astron. Astrophys. 5, 571.
Knipe, G. F. F.:   1965, Astrophys. J., 142, 1068.
Kopal, Z.:   1955, Ann. d'Astrophys., 18, p. 379.
Kopal, Z.:   1959, *Close Binary Systems*, Chapman and Hall Ltd., London (Chapter III).
Kopal, Z.:   1968, Astrophys. Space Sci., 2, 23.
Kraft, R.:   1969, in *Stellar Astronomy*, Volume II (ed. H. Y. Chiu et al.) Gordon and Breach, New York-London-Paris, p. 35.
Kukarkin, R. V., Kholopov, P. N., Efremov, Yu. N., Kukarkina, N. O., Kurochkin, N. E., Medvedeva, G. I., Perova, N. B., Fedorovich, V. P. and Frolov, M. S.:   1969, *General Catalogue of Variable Stars*, Moscow.
Lucy, L. B.:   1968, Astrophys. J., 151, 1123.
Lucy, L. B.:   1976, Astrophys. J., 205, 208.
Lucy, L. B. and Ricco, E.:   1979, Astron. J., 84, 401.
Mestel, L.:   1968, Mon. Not. R. astr. Soc., 138, 359.
Mochnacki, S. W.:   1981, Astrophys. J., 245, 650.
Plavec, M.:   1982 (This volume).
Popper, D. M.:   1980, Ann. Rev. Astron. Astrophys., 18, 115.
Refsdal, S., Roth, M. L. and Weigert, A.:   1974, Astron. Astrophys., 36, 113.
Ruciński, S. M., Pringle, J. E. and Whelan, J. A. J.:   1982 (This volume).
Shapley, H.:   1948, Harvard Obs. Mono. No. 7, Cambridge, Mass. p. 249.
Svechnikov, M. A.:   1969, *Catalogue of Orbital Elements, Masses and Luminosities of Eclipsing Binary Stars* (Sverdlovsk).
Tutukov, A. V. and Yungelson, L. R.:   1980, in IAU Symp. No. 88, *Close Binary Stars: Observations and Interpretation* (ed. M. Plavec et al. Reidel, p. 15.
Van't Veer, F.:   1975, Astron. Astrophys., 40, 167.
Van't Veer, F.:   1976, in IAU Symp. No. 73, *Structure and Evolution of Close Binary Systems*, Reidel, p. 343.
Vilhu, O.:   1981, Astrophys. Space Sci., 78, 401.
Wilson, R. E. and Rafert, J. B.:   1981, Astrophys. Space Sci., 76, 23.
Yamasaki, A.:   1975, Astrophys. Space Sci., 34, 413.

# Part IV

# Cataclysmic Binaries and their Role in Stellar Evolution

# BINARY STARS AND SS433

Remo Ruffini

Istituto di Fisica "G.Marconi" - Università di Roma, Italy

ABSTRACT

Some of the most unique experimental features of the source SS433 are
outlined as well as some implications of the theoretical models of this
source.

The fact that as many as $10^9$ stars in our galaxy may be members of mul-
tipole systems make the theoretical study of binary sources one of the
most important in the entire field of astronomy and astrophysics. Moreover, since all type of stellar population are member of binary systems
the study of such binaries gives as well basic information about the
Hertzsprung Russell diagram and evolution of a single star. In particu
lar the knowledge of the binary parameters, such as the orbital period
and the velocity of each one of the component, allows to infer the
value of the star masses of their radii and their luminosity. Beyond
any doubt the entire study and discovery of white dwarfs can be considered a clear success, a biproduct of this type of research. In order
to pursue the theoretical analysis of these binary systems our entire
field of applied mathematics has been developed reaching classical re
sults on the equilibrium configurations of self gravitating fluids in
rotation.[1,2] All these classical works have become in recent years
of paramount importance for the understanding of an entire new branch
of high energy astrophysics: the binary X-ray sources. Unlike usual
binaries, in this case, one of the star is a gravitationally collapsed
star; either a neutron star or a black hole. The matter falls from the

*Z. Kopal and J. Rahe (eds.), Binary and Multiple Stars as Tracers of Stellar Evolution, 373–388.*

normal star into tne deep potential well of the collapsed companion
star. Under these circumstances a large emission of X-rays (dE/dt $\simeq 10^{38}$
erg/sec) can occur.[3] The development of X-ray astronomy by the group
of Riccardo Giacconi has allowed the identification of a large variety
of such binary X-ray sources. It has been possible for the first time
to measure the masses of neutron stars[4] and to meet the necessary
conditions required for the positive identification of a black hole in
Cygnus XI.[5]

A second field of research which has also shown remarkable progress in
recent years thanks to the ample use of radiotelescopes,[6] radars and
satellites[7] and atomic clocks,[8] is tnat of testing general relativistic effects within the weak gravitational fields of the solar system.
Specifically conceived and directed experiments have tested such effects
as time delay in the propagation of signals in a gravitational field,
light deflection, gravitational redshifts and planetary motions[9] to
such accuracy that general relativity has become the most viable theory
of gravity consistent with experimental tests.

What one would like to have in the future is a mixture of these two
fields of research, namely, clean and precise tests of general relativistic effects like those in the solar system, but in the strong gravitational field of a gravitationally collapsed object. Such a goal more
than likely cannot be achieved with binary X-ray sources since there the
most relativistic region of the gravitational field is occupied by matter
accreting from the normal companion star.

An important step towards this goal has been achieved by the discovery
of the binary pulsar PSR 1913+16. This system has given the first clear
evidence of the existence of gravitational waves in nature.[10] Progress
is also currently being made in developing suitable experiments for the
detection of gravitational waves on the earth.[11] Other important systems
in tnis context may be the X-ray burst sources in globular clusters or
in tne galactic bulge[12] and possibly SS433. In the following, we will
limit our attention to the system SS433.

SS433 was originally observed in the sixties by Stephenson and Sanduleak
in a search for stars with strong hydrogen emission lines and was the
433rd object in their list.[13] In 1978, it was identified[14,15] with a
point-like radio source located at the center of the supernova remnant
W50, initially discovered by Holden and Caswell;[16] radio studies done
by several groups[17] have since given a very detailed map of W50 and of
SS433. In 1978, Clark and Murdin[15] had proposed the identification of
SS433 with the X-ray source A1909+04 discovered in 1975 by the Ariel 5
satellite.[18] The total flux in the optical region was estimated to be

$L_{opt} \simeq 10^{37}$ erg/sec, the one in radio to be $L_{radio} \simeq 10^{31 \div 32}$ erg/sec, and the one in X-rays to be $L_x \simeq 10^{35}$ erg/sec. Except for the anomalous abundance and strength of the emission lines, nothing peculiar had been observed until that time, not even in the energetics of the system, with the possible exception of a low value of the X-ray flux compared to the one for binary X-ray sources.

What has made this system unique is the spectroscopic analysis of the optical spectrum done by Ciatti, Mammano and Vittone[19] of the Asiago Observatory; they showed the existence of strong anomalous emission lines moving from night to night by an amount $\Delta\lambda$ 50 Å! These lines were attributed at that time by Mammano and co-workers to a magnetic splitting of the $H_\alpha$ line by a strong magnetic field ($10^{7-8}$ Gauss). Further investigation by various research groups throughout the world has shown some regular patterns in the shift of the lines as a function of time and finally an overall periodicity of $P \simeq 164$ days with a very specific pattern (Fig. 1).[20,21] Analogous moving emission lines have been observed around $H_\beta$, $H_\gamma$ and in higher series of the hydrogen spectrum.[22,23]

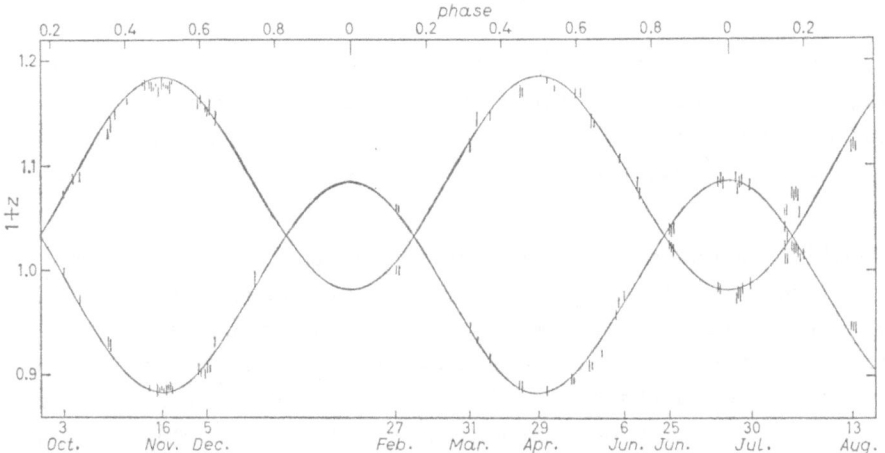

Figure 1. The 164 day modulation of the shifted lines of SS433 during 1978-1979. For details see refs. (18), (19) and (29).

Before discussing theoretical models for SS433, let us review some of the points on which everyone agrees.

(1) SS433 is a relativistic system. The widely accepted hypothesis of kinematical (and possibly gravitational) effects can be clearly inferred

by the magnitude of the shifts in Fig.1. It is well known that the frequency shift observed by an observer with four-velocity $u_{obs}$ emitted by a source with four-velocity $u_{em}$ is simply given by:

$$\nu_{obs}/\nu_{em} = (u^{\alpha} k_{\alpha})_{obs}/(u^{\alpha} k_{\alpha})_{em} \tag{1}$$

where k is the wave four-vector propagating from the emitter to the observer. If the shifts are due uniquely to the kinematic effect of two motions of equal speed in opposite directions, we have:

$$\lambda_{\substack{red \\ blue}}/\lambda_{o} = (1 \pm v/c \cos \psi)(1-v^2/c^2)^{-1/2} \tag{2}$$

and the fact that $(\lambda_{red} + \lambda_{blue})/2 \lambda_{o} \simeq 1.036$ is a direct measure of the $\gamma$ factor of the special relativistic Doppler formula. If the shifts are due to a general relativistic effect (see below), there will be an additional gravitational redshift. Therefore, for the first time, either in "bulk" is observed to be moving inside our own galaxy at a relativistic speed or matter in "bulk" is observed to be moving around a gravitationally collapsed object with high regularity.

(2) The relativistic effects are modulated in time with a period $T_p \simeq 164$ days.

(3) SS433 has to be a gravitationally collapsed object. Inference of this comes mainly from its association with W50 and on the general grounds of the total energetics of the system. The proof of this last point, however, will be only possible by consistency with a detailed theoretical model.

M. Milgrom[24] in one of the earliest theoretical papers on this subject suggested a variety of possible models explaining the shift of the lines and their time variation in terms of Doppler and possibly gravitational effects. It is interesting to note that at the time of the Milgrom work, the 164 day period in the shift of the moving lines had still not been discovered. Two of the models proposed by Milgrom are still being actively pursued today: the two jet model in which material is ejected in two highly collimated relativistic beams (this model has been proposed independently by Fabian and Rees[25]) and a ring model of material orbiting a collapsed object.

The widely publicized two jet model (Fig. 2) has been discussed by Fabian and Rees,[25] Milgrom,[26] Abell and Margon,[21] Katz,[27] Martin and Rees,[28] and Maraschi and Treves.[29] In this model, matter is moving at an almost relativistic speed $v \simeq 0.26c$ in two highly collimated jets precessing around a fixed axis. This idea has found some support by X-ray and radio observations[17] but, interesting as it is, it fails to be a definite model in more than one respect.

(1) The origin of the acceleration process of matter in these beams and the mechanism by which the speed of the matter in the beams should be kept at an almost constant velocity are not yet determined.

(2) The physical reasons by which the beams are collimated within an angle $\lesssim 2°$ are again not determined.

(3) The clock mechanism characterizing the 164 day modulation is also unclear.

(4) The most severe constraint comes, in our opinion, from the overall energy balance in the jets. Every photon of the shifted lines should be emitted by an atom of mass, m, moving at a velocity v=0.26c. The kinetic power of the jets is then given by:

$$W_{kin} \simeq G \cdot V \cdot N_{ph} \cdot 1/2 \ mv^2 \cdot P^{-1} \tag{3}$$

where the number $M_{ph}$ of $H_\alpha$ photons emitted in the jets per second is easily computed by the observed flux in the moving lines ($L_{m\ell} \simeq 10^{34}$ erg/sec), $V \geq 1$ is given by (number of atoms in the jets)/(number of atoms emitting $H_\alpha$), $G \geq 1$ takes into account the geometry of the jets and the optical depth of the emitting region and $P \geq 1$ is the number of re-emissions in each hydrogen atom. If one assumes $V \cdot G \cdot P^{-1} = 1$, then $W_{kin} \simeq 10^{41}$ erg/sec. This is an extremely high value for a galactic object.

Let us now consider the other possibility: a ring model. Although the model is still not unique, there are a variety of points which can be explicitly explained, some predictions of the model which can be tested and quantitative estimates and computations which can be done in this theoretical framework.

In the ring model[12,30] the observed shifts in $H_\alpha$, $H_\beta$, ... lines are explained in terms of a combination of gravitational and Doppler effects associated with matter orbiting a black hole. The mechanism of acceleration of matter up to the observed relativistic velocities is simply accretion through (almost) Keplerian orbits. In order to fit the data the emission has to occur only from two opposite spots in the precessing ring[31] (Fig. 3). If we assume, for simplicity, that photons propagate along straight lines, we can fit the observed data very satisfactorily (see Fig. 1) and the predicted behavior is virtually identical to the one obtained in the two jet model.[31] The radius inferred for the orbit is then r = 50M (wee use c = G = 1) implying that the central object is either a compact neutron star or a black hole. Since the orbits have to be almost Keplerian and unperturbed by electromagnetic fields, we are going to assume in the following that the central object is indeed a black hole. We have proposed some experimental tests to check this (see below).

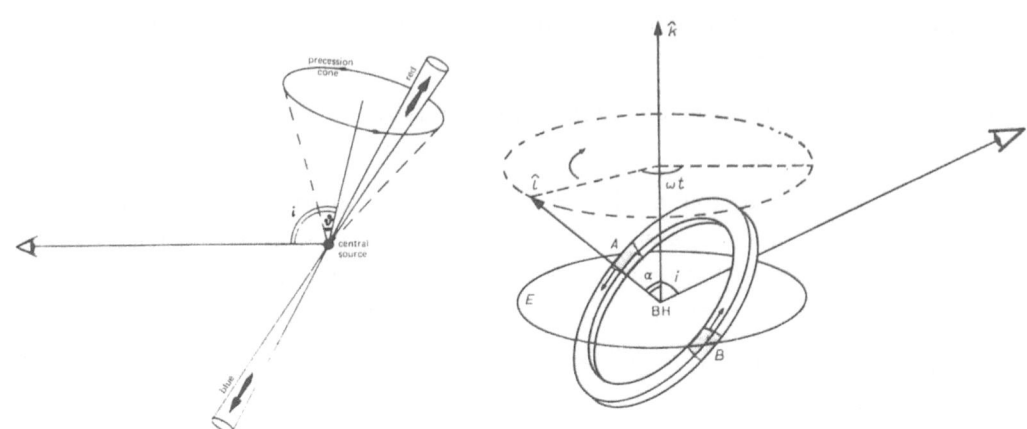

Figure 2. Two jets models        Figure 3. Ring models for SS433
for SS433

One of the main ideas in the development of this ring model has been to include as many relativistic effects as possible in the treatment with a double purpose: (a) to recognize the self-consistency of the model, and (b) in case of experimental configuration, to clearly point out the most unique features of SS433 from the point of view of relativistic astrophysics.

The 164 day modulation is explained in terms of an additional general relativistic effect: the Lense-Thirring-Wilkins effect.[30] If a black hole of mass M is endowed with a specific angular momentum a = L/M, this effect predicts a precession of tne angular momentum $\hat{\ell}$ of the ring with respect to the angular momentum $\hat{K}$ of the black hole by an angle of:

$$\Delta \Omega = 4\Pi \ (a/M)(M/r)^{3/2} \qquad (4)$$

per revolution, where r is the radius of the emitting ring previously determined by the amplitude of the shifts. This interpretation of the 164 day periodicity, if confirmed, would be the first measurement of the specific angular momentum of the black hole:

$$a/M = 1.2 \times 10^{-7} \ (M/M_{\odot}) \qquad (5)$$

Until this point there has been no need to fix the mass of the black hole, since all the results given above scale with the black hole mass. Shaham et al. and Terlevich and Pringle[30] have assumed a black hole

mass of $\sim 10^6$ M$_\odot$ in order to explain the energetics of the observed opti_
cal radiation from the moving lines of SS433:

$$L_{ml} = \alpha\, A\, \sigma\, T^4 \tag{6}$$

where A is the surface area of the line emitting regions, $\alpha \simeq 1$ and T
is a suitable temperature for the emission of Balmer lines (T = $10^{4\,\circ}$K).
Eq. (6) has been obtained under the assumption that the emission occurs
by a thermal process: as pointed out in ref. (33), this is not necessari_
ly the case. The existence of such a large mass for SS433 can hardly be
explained from an astrophysical point of view. For this reason, we assume
in the following that the mass of SS433 is around 10 M$_\odot$ , and we overcome the constraint given by Eq. (6) by choosing a different radiation
mechanism. Two of the points still to be explored are: (a) why the hydro_
gen lines are emitted only in a ring, and (b) why only two spots on the
ring are observed to emit.

In order to give an answer to these questions, it is necessary to make a
model of the emission process of the shifted lines. Since the thermal
radiation leads to the constraint given by Eq. (6), namely, to a mass of
the gravitationally collapsed object $M \geq 3 \times 10^5$ M$_\odot$ , we have explored
the possibility that cooperative or stimulated emission takes place. The
phenomenon of cooperative or stimulated emission, well known from labora_
tory physics, can occur only under very restrictive conditions in any
astrophysical setting. The fact that a variety of astrophysical systems
demonstrating this phenomenon have been found[34] clearly shows that
these necessary conditions are fulfilled at least in some cases.

One of the major novelties in an astrophysical setting is that different
oscillators will in general have large relative velocities and find them_
selves separated by large differences of gravitational potential. As a
consequence, due both to Doppler shifts (transverse and parallel) and
gravitational red or blue shifts, the frequencies of the individual
oscillators will differ by an amount larger than the intrinsic width
allowed by the cooperative emission process.[33] Some considerations
concerning the conditions of population inversion necessary in order to
have a cooperative or stimulated emission process at work, have been ex_
plored in refs. (36) and (37). Here we focus our attention on the neces_
sary conditions that the gravitational potential and the velocity field
of matter have to fulfill in order to have such an emission mechanism
working and being observable from infinity.

A first necessary condition in order to have cooperative emission
between an oscillator at a point P and an identical one at a point P' is
that the two oscillator frequencies related by Eq. (1) fulfill the ine-
quality:

$$|\nu_p - \nu_{p'}| \leq \Delta\nu \qquad\qquad\qquad (7)$$

where $\Delta\nu$ is the width of the cooperative or stimulated emission process.

The elucidate the meaning of this constraint, we present in Fig. 4 the case of a disk of matter in circular orbit around a Schwarzschild black hole of mass M. The dashed region shows the locus of corotating points P' in "frequency contact" with the corotating point P within the bandwidth $\Delta\nu/\nu \leq 10^{-2}$. Details are given in ref. (35). We want to point out here that, independent of the numerical value of the width of the cooperative emission process and of the location of the emitting point P, every atom P can be in "frequency contact" with an entire ring-like region in the disk, having a mean radius equal to the distance of the point P from the black hole, and with two almost radially pointing columns. The lasing or masing process can only occur in these regions of "frequency contact" and in directions in which a critical size for the emitting region is reached. This size is dictated by the cooperative mechanism (radiation flux, density of particles, etc.).

A further necessary condition must be fulfilled in order that cooperative or stimulated emission generated in the system be observed in the shape of an emission line by an external observer at point O. All the emitting points need to have constant shifts, with the balancing of Doppler and gravitational effects with respect to the observer at the point O (clearly this equality has to be fulfilled only within the linewidth observed, see Eq. (7)). We define the "regions of constant shift" with respect to the observer O at infinity to be the sets of points P satisfying the following condition:

$$\nu_p/\nu_o = (u^\alpha k_\alpha)_p / (u^\alpha k_\alpha)_o = \text{constant} \qquad\qquad (8)$$

For the sake of example, we again consider a Keplerian ring around a Schwarzschild black hole;[35] the results for the case of an observer at infinity lying on the disk plane, are shown in Fig. 5.

It is then clear that in order to have the mechanism of cooperative emission working and observable, both necessary conditions given above have to be fulfilled. Stimulated or cooperative emission will be observed only from the intersection of the two sets, namely, regions of "constant shift" and regions in "frequency contact". It is likely that only intersection regions of large enough depth along the line of sight can lead to observable phenomena of amplified emission. The astrophysical setting for the simultaneous fulfillment of the two necessary conditions given above will only occur along selected directions. It is also a matter of course that observations of the same astrophysical system

from different directions of sight will lead to different laser or maser
patterns.

In the simple case of the Keplerian disk seen edge-on shown in Figs.  4
and 5, the intersection of the two conditions with the above condition
of "maximum depth" is found to select two emitting regions in each ring
of matter constituting the disk.[35] Going back to the model of  SS433,
this result is expected to play a fundamental role in the characterization

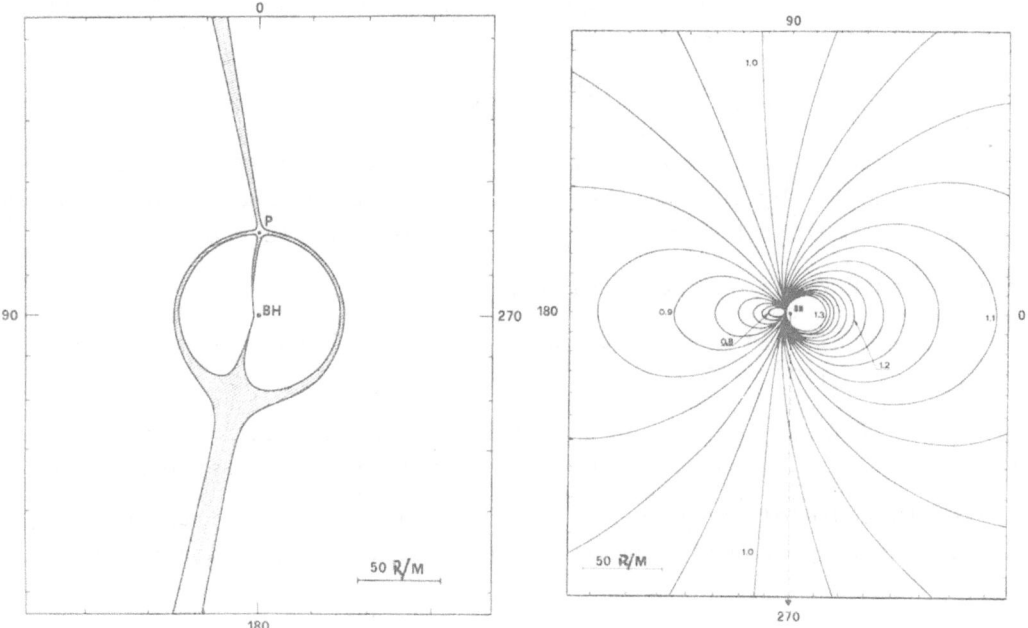

Figure 4. The dashed region is
the set of points in a Keplerian
disk in "frequency contact" with
the point P orbiting the black
hole at a distance r=50M.  This
region consists of points having
frequencies within a bandwidth
$\Delta\nu/\nu \leq 10^{-2}$ about the frequency
$\nu$ of the point P. The disk is
rotating counter-clockwise (ref.
(35)).

Figure 5. Lines of constant shift
for the Keplerian disk of Fig. 4,
seen edge-on (the line of sight
coincides with the dashed line).
Selected numerical values of (1+z)
are identicated on the redshifted
constant shift lines (z > 0)  and
on the flueshifted lines (z < 0)
(ref. (35)).

of the two emitting regions during the precessing of the ring plane.[31]

In order to give an astrophysical meaning to these considerations and
identify the source of energy of the system, we consider the ring as a
part of an accretion disk. The structure of this disk should, however,
be very different from the ones usually described in the literature for
accretion processes in binary X-ray sources.[38] One of the major diffe
rences is that the present disk should be at very low temperature near
the black hole in order to have, at r = 50 M, emission of the Balmer
lines of hydrogen. This is in clear contrast to the traditional models
which assume that the X-rays are emitted by matter accreting near  the
surface of the black hole. Traditionally, the accretion occurs from  a
companion star overflowing its Roche lobe; the ensuing X-ray luminosity
is typically near the critical value

$$L_{crit} = 4 \pi cGM/k_T \simeq 1.3 \times 10^{38} M/M_\odot \text{ erg/sec} \qquad (9)$$

where $k_T$ is the opacity due to Thomson scattering. In SS433, the accretion rate implies by the observed  fluxes is not necessarily high enough
to require a violent Roche-lobe flow from the possible companion star.[39]

The main assumptions of the model are:[12]

(a) The X-ray flux is not generated near the glack hole but in a shell
at redius $r_0 = (10^3 + 10^4)$ M.

(b) Since $L < L_{crit}$ for $r < r_0$ the distribution of matter is assumed to
the disk-like with an angular velocity very near to the Keplerian one.

(c) The structure of the magnetic field is almost toroidal and exerts
minimal viscosity, implying small accretion rates in the inner part of
the disk.

(d) The temperature of the disk is assumed to be monotonically increasing outward starting from zero temperature at the black hole surface[38]
all the way up to $T_0 \simeq 10^{6 \pm 7}$ °K at $r \simeq r_0$.

(e) Finally, in a region $r \gg r_0$, the accreting material is heated up
($T \simeq 10^4$ °K); from this region the unshifted optical lines and the optical and infrared continuum should be emitted.

It is important also to stress that, unlike the traditional models, in
the Fang Li Zhi-Ruffini model, the accretion rate M is not considered
to be constant in the disk, but is a function of the radial coordinate.
The loss of mass will occur from the surface of the disk either as "co-
ronal outflow" or in jets of material of high temperature and low densi
ty.[40] This material should be responsible for the observed jet structure in radio emission.

An attempt to integrate the full equations in order to describe the structure of the disk has been made by W. Stoeger.[40] But independently of the details of the complete treatment, some important and general results can be obtained by the study of some idealized situations in which analytical treatment can be applied. It has been shown in fact by Rosner, Ruffini and Vaiana[41] and, independently, by Fang Li Zhi[42] that the narrowness of the recombination region in the inner part of the disk can be explained in a natural way and quantitatively determined both in a magnetized and an α -viscosity law approach to the disk structure (see also ref. (36)). The results clearly show a temperature gradient in the inner part of the disk and the possible existence of a recombination layer at $r \simeq 50$ M. This is the reason why mainly a ring of matter should be visible in the optical region as expected from the kinematical analysis given above. The sharpness of the temperature transition can be also an important ingredient in generating both the inversion of population levels and the pumping mechanism for the cooperative emission process.[36,37]

We would like now to point out some possible general relativistic tests which can be checked on SS433 and could help in establishing the validity of the ring models. It has been pointed out that both the jet model and the disk model can explain the observed features of SS433 at least from a kinematical point of view. The correlation of simultaneous variations in the blue and redshifted lines, especially on short time scales,[20] finds a very natural explanation in the disk model as due to small changes in the position of the emitting zones. Quite apart from these features, there is a "gedanken" process which, in principle, could give a very important test of the strong field treatment of general relativity and, at the same time, of the correctness of the ring geometry in SS433.

In the fit of Fig. 1, the observed photons are supposed to propagate along straight lines from the emissing regions to the observer.[31] If we generalize this treatment, taking into proper account the light deflection by the central object, we have at least two additional paths to the observer in addition to the direct ones (see Fig. 6). These paths originate from photons being emitted from the points A and B and arriving at infinity after a strong deflection near the unstable orbit $r = 3$ M (paths undergoing a total deflection $> 180°$ have not been considered, since they are expected to give rise to negligible intensity). The complete relativistic treatment,[43] including the numerical integration of the null geodesic equation:

$$\frac{d^2 u}{d\phi^2} = 3u^2 - u \qquad (10)$$

(where u = M/r and  φ is a polar angle in the photon orbital plane),
induces in the predicted behavior of the shifted lines the deformation
given by the solid curves of Fig. 7; of course, this effect should not
occur in the jet model since in that case the shifted lines are supposed
to originate far from the central object (dashed curves of Fig. 7). The
predicted asymmetry between the blueshifted and the redshifted lines
could be checked experimentally.

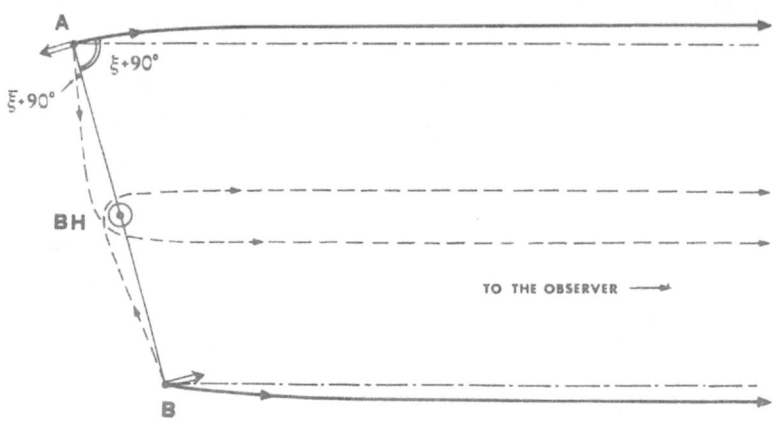

Figure 6. Selected trajectories of a photon in a Schwarzschild
geometry reaching an asymptotic observer. The solid curves
refer to the trajectories going from the emitting regions A
and B (Fig. 3)  directly to the observer. The dashed curves
refer to trajectories being highly deflected by the black hole
near the unstable orbit r = 3 M, represented here by the cir-
cle around the black hole. The dot-dashed straight lines refer
to the rectilinear photon propagation used to obtain the fit
of Fig. 1 (ref. (35)). Clearly the observed shifts of photons
propagating along the direct or the highly deflected paths are
very different (Fig. 7).

The two inner curves, with smaller shifts and complicated modulations,
originate from photons reaching the observer along the two highly deflec
ted paths in Fig. 6. The detection, even if sporadic, of these shifted
lines (they are expected to be much fainter than the other shifted lines)
could constitute a fundamental check of the strong field effects of gene
ral relativity.

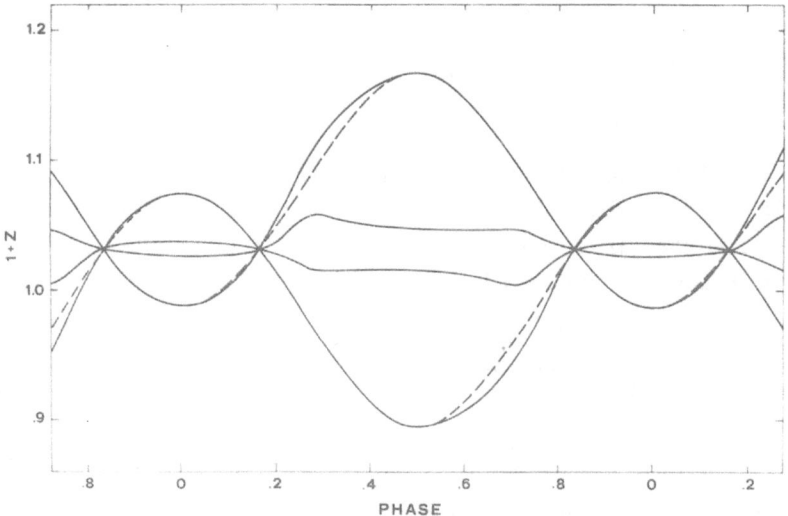

Figure 7. The predicted modification of the sinusoidal
behavior of the shifted lines, in a ring model, due to
general relativistic effects are here plotted as a fun
ction of pnase. The dashed curves represent the corres
ponding behavior obtained within the jet model. The two
inner curves, with smaller shifts and complicated modu
lations, originate from photons being highly deflected
by the black hole near r   3 M. The opposite modulation
(phase increasing from left to right) can also occur
for the opposite sign of the black hole spin (ref.(43)).

Another relativistic effect is expected in SS433 in the case in which
the precession of the emitting ring has to be ascribed to the Lense-
Thirring coupling between the angular momentum of the emitting ring and
the spin of the black hole. If matter is accreting into the black hole,
the spin of the black hole is expected to change and, in turn, the pre
cession period due to the Lense-Thirring effect, as given by Eq. (4),
is going to cnange. Following the above assumption that the dissipation
process in the inner part of the disk are small and that matter moves in
almost Keplerian orbits, we can infer that the intalling mass   $M_{in}$ will
carry in tne angular momentum corresponding to the last stable orbit:[43]

$$\Phi_{LSO} \simeq \Delta M_{in} (Mr_{LSO})^{1/2} \cos \alpha' (1-3M/r_{LSO})^{-1/2} \times$$

$$\left[ 1-3 \cos \alpha' (a/M)(M/r_{LSO})(1-2M/r_{LSO})(1-3M/r_{LSO})^{-1} \right] \qquad (11)$$

where the radius $r_{LSO}$ is given by:

$$r_{LSO} \simeq 6M \left[ 1 - \frac{2\sqrt{2}}{3\sqrt{3}} \left( \frac{a}{M} \right) \cos \alpha' \right] \tag{12}$$

and the angle $0 \leq \alpha'(r=r_{LSO}) \leq \alpha(r=50M)$ (see Fig. 3), since the effective potential in the $\theta$-direction tends to constrain the orbital motion close to the equatorial plane of the black hole. The corresponding change in the Lense-Thirring period is given by (L is the spin of the black hole)

$$\frac{\Delta T_P}{T_P} \simeq \frac{\Delta a}{a} + \frac{1}{2} \frac{\Delta M}{M} \simeq \frac{\Delta L}{L} \simeq \frac{\dot{\Phi}_{LSO}}{L} \tag{13}$$

since, for the orbits considered here, the second term can be neglected.

Then from any observed change in the precession period $T_P$, the corresponding accretion rate $\dot{M}_{in}$ into the black hole can be inferred:

$$\dot{M}_{in} \simeq (3.2 \div 5.0) \times 10^{-10} \left( \frac{M}{M_\odot} \right)^2 \left[ \frac{|\dot{T}_P|}{10^{-2} T_P yr^{-1}} \right] M_\odot yr^{-1} \tag{14}$$

This result could be of great relevance in order to estimate important parameters in the models of SS433, e.g. the ratio of matter accreting versus the one ejected in the Fang Li Zhi-Ruffini model.

In all the above considerations, the radius r at which the shifted lines are emitted is assumed to be constant in time. Viceversa, a change in the radius r of the emitting regions will also produce a change of the observable precession period. In this case, however, the orbital velocity of the emitting matter will also change the amplitude of the frequency separation in the shifted lines.

The recent discovery of a change in the 164 day period clearly agrees with the picture presented above. From the observed value $\dot{T}_P \simeq 0.01$ days/day, we can directly infer an accretion rate, $\dot{M}_{in} \simeq (0.7 \div 1.1) \times 10^{-9} (M/M_\odot)^2 M_\odot yr^{-1}$, implying for masses $\lesssim 10\,M$ a subcritical accretion regime as in the hypothesis of our model. It goes without saying that confirmation of our explanation and the eventual observation of the other predicted relativistic effects would represent an extremely important test of general relativity and of the existence of a black hole in SS433.

REFERENCES

1.  See e.g. Kopal, Z. "Close binary systems" Chapman and Hall, London 1959 and references therein.

2.  See e.g. Plavec, M. "Mass éxchange and Evolution of Close binaries" in Advances in Astronomy and Astrophysics Vo.6-1968 and reference therein.

3.  Giacconi, R., Ruffini, R.: "Physics and astrophysics of neutron stars and black holes", Proc. International School "Enrico Fermi" (Varenna, Italy), 1978, North Holland, Amsterdam.

4.  See e.g.: Bahcall, J.N.: 1978, Am. Rev. As. Ap. 16, 241.

5.  Rhoades, C.E., Ruffini, R.: 1974, Phys. Rev. Lett. 32, 324.

6.  Fomalont, E.B., Sramek, R.A.: 1976, Phys. Rev. Lett. 36, 1475.

7.  Shapiro, I.I., et al.: 1977, J. Cheophys. Res. 82, 4329.

8.  Williams, J.G., et al.: 1976, Phys. Rev. Lett. 36, 551; Shapiro, I.I., Counselman, C.C., King, R.W.: 1976, Phys. Rev. Lett. 36, 555.

9.  See e.g.: Will, C.M. in "General Relativity: an Einstein centenary survey", Hawking, S.W. and Israel, W., editors, 1979, Cambridge University Press.

10. Taylor, J.H., Fowler, L.A., McCulloch, P.M.: 1979, Nature 277, 437.

11. See e.g.: Amaldi, E., et al.:

12. Ruffini, R.: 1979, Nuovo Cimento Lett. 26, 239.

13. Stephenson, C., Sanduleak, N.: 1977, Ap. J. Suppl. 33, 439.

14. Seaquist, E., et al.: 1978, IAU Circular n.3256.

15. Clark, D., Murdin, P.: 1978, Nature 276, 44.

16. Holden, D.J., Caswell, J.L.: 1969, MNRAS 143, 407.

17. Seward, P., Grindlay, J., Seaquist, E., Gilmore, W.: 1980, Nature 287, 806;
    Hjellming, R.M., Johnston, K.J.: 1981, Ap. J. 246, L141;
    Geldzahler, B., Pauls, T., Salter, C.: 1980, As. Ap. 84, 237 and references therein.

18. Seward, F., Page, C., Turner, M., Pounds, K.: 1976, MNRAS 175, 39P.

19. Ciatti, F., Mammano, A., Vittone, A.: 1978, IAU Circular, n.3305.

20. Ciatti, F., Mammano, A., Vittone, A.: 1980, As. Ap. 85, 14.

21. Margon, B., et al.: 1979, Ap. J. 230, L41;
    Bedogn, R., et al.: 1980, As. Ap. 84, L4.
    Abell, G., Margon, B.: 1979, Nature, 279, 701.
    Margon, B., et al.: 1979, Ap. J. 233, L63.

22. Margon, B., Grandi, S.A., Downes, R.A.: 1980, Ap. J. 241, 306 and references therein.

23. Allen, D.: 1979, Nature 281, 284;
    McAlary, C., McLaren, R.: 1980, Ap. J. 240, 853.

24. Milgrom, M.: 1979, As. Ap. 78, L9.

25. Fabian, A.C., Rees, M.J.: 1979, MNRAS 187, 13P.

26. Milgrom, M.: 1979, As. Ap. 78, L17.

27. Katz, J.I.: 1980, Ap.J. 236, L127.

28. Martin, P.G., Rees, M.J.: 1979, MNRAS 189, 19P.

29. Maraschi, L., Treves, A.: 1979, preprint.

30. Amitai-Milchrub, A., Piran, T., Shaham, J.: 1979, Nature 279, 505;
    Terlevich, R., Pringle, J.: 1979, Nature 278, 219.

31. Ruffini, R., Stella, L.: 1980, Nuovo Cimento Lett. 27, 529.

32. Lense, J., Thirring, H.: 1918, Phys. Z. 19, 156;
    Wilkins, D.: 1972, Phys. Rev. D5, 814.

33. Ruffini, R.: "Gravitationally collapsed objects", 1979, Proc. Second
    M. Grossman Meeting, in press.

34. This phenomenon has been widely observed in radio wave lengths
    (Maser). See e.g.: Moran, J.M., Radio observations of galactic masers,
    in: Frontiers of Astrophysics, ed. Avrett, E.H. (Harvard U.P., 1976).
    The multipole fragmentation of the maser sources observed in Astro-
    physical systems (see e.g.: Genzel, R., et al.: 1978, As. Ap. 66,
    13 and references therein), could be originated by the fulfillment
    of the necessary conditions presented here and in ref. (33)).

35. Ruffini, R., Stella, L.: 1980, Phys. Lett. 93B, 107.

36. Fang Li Zhi, Ruffini, R., Stella, L.: "SS433: Background for a rela
    tivistic model", 1981, Vistas in Astronomy, in press.

37. Stoeger, W.: 1981, preprint.

38. See e.g.: Lightman, A.P., Shapiro, S.L., Rees, M.J.: 1978, in
    ref. (1), and references therein.

39. Crampton, D., Cowley, A.P., Hutchings, J.B.: 1980, Ap. J. 235, L131.

40. Stoeger, W.: 1981, preprint.

41. Rosner, R., Ruffini, R., Vaiana, G.: 1980, preprint.

42. Fang Li Zhi: 1981, MNRAS 194, 177.

43. Ruffini, R., Song, D.J., Stella, L.: 1981, As. Ap., in press.

44. Collins, II G.W., Newsom, G.H.: 1980, IAU Circular n.3547;
    Margon, B., Anderson, S.: 1981, IAU Circular n.3626.

# OBSERVATIONS OF SECULAR CHANGES IN THE KINEMATIC MODEL OF SS433

G. W. Collins, II, and G. H. Newsom
The Perkins Observatory and The Ohio State University

ABSTRACT

In this paper we present evidence that several of the defining parameters of the Kinematic Model for SS433 are not constant but rather exhibit long term systematic changes. Recent data confirm the existence of the previously reported decrease in the precessional period. The value for this period change, when combined with the observed change in the period of the synodic spectral variations, implies that the orbital period is not significantly changing on a time scale less than 1000 years.

In addition we find mounting evidence for a statistically significant $(4\sigma)$ secular change in the cone angle $\theta$ at a rate of about $-1.5 \times 10^{-3}$ deg/day. However, the surprisingly short time scales implied by the observed values of $\dot{P}$ and $\dot{\theta}$ when combined with estimates of the system age suggest the possible existence of detectable higher time derivatives. This view is supported by the most recent data which suggest a value for $\ddot{P} \sim 10^{-5}$ (days)$^{-1}$. It is possible to understand these secular changes in terms of the motions to be expected from an object exhibiting classical precession in response to an external torque.

## INTRODUCTION

In the model we proposed for SS433 (Collins and Newsom 1979; Collins et al. 1980, 1981), a large magnetically-distorted star precesses due to the classical gravitational torque of a companion member in a binary system. Ionized matter escaping from the distorted star reaches high speed as it co-rotates out to a large distance, where the matter becomes compressed and emits the characteristic "moving" spectral lines as it encounters material trapped in the stellar magnetosphere. A consequence of this model was that deviations from the five-parameter kinematic model might be observable in the wavelengths of the moving lines in the form of a 6-day period (Collins et al. 1981).

Z. Kopal and J. Rahe (eds.), Binary and Multiple Stars as Tracers of Stellar Evolution, 389–397.

Following the discovery that this period is indeed present in the moving lines (Newsom and Collins 1981), we searched for other periodic or secular deviations from the kinematic model. By the end of the 1980 observing season, we were surprised to find that the precessional "164-day" period had been decreasing at the remarkably rapid rate of about 1% (Collins and Newsom 1981). It should be noted that Ciatti et al. (1981) also suggested the possibility that the period was decreasing.

This result has proven to be controversial. Following our initial announcement (Collins and Newsom 1980), Margon (1980) concluded that, if the period is decreasing, the rate of decrease is not greater than 0.4% at the 95% confidence level. If our value is confirmed and if the change would continue at this rate, the period would reach zero in half a century. Yet the length of the X-ray jets (Seward et al. 1980) and the presence of the radio "lobes" in W 50 (Geldzahler et al. 1980) imply that high-speed jets (if not the moving lines) have been in existence for at least $10^5$ years. Unless we are seeing the final stages of a long-lasting phenomenon, the rate of change of the period would itself be expected to decrease. To confirm the large value of $\dot{P}$ and search for higher-order derivatives in the period, plus secular changes in other parameters of the system that may accompany the period change, we have continued our analysis of the wavelength data. Since some of these parameters have been found to be subject to change, the model parameters are valid only for a specific epoch. For this study this epoch is JD 2444167.5.

OBSERVATIONS AND ANALYSIS

Most of the Doppler shifts used in our analysis are listed in Collins and Newsom (1981) and Wagner et al. (1981). The sources of data and number of data points from each source are as follows: Grandi (1980) 134; Blair (1981) 111; Koski et al. (1980) 36; Crampton (1980) 35; Ohio State University (Newsom et al. 1980; Wagner et al. 1980; Peterson and Crenshaw 1981; Wagner et al. 1981) 27; Liebert et al. (1979) 6; Leibowtz and Mazeh (1979) 5; Ford (1981) 4; Liller (1979) 2; Seaquist et al. (1979) 2; Seaquist (1979) 1. Our 1981 data are all more compatible with a large negative $\dot{P}$ than with a constant period. However, final confirmation that $\dot{P}$ is indeed large and negative came from observations during five nights in June 1981 obtained at the McGraw-Hill Observatory and kindly provided to us by Blair (1981). With these data included, the best current value of $\dot{P}$ is $-0.009 \pm .001$ days/day, as shown by the solid line in Figure 1. More indirect support for the decrease in the precessional period was provided by the observations of Wagner et al. (1981) showing that the 6 day period was also decreasing. Since the 6 day period represents half the synodic period derived from the 13.1-day orbital period and the 164-day precessional period, a change in the precessional period would cause a change in the 6 day period. The rate of change observed in the 6 day period,$-(3.0 \pm 1.1) \times 10^{-5}$ days/day, is very close to the value expected if the orbital period has not changed but the precessional period is decreasing at the rate we had previously found. The change in the precession thus does not appear to result from orbital changes in the binary.

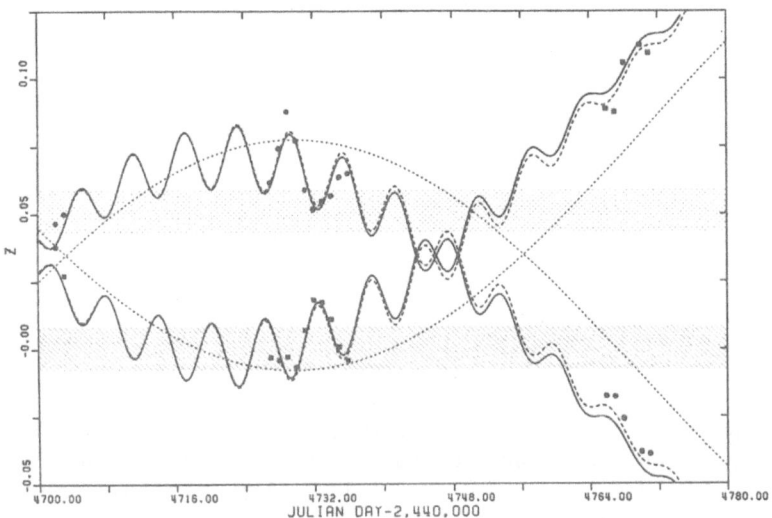

Fig 1. Doppler shift vs time for J.D. 2444700-4780. The solid line includes first derivatives of both the 6 and 164-day periods, while the dashed line also includes the second derivative of the 164-day period. The dotted line is the ephemeris of Margon (1980) which has a constant 164-day period. The ten data points on the right are from Blair (1981) and provide a crucial test of changes in the 164-day period. Shaded areas designate wavelengths at which moving H-α lines are blended with unshifted H-α or atmospheric absorption bands.

When we include the second derivative of the precessional period in our fit to the data, a positive value emerges, as would be expected from the fact that, as more recent data have been added to our fit, the magnitude of $\dot{P}$ has steadily decreased. The presently-available span of observations is not yet adequate to confirm that $\ddot{P}$ is statistically significant, but observations during November, 1981 will hopefully resolve this question. Our best fit, with $\ddot{P} = 1.2 \times 10^{-5}$, is shown as the dashed line in Figure 1. For comparison, the ephemeris of Margon (1980) is shown as a dotted line.

If we regard the angle of inclination of the system as constant, then the two remaining variables in the five-parameter fit subject to secular variation are the cone angle θ and the speed of the beams (represented here by the relativistic time dilation factor, γ ). When a first-order expansion of these two variables was included in the

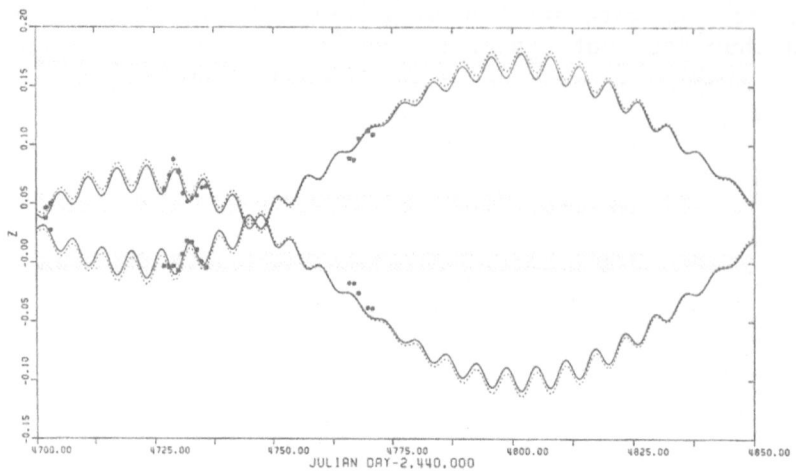

**Fig. 2.** Doppler shift vs time during spring and summer 1981. First derivatives of both the 6 and 164-day periods are included in both the dotted and solid lines, while the solid line also includes the first derivative in the cone angle $\theta$.

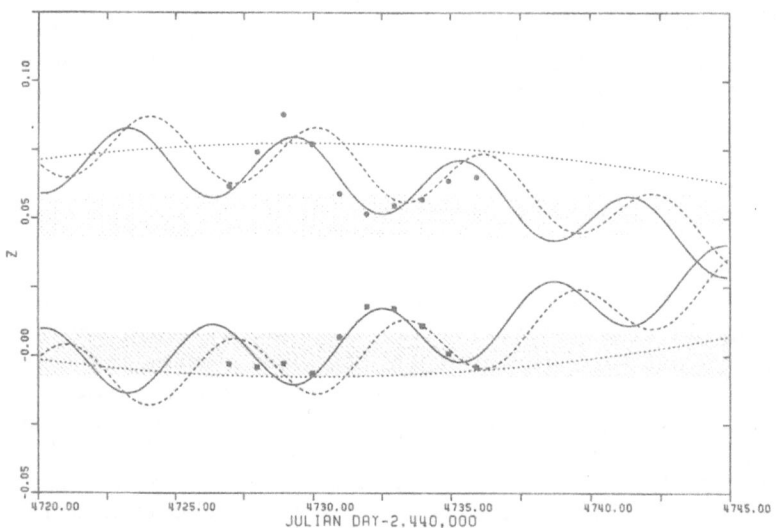

**Fig. 3.** Doppler shift vs time for J. D. 2444720-4745. The solid and dashed lines both include the first derivative of the 164-day period, while the solid line also includes the first derivative in the 6-day period and cone angle $\theta$. The dotted line is the ephemeris of Margon (1980).

analysis, a remarkably large value of $\dot{\theta}$ was found, amounting to $-(1.5 \pm 0.5) \times 10^{-3}$ degrees/day. During the course of spectroscopic monitoring of SS433, the cone angle has decreased by somewhat less than $2^{\circ}$, resulting in a small but steady decrease in the amplitude of the radial velocity curve. Figure 2 shows a recent cycle with and without including the variation in $\theta$ as shown as solid and dotted lines respectively. The combined effects of $\dot{\theta}$ and the decrease in the 6 day period are illustrated in Figure 3 by the solid line. The dashed line shows the best fit to the data if these two effects are excluded, and the fit is considerably poorer. The dotted line is the ephemeris of Margon (1980).

In the midst of these changes, however, the value of $\gamma$ appears unchanged. Our best fit to $\dot{\gamma}$ is $(0.6 \pm 1.2) \times 10^{-6}$/day. This result implies that there are no <u>secular</u> changes in the speed of the emitting gas greater than 2 km/s/day.

If we assume that the decrease in the cone angle derived from motions of the emitting regions is equal to a change in the orientation of the rotation axis about a precession axis (a plausible but not well-established assumption), then the discoveries reported above can be used to help understand the dynamics in the binary system that appears to drive the high-speed beams.

## SOME IMPLICATIONS OF THE SECULAR CHANGES OF THE SYSTEM PARAMETERS

One of the hallmarks of geodetic precession is that for most plausible objects, the direction of the precession is opposite that of the orbital motion. If this is indeed the case, any phenomena resulting from the 'beating' of the orbital period with the long period will produce phenomena with a synodic period less than the orbital period. On this basis Collins <u>et al</u>. (1981) suggested that a short period fluctuation exhibiting half the synodic period might be present in the moving line spectra and that if the precession were indeed retrograde, the observed period should be 6 days. The discovery of periodic variations exhibiting a period of 6.06 days (Newsom and Collins 1981) we take as confirmation of the retrograde nature of the precession.

Let us now investigate the extent to which the observed secular changes are consistent with the picture of a geodetically precessing object. Collins <u>et al</u>. (1981) have shown that under a wide variety of circumstances the precessional frequency $\omega_o$ and spin frequency $\omega_z$ of an object exhibiting both geodetic and forced precession is given by:

$$\omega_o^2 \cos\,\theta\,-(I_z\,\omega_z\,\omega_o/I_x)+ 3\,V_o \cos\theta/I_x = 0 , \qquad (1)$$

where $\theta$ is cone angle of the precession and $V_o$ is the perturbing potential resulting from the presence of the companion. The perturbing potential is:

$$V_o = - \omega_b^2 \; \mu \; (I_z - I_x)/2 \tag{2}$$

where $\omega_b$ is the angular frequency of the binary motion, $\mu$ is the reduced mass of the system and $I_z$ and $I_x$ are the moments of inertia of the precessing object about the body axis and an orthogonal axis respectively. It is worth noting that for any oblate object $I_z > I_x$ which implies $V_o < 0$.

Equation (1) may be rewritten in terms of the dimensionless parameters r and q so that

$$(1+q) \cos \theta - r = 0 \; ,$$

where        $q = 3 \; V_o/I_x \; \omega_o^2 \; ,$                                        (3)

and        $r = I_z \; \omega_z/(I_x \; \omega_o) \; .$

Although we initially suggested (Collins et al. 1981) that magnetic torques on the system implied by the high velocity mass loss should reduce $\omega_z$ to zero, it is clear that tidal acceleration by the secondary would tend to oppose this result. Thus it is not unreasonable to suspect that $\omega_z$, although small, may be variable. Such variation would result in a secular change in both $\omega_o$ and $\theta$.

Just as it is clear that equations 3 cannot uniquely specify r and q, so it is clear that time derivatives of equations 3 which describe the relationship between the secular changes cannot yield a unique result. The additional physics implying those changes is not included in the equilibrium model giving rise to equations 3. Nevertheless, the time derivatives of equations 3 do place a constraint on the angular spin and its rate of change. Thus

$$\left( \frac{\dot{\omega}_z}{\omega_z} \right)_{t_o} = \left( \frac{\dot{\omega}_o}{\omega_o} \right)_{t_o} \left[ \frac{2 \cos \theta}{r} - 1 \right]_{t_o} - (\dot{\theta} \; \tan \theta)_{t_o} \; , \tag{4}$$

where $t_o$ represents the epoch for which the system parameters have been determined.

If $(\dot{\omega}_z/\omega_z) < 0$, the precessing object is currently losing spin and we would expect $(\dot{\omega}_o/\omega_o) > 0$. The constraint imposed on r for this to be the case is

$$\frac{2 \cos \theta}{r} < 1 + (\dot{\theta} \tan \theta)/(\dot{\omega}_o/\omega_o) \ . \tag{5}$$

For the observed values of $\theta$, $\dot{\theta}$, $P_o$ and $\dot{P}_o$ (hence $\omega_o$ and $\dot{\omega}_o$) and the definition of r, equation 5 would imply that

$$\frac{I_x \omega_o}{I_z \omega_z} < 0.47 \ . \tag{6}$$

Since the precession is retrograde (i.e. $\omega_o < 0$), any prograde spin ($\omega_z > 0$) is compatible with equation 6. Thus the observed secular changes are consistent with a geodetically precessing object exhibiting a small amount of prograde rotation which is decreasing at the moment. This decrease leads to an increase in the precession frequency and a decrease in the magnitude of the cone angle as the external torque attempts to align the body axis of the object with the orbital axis.

Admittedly the above interpretation, at this point, is not unique but it is consistent and suggestive. For instance one can explore further aspects of the dynamics of the system for additional constraints on q and r. Our preliminary investigations indicate indeed that such additional constraints do exist. We have found (Collins & Newsom 1982) that the object should exhibit small amplitude nutational variations in the cone angle with a period very roughly of the order of 100 days. Determination of this period, for which we feel some evidence already exists, would enable a unique determination of q, r and $\dot{r}$.

Finally it is worth noting the effect that the dynamical constraints on q and r have on the nature of the precessing object itself. If the object is rapidly spinning (i.e., $\omega_z \gg \omega_o$), then it is clear from the definition of r that $r \gg 1$. Thus equation 4 becomes

$$(\frac{\dot{\omega}_z}{\omega_z})_{t_o} = - [ \frac{\dot{\omega}_o}{\omega_o} + (\dot{\theta} \tan \theta) ]_{t_o} \ . \tag{7}$$

For the measured values of $\dot{P}_o$ and $\dot{\theta}$ we can conclude that $(\dot{\omega}_z/\omega_z)_{t_o} \sim$ $-4.9 \times 10^{-5}$ day$^{-1} < 0$ and the object is spinning down. Thus the observed secular changes require that any rapidly spinning object subjected to forced geodetic precession is undergoing a loss of spin regardless of the direction of rotation.

It seems likely that any disk model for the precessing object will require $\omega_z > \omega_b \gg \omega_o$. Thus it would appear that any disk model will have to incorporate a physical mechanism which allows the disk to substantially slow down (and probably later speed up). Within the framework of standard disk models this would seem very difficult to accomplish. Although one may argue that the classical dynamics approach exhibited here is inapplicable to disks, the fact remains that the observed secular changes must be incorporated into any model of this system and they are natural consequences of the classical description.

Additional observation of this object should serve to further quantify the secular changes we have described here as well as establish the existence of nutation and period acceleration (i.e. $\ddot{P}_o$). Within the classical picture, we can expect such quantification to uniquely specify q and r and thereby illuminate the specific nature of the binary. It seems reasonable to suggest that, although this may well be the most enigmatic object to be discovered in twenty years, we can understand it and thereby learn something of the evolution of a truly unusual binary system.

We would like to thank the observers who provided us directly with data. Particular thanks are due W. P. Blair, whose results from June 1981 were definitive in establishing the values quoted in this paper. We also would like to thank Z. Kopal for suggesting we make this contribution. Since the original preparation of this work, Margon has reversed his position on the value of $\dot{P}$ (Margon 1980), and he now finds a value (Margon et al. 1981) that confirms our original announcement (Collins and Newsom 1980) and is in precise agreement with the value quoted here.

REFERENCES

Blair, W.P.: 1981. Private communication.

Ciatti, F., Mammano, A., and Vittone, A.: 1981, Vistas in Astron. (in press).

Collins, G. W., II, and Newsom, G. H.: 1979, Nature 280, pp. 474-475.

Collins, G. W., II, and Newsom, G. H.: 1980, IAU Circ. No. 3547.

Collins, G. W., II, Newsom, G. H., and Boyd, R. N.: 1980 in M. Plavec, D. M. Popper, and R. K. Ulrich (eds.), "Close Binary Stars: Observations and Interpretation", IAU Symp. 88, pp. 375-379.

Collins, G. W., II, and Newsom, G. H.: 1982, Astrophys. Space Sci. 81, pp. 199-208.

Collins, G. W., II, Newsom, G. H., and Boyd, R. N.: 1981, Astrophys. Space Sci 76, pp. 417-440.

Collins, G. W., II, and Newsom, G. H.: 1982, in preparation.

Crampton, D.: 1980. Private communication.

Ford, H.: 1981. Private communication.

Geldzahler, B. J., Pauls, T., and Salter, C. J.: 1980, Astron. Astrophys., 84, 237-244.

Grandi, S. A.: 1980. Private communication.

Koski, A., Burbidge, E. M., and Smith, H. E.: 1980. Private communication.

Leibowitz, E. M., and Mazeh, T.: 1979, IAU Circ. No. 3367.

Liebert, J., Angel, J. R. P., Hege, E. K., Martin, P. G., and Blair, W. P.: 1979, Nature 279, pp. 384-387.

Liller, W.: 1979. Private communication.

Margon, B.: 1980. Paper delivered at the Tenth Texas Symposium on Relativistic Astrophys, December, Baltimore, Maryland.

Margon, B., Anderson, S., Grandi, S., and Downes, R.: 1981, IAU Circ. No. 3626.

Newsom, G. H., Jenkner, H., and Wagner, R. M.: 1980, Astron. J. 85, 1229-1231.

Newsom, G. H., and Collins, G. W., II: 1981, Astron. J. 86, 1250-1258.

Peterson, B. M., and Crenshaw, D. M.: 1981. Private communication.

Seaquist, E. R.: 1979. Private communication.

Seaquist, E. R., Garrison, R. F., Gregory, P. C., Taylor, A. R., and Crane, P. C.: 1979, Astron. J. 84, pp. 1037-1041.

Seward, F., Grindlay, J., Seaquist, E., and Gilmore, W.: 1980, Nature, 287, pp. 806-808.

Wagner, R. M., Byard, P. L., Foltz, C. B., and Peterson, B. M.: 1980. Private communication.

Wagner, R. M., Newsom, G. H., Foltz, C. B., and Byard, P. L.: 1981, Astron. J. (in press).

# TIME RESOLVED CIRCULAR POLARIMETRY OF WHITE DWARF PULSARS

W. Krzemiński
N. Copernicus Astronomical Centre, Polish Academy
of Sciences, and Department of Astronomy, University of Western Ontario

J. D. Landstreet, and I. Thompson
Department of Astronomy, University of Western
Ontario

There are two recognized subsets of cataclysmic binaries that contain white dwarfs: the AM Her stars, whose
white dwarf components are sufficiently magnetized to en-
sure synchronous rotation with the orbital period, and the
DQ Her stars, which do not maintain synchronism and probably have been spun up by mass accretion. Both groups of
stars, recognized also as X-ray sources, are important as
probes of the accretion process, and the radiation mechanisms in the vicinity of the white dwarf. So far, we know
five objects belonging to the former group (AM Her, AN UMa,
VV Pup, 2A 0311-227, and PG 1550+191), and seven falling
into the latter: WZ Sge, AE Aqr, V533 Her, DQ Her, V1223
Sgr, H2252-035, and EX Hya, with the corresponding rotation periods of their white dwarf primaries of 28, 33, 64,
71, 794, 805, and 4020 s, respectively. While the main
observational and theoretical efforts have been focussed
on the AM Her class, the DQ Her binaries have been investigated to much lesser extent. This is probably because
the DQ Her stars have been recognized only very recently
as a homogeneous class.

The DQ Her-type stars are characterized primarily by
the highly coherent oscillations in the optical region,
with a quality factor $Q = 1/|\dot{P}| \simeq 10^{12}$. All but one (DQ Her
itself) are X-ray emitters with the X-ray flux being modulated on the fundamental period of optical oscillations.
The observed oscillations of DQ Her stars are most readily
explained by invoking an "oblique rotator" or a "white
dwarf pulsar" model with the accretion induced hot spots
at the surface of a rapidly rotating, magnetic white dwarf
(Lamb 1974). In analogy to AM Her binaries, at least in
active state, in DQ Her stars one might expect to observe

399

*Z. Kopal and J. Rahe (eds.), Binary and Multiple Stars as Tracers of Stellar Evolution, 399–401.*

some continuum polarization being related to the accretion
column phenomena. It has been shown (Lamb and Masters 1979;
Chanmugam and Dulk 1981) that the optically polarized light
in AM Her objects arises as a result of high harmonic
(m $\simeq$ 5) optically thick cyclotron emission from the shock-
heated region above a magnetic pole of accreting white
dwarf. In lower fields, $\lesssim$ 10 Mgauss, the polarized cyclotron emission is shifted towards infrared, while in high
fields, $\gtrsim$ 100 Mgauss, it moves to the UV domain (Lamb 1979).
Because of the existence of large accretion disks around
magnetic primaries in the DQ Her stars, and because of
rapid rotation of these primaries, one should not expect
magnetic fields in these objects to be as large as those
observed in AM Her stars. Furthermore, in few cases (eg.
H2252-035) one is observing the dominant optical modulation
at a period arising as a beat phenomenon between the fundamental oscillation period (i.e. white dwarf rotation)
and the orbital period. This is interpreted that the significant fraction of the pulsed optical light arises not
from the white dwarf directly, but from the reprocessing
of the white dwarf's pulsed radiation in the surrounding
accretion disk (Patterson 1980).

The detection of both the circular and linear polarization changes synchronous with the rotation of the underlying white dwarf pulsars in DQ Her binaries would allow
us to constrain class of models which have been suggested
for these objects. Of particular importance would be the
determination of the strength of a pulsar magnetic field.
The early results for DQ Her (Swedlund et al. 1974; Kemp
et al. 1974), although being on a threshold of detection,
seemed to be encouraging: both the circular and linear
polarization variations synchronous with the white dwarf
rotation have been found. Being stimulated by these early
works, and by the subsequent discovery of more members of
the class, during later part of 1980 and in early 1981
we have been carrying out a limited survey of circular
polarization properties of five objects: AE Aqr, V533 Her,
H2252-035, V1223 Sgr, and EX Hya in the blue spectral region. A Pockel cell polarimeter (Angel and Landstreet 1970)
attached to either the 122-cm telescope of the University
of Western Ontario or the 1-m Las Campanas telescope has
been utilized. For rapid rotators (AE Aqr, V533 Her) we
used a synchronous data averaging.

This pilot survey has shown that for all objects but
V533 Her the amplitude of the periodic component of the
circular polarization is not greater than 0.10 - 0.15 per
cent in the blue band. In case of V533 Her a $\sim$ 0.5% sinusoidal changes in a circular polarization have been detected on two nights. This might mean that in DQ Her stars the

dilution by an unpolarized background is very large, and furthermore, that this dilution is variable (the case of V533 Her).

The detailed account on both the observational and theoretical aspects of the circular polarimetry of DQ Her stars will appear in the Monthly Notices of the Royal Astronomical Society.


References

Angel, J. R. P., and Landstreet, J. D.: 1970, Astrophys. J. Letters 160, L147.
Chanmugam, G., and Dulk, G. A.: 1981, Astrophys. J. 244, 569.
Kemp, J. C., Swedlund, J. B., and Wolstencroft, R. D.: 1974, Astrophys. J. Letters 193, L15.
Lamb, D. Q.: 1974, Astrophys. J. Letters 192, L129.
Lamb, D. Q.: 1979, in "Compact Galactic X-ray Sources", pp. 27-47.
Lamb, D. Q., and Masters, A. R.: 1979, Astrophys. J. Letters 234, L117.
Patterson, J.: 1980, Astrophys. J. 241, 235.
Swedlund, J. B., Kemp, J. C., and Wolstencroft, R. D.: 1974, Astrophys. J. Letters 193, L11.

# THE MASS DISTRIBUTION OF WHITE DWARFS AND CENTRAL STARS OF PLANETARY NEBULAE

Volker Weidemann
Institut f. Theoretische Physik und Sternwarte
Universität Kiel, Fed. Rep. Germany

Recent results on mass distributions for white dwarfs and planetary nebulae are presented and compared with current theoretical predictions. Whereas single star evolution leads to final masses predominantely in a narrow interval around $0.6\,\mathcal{M}_c$ which can be explained by current mass loss schemes degenerate stars in binaries present a larger range of masses. The average mass of the primaries in cataclysmic binaries seems to be more around 0.7 than $1\,\mathcal{M}_c$.

Analysis of spectroscopic and photometric data for DA white dwarfs (i.e. the majority) during the last years have yielded the interesting result that the mass distribution is rather narrow, around $0.6\,\mathcal{M}_c$. This is apparent as well from the HR-diagram (Weidemann, 1978) if one assigns weights to parallaxes, resulting in a mass distribution $\mathcal{M}(R)$ - which is obtained from the radius distribution via the mass-radius relation - confined to $\mathcal{M}(R) = 0.58 \pm 0.10\,\mathcal{M}_0$ (Koester, Schulz, Weidemann, 1979, -KSW-, Fig. 9b) as also from the two-color diagrams in the Strömgren (Graham, 1972, Wegner, 1979) or multichannel system (Greenstein, 1976) which enable sensitive surface gravity determinations (Weidemann, 1971, KSW, Figs. 2, 4 and 6, Shipman and Sass, 1980). The corresponding mass distribution, $\mathcal{M}(g)$ for 122 DA white dwarfs KSW, Fig. 8b) has recently been confirmed by a careful study of 20 DA stars by Schulz and Wegner (1981). The (15 - 30%) non-DA white dwarfs (spectral type DB, DC, C2), although less certain, seem to occupy the same narrow mass range (Koester et al. 1981),(Wickramasinghe, 1981, Weidemann, 1981a).

The situation is similar for the nuclei of planetary nebulae (NPN). Consideration of evolutionary tracks in the HR-diagram by Weidemann (1977b, Fig. 2) and by Renzini (1979, Fig. 1) suggested a new method of mass determination in which the time scale of evolution as measured by nebular expansion is compared with luminosities of NPN as predicted by evolutionary tracks for different masses (Schönberner and Weidemann, 1981). These tracks differentiate very effectively between NPN masses, essentially since the core mass in the pre-PN stages at the asymptotic (second) giant branch is highly sensitive to luminosity (core mass-

403

*Z. Kopal and J. Rahe (eds.), Binary and Multiple Stars as Tracers of Stellar Evolution, 403–408.*
*Copyright © 1982 by D. Reidel Publishing Company.*

luminosity relation), (Schönberner and Weidemann, 1981, Fig. 6).
Comparison with a local ensemble - less biased by observational
selection - confirms the narrow distribution with few if any massive
NPN at very low luminosities (Schönberner, 1981). The NPN distribution
cuts off at $\mathcal{M} = 0.55\ \mathcal{M}_O$ - not surprisingly, since stars with smaller
core masses do not reach the asymptotic giant branch and will not
produce PN (Sweigart et al. 1974, Weidemann, 1975). However it was
more surprising that it falls also off very steeply for higher masses,
$\mathcal{M} > 0.64\ \mathcal{M}_\odot$ . This contradicts canonical stellar evolution which
predicts dredgeup and PN enrichment not below progenitor masses of
$3\,\mathcal{M}_\odot$ or core masses above $0.7\,\mathcal{M}_\odot$. Schönberner's material shows PN of
all enrichment classes (Schönberner, 1981, Fig. 13/14).
Thus single stars seem in general not to produce larger degenerate
core masses than about $0.65\ \mathcal{M}_O$ . This conclusion has been supported by
recent downward revisions of Mira (pre-PN) luminosities (Robertson and
Feast, 1981, Willson, 1981) and by the C star luminosity distribution
in the Magellanic clouds (Richer, 1981), which also show enrichment
at small core masses - a fact which has lead Iben (1981) to dicuss
the "carbon star mystery".
The PN results have independently confirmed by NPN spectroscopy and
NLTE analysis at Kiel (Mendez et al., 1981, Kudritzki et al., 1981),
again by a comparison with Schönberner tracks in $g\text{-}T_{eff}$ diagrams.

In summary the observational evidence presented points to a very flat
initial-final mass relation (Weidemann, 1981 b, Fig. 1) for which
progenitors with masses up to the limit of the degenerate core range
( $\sim 8\ \mathcal{M}_\odot$) may produce white dwarfs (Weidemann, 1977a, 1979). Koester
and Reimers (1981) have recently confirmed at least partly the prediction of Romanishin and Angel (1980) that white dwarfs are present in
young clusters with turn-off masses larger than $4\,\mathcal{M}_\odot$·Wegner (1981)
reaches are similar conclusion by consideration of white dwarfs in
binaries  whose age and progenitor masses are estimated from kinematical properties of the system.
On the theoretical side the only mechanism proposed yielding the
needed high mass loss rates is shock ejection in late giant stages
(Barkat and Tuchman, 1980). It must be combined with a fairly large
steady mass loss rate (Reimers factor $\eta \sim 1.4$) in order to reproduce
the flat empirical initial-final mass relation. Schönberner (1981,
Fig. 10) has demonstrated that the Barkat-Tuchman PN ejection line
indeed predicts the sharply peaked NPN mass distribution observed.

We now turn to the topic of this Conference: evolution in binary and
multiple stars and ask how our picture will be changed. White dwarfs
occur in quite a variety of binary stars which we shall consider
in turn.
a) Wide binaries, comprising astrometric and visual binaries, and
common proper motion pairs. There are 7 white dwarfs in binaries
within 10 pc (compared with 13 single WDs) with separations between
16 and 100 AU - thus there will be many more at larger distances which
have not get been found, staying too close to be visible and having
too long periods for spectroscopic detection. However there is the

possibility to detect the hotter ones in the ultraviolet.
About 20 % of all WD listed are wide binaries or cpm objects with
separations between 100 and typically several thousand AU. We do not
expect them to have been influenced during evolution by the existence
of the companion (red giant radii remain smaller than 10 AU) and thus
they may be counted as single stars. Luyten (1969) has given lists and
also proposed 15 candidates for double white dwarfs, from which 5 have
been confirmed up to now. Although it will be difficult in many cases
to detect duplicity of degenerates the number of double WDs is surprisingly small. A last category in this group comprises the unseen
companions - remember that our classical WD Sirius B has been an
unseen companion first! - about which we obtained in formation at this
Conference by Prof. Van de Kamp and Dr. Abhyankar. In some cases
there are suspected white dwarf companions with masses between 0.5 and
$1 \mathcal{M}_{\odot}$, worth an effort of search with the IUE or the Space Telescope.
b) Close binaries. We distinguish three categories:
b1) degenerates in detached systems, partly found by eclipses, partly
by heating, EUV or X radiation. We list as examples (periods in
brackets) F 24 ($4^d$), V 471 Tau = BD+$16^{\circ}516$ ($0.5^d$),GK Vir = PG 1413+01
($0.34^d$), and the polars AM Her ($0.13^d$), EF Eri = 2A 0311-227 ($0.06^d$)
with separation between 1.5 and 6 $R_{\odot}$, and thus products of common
envelope evolution (Paczynski, 1981) but now with comparatively little
interaction.
b2) the cataclysmic variables, with degenerate primaries, Roche lobe
overflow of the secondaries and accretion disks, periods from typically half a day down to about 80 minutes with separations of the order
of 1 $R_{\odot}$. Much effort has been devoted to a better determination of
the physical parameters of these systems, and better understanding
is at hand. (Reference to the contributions by Ritter, N. Vogt,
and others at this Conference). Whereas earlier mass determinations
have yielded white dwarf masses around $1 \mathcal{M}_{\odot}$ (see Robinson , 1976)
recent data, collected by Ritter (1980) give often smaller masses. I
have studied the literature up to today and summarize the present
situation in Fig. 1, where the estimated uncertainties are indicated
by the size and width of the quadrangles in each case. Summed up over
the mass intervals I obtain the mass distribution of Fig. 2. It is
evident that the average mass of the degenerates in cataclysmics is
around $0.7 \mathcal{M}_{\odot}$ rather than $1 \mathcal{M}_{\odot}$.

The new result appears more reasonable in view of what has been
outlined for single star evolution: it may even be possible that the
WD masses are actually equal. Of course there are cases in which mass
transfer stops the core evolution before it reaches the critical
values for single star mass loss, however it is difficult to imagine
how mass exchange might increase the core masses beyond those reached
by single star evolution. The fact that furthermore the WD masses in
cataclysmics are not dependent on the period (Fig. 1) - i.e. that
there seem to be no significant differences for WD masses above and
below the period gap, or between post-novae and dwarf novae - speaks
also for a general integrated scheme of evolution like that outlined
by Vogt at this Meeting. In any cases I strongly recommend not to use

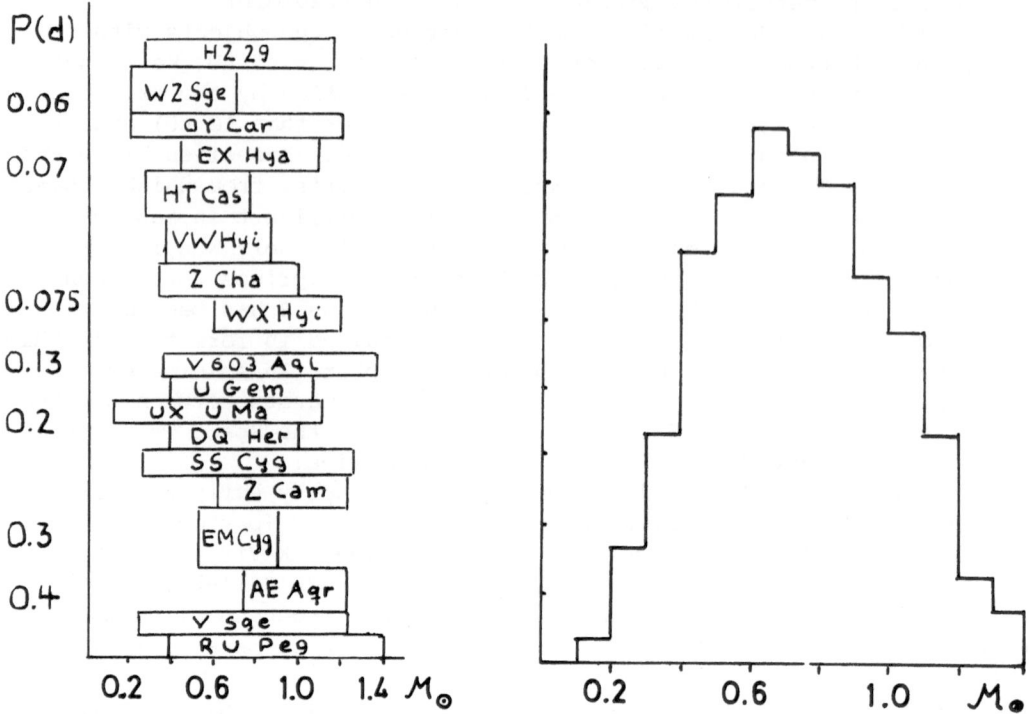

Fig. 1. Masses of degenerate primaries for 18 cataclysmic variables,
ordering according to period (P) in
days (d), see text

Fig. 2. Mass distribution of degenerate primaries in CB's, derived
from Fig. 1

$1\,\mathcal{M}_\odot$ any more as the typical WD mass value in all kind of model calculations for novae and dwarf novae. If, on the other hand, the average
WD mass in CBs will turn out to be really higher than in the single
star case - or if there are individual cases with definitely higher
masses - we are forced to look for very massive progenitors in order
to build up the higher core masses in red giant stages. Examples are
given by Law and Ritter at this Conference: they need a helium
primary star of 0.8 to 3.3 $\mathcal{M}_\odot$ to begin with, and total masses above
10 $\mathcal{M}_\odot$ in order to obtain higher mass white dwarfs in CBs. So, if WD
masses are high, progenitor masses must have been even higher, such as
to make the formation of progenitor systems a comparatively rare event
in the galaxy. The WD mass is thus indeed crucial to the question of
the origin of CBs! If massive progenitor systems are acceptable
statistically or not depends of course on the longevity of the CBs, a
problem which has not yet been solved, although predictions are being
made (Taam et al., 1980, Paczynski and Sienkiewicz, 1980, Rappaport
et al., 1981)
b3) the last category  comprise the two extremely short period binaries GP Com = G 61-29 (46[m]) and HZ 29 (16[m]) which both show only
helium in the system and which are probably double degenerates with a
small incipient black dwarf secondary of 0.02 resp. 0.04 $\mathcal{M}_\odot$. The

secondary will be soon eroded and the system will appear as a single white dwarf of spectral type DB (Nather et al., 1981). The fraction of the DB stars which is formed in this way depends on the lifetime and space density of these systems which are both very uncertain.

It is however interesting to note that according to current theories low period CB evolution with periods increasing again from a minimum around $80^m$ will also lead to erosion of the secondary with a DA white dwarf remaining. So there would be no way for evolution via the normal CB stage to these strange systems. We further want to emphasize that within the present schemes CB evolution will not lead to mass increase of the degenerate primary and thus not to supernovae (type I) events caused by collapse on reaching the Chandrasekhar mass limit.

If one finally takes into account what has been discussed by Webbink (1979) about possible channels of binary evolution, noting that all contact binaries on the main sequence evolve into coalescence and thereby finally follow single star evolution one is lead to the conclusion that the ultimate fate even in the cases of binary evolution with mass exchange does not differ so much from that of single stars after all! I have estimated in my Rochester Lecture (1979) that about 70 % of all stars in the main progenitor range ($1 < M < 8 M_\odot$) follow single star evolution. The remaining 30 % may undergo mass exchange and common envelope evolution. Paczynski (1981) expressed the view that detached systems like V 471 Tau might be the normal result of common envelope evolution and that cataclysmics are formed only if they lose additional mass and angular momentum. This depends very much on the fate of the detached systems (b1): will they survive at the present separations or get in contact again by magnetic braking or stellar wind leading them into the CB channel? We are not yet at the end of the road!

## References

Barkat, Z., Tuchman, Y.: 1980, Astrophys. J. 237, 105 - 110
Graham, J.A.: 1972, Astron. J. 77, 144-149
Greenstein, J.L.: 1976, Astron. J. 81, 323 - 338
Iben, Jr., I.: 1981, Astrophys. J. 246, 278-291
Koester, D., Schulz, H. Weidemann, V.: 1979, Astron.Astrophys. 76, 262 - 275
Koester, D., Reimers, D.: 1981, Astron. Astrophys. 99, L8 - 11
Koester, D., Schulz, H., Wegner, G.: 1981, Astron.Astrophys. in press.
Kudritzki, R.P., Méndez, R.H., Simon, K.P.: 1981, Astron.Astrophys. 99, L15 - 17
Méndez,R.H., Kudritzki, R.P., Gruschinske, J., Simon, K.P.: 1981, Astron. Astrophys. in press.
Nather, R.E., Robinson, E.L., Stover, R.J.: 1981, Astrophys. J. 244, 269 - 279
Paczyński, B.: 1981, Acta Astronomica 31, 1 - 11

Paczyński, B., Sienkiewicz, R.: 1981, preprint
Rappaport, S., Joss, P.C., Webbink, R.E.: 1981, preprint
Renzini, A.: Stars and Star Systems, IAU Regional Meeting, Uppsala,
          B.E. Westerlund, Ed. Reidel, Dordrecht, pp. 155 - 169
Richer, H.: 1981, Astrophys. J. 243, 744 - 755
Ritter, H.: 1980, private communication
Robertson, B.S.C., Feast, M.W.: 1981, preprint
Robinson, E.L.: 1976, Ann. Rev. Astr. Astroph. 14, 119 - 142
Romanishin, W., Angel, J.R.P.: 1980, Astrophys. J. 235, 992 - 998
Schönberner, D.: 1981, Astron. Astrophys., in press.
Schönberner, D., Weidemann, V.: Physical Processes in Red Giants
          I. Iben Jr., A. Renzini: Eds. D. Reidel, Dordrecht,
          pp. 463 - 468
Schulz, H., Wegner, G.: 1981, Astron. Astrophys. 94, 272 - 279
Shipman, H.L., Sass, C.A.: 1980, Astrophys. J. 235, 177 - 185
Sweigart, A.V., Mengel, J.G., Demarque, P.: 1974, Astron. Astrophys.
          30, 13 - 19
Taam, R.E., Flannery, B.P., Faulkner, J.: 1980, Astrophys. J. 239,
          1017 - 1023
Webbink, R.E.: 1979, IAU Coll. 53, White Dwarfs and Variable Degenerate
          Stars, H. Van Horn, V. Weidemann Eds. Univ. Rochester,
          pp. 426 - 447
Wegner, G.: 1979, Astron. J. 84, 1384 - 1394
Wegner, G.: 1981, Astron. J. 86, 264 - 270
Weidemann, V.: 1971, IAU Symp. 42, White Dwarfs, W.J. Luyten Ed.
          D. Reidel, Dordrecht, pp. 81 - 96
Weidemann, V.: 1975, Problems in Stellar Atmospheres and Envelopes,
          B. Baschek, W.H. Kegel, G. Traving, Eds. Springer,
          Berlin, pp. 173 - 203
Weidemann, V.: 1977a, Astron. Astrophys. 59, 411 - 418
Weidemann, V.: 1977b, Astron. Astrophys. 61, L 27 - 30
Weidemann, V.: 1978, IAU Symp. 80, The HR Diagram, A.G.D. Philips,
          D.S. Hayes, Eds. D. Reidel, Dordrecht, pp. 121 - 124
Weidemann, V.: 1979, IAU Coll. 53, White Dwarfs and Variable
          Degenerate Stars, H.M. Van Horn, V. Weidemann
          Eds. Univ. Rochester, pp. 206 - 222
Weidemann, V.: 1981a, Lecture at 4th Europ. Workshop on White Dwarfs,
          Paris, 1981
Weidemann, V.: 1981b, IAU Coll. 59, Effects of Mass Loss on Stellar
          Evolution, C. Chiosi, R. Stalio Eds. D. Reidel,
          Dordrecht, pp. 339 - 344
Wickramasinghe, D.T.: Contributed paper at the 4th Europ. Workshop on
          White Dwarfs, Paris, 1981
Willson, L.A.: 1981, Physical Processes in Red Giants
          I. Iben Jr., A. Renzini, Eds. Reidel, Dordrecht,
          pp. 225 - 230

ABUNDANCE PECULIARITIES IN WHITE DWARF ATMOSPHERES AS A RESULT OF BINARY
 EVOLUTION?

I.Bues
Dr.Remeis Sternwarte Bamberg,Astron.Inst.Univ.Erlangen-
Nürnberg,G.F.R.

We investigate metal abundances in helium-rich white dwarf atmospheres
and check the possibility of variations due to binary evolution.

I.    INTRODUCTION

From the spectroscopically classified sample of white dwarfs,
continuously enlarged by Greenstein,it is well known that most of these
objects belong to the group of extremely hydrogen-rich DA-stars.If we
look for binaries,however,we find only 12 per cent being members of
wide binary systems compared to 20 per cent of the whole sample. The two
close binary systems - the ultrashort period variables AM CVn and G 61-
29 - show only helium lines in their spectra.

According to Nather et al.(1979) these systems might be the pre-
cursors of very low mass ($M < 0.4$ $M_\odot$ ) single helium-rich DB stars with
a helium core since the secondary ($M \approx 0.02$ $M_\odot$) is likely to be completely dissolved. At least a certain fraction of the DB-stars and their cooler descendants of spectral type DC,$C_2$, DF and DG should be the result
of a process like that.If compared to other stars of the same spectral
type, they should have a smaller log g and a decreased metal abundance.

Wide binary systems containing a non-DA white dwarf are more
massive  ($M \approx 1M_\odot$) than the above mentioned systems with a typical separation of 270 AU. Both components should have undergone the evolution of
single stars with a certain amount of mass loss from the system. We expect degenerate stars with cores of elements heavier than helium ($M >$
$0.5$ $M_\odot$) and at least traces of heavier elements in the outer helium-rich
atmosphere.

With model atmospheres fitted to observations of helium-rich white
dwarfs in the range of $T_{eff}$ where metal lines occur in the visible and
UV spectrum we try to see a difference in mass and/or metal abundances
between single stars and binary components.

*Z. Kopal and J. Rahe (eds.), Binary and Multiple Stars as Tracers of Stellar Evolution, 409–411.*
*Copyright © 1982 by D. Reidel Publishing Company.*

## II.   MODEL ATMOSPHERES AND RESULTS

Flux-constant helium-rich model atmospheres including convection have been computed in the range $12000 > T_{eff} > 6000$ K with abundance ratio of $10^6 > He/H > 10^3$, $10^8 > He/C > 10^3$, $10^3 > C/O > 1/3$, all heavier elements reduced by a factor of $10^3$ compared to the solar value. Special emphasis has been taken for the treatment of all molecules, not only their dissociative equilibria but also their absorption of the flux due to strong bands.
(Details of the computations and some model atmospheres will be published in Astronomy & Astrophysics)

The resulting fluxes have been compared to observed spectra in the visible and UV taken by Wegner (1973,1980,1981),scannerobservations by Oke(1974)and IUE observations by Weidemann et al.(1980,1981). Our results have been compiled in Tables I and II.

TableI : Cool helium-rich white dwarfs (single stars)

| | $T_{eff}$ | log g | $M/M_\odot$ | He/H | He/C | He/O | He/Ca |
|---|---|---|---|---|---|---|---|
| vMa2 | 5800 | 7.42 | 0.31 | $>10^5$ | $>10^7$ | $>10^7$ | $5 \cdot 10^{10}$ |
| R640 | 8300 | 7.5 | 0.45 | $10^{3.6}$ | $>10^5$ | $>10^5$ | $4 \cdot 10^8$ |
| L145-141 | 7500 | 7.8 | 0.40 | $>10^5$ | $10^6$ | $>10^8$ | $>10^{10}$ |
| EG148 | 7500 | 7.5 | | $>10^5$ | $5 \cdot 10^6$ | $>10^8$ | $>10^{10}$ |
| L97-3 | 9800 | 7.5 | | $>10^5$ | $2 \cdot 10^5$ | $>10^7$ | $>10^7$ |
| G257-38 | 7000 | 7.5 | | $>10^5$ | $5 \cdot 10^5$ | $>10^7$ | $>10^{10}$ |

Table II: Cool helium-rich white dwarfs (wide binaries)

| | $T_{eff}$ | log g | $M/M_\odot$ | He/H | He/C | He/O | He/Ca |
|---|---|---|---|---|---|---|---|
| BPM27606 | 8200 | 7.8 | 0.82 | $10^4$ | $5 \cdot 10^4$ | $> 5 \cdot 10^5$ | $>10^8$ |
| G47-18 | 8800 | 7.5 | 0.75 | $10^5$ | $3 \cdot 10^4$ | $> 3 \cdot 10^5$ | $>10^8$ |
| LDS678B | 10600 | 7.8 | 0.7 | $10^4$ | $2 \cdot 10^4$ | $>10^5$ | $>10^7$ |
| G218-8 | 10300 | 7.8 | 0.65 | $>10^5$ | $10^4$ | $>10^5$ | $>10^7$ |
| L745-46A | 7500 | 8.0 | 0.8 | $>10^5$ | $>10^8$ | $>10^8$ | $4 \cdot 10^9$ |

For vMa2, the values obtained by Grenfell (1974) have been included.
Due to several improvements of the model atmosphere program the results of stars included already in the analysis of Bues (1973) and

Bues (1979) have been slightly changed. For R640, the effective temperature is lower than earlier work by Liebert (1977) and Cottrell and
Greenstein (1980).

## III.  DISCUSSION

Although the analyzed sample of cool helium-rich white dwarfs is
very small the difference in masses and abundances seems to be significant. If more parallaxes of single white dwarfs could be measured,
the mass determination would be improved. Shipman (1979) derived statistical parallaxes from scanner observations of a large sample of white
dwarfs.His mass determination of DB stars (his Table 6) compare very
well with our result:stars in binary systems have larger masses than
single ones.

Concerning the purity of helium in the atmospheres,single degenerates are favoured against binary components independent of spectral
type.There is no systematic difference in total metal content between
stars with carbon or calcium determined spectra.

So we may conclude that the mechanism proposed by Nather et al.
cannot work for helium-rich binary components but might have been
effective for the precursor of vMa2 or L145-141.Cool atmospheres of
white dwarfs consisting of helium only have not been observed,the
formerly classified DC stars reveal strong features due to metal lines
in the far UV region.

### REFERENCES

Bues, I., 1973, Astron. Astrophys. 28, 181
Bues, I., 1979, IAU Coll. 53, Rochester University Press, p. 186
Bues, I., 1981, preprint
Cottrell, P.L., Greenstein, J.L., 1980, Ap. J. 238, 941
Grenfell, T.C., 1974, Astron. Astrophys. 31, 303
Liebert, J.W., 1977, Astron. Astrophys. 60, 101
Nather, R.E., Robinson, E.L., Stover, R.J., 1979, IAU Coll. 53,p. 453
Oke, J.B., 1974, Ap.J. Suppl. 27, 21
Shipman, H.L., 1979, Ap.J. 228, 240
Wegner, G., 1973, Monthly Not. roy.Astron. Soc. 163, 381
Wegner, G., 1981a, Ap.J. 245, L27
Wegner, G., 1981b, Ap.J. 248, L129
Weidemann, V., Koester, D., Vauclair, G., 1980, Astr.Astrophys.83, L13
Weidemann, V., Koester, D., Vauclair, G., 1981, Astr.Astrophys.95, L9

# THE MINIMUM ORBITAL PERIOD OF HYDROGEN-RICH CATACLYSMIC BINARIES

R. Sienkiewicz and B. Paczyński
N. Copernicus Astronomical Center
Polish Academy of Sciences
ul. Bartycka 18, 00-716 Warszawa, Poland

This communication summarizes the investigations of the relationship between the minimum binary period and the evolution of very short period hydrogen-rich cataclysmic binaries /CB's/. In the first work /Paczyński, 1981/ some ideas were developed concerning an influence of gravitational radiation on the evolution of CB's, and some quantitative predictions were done. In the next paper /Paczyński and Sienkiewicz,1981, hereafter referred to as PS/ more detailed numerical calculations were performed. We considered a widely accepted picture of such binaries. They are believed to have a white dwarf primary component surrounded with an accretion disk. The secondary component, probably the lower main sequence star or degenerate dwarf, overflows its Roche lobe, and matter flows from the inner Lagrangian point towards the disk. In all known cases but one, the binary orbits are circular. No hydrogen-rich CB's are observed with periods below 80 minutes. Because the nuclear evolution cannot be of any importance for the secondary, we assumed that angular momentum loss due to gravitational radiation is the driving force which causes the mass transfer and a cataclysmic activity. We followed the evolution of such system with a very simple and fast code, and the best input physics actually available /PS, Sienkiewicz, 1981/. We have calculated a large number of evolutionary tracks for many values of a total mass of a binary and the initial mass of the secondary.
Because of a transfer of matter, the mass of the secondary decreases and the orbital period initially decreases and later increases passing through minimum while the secondary proceeds from the lower main sequence stage towards a degenerate condition. It should be pointed out, that during this stage the secondary is out of thermal equilibrium.
A value of the minimum orbital period depends on the total mass of CB, an adopted metal content and physics of opacity sources /Paczyński and Sienkiewicz, 1982/. Assuming moderate opacities and gravitational radiation as the only sink of

*Z. Kopal and J. Rahe (eds.), Binary and Multiple Stars as Tracers of Stellar Evolution, 413–414.*

angular momentum, we obtained values of the orbital minimum
period reasonably close to the observed 80 minutes cutoff.
If the rate of angular momentum loss is increased, than the
minimum period increases too /Paczyński and Sienkiewicz,1982/.
To determine more precisely a value of the minimum orbital
period, it is necessary to know more about the metal abundance
in CB´s, as well as about opacity sources in the outer layers
of very low luminosity stars.

REFERENCES

Paczyński, B.: 1981, Acta Astron.,31, pp.1-12.
Paczyński, B., and Sienkiewicz, R.: 1981, Astrophys. J.
                Letters,248, pp. L27-30.
Paczyński, B., and Sienkiewicz, R.: 1982, in preparation.
Sienkiewicz, R.: 1981, Acta Astron., in press.

# REMARKS ON THE EVOLUTIONARY STATUS OF CATACLYSMIC VARIABLES

N. Vogt
European Southern Observatory
Santiago, Chile 1/

The basic binary parameters (masses, system dimensions) of all cataclysmic variables are essentially identical, but there is a great variety in the outburst behaviour. Since there is no evident physical reason for this, it is suggested that nova and dwarf nova variability are periodically repeating states of activity of the same binary. After the nova eruption first follows a postnova state (very low disc mass), later the dwarf nova states BV Pup, U Gem and Z Cam (slowly increasing disc mass and outburst frequency, finally standstills), and finally the UX UMa state (permanent standstill) which is terminated by a new nova outburst. The mass of the disc increases continuously during this cycle. Also the mean mass transfer and accretion rates vary slowly from $10^{18}$ g s$^{-1}$ at beginning and end of the cycle to $10^{16}$ g s$^{-1}$ at mid-cycle, as a U Gem star. The nova outburst is understood in terms of a thermonuclear runaway near the surface of the white dwarf, the dwarf nova outburst is due to intermittent accretion of gas which was accumulated in the outer disc during quiescence (several arguments in favor of this model are given). The time interval between consecutive nova outbursts is of the order of $10^{5}$ years. The secular evolution of a cataclysmic binary is characterized by a period decrease from $10^{h}$ to $1^{h}5$ in a time scale of $10^{10}$ years. The star probably passes the period gaps in the ultra-short period domain in form of no-contact configurations.

1/ Present address: Universitäts-Sternwarte, 8000 München 80, F.R.G.

*Z. Kopal and J. Rahe (eds.), Binary and Multiple Stars as Tracers of Stellar Evolution, 415.*
*Copyright © 1982 by D. Reidel Publishing Company.*

# SUPERNOVAE IN BINARY SYSTEMS : PRODUCTION OF RUN-AWAY STARS AND PULSARS

Jean-Pierre De Cuyper
Astrofysisch Instituut, Vrije Universiteit Brussel, Belgium.

ABSTRACT

The effects of an instantaneous asymmetric supernova explosion in an eccentric binary system are analyzed, taking into account the mass loss out of the system, the influence of the impact of the supernova shell on the companion star and the extra "kick" velocity which a collapsed star might receive in an asymmetric supernova explosion. For a random orientation in space of this asymmetric kick velocity, the survival probability and the runaway velocities are derived and their properties discussed for an explosion occurring at a given position in the initial keplerian orbit and the mean and extreme values of these quantities over one orbit are derived. As an example, the outcome of a possible supernova explosion in the ten best known WR+OB binaries is studied and a comparison is made with the observed run-away OB stars, radio pulsars and binary X-ray pulsars.

## 1. INTRODUCTION

The effects of supernovae occurring in binaries were first studied by Blaauw (1960) in the context of run-away OB stars. In those days one was not yet aware of the great effects of mass exchange and mass transfer on the evolution of components of close binary systems and the idea that the most massive component exploded first as a supernova was invoked to explain the observed run-away OB stars as being the released companions of stars which had gone through the supernova stage. The main effect studied was the mass loss from the initially circular system, taking into account numerically the details of the mass distribution in the ejected shell (Boersma, 1960). Later on the mass exchange in a binary system was included and the effects of the explosion of the less massive star were studied in the same way (van den Heuvel, 1968; De Cuyper, de Loore and van den Heuvel, 1977). Analytic approximations were given by Savedorff and Vila (1964) and by Hut and Verhulst (1981). The extention to initial eccentric orbits was given by Hadjidemetriou (1966), which studied the time dependent mass loss problem numerically

417

Z. Kopal and J. Rahe (eds.), Binary and Multiple Stars as Tracers of Stellar Evolution, 417–443.
Copyright © 1982 by D. Reidel Publishing Company.

and gave an analytic solution for the instantaneous case. Analytic solutions for initially circular orbits in case of an instantaneous explosion
were also given by Sofia (1967), Gott (1972) and Mitalas (1976).

The effects of the interaction of the supernova shell with the
companion star were found to be unable (unless under very special conditions) to make systems unbound, which were revolving initially in circular orbits, and which had undergone a mass loss of less than half of the
initial total mass (Colgate, 1970; McCluskey and Kondo, 1971; Cheng,
1974; Sutantyo, 1974,1975; Khabazin, 1975). Wheeler, Lecar and McKee
(1975) made an analytic approximation taking into account the effects of
the internal structure on the reduction of the effective cross section
of the star and on the mass ablated by the subsequent heating. A fully
hydrodynamical treatment of the impact problem was given by Fryxell and
Arnett (1981).

An acceleration of the collapsed star due to an asymmetry in the
explosion process was taken into account numerically by De Cuyper (1974);
de Loore, De Grève, van den Heuvel and De Cuyper (1975) for a time dependent explosion in a circular orbit. The analytic treatment of an instantaneous asymmetric explosion neglecting the effects of impact was
given by Sutantyo (1978).

We will generalize here the problem to an instantaneous asymmetric
explosion occurring at a given position in an initially eccentric orbit
taking into account the mass loss out of the system, the effects of the
interaction of the supernova shell with the companion star and the asymmetric kick velocity the collapsed star gets due to the asymmetry of the
explosion process.
In section 2 the supernova explosion of a component of a binary is
formulated, using observational data and theoretical prospections, as a
celestial mechanical problem and the effects of an instantaneous asymmetric supernova explosion on the orbital parameters are derived analyt-
ically.
In section 3 the survival condition of the post-supernova system is
analyzed and the survival probability for a given probability distribution of the asymmetric velocity parameter, the collapsed star gets as a
result of the asymmetry of the explosion, is defined for an explosion
occurring in a given point of the initial orbit and for its mean value
over one orbit. Analytic expressions for the run-away velocities of the
remaining bound system or of both remaining components in case of disruption with respect to the center of mass of the pre-supernova system
are given in section 4.
These formulae are used to study the outcome of the first supernova
explosion in massive close binaries. In section 5 an evaluation of the
explosion parameters is made using observational data and theoretical
model results. In section 6 the initial systems are derived from the
best observed WR+OB binaries and the outcome of the supernova explosion
of the Wolf-Rayet star at the end of its nuclear evolution is analyzed.
The results are discussed together with the concluding remarks in
section 7.

## 2. THE EFFECTS OF AN INSTANTANEOUS ASYMMETRIC SUPERNOVA EXPLOSION ON THE ORBITAL PARAMETERS

### 2.1. Formulation of the problem

One of the major events in the evolution of a binary system is the supernova explosion of one of its components. The supernova itself is characterized by the implosion of a gravitationally unstable stellar core forming a neutron star (or possibly a black hole), the ejection of the stellar envelope with an observed velocity of the order of 1-3 $10^4$ km.s$^{-1}$ (Shklovsky, 1968; Schatzman, 1965; Zwicky, 1965; Minkowski, 1969). In case the explosion itself is not fully symmetric internal forces will accelerate the exploding star. Hence the collapsed star may get an extra asymmetric velocity (Shklovsky, 1970; Fryxell, 1979).

In a binary, part of the ejected supernova shell will impact on the companion star. This inelastic collision strips off the outer edges of the companion star (as seen from the supernova) and accelerates the remaining part by direct momentum transfer and subsequent anisotropic ablation of some mixed stellar and impacting material due to the heating behind the shockfront, which forms a bow shock around the stellar core (Fryxell and Arnett, 1981).
When the supernova shell has past the companion star it no longer exerts any significant attraction on the remaining system. This is due to the fact that the mass in the ejected shell is nearly isotropically distributed so that its inside gravitational field can be neglected.
As a consequence of this the orbital motion of the remaining components changes. First of all the mass loss out of the system will decrease the gravitational binding energy. Secondly the orbital velocity change, due to the effects of the impact and the asymmetric explosion, will modify the orbital kinetic energy.
As the orbital velocities ($\sim 10^2$ km.s$^{-1}$) are some orders of magnitude smaller than the supernova ejection velocities ($\sim 10^4$ km.s$^{-1}$) and the typical thickness of the supernova shell is about one third of its expansion distance (Colgate, 1970), the time needed for the supernova shell to pass the companion star is much shorter than the initial orbital period. Hence we can neglect the time dependence of the mass loss out of the system and assume that the explosion occurs instantaneously.

### 2.2. The pre-supernova system

We consider a binary system with component masses $M_1^0$ (the pre-supernova star) and $M_2^0$ (the companion star) revolving around their common center of gravity $C_g^0$ in a keplerian orbit with period $P^0$, eccentricity $e^0$ and barycentric semi-major axes $a_1^0$ and $a_2^0$, respectively. The semi-major axis of the relative orbit one component describes around its companion is denoted $a^0$.

At the instant of the explosion the barycentric distances and velocities: $r_1$, respectively $r_2$, and $\bar{v}_1^0$, respectively $\bar{v}_2^0$; the separation between both components $r$ and the relative velocity $\bar{v}^0$ of the companion

star with respect to the pre-supernova star, can be determined from the
eccentric anomaly $E^o$ (figure 1a,b) (Whittaker, 1944; Roy, 1978) as :

$$r = (1 - e^o \cdot \cos E^o) \cdot a^o \qquad (2.1)$$

with:     $M_1^o \cdot r_1 = M_2^o \cdot r_2 = \mu^o \cdot r \qquad (2.2)$

where:     $\mu^o = \dfrac{M_1^o \cdot M_2^o}{M_1^o + M_2^o} \qquad (2.3)$

denotes the reduced mass of the initial system,

and     $\bar{v}^{o2} = \left[ \dfrac{1 + e^o \cdot \cos E^o}{1 - e^o \cdot \cos E^o} \right] \cdot v_c^{o2} \qquad (2.4)$

with radial and tangental components given by, respectively:

$$v_r^o = \frac{e^o \cdot \sin E^o}{1 - e^o \cdot \cos E^o} \cdot v_c^o \qquad (2.5)$$

$$v_t^o = \frac{\sqrt{1 - e^{o2}}}{1 - e^o \cdot \cos E^o} \cdot v_c^o \qquad (2.6)$$

Making use of the pre-supernova momentum relation with respect to the
center of gravity $C_g^o$ one finds:

$$- M_1^o \cdot \bar{v}_1^o = M_2^o \cdot \bar{v}_2^o = \mu^o \cdot \bar{v}^o \qquad (2.7)$$

whereas:     $v_c^{o2} = G \dfrac{M_1^o + M_2^o}{a^o} = - \dfrac{2 \cdot E^o}{\mu^o} \qquad (2.8)$

(with G the universal constant of gravity), denotes the square of the
constant relative velocity the pre-supernova system would have if it
revolved in a circular orbit of radius $a^o$. More generally this quantity
is identical to the mean over one orbit of the square of the relative
orbital velocity of the pre-supernova system. It can be expressed as a
function of the total orbital energy:

$$E^o = -G \frac{M_1^o \cdot M_2^o}{2 \, a^o} \qquad (2.9)$$

and the reduced mass $\mu^o$ of the initial system and is therefore independent of the eccentricity.

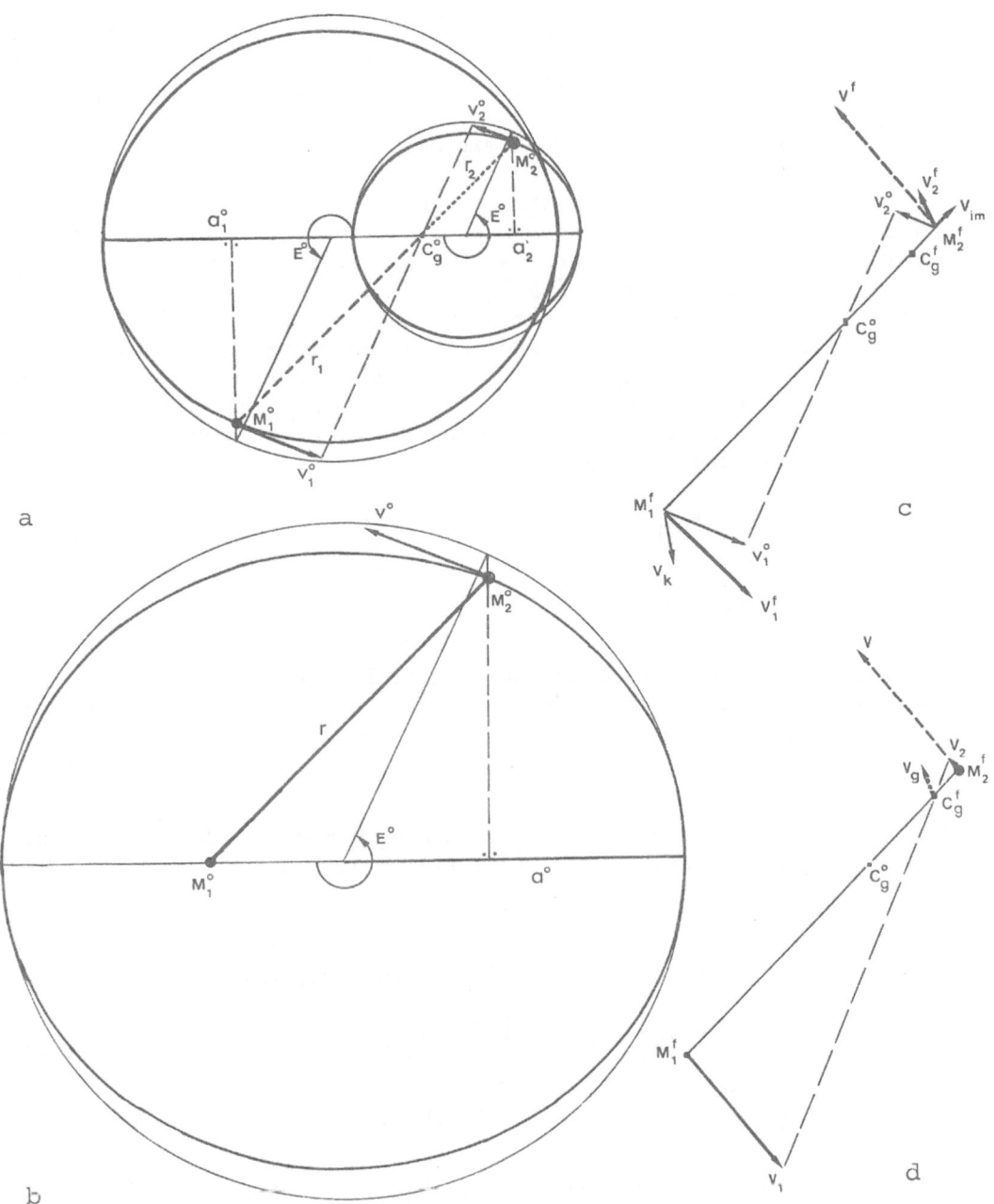

Figure 1. The orbital change due to an asymmetric instantaneous supernova explosion in an eccentric binary ($e^o$=.4). Part (a) and (b) show
the position and relative velocities at the instant of the supernova
event in the initial barycentric and relative orbits, respectively.
The velocity change with respect to $C_g^o$ is given in part (c). The new
barycentric velocities with respect to $C_g^f$ are given in part (d) together
with the run-away velocity $\bar{v}_g$ of $C_g^f$ with respect to $C_g^o$.

We will make the problem dimensionless by expressing the distances in
units of the semi-major axis of the initial relative orbit $a^o$ and the
velocities in units of the initial circular orbital velocity $v_c^o$.

## 2.3. The instantaneous supernova explosion

The instantaneous asymmetric supernova explosion of a component of
a binary system will change the masses and velocities of both components.
The collapsed star, a neutron star (or possibly a black hole) of mass $M_1^f$,
receives an extra kick velocity $\bar{v}_k$, as a consequence of the assumed asymmetry of the explosion. The companion star loses a fraction of its mass
by stripping and ablation and receives an extra radial velocity $\bar{v}_{im}$, due
to the effects of the impact of the supernova shell. Hence, immediately
after the supernova event (assumed to be of negligible duration) the
collapsed star and its remaining companion of mass $M_2^f$ revolve with barycentric velocities (cf. figure 1c) :

$$\bar{v}_1^f = \bar{v}_1^o + \bar{v}_k \tag{2.10}$$

and

$$\bar{v}_2^f = \bar{v}_2^o + \bar{v}_{im} \tag{2.11}$$

respectively, with respect to their initial center of gravity $C_g^o$, while
the companion star has a relative velocity versus the collapsed star:

$$\bar{v}^f = \bar{v}^o + \bar{v}_{im} - \bar{v}_k \tag{2.12}$$

The new center of mass $C_g^f$ of the remaining components has a run-away
velocity $\bar{v}_g$ with respect to $C_g^o$ (cf. figure 1d).

## 2.4. The final orbital parameters

From the vis-viva integral of the final relative orbit:

$$v^{f2} = G \cdot (M_1^f + M_2^f) \cdot \left[ \frac{2}{r} - \frac{1}{a^f} \right] \tag{2.13}$$

one finds, using eq. (2.1) and (2.8) that:

$$\frac{a^o}{a^f} = \frac{2}{1 - e^o \cdot \cos E^o} - \frac{1}{\alpha} \cdot \left[ \frac{\bar{v}^f}{\bar{v}_c^o} \right]^2 \tag{2.14}$$

where:     $\alpha = \dfrac{M_1^f + M_2^f}{M_1^o + M_2^o}$ \hfill (2.15)

denotes the fractional mass of the remaining system.
Hence the ratio of the semi-major axis of the pre- and post-supernova
relative orbit is a function of the ratio of the final to the initial

total mass of the system, the ratio of the post explosion relative velocity to the initial circular velocity and the relative separation in the
initial orbit at the time of the supernova explosion.  Using Kepler's
third law we can express the ratio of the initial and final orbital
period as a function of the ratios of the pre- and post-supernova total
mass and relative semi-major axis, as:

$$\left[\frac{P^o}{P^f}\right]^2 = \alpha \cdot \left[\frac{a^o}{a^f}\right]^3 \tag{2.16}$$

Hence the changes in the relative semi-major axis and orbital period are
independent of the individual mass loss of each component, but depend
only on the total fraction of mass loss.

The ratio of the total orbital energy of the pre- and post-supernova
relative orbit is given (cf. eq. 2.9) as a function of the ratios of the
initial and final mass of each component and relative semi-major axis
as:

$$\frac{E^f}{E^o} = \frac{M_1^f \cdot M_2^f}{M_1^o \cdot M_2^o} \cdot \frac{a^o}{a^f} \tag{2.17}$$

For the final eccentricity we find using Kepler's second law the equality:

$$1 - e^{f2} = \frac{(r \cdot v_t^f)^2}{G \cdot (M_1^f + M_2^f) \cdot a^f} \tag{2.18}$$

which gives using eq. (2.1) and (2.8):

$$1 - e^{f2} = \frac{(1 - e^o \cdot \cos E^o)^2}{\alpha} \cdot \frac{a^o}{a^f} \cdot \left[\frac{v_t^f}{v_c^o}\right]^2 \tag{2.19}$$

The orbital angular momentum of the post explosion relative orbit is
defined as:

$$\bar{h}^f = \mu^f \; \bar{r} \wedge \bar{v}^f \tag{2.20}$$

Hence the ratio of the magnitude of the initial and final orbital angular
momentum is given as a function of the ratio of the reduced mass of the
pre- and post-supernova system and of the ratio of the tangental component
of  the initial and final relative orbital velocity, and using eq. (2.6)
becomes:

$$\left\|\frac{\bar{h}^f}{\bar{h}^o}\right\| = \frac{\mu^f \cdot v_t^f}{\mu^o \cdot v_t^o} = \frac{1 - e^o \cdot \cos E^o}{\sqrt{1 - e^{o2}}} \cdot \frac{\mu^f}{\mu^o} \cdot \frac{v_t^f}{v_c^o} \tag{2.21}$$

## 3. THE SURVIVAL PROBABILITY

### 3.1. The survival condition

The condition for the system to remain bound, after the supernova explosion of one of its components, is that the final orbit is elliptic; i.e. the relative semi-major axis $a^f$ must be positive. Hence, the condition for survival may be written using eq. (2.14) as:

$$\beta^2 \geq \left[\frac{\overline{v}^{-f}}{\overline{v}^{-o}}\right]^2 \tag{3.1}$$

with $\beta$ depending on the fractional mass of the remaining system and the relative separation at the time of the explosion:

$$\beta = \left\{\frac{2\alpha}{1 - e^o \cdot \cos E^o}\right\}^{1/2} \tag{3.2}$$

In carthesian coordinates, for example with the X-axis connecting the exploding star to its companion at the instant of the explosion and the Y-axis lying in the orbital plane, we find that:

$$v^{f2} = (v_{k_x} - v_r^o - v_{im})^2 + (v_{k_y} - v_t^o)^2 + v_{k_z}^2 \tag{3.3}$$

where subscripts x, y, z indicate the components along the X, Y, Z axes. From this we can state:

*The survival condition* for the remaining system requires that in the velocity space the endpoint of the kick velocity vector $\overline{v}_k$, the collapsar gets as a result of the asymmetry of the supernova explosion, is situated inside the sphere $S$ of radius $\beta.v_c^o$ and center coinciding with the endpoint of the velocity vector $\overline{v}^o + \overline{v}_{im}$.

An explosion for which the endpoint of the vector $\overline{v}_k$ is situated on the surface of the sphere $S$ gives a parabolic final orbit. Hereto the velocity vector $\overline{v}_k$ of magnitude $K.v_c^o$, should make an angle $\theta_S$ with respect to $\overline{v}^o + \overline{v}_{im}$, given by:

$$\cos \theta_S = \frac{K^2 + A^2 - \beta^2}{2.K.A} \tag{3.4}$$

where the value of $\beta$ needed for this ranges from $\beta_{min} \leq \beta \leq \beta_{max}$, with:

$$\beta_{min} = |K - A| \tag{3.5}$$

$$\beta_{max} = K + A \tag{3.6}$$

and
$$A = \left\{\left[\frac{e^o \cdot \sin E^o}{1 - e^o \cdot \cos E^o} + I\right]^2 + \frac{1 - e^{o2}}{(1 - e^o \cdot \cos E^o)^2}\right\}^{1/2} \tag{3.7}$$

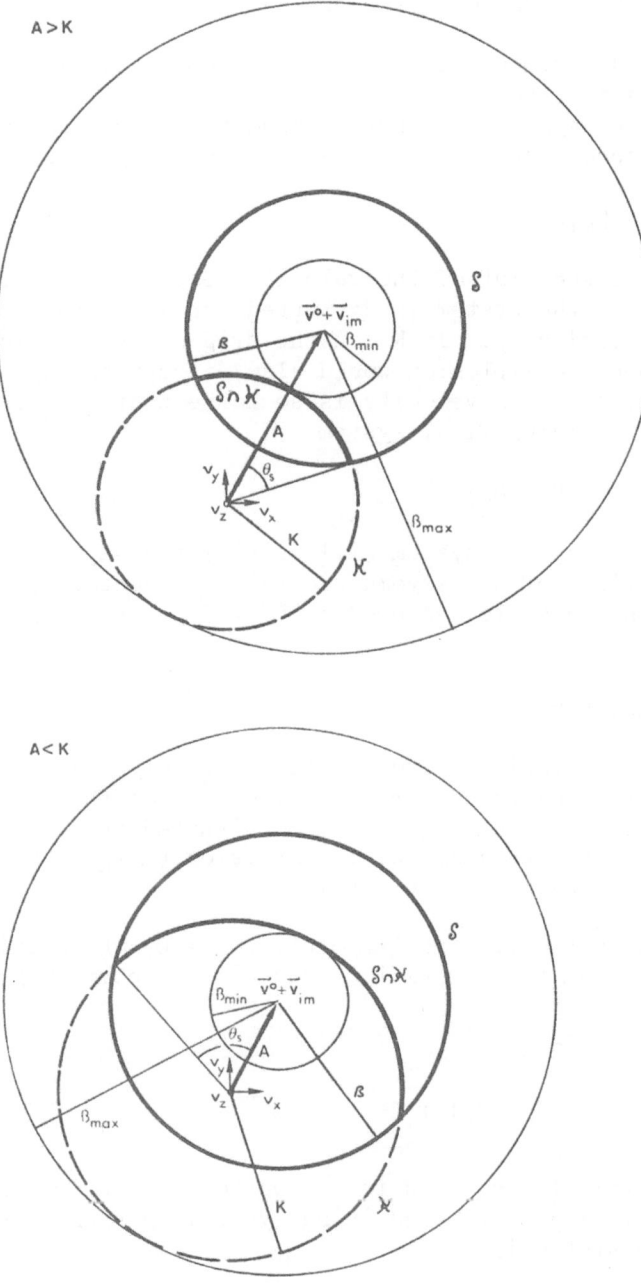

Figure 2.  The intersection of the spheres $S$ and $K$ together with the critical values of the survival parameter $\beta$ for the two possible cases $A > K$ and $A < K$.

where K, I and A denote the magnitude of $\bar{v}_k$, $\bar{v}_{im}$ and $\bar{v}^0 + \bar{v}_{im}$, respectively, in units of the circular orbital velocity $v_c^0$.

If $\beta \leq A$ a symmetric explosion disrupts the system whereas if $\beta > A$ the opposite is true.
Hence for the asymmetric survival condition three cases are to be distinguished according to the value of $\beta$ (cf. figure 2), i.e.:

<u>case a</u> : $\beta < \beta_{min}$

Here the endpoint of the velocity vector $\bar{v}_k$ is situated outside the sphere $S$, i.e. the system is disrupted for any direction of the asymmetric kick velocity $\bar{v}_k$. If $K < A$ the origin lies outside the sphere $S$, hence a symmetric explosion would also disrupt the system. If $K > A$ the magnitude of the kick velocity is so large that it even disrupts the retrograde revolving final system.

<u>case b</u> : $\beta_{min} \leq \beta \leq \beta_{max}$

In this case the system will survive the supernova explosion for those directions of the asymmetric velocity vector $\bar{v}_k$, which are situated inside the solid angle centered at $\bar{v}^0 + \bar{v}_{im}$ with semi-apex angle $\theta_s$ given by eq. (3.4).

<u>case c</u> : $\beta_{max} < \beta$

Here the velocity vector $\bar{v}_k$ lies entirely inside the sphere $S$, i.e. the system remains bound for any direction of the asymmetric velocity $\bar{v}_k$. As $\beta_{max} \geq A$, the origin is also situated inside the sphere $S$, so the system would survive a symmetric explosion ($\| \bar{v}_k \| = 0$) occurring at the same position in the initial orbit.

For an initially circular orbit ($e^0 = 0$) the magnitude of the orbital velocity and the separation between both components remains constant, hence:

$$\beta = \sqrt{2.\alpha} \qquad\qquad\qquad (3.8)$$

$$\beta_{\substack{min \\ max}} = A = \sqrt{1 + I^2} \qquad\qquad\qquad (3.9)$$

Here the survival condition is a function of the fractional mass and relative velocity of the remaining system. A symmetric explosion would disrupt the system if:

$$\alpha \leq \frac{1 + I^2}{2} \qquad\qquad\qquad (3.10)$$

which is always fulfilled if more than half of the mass leaves the system. In terms of energy this is due to the fact that in a circular orbit the total orbital energy is one half of the constant gravitational energy. Taking away half of the total mass without changing the kinetic energy will unbound the system.

3.2. The survival probability for an explosion occurring in a given
     point of the initial orbit

     Assuming a certain probability disruption of the asymmetric velocity
vector $\bar{v}_k$, one can define the survival probability for an instantaneous
supernova explosion occurring in a given point with eccentric anomaly $E^o$
of the initial orbit (of eccentricity $e^o$ and circular velocity $v_c^o$), for
a mass loss parameter $\alpha$ and impact parameter $I$.

     As the supernova explosion itself is supposed to be unaffected by
the presence of a companion star, we restrict ourselves here to a random
orientation of the asymmetric velocity vector $\bar{v}_k$ with given magnitude
$k.v_c^o$. Hence the endpoint of the asymmetric velocity vector $\bar{v}_k$ is randomly situated on the sphere $K$, centered at the origin with radius $k.v_c^o$.
From the survival condition it follows that the corresponding survival
probability $P$ $(k;E^o)$ is equivalent to the fraction of the surface of the
sphere $K$ that is located inside the sphere $S$.
For $\beta_{min} \leq \beta \leq \beta_{max}$ the part of the surface of $K$ inside $S$ is a polar cap
centered at $\bar{v}^o + \bar{v}_{im}$ with semi-apex angle $\theta_s$, given by eq. (3.4). The
survival probability is thus given by:

$$P(k;E^o) = \frac{1 - \cos \theta_s}{2} = \frac{\beta^2 - [k-A]^2}{4.k.A} = \frac{\beta^2 - \beta_{min}^2}{4.k.A} \qquad (3.11)$$

If $\beta < A$ the origin lies outside the sphere $S$ and $P(k;E^o) < 1/2$.
The survival probability $P(k;E^o)$ of the remaining system after an instantaneous supernova explosion at a given point of eccentric anomaly $E^o$, for
a randomly orientated asymmetric velocity $\bar{v}_k$ of given magnitude $k.v_c^o$, can
be summarized as in Table I.

Table I

|  | $P(k;E^o)$ |
|---|---|
| $\beta < \beta_{min}$ | $0$ |
| $\beta_{min} \leq \beta \leq \beta_{max}$ | $\dfrac{\beta^2 - \beta_{min}^2}{4.k.A}$ |
| $\beta_{max} < \beta$ | $1$ |

For a circular pre-supernova orbit the dependence of the survival probability $P$ on the total mass loss and magnitude of the asymmetric kick velocity is given in figure 3 for two values of the impact velocity.

3.3. The mean survival probability over one orbit

     As the supernova explosion may equally likely occur at any instant
of time during the revolution in the initial orbit, we define the mean
survival probability $<P(k)>$ as the weighted average over one orbital

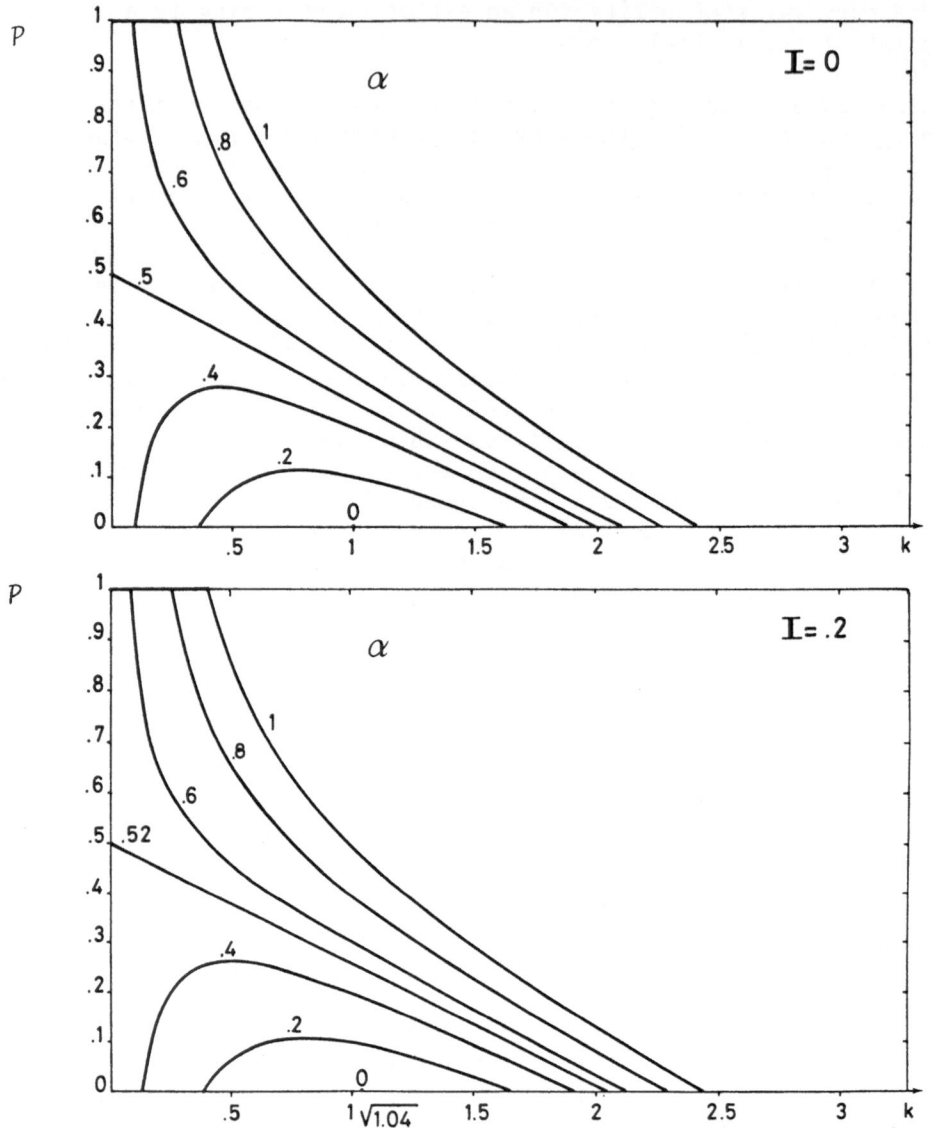

Figure 3.  The dependence of the survival probability $P$ on the magnitude
of the asymmetric kick velocity for different constant values of the
mass loss parameter and two values of the impact parameter in case of a
circular pre-supernova orbit.

period of the survival probability $P(k;E^{0})$ for an explosion that occurs
in a given point of the initial orbit, with equal weight per unit of
time; i.e. using Kepler's law of areas as weighting function:

$$<P(k)> = \frac{1}{P^{0}} \int_{o}^{P^{0}} P(k;E^{0}) \, dt \qquad\qquad (3.13)$$

Differentiating Kepler's parametric representation:

$$\frac{2.\pi.t}{P^o} = E^o - e^o . \sin E^o \tag{3.14}$$

the time integral defining $<P(k)>$ can be replaced by an integral over the position in the initial orbit in function of the eccentric anomaly as:

$$<P(k)> = \frac{1}{\pi} \int_0^\pi P(k;E^o) . (1 - e^o . \cos E^o) \, dE^o \tag{3.15}$$

## 4. THE RUN-AWAY VELOCITIES

### 4.1. The run-away velocity of the new center of gravity

The run-away velocity of the center of gravity $C_g^f$ of the remaining system, after an instantaneous supernova explosion, with respect to the center of gravity $C_g^o$ of the pre-supernova binary is given by the momentum equation of the post-supernova components in the initial center of mass system as a function of the velocities, at the instant after the instantaneous supernova explosion, and the masses of the remaining components. This equation takes the form using eq. (2.10, 2.11):

$$(M_1^f + M_2^f) . \bar{v}_g = M_1^f . \bar{v}_1^o + M_2^f . \bar{v}_2^o + M_2^f . \bar{v}_{im} + M_1^f . \bar{v}_k \tag{4.1}$$

The first two terms on the right-hand side denote the momentum due to the mass loss from the system, the third term represents the momentum imparted by the supernova shell to the companion and the last term gives the momentum contributed by the asymmetry of the supernova explosion. Making use of the pre-supernova momentum relation with respect to $C_g^o$ (cf. eq. 2.7) this can be written as:

$$\bar{v}_g = \frac{1}{M_1^f + M_2^f} \left\{ \left[ \frac{M_1^o . M_2^f - M_2^o . M_1^f}{M_1^o + M_2^o} \right] . \bar{v}^o + M_2^f . \bar{v}_{im} + M_1^f . \bar{v}_k \right\} \tag{4.2}$$

or:

$$\bar{v}_g = \frac{1}{1 + q^f} \left\{ \left[ \frac{q^f - q^o}{1 + q^o} \right] . \bar{v}^o + q^f . \bar{v}_{im} + \bar{v}_k \right\} \tag{4.3}$$

with:-

$$q^o = \frac{M_2^o}{M_1^o} \tag{4.4}$$

and:

$$q^f = \frac{M_2^f}{M_1^f} \tag{4.5}$$

denoting the mass ratio of the secondary to the primary component before
and after the supernova explosion, respectively.

Hence the extreme values of the magnitude of $\bar{v}_g$ as a function of the
asymmetric velocity vector $\bar{v}_k$ with given magnitude $k.v_c^o$ coincide with $\bar{v}_k$
directed along or opposite to the vector $\left[\dfrac{q^f - q^o}{1 + q^o}\right].\bar{v}^o + q^f.\bar{v}_{im}$, and are
given respectively by:

$$v_{g_{min}} = \frac{1}{1 + q^f} (k - B) \tag{4.6}$$

$$v_{g_{max}} = \frac{1}{1 + q^f} (k + B) \tag{4.7}$$

with:

$$B = \left\{ \left[ \left( \frac{q^f - q^o}{1 + q^o} \right) . \frac{e^o . \sin E^o}{1 - e^o . \cos E^o} + q^f . I \right]^2 + \left( \frac{q^f - q^o}{1 + q^o} \right)^2 \frac{1 - e^{o2}}{(1 - e^o . \cos E^o)^2} \right\}^{1/2} \tag{4.8}$$

denoting the magnitude of the vector $\left[\dfrac{q^f - q^o}{1 + q^o}\right].\bar{v}^o + q^f.\bar{v}_{im}$ in units of
the circular orbital velocity $v_c^o$.

## 4.2. The run-away velocity of the bound systems

If the system survives the supernova explosion of one of its components, the newly formed remnant and its remaining companion revolve in
elliptic orbits around their center of gravity $C_g^f$. This center of gravity will possess a run-away velocity $\bar{v}_b$ with respect to the center of
gravity of the initial system as given by eq. (4.3) of the preceeding
section. Here the extreme values of the magnitude of $\bar{v}_b$ as a function
of the asymmetric velocity vector $\bar{v}_k$ with given magnitude $k.v_c^o$ also
depend on the survival condition, i.e. on the extreme values of the
distance of the endpoint of the vector: $-\left[\dfrac{q^f - q^o}{1 + q^o}\right].\bar{v}^o - q^f.\bar{v}_{im}$, to the
part of the sphere $K$ inside the sphere $S$, deduction of which will be
given elsewhere.

## 4.3. The run-away velocities of the disrupted components

The magnitude of the relative velocity at infinity $\bar{v}^\infty$ of the remaining components of an unbound system is given by the vis-viva integral as:

$$\bar{v}^{\infty 2} = -G \frac{(M_1^f + M_2^f)}{a^f} \tag{4.9}$$

Using the definition of the initial circular velocity $v_c^o$ (eq. 2.8) together with the expression of the semi-major axes of the relative orbit
(eq. 2.14) this becomes:

$$\bar{v}^{\infty 2} = -\alpha \cdot \frac{a^o}{a^f} \cdot v_c^{o2} = \bar{v}^{f2} - \beta^2 \cdot v_c^{o2} \tag{4.10}$$

Hence the run-away velocities of both components in case of disruption
are, respectively :

$$\bar{v}_1^{\infty} = \bar{v}_g - \left[ \frac{q^f}{1 + q^f} \right] \cdot \bar{v}^{\infty} \tag{4.11}$$

$$\bar{v}_2^{\infty} = \bar{v}_g + \left[ \frac{1}{1 + q^f} \right] \cdot \bar{v}^{\infty} \tag{4.12}$$

where we used the momentum equation at infinity of the post-supernova
system with respect to the new center of gravity $C_g^f$ :

$$-M_1^f \cdot \bar{v}_1^{\infty} = M_2^f \cdot \bar{v}_2^{\infty} = \mu^f \cdot \bar{v}^{\infty} \tag{4.13}$$

## 5. EVALUATION OF THE EXPLOSION PARAMETERS

Observations of white dwarfs in galactic clusters and theoretical
stellar models indicate that for single stars the lower mass limit for
becoming a supernova is around 6-8 $M_\odot$ (Gunn and Ostriker, 1970; van den
Heuvel, 1975; Sugimoto and Nomoto, 1980). On the other hand, a component
of a close binary loses before the end of its evolution more than two
thirds of its mass by mass transfer to its companion and mass loss out
of the system. As a consequence of this mass loss a much larger initial
mass of a component of a binary is required for direct collapse of the
stellar core to a relativistic star (van den Heuvel, 1981). Theoretical
computations (van den Heuvel, 1974; de Loore and De Grève, 1975,1976) indicate that the lower mass limit for a component of a close binary to
become a supernova by direct core collapse is at least 12-15 $M_\odot$. Mass
loss by stellar wind during the main sequence and especially during the
helium burning stage (presumably identified with Wolf-Rayet stars
(Paczynski, 1971)) gives a lower limit of 15-20 $M_\odot$. Such stars leave
helium cores more massive than 4 $M_\odot$. Helium stars of smaller masses
evolve in close binaries through a second stage of mass exchange and
are expected to end up as white dwarfs with masses below the Chandrasekhar limit (Arnett, 1974; Tutukov and Yungelson, 1973; van den Heuvel and
Heise, 1972; de Loore and De Grève, 1975,1976). More massive helium components of a binary are expected to explode as supernovae (Arnett, 1975,
1978).

We assume here that the core of a supernova star will form a neutron
star of mass 1.5 $M_\odot$ (Weaver and Woosley, 1978).

The initial stellar mass function (in the mass range 1 to 100 $M_\odot$) may be approximated by :

$$\psi(M) = C M^{-2.55}$$

(Limber, 1960; Taff, 1974; Miller and Scalo, 1979). As during the last $10^9$ years the stellar birth rate in the galaxy was practically constant (Schmidt, 1959; Miller and Scalo, 1979) the number of the short lived stars of mass larger than 2 $M_O$ will be close to a steady state in which the birth rate equals the death rate. Assuming that 50% of the new born stars are in close binaries (Garmany, Conti and Massey, 1980; Abt and Levy, 1978) the ratio $\xi$ of the birth rate of collapsing stars in close binaries and of collapsing single stars is given by :

$$\xi \simeq .5 \int_{15}^{\infty} \psi(M) \, dM \, / \int_{5}^{\infty} \psi(M) \, dM = .09$$

Radio pulsars are thought to be neutron stars formed during a supernova explosion (Ruderman, 1972), hence most of them must have originated from low velocity single stars. As they are high velocity objects (up to $6 \ 10^2$ km.s$^{-1}$)(Taylor and Manchester, 1977,1981; Gullahorn and Rankin, 1978; Lyne, 1981) their acceleration must be due to their formation process (Shklovsky, 1970; Iben, 1972; Buchler, 1973) or their relativistic nature (Harrison and Tademaru, 1977). The radio pulsars with measured run-away velocities belong to the maximum part of the velocity distribution as they are easiest measurable. Hanson (1979) in a statistical study estimated the mean of the run-away velocity distribution to be about 100 km.s$^{-1}$.

We will assume here that the observed run-away velocities of radio pulsars originating from single stars are due to the asymmetry of the supernova explosion which created them and to be 75-150 km.s$^{-1}$

For collapsed stars formed in binaries the run-away velocities depend on the initial binary conditions. In a binary system mass exchange will cause the originally most massive star (primary) to become the less massive one. As the evolution of this primary remains faster than the evolution of the now more massive secondary, it will as first become a supernova while its companion is still a main sequence star (van den Heuvel and Heise, 1972). Hence during the first supernova explosion in a binary less than half of the total mass will leave the system, which cannot enhance the disruption of the remaining system. According to two-dimensional hydrodynamical calculations by Fryxell and Arnett (1981) the momentum imparted by the supernova shell to a companion, of polytropic structure with polytropic index n=3, decreases from 80% to 35% of the momentum incident on the geometric cross section of the star as the kinetic energy of the incident matter increases from 20% to 60% of the binding energy of the companion star. They find a fairly good agreement between their hydrodynamical results and the analytic model of Wheeler, Lecar and McKee (1975) for the evaluation of the mass loss due

to stripping and ablation and the reduction of the effective cross
section in case that the ratio $\psi$ of the momentum of the impacting matter
to the momentum which the undisturbed companion would have if it moved
with outer escape velocity (as defined below) ranges from:

$$\psi = 6.10^{-3} - 4.10^{-2} \tag{5.1}$$

As the effects of the impact may become important only for larger values
of $\psi$ we make the following extrapolation.

The mass loss from the companion due to stripping and ablation is
assumed to be given by the results for polytropic models of Wheeler et
al. (1975) which can be fitted by a simple function of $\text{Log}_{10}\,\psi$ as:

$$M_2^f = M_2^o \cdot \left[ 1 - \left\{ \frac{\text{Log}\,\psi + 3}{4} \right\}^c \right] \quad , c = \begin{cases} 4.3 \text{ if } \psi > .01 \\ 3.3 \text{ if } \psi \leq .01 \end{cases} \tag{5.2}$$

where:

$$\psi = F_{in} \cdot \left[ \frac{V_{SN\ shell}}{v_{es}} - 1 \right] \tag{5.3}$$

with:

$$F_{in} = \frac{(M_1^o - M_1^f)}{4.M_2^o} \cdot \left[ \frac{R_2^o}{r} \right]^2 \tag{5.4}$$

and

$$v_{es} = \left\{ 2.G. \frac{M_2^o}{R_2^o} \right\}^{1/2} \tag{5.5}$$

$F_{in}$ denotes the ratio of the mass of the supernova shell which interacts
with the companion to the total mass of this companion; $v_{es}$ stands for
the magnitude of the escape velocity in the outer part of the companion;
$V_{SN\ shell}$ denotes the ejection velocity of the supernova shell. Observations give values between $0.5\text{-}3.10^4$ km.s$^{-1}$ (Shklovsky, 1968; Schatzman,
1965; Zwicky, 1965; Minkowski, 1969).

We assume here that $V_{SN\ shell} = 10^4$ km.s$^{-1}$.
$R_2^o$ represents the pre-supernova radius of the companion star, which is
still on the main sequence.

We assume here that $R_2^o$ equals two times the radius on the zero age
main sequence given by Plavec (1968) in solar units:

$$R_2^o = 2\ 10^{(0.63\ \log M_2^o - 0.08)} \tag{5.6}$$

As to the impacted momentum we assume it equals 30% of the incident
momentum, hence

$$v_{im} = 0.3\ F_{in} \cdot V_{SN\ shell} \tag{5.7}$$

This way the effects of the interaction of the  supernova shell with the companion star will be overestimated if they play a role in the faith of the post-supernova binary ($\psi$ > .1).


# 6. THE OUTCOME OF THE FIRST SUPERNOVA EXPLOSION IN MASSIVE BINARIES

## 6.1. The data

Out of the best known data on Wolf-Rayet + OB binaries we selected our pre-supernova systems by assuming a value for the inclination i of the orbital plane.  Except for V444 Cyg  the only system with an accurately measured value of i = 78°4 (Münch  1950)  For the other systems the choice of the inclination was inferred from the eclipse condition taking into account the radius of the Wolf-Rayet envelope  the spectral type of the OB star and the Roche radii of both components.  The value retained yields the best mass-spectral type relation for the OB star and a minimum mass of 5 $M_O$ for the Wolf-Rayet star.  For the two systems with highly eccentric orbits, i.e. $\gamma^2$Vel ($e^o \simeq 0.40$) and HD 90657 ($e^o = 0.42$) this condition was tested at periastron.

Table II .  The pre-supernova data.

| | System (★ eclipses) | Spectral type | $P^o$ (days) | $\frac{M_{WR} \cdot \sin^3 i}{M_{OB} \cdot \sin^3 i}$ ($M_\odot$) | $q^o$ | i | $\frac{M_{WR}}{M_{OB}}$ | $v_c^o$ (km.s$^{-1}$) |
|---|---|---|---|---|---|---|---|---|
| a | CX Cep | WN5 O | 2.12 | 5.3 12. | 2.3 | 55° | 9.6 22. | 524 |
| b | HD 193576= V444 Cyg ★ | WN6 O6I | 4.21 | 9.5 24. | 2.4 | 78°4 | 11. 26. | 439 |
| c | HD  94546 | WN4 O | 4.90 | 8. 23. | 2.9 | 60° | 12. 35. | 452 |
| d | HD  90657 | WN5 O6 | 6.42 $e^o$=.42 | 7.6 17. | 2.2 | 50° | 17. 37. | 432 |
| e | HD 211853= GP Cep  ★ | WN6 O6I | 6.69 | 7.6 20. | 2.6 | 75° | 8.5 22. | 353 |
| f | HD 152270 | WC7 O5-8 | 8.89 | 1.8 4.9 | 2.7 | 30° | 9.3 26. | 337 |
| g | HD 186943 | WN4 O9V | 9.55 | 9. 18. | 2.0 | 60° | 9.4 19. | 306 |
| h | HD 168206= CV Cer  (★) | WC8 O8-9 | 29.7 | 11. 22. | 2.0 | 75° | 12. 25. | 229 |
| i | HD  68273= $\gamma^2$ Vel | WC8 O9I | 78.5 $e^o$=.4 | 17. 32. | 1.9 | 70° | 20. 39. | 194 |
| j | HD 190918 | WN4 O9I | 85.0 | 0.20 0.77 | 3.8 | 20° | 5.9 22. | 147 |

(a) Massey and Conti (1980)      (f) Seggewiss (1974)
(b) Münch (1950)                 (g) Massey (1981)
(c) Niemela (1980)               (h) Massey and Niemela (1980)
(d) Niemela (1976)               (i) Niemela and Sahade (1980)
(e) Hiltner (1945) and Massey (1981)   (j) Wilson (1949)

The assumption is made that these data, as given in Table II, are comparable with the pre-supernova binary parameters at the time the Wolf-Rayet star becomes a supernova

## 6.2. The results

Table III shows the results of a supernova event leaving a collapsed star of mass $M_p = 1.5\ M_\odot$ for an assumed value of $V_{SN\ shell} = 10^4\ km.s^{-1}$ and a random orientated kick velocity $\bar{v}_k$ of 75 km.s$^{-1}$ (case a) and 150 km.s$^{-1}$ (case b), respectively. In case of an eccentric orbit the results for the supernova explosion occurring at periastron and at apastron are given. In the first three columns the pre-supernova parameters are shown. The initial mass of the pre-supernova star $M_{WR}$ and its companion $M_{OB}^o$ are given in the first column in solar units, in the second column the initial period $P^o$, expressed in days, and the initial eccentricity $e^o$, if non zero, is shown. Column three gives the barycentric velocities at the instant before the explosion of both components $v_{WR}^o$ and $v_{OB}^o$. If $e^o \neq 0$ the upper and lower case correspond to an explosion occurring at periastron and at apastron, respectively. Column four gives the mass loss parameter $\alpha$ and if $e^o \neq 0$ the survival parameter $\beta^2$ is also given. The effects of the impact are shown in the subsequent columns. The fifth column shows the magnitude of the impact velocity $v_{im}$ and the value of the momentum parameter $\psi$ and column six the final mass of the companion $M_{OB}^f$. In column seven the run-away velocity of the bound system $v_{b\ sym}$ or of the disrupted components $v_{p\ sym}^\infty$ and $v_{OB\ sym}^\infty$ are given for the symmetric supernova explosion ($\|\bar{v}_k\| = 0$). In case of an asymmetric explosion the survival probability $P$ and the extreme values of the run-away velocity of the bound system $v_{b}$, of the disrupted companion star $v_{OB}^\infty$, and of the single collapsed star $v_p^\infty$ are shown respectively in columns eight, nine and ten for case a ($\|\bar{v}_k\| = 75\ km.s^{-1}$) and in colums eleven, twelve and thirteen for case b ($\|\bar{v}_k\| = 150\ km.s^{-1}$). In case of an eccentric initial orbit the value of the mean survival probability $<P>$ is also shown in columns seven, eight and eleven, respectively.

## 6.3. The survival probability

A massive close binary revolving in a nearly circular initial orbit survives the symmetric supernova explosion of its most evolved and hence less massive component. If the initial revolution is eccentric the system is disrupted in case the symmetric explosion occurs at the less probable positions around periastron in the initial orbit. The mean survival probability over one revolution for the short periodic system (d) is $<P_{sym}> = .80$, due to the influence of the impact especially around periastron; whereas for the long periodic system (i) : $<P_{sym}> = .92$.

In case the supernova explosion is asymmetric the survival condition also depends on the ratio of the asymmetric kick velocity to the initial circular velocity. As $v_c^o$ decreases for increasing values of $a^o$ (equation 2.8), the influence of the asymmetric kick velocity will increase with $P^o$.

Originally circular systems with an initial orbital period up to about

Table III.  The survival probability and the run-away velocities.

| | $M_{WR}$ | $P^o$ ($e^o$) | $v^o_{WR}$ / $v^o_{OB}$ | $\alpha$ ($\beta^2$) | $v_{im}$ / $\psi$ | $M_2^f$ | $v_b$ sym/or [$v^\infty_P$ sym / $v^\infty_{OB}$ sym] ($<P_{sym}>$) | case a: $\|\bar{v}_k\|=75$ km.s$^{-1}$ $P$ ($<P>$) | $v_b$ | $v^\infty_{OB}$ | $v^\infty_P$ | case b: $\|\bar{v}_k\|=150$ km.s$^{-1}$ $P$ ($<P>$) | $v_p$ | $v^\infty_{OB}$ | $v^\infty_P$ |
|---|---|---|---|---|---|---|---|---|---|---|---|---|---|---|---|
| a | 9.6 | 2.12 | 365 | .66 | 78. | 19.3 | 141 | .98 | 135 | 136 | 142 | .68 | 134 | 135 | 116 |
|   | 22. |      | 159 |     | .28 |      |     |     | 147 | 137 | 151 |     | 152 | 145 | 263 |
| b | 11. | 4.21 | 309 | .70 | 34. | 24.4 | 110 | 1.  | 105 |     |     | .70 | 105 | 106 | 79  |
|   | 26. |      | 131 |     | .12 |      |     |     | 115 |     |     |     | 119 | 113 | 221 |
| c | 12. | 4.90 | 337 | .74 | 29. | 33.3 | 100 | 1.  | 96  |     |     | .77 | 95  | 96  | 81  |
|   | 35. |      | 115 |     | .09 |      |     |     | 103 |     |     |     | 106 | 100 | 205 |
| d | 17. | 6.46 | 463 | .64 | 60. | 32.8 | [221 | .20 | 200 | 200 | 180 | .31 | 199 | 201 | 143 |
|   | 37. | (.42)| 213 | (1.1)| .26 |     | 204] | (.80) | 202 | 208 | 295 | (.71) | 205 | 212 | 385 |
|   |     |      |     |     |     |      | (.80) |     |     |     |     |     |     |     |     |
|   |     |      | 189 | .70 | 13. | 36.2 | 77  | 1.  | 74  |     |     | 1.  | 71  |     |     |
|   |     |      | 87  | (.49)| .04|      |     |     | 80  |     |     |     | 83  |     |     |
| e | 8.5 | 6.69 | 255 | .75 | 15  | 21.4 | 77  | 1.  | 71  |     |     | .68 | 72  | 72  | 44  |
|   | 22. |      | 98  |     | .05 |      |     |     | 82  |     |     |     | 86  | 81  | 202 |
| f | 9.3 | 8.89 | 248 | .77 | 11. | 25.5 | 71  | 1.  | 66  |     |     | .68 | 67  | 67  | 39  |
|   | 26. |      | 89  |     | .04 |      |     |     | 75  |     |     |     | 79  | 75  | 199 |
| g | 9.4 | 9.55 | 205 | .71 | 11. | 18.6 | 79  | .86 | 75  | 75  | 66  | .59 | 77  | 75  | 25  |
|   | 19. |      | 101 |     | .04 |      |     |     | 85  | 78  | 100 |     | 90  | 87  | 204 |
| h | 12. | 29.7 | 155 | .72 | 3.  | 24.8 | 61  | .75 | 59  | 59  | 40  | .50 | 62  | 58  | 1   |
|   | 25. |      | 74  |     | .01 |      |     |     | 66  | 63  | 98  |     | 70  | 68  | 201 |
| i | 20. | 78.5 | 195 | .68 | 3.  | 38.6 | [88 | .42 | 89  | 89  | 46  | .36 | 91  | 89  | 5   |
|   | 39. | (.4) | 100 | (1.1)| .01|     | 90] | (.69) | 92  | 94  | 152 | (.43) | 95  | 96  | 246 |
|   |     |      |     |     |     |      | (.92)|     |     |     |     |     |     |     |     |
|   |     |      | 84  | .69 | 1.  | 39.0 | 38  | .89 | 36  | 36  | 27  | .48 | 39  | 35  | 20  |
|   |     |      | 43  | (4.9)| 0. |      |     |     | 41  | 37  | 55  |     | 44  | 41  | 162 |
| j | 5.9 | 85.0 | 116 | .84 | 0.  | 22.0 | 22  | .71 | 20  | 19  | 9   | .41 | 25  | 18  | 22  |
|   | 22. |      | 31  |     | 0.  |      |     |     | 26  | 24  | 94  |     | 31  | 28  | 202 |

one week stay together for case a and have a survival probability of
$.7 \leq P \leq .8$ for case b.  Except the shortest system (a) which has a
relatively small value for $\alpha$ = .66 and is disrupted for a kick directed
along $\bar{v}^o + \bar{v}_{im}$ of minimum magnitude $v_{k\ min}$ = 78 km.s$^{-1}$ if $v_{im}$ = 0, with
$P$ = 1 (case a) and $P$ = .71 (case b), respectively.  Whereas here $v_{im}$ =
78 km.s$^{-1}$ so that $v_{k\ min}$ = 73 km.s$^{-1}$, with $P$ = .98 (case a) and $P$ = .68
(case b), respectively.  Hence the impact lowers the survival probability here with a few percent.
The short periodic system (d) with an initially eccentric orbit is disrupted for some directions  of the asymmetric kick velocity if the explosion occurs near periastron in the initial orbit, with $P_{per}$ = .20
(case a) and $P_{per}$ = .31 (case b), respectively; but stays bound for an
explosion occurring around apastron, so the mean survival probability
over one orbit is as high as $<P>$ = .80 (case a), i.e. equal to the symmetric value, and $<P>$ = .71 (case b), respectively.  This last value lies
in the range of the survival probabilities of the circular short periodic systems.

For originally circular systems having an initial orbital period of the order of months the survival probability lies in the range $.7 \leq P \leq .9$ (case a) and $.4 \leq P \leq .6$ (case b), respectively. The long periodic system (i) with an initially eccentric orbit has a survival probability around periastron as low as $P_{per} = .42$ (case a) and $P_{per} = .36$ (case b), respectively and around apastron up to $P_{ap} = .89$ (case a) and $P_{ap} = .48$ (case b), respectively; with a mean survival probability of $<P> = .69$ (case a) and $<P> = .43$ (case b), respectively, both in the range of the corresponding survival probabilities for long periodic circular systems.

## 6.4. The run-away velocities

The magnitudes of the run-away velocities of the OB stars with a collapsed companion $\bar{v}_b$ and of the disrupted OB stars $\bar{v}_{OB}^{\infty}$ are comparable. They are slightly smaller than the barycentric velocity $\bar{v}_{OB}^{o}$ of the OB star at the instant of the supernova explosion and their asymmetric values enclose the symmetric one : $\bar{v}_b$ sym or $\bar{v}_{OB}^{\infty}$ sym. According to the pre-supernova momentum relation with respect to the initial center of gravity $C_g^o$ (eq. 2.7), these run-away velocities are inversely proportional to the pre-supernova mass ratio $q^o$. They decrease for increasing values of the initial orbital period. The possible values of $\bar{v}_b$ are slightly larger than those of $\bar{v}_{OB}^{\infty}$ (except for an explosion occurring near periastron in an elliptic initial orbit where the opposite is true) and their maximum values are most probable (cf. figure 4).

Originally circular systems with an orbital period up to one week create high velocity OB stars (single or with a collapsed companion) with run-away velocities between 75-150 km.s$^{-1}$. If the initial orbit is eccentric, as in the short periodic system (d), a supernova explosion occurring at the less probable positions near periastron in the initial orbit releases run-away OB stars faster than 200 km.s$^{-1}$.
For originally long periodic circular systems the post-supernova system, consisting of an OB star and a collapsed companion, and the disrupted single OB stars have run-away velocities in the range 60-70 km.s$^{-1}$ if the initial orbital period is about one month and smaller than 25 km.s$^{-1}$ if the initial orbital period is longer than a few months. If the initial orbit is eccentric and the explosion occurs near periastron somewhat faster run-away OB stars are released.

In case a nearly circular system is disrupted a single pulsar is formed with a run-away velocity up to $v_p^{\infty} \leq 100$ km.s$^{-1}$ (case a) and $v_p^{\infty} \leq 200$ km.s$^{-1}$ (case b), respectively; independent of the initial orbital period. Few of them have a negligible run-away velocity (cf. figure 4). Except system (a) where $140$ km.s$^{-1} \leq v_p^{\infty} \leq 150$ km.s$^{-1}$ (case a) and $115$ km.s$^{-1} \leq v_p^{\infty} \leq 265$ km.s$^{-1}$ (case b), respectively due to the relative large mass loss and the high value of the impact velocity.
In case the supernova explosion disrupts an initially eccentric revolving system the run-away velocity of the collapsed star is comparable with the value for circular systems. Only in the less probable cases the supernova explosion occurs near periastron in the initial orbit, much higher

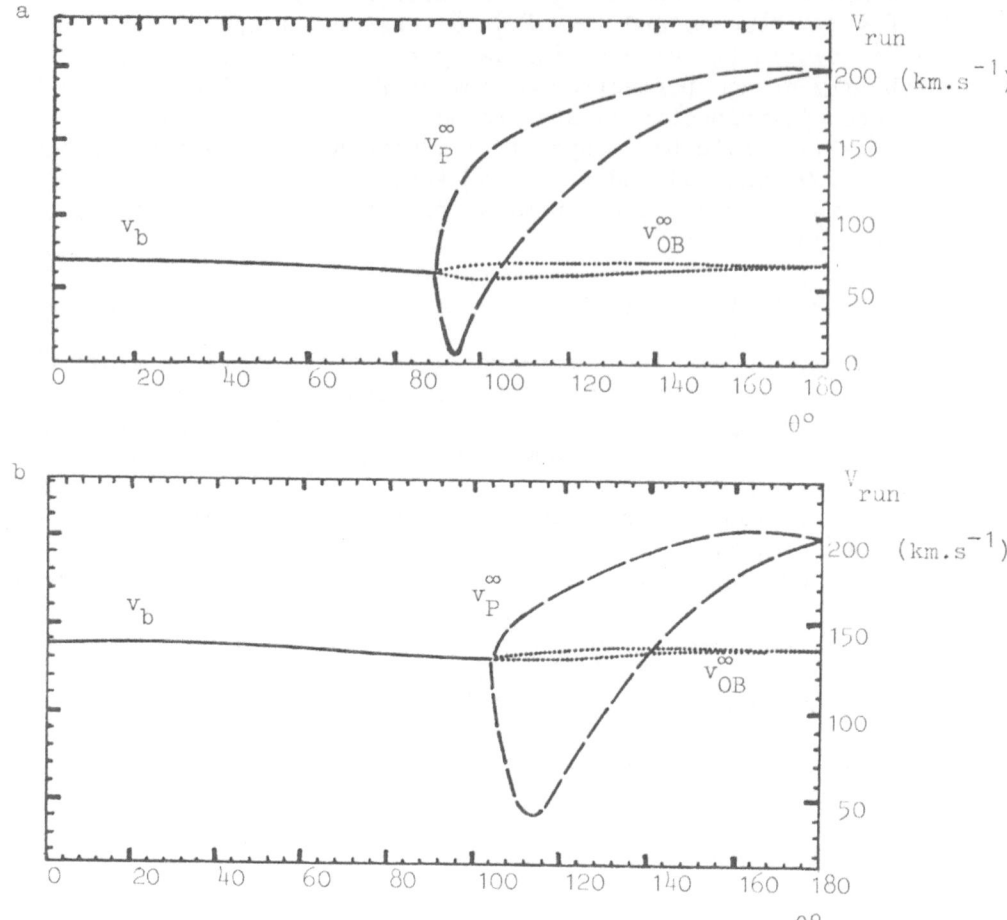

Figure 4.   The run-away velocity range as a function of the
angle θ between the asymmetric kick velocity $\bar{v}_k$ and the vector
$\bar{v}^0 + \bar{v}_{im}$.  The outcome for a case (b) supernova is shown for
the system (h) in part (a) and for the system (d) with $E^0 =$
$\pi/2$ in part (b).

velocity pulsars are created.  For example the symmetric explosion
occurring near periastron in the short periodic system (d) releases a
run-away pulsar with $v_p^\infty$ sym = 220 km.s$^{-1}$; for an asymmetric explosion
near periastron : 180 km.s$^{-1}$ $\leq v_p^\infty \leq$ 295 km.s$^{-1}$ (case a) and 140 km.s$^{-1}$
$\leq v_p^\infty \leq$ 385 km.s$^{-1}$ (case b), respectively.  For an explosion near periastron in the long periodic system (i) : $v_p^\infty$ sym = 90 km.s$^{-1}$, $v_p^\infty \leq$ 150
km.s$^{-1}$ (case a) and $v_p^\infty \leq$ 245 km.s$^{-1}$ (case b), respectively.

## 7. DISCUSSION AND CONCLUDING REMARKS

In view of the above we can conclude that the main effect of the impact of the ejected supernova shell on the companion star is the induced mass loss out of the system. Its magnitude is roughly inversely proportional to the square of the separation at the time of the supernova explosion. But, even for the shortest initial orbital periods considered or near periastron in eccentric orbits, the influence of the impact on the survival probability and the run-away velocities is marginal. Though $\bar{v}_{im}$ enters the equation (3.4) of $\bar{v}_g$ multiplied with $q^f = 15$-$26$.

The asymmetric kick velocity $\bar{v}_k$, the collapsed star is supposed to receive due to the asymmetry of the supernova explosion, lowers the survival probability of the remaining system and its importance is inversely proportional to the initial circular velocity $v_c^o$; i.e. increases for increasing values of the initial orbital period. The run-away velocity of the disrupted collapsed stars depends only on the magnitude of the asymmetric kick velocity. Whereas its influence on the run-away velocities of the remaining systems and of the disrupted OB stars is found to be marginal.

An initially eccentric orbit is mostly disrupted if the explosion occurs in the less probable positions near periastron. As most of the short periodic systems are expected to be circularized by tidal forces, OB stars with run-away velocities larger than 160 km.s$^{-1}$ should be very rare. Initially eccentric systems with long orbital periods may be quite common. The range of their run-away velocities is somewhat larger than in the circular case, but their mean survival probability equals the circular value.

If the supernova explosion is symmetric all the run-away OB stars originating from initially circular systems will have a collapsed companion. Single run-away OB stars are formed only in case the supernova explosion occurs at the less probable positions near periastron in an initially eccentric orbit. In case of an asymmetric supernova explosion however the mean survival probability and the run-away velocity of the OB star (be it single or with a collapsed companion) are related. They both decrease for increasing values of the initial orbital period.
Hence most of the high velocity OB stars ($\geq 60$ km.s$^{-1}$) have a collapsed companion. They originate from systems, with an initial orbital period up to one month, which have a large survival probability. Whereas systems with an initial orbital period of a few months create low velocity OB stars and about half of them are disrupted. Single run-away OB stars are expected to be less numerous and most of them have small run-away velocities.

In a bound system the collapsed star can show up as an X-ray pulsar, in the very rare case a suitable mode of mass transfer is available (cf. Davidson and Ostriker, 1973; Illarionov and Sunyaev, 1975; Savonije,

1979,1978). From observations of massive X-ray binaries we know that most of them have a high z-latitude, indicating that it are run-away objects; the observed orbital periods are up to about one month and they revolve in nearly circular orbits. The theoretically expected number of massive X-ray binaries in the galaxy (van den Heuvel, 1974) is in agreement with the observed number in case their survival probability is high. This supports our results that a supernova explosion in a massive close binary, with an initial orbital period up to a few months, creates run-away OB stars most of which have a collapsed companion. Hence an asymmetric supernova explosion in a massive close binary with initial period up to one year cannot lower the survival probability by orders of magnitude (cf. De Cuyper, De Grève, de Loore and van den Heuvel, 1976).

If the collapsed star is not directly observable, most of the bound systems will be regarded as single stars. This is due to the small barycentric velocity of the much heavier OB star in its final orbit, to the marginal effects the collapsed star has on its photometric light variations (unless the side towards the pulsar companion is heated by electromagnetic radiation from the pulsar beams) and to the observational limits on these quantities for OB stars.

In spite of the above, Stone (1979, 1981a,b) derived a mass-velocity relation for O-stars from a theoretical evolution model. Starting from a "standard" zero-age-main-sequence WR+OB binary with fixed initial orbital period and mass ratio, equal to their so called "most probable" value, only the initial total mass of the system was used as a free parameter. The system was assumed to revolve in a circular orbit and the period change due to mass and momentum loss and exchange was calculated neglecting the rotational energy and angular momentum of both components. The observational evidence given for the supposed increase of the run-away velocity of O-stars with their mass is weak. It depends on only three high run-away velocities of stars heavier than 60 $M_\odot$, which may be due to selection effects. Whereas the given run-away velocities of stars smaller than 60 $M_\odot$ are uncorrelated.

Our results show no evidence for a correlation between the mass of the remaining companion and its run-away velocity, in agreement with the observed run-away OB-star velocities (Cruz-González et al., 1974; Stone, 1979).

If the system is disrupted a single pulsar is released with maximum run-away velocity merely depending on the magnitude of the asymmetric kick velocity. We started by deriving this magnitude from observations of single radio pulsars. The run-away velocities calculated for pulsars escaping from massive binaries are spread around the magnitude of the asymmetric kick velocity used. But even for $\|\bar{v}_k\| = 150$ km.s$^{-1}$ they are not larger than 200 km.s$^{-1}$ for initially circular systems and about 400 km.s$^{-1}$ for supernovae occurring near periastron in initially short periodic eccentric orbits. The highest run-away velocities observed

($\lesssim$ 650 km.s$^{-1}$) are probably the outcome of the second supernova explosion in a massive close binary which survived the supernova explosion of its less massive companion.  Here the effect of the second supernova depends critically on the outcome of the instable spiral-in of the collapsed star into the envelope of its more massive companion.
Hence the supposed asymmetric supernova explosions reproduce the observed run-away velocities of pulsars.


ACKNOWLEDGEMENT

The author wishes to thank W.D. Arnett for discussion, B. Fryxell for correspondence and E.P.J. van den Heuvel for suggesting the subject and reading the manuscript.


REFERENCES

Abt, H.A. and Levy, S.G. 1978, Ap.J.Suppl. 36, 241
Arnett, W.D. 1974, I.A.U.Symp. N°66 : "Late Stages of Stellar Evolution".
    Ed. R.J. Taylor, Reidel, Dordrecht, Netherlands, p.1
Arnett, W.D. 1975, Ap.J. 195, 727
Arnett, W.D. 1978, International School of Physics, "Enrico Fermi"
    Course LXV : "Physics and Astrophysics of Neutron Stars and Black
    Holes", Ed. R. Giacconi and R. Ruffini, North Holland Publ. Comp.,
    Amsterdam, Netherlands, p.356
Blaauw, A. 1960, Bull.Astr.Inst.Netherlands 15, 265
Boersma, J. 1960, Bull.Astr.Inst.Netherlands 15, 291
Buchler, J.-R. 1973, in "Explosive Nucleosynthesis", Ed. W.D. Arnett
    and D.N. Schramm, Austin Univ. of Texas Press, p.229
Cheng, A. 1974, Ap.Space Sci. 31, 49
Colgate, S.A. 1970, Nature 225, p.247
Cruz-González, D., Recillas-Cruz, E., Costero, R., Peimbert, M., and
    Torres-Peimbert, S. 1974, Rev. Mexicana Astr.Ap. 1, 211
Davidson, K. and Ostriker, J.P. 1973, Ap.J. 179, 585
De Cuyper, J.-P. 1974, Master Thesis, Vrije Universiteit Brussel, Belgium
De Cuyper, J.-P., De Grève, J.P., de Loore, C. and van den Heuvel, E.P.J.
    1977, Astr.Ap. 52, 315
De Cuyper, J.-P., de Loore, C. and van den Heuvel, E.P.J. 1977, Astr.Ap.
    Suppl. 30, 93
de Loore, C. and De Grève, J.P. 1975, Ap. Space Sci. 35, 291
de Loore, C. and De Grève, J.P. 1976, I.A.U.Symp. N°73, "Structure and
    Evolution of Close binary Systems", p.27
de Loore, C., De Grève, J.P., van den Heuvel, E.P.J. and De Cuyper, J.P.
    1975, Mem.Soc.Astron.Italiana 45, 893
Fryxell, B.A. 1979, Ap.J. 234, 641
Fryxell, B.A. and Arnett, W.D. 1981, Ap.J. 243, 994
Garmany, C.D., Conti, P.S. and Massey, P. 1980, Ap.J. 242, 1063
Gott, J.R. 1972, Ap.J. 173, 227
Gullahorn, G.E. and Rankin, J.M. 1978, Ap.J. 225, 963
Gunn, J.E. and Ostriker, J.P. 1970, Ap.J. 160, 979
Hadjidemetriou, J.D. 1966, Z.Ap. 63, 116

Hanson, R.B. 1979, Mont.Not.Roy.Astr.Soc. 186, 357

Harrison, E.R. and Tademaru, E. 1975, Ap.J. 200, 145

Hiltner, W.A. 1945, Ap.J. 101, 356

Hut, P. and Verhulst, F. 1981, Astr.Ap. 101, 134

Iben, I.Jr. 1972, Ap.J. 178, 433

Illarionov, A.F. and Sunyaev, R.A. 1975, Astr.Ap. 39, 185

Khabazin, Yu.G. 1975, Astr.Zh. 52, 57 - Soviet Astron. A.J. 19, 34

Limber, D.N. 1960, Ap.J. 131, 168

Lyne, E. 1981, IAU Symp. N°95 "Pulsars", Ed. W. Sieber and R. Wielebinski
    Reidel, Dordrecht, the Netherlands

Massey, P. 1981, Ap.J. 244, 157

Massey, P. and Conti, P.S. 1981, Ap.J. 244, 169

Massey, P. and Niemela, V.S. 1981, submitted to Ap.J.

McCluskey, G.E. and Kondo, Y. 1971, Ap. Space Sci. 10, 464

Miller, G.E. and Scalo, J.M. 1979, Ap.J.Suppl. 41, 513

Minkowski, R. 1969, in "Supernovae and their Remnants", Ed. P.J.
    Brancario and A.G.W. Cameron

Mitalas, R. 1976, Astr.Ap. 46, 323

Munch, G. 1950, Ap.J. 112, 266

Niemela, V.S. 1976, Ap. Space Sci. 45, 191

Niemela, V.S. 1980, IAU Symp. N°88 "Close Binary Stars", Ed. M. Plavec,
    D.M. Popper and R.K. Ulrich, Reidel, Dordrecht, Netherlands, p.177

Niemela, V.S. and Sahade, J. 1980, Ap.J. 238, 244

Paczynski, B.E. 1971, Ann.Rev.Astr.Ap. 9, 183

Plavec, M. 1968, Astr.Ap. 6, 201

Roy, A.E. 1978, "Orbital Motion", Adam Hilger Ltd., Bristol, Great-
    Britain

Ruderman, M. 1972, Ann.Rev.Astr.Ap. 10, 427

Savedorff, M.P. and Vila, S. 1964, Astr.J. 69, 241

Savonije, G.J. 1978, Astr.Ap. 62, 317

Savonije, G.J. 1979, Astr.Ap. 71, 352

Schatzman, E. 1965, in "Stellar Structure", Ed. L.H. Aller and D.B.
    McLaughlin, Univ. Chicago Press, Chicago, Gordon and Breach, New
    York, USA

Seggewiss, W. 1974, Astr.Ap. 31, 211

Shklowsky, I.S. 1968, "Supernovae", Interscience, New York, USA

Shklowsky, I.S. 1970, Astr.Zh. 46, 715 - Soviet Astron. A.J. 13, 562

Schmidt, M. 1959, Ap.J. 129, 243

Sofia, S. 1967, Ap.J. 149, 1

Stone, R.C. 1979, Ap.J. 232, 520

Stone, R.C. 1981 a, Astron. J. 86, 544

Stone, R.C. 1981 b, submitted to Ap.J.

Sugimoto, D. and Nomoto, K. 1980, Space Sci.Rev. 25, 155

Sutantyo, W. 1974, Astr.Ap. 31, 339

Sutantyo, W. 1975, Astr.Ap. 41, 47

Sutantyo, W. 1978, Ap. Space Sci. 54, 479

Taff, L. 1974, Astron.J. 79, 1280

Taylor,J.H. and Manchester,R.N. 1977, Ap.J. 215, 885

Taylor,J.H.and Manchester,R.N.1981, IAU Symp. N°95 "Pulsars", Ed.
    W. Sieber and R. Wielebinski, Reidel, Dordrecht, the Netherlands

Tutukov, A.V. and Yungelson, L.R. 1973, Nauch.Inform.Moscow 27, 86

van den Heuvel, E.P.J. 1968, Bul.Astr.Inst.Netherlands 19, 326 & 432 van den Heuvel, E.P.J. 1974, Proc. 16th Solvay Conf. on "Astrophysics and Gravitation", Univ. of Brussels Press, Brussels, Belgium, p.119

van den Heuvel, E.P.J. 1975, Ap.J. 198, L.109

van den Heuvel, E.P.J. 1981, IAU Symp. N°93 "Fundamental Problems in Stellar Evolution", Ed. D. Sugimoto, D.G. Lamb and D.N. Schramm, Reidel, Dordrecht, the Netherlands

van den Heuvel, E.P.J. and Heise, J. 1972, Nature 239, 67

Weaver, A. and Woosley, S.E. 1979, Proc. 9th Texas Symp. on "Relativistic Astrophysics"

Wheeler, J.C., Lecar, M. and McKee, C.F. 1975, Ap.J. 200, 145

Whittaker, E.T. 1944, "Analytical Dynamics", Dover Publ.Comp., New York, USA

Wilson, O.C. 1949, Ap.J. 95, 402

Zwicky, F. 1965, in "Stellar Structure", Ed. L.H. Aller and D.B. McLaughlin, Univ. of Chicago Press, Chicago, Gordon and Breach, New York, USA

# AN OPTICAL BURST IN Sco X-1

H.Mauder

Astronomisches Institut
Universität Tübingen

Sco X-1 is known to show periods of correlated X-ray and optical
activity on a timescale of minutes to days, see e.g. Ilovaisky
et al.,1980. The rapid optical variations show typically an amplitude of about 5%. During the 1979 campaign on Sco X-1, blue band
observations were obtained on several nights with the ESO 1 m
photometric telescope with a time resolution of 2 seconds.

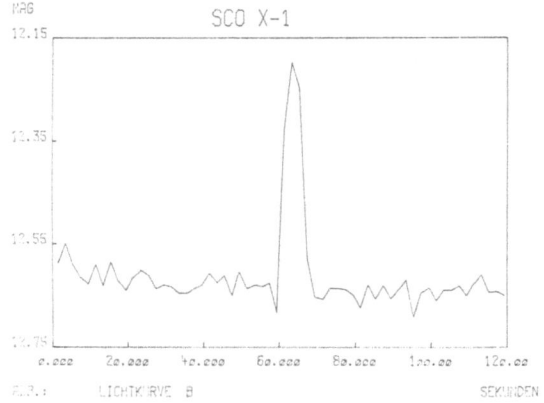

Figure:

An optical burst
in Sco X-1

On March 13th,1979, Sco X-1 was in the normal active state. However,
at $7^h$ $29^m$ $49^s$UT a burst event was observed, as is shown in the figure. This optical burst is very similar to the optical bursts of
X-ray bursters, lasting for only about 10 seconds with an amplitude
of almost 0.5 mag. Thus it seems possible that Sco X-1 shows at
least occasionally the characteristics of normal X-ray bursters.

Reference:

S.A.Ilovaisky,C.Chevalier,N.E.White,K.O.Mason,P.W.Sanford,J.P.Del-
vaille and H.W.Schnopper,1980, MNRAS 191,81

445

*Z. Kopal and J. Rahe (eds.), Binary and Multiple Stars as Tracers of Stellar Evolution, 445.*
*Copyright © 1982 by D. Reidel Publishing Company.*

# THE FORMATION OF MASSIVE WHITE DWARFS IN CATACLYSMIC BINARIES

Wai-Yuen Law and Hans Ritter
Max-Planck-Institut für Physik und Astrophysik
8046 Garching, Federal Republic of Germany

## Abstract

In contrast to the mass spectrum of single white dwarfs which has a single narrow peak at $\sim 0.6$ $M_\odot$, the observed mass spectrum of white dwarfs of cataclysmic binaries (CB's) shows a rather uniform distribution of the masses in the range $\sim 0.3$ $M_\odot$ to $\sim 1.3$ $M_\odot$. The formation of CB's with white dwarfs of less than about 0.8 $M_\odot$ can be understood as the result of a binary evolution according to low mass Case B or Case C with a subsequent spiraling-in in a common envelope. On the other hand the formation of massive white dwarfs of $M \gtrsim 1$ $M_\odot$ can be explained as the result of a massive Case B mass transfer yielding a helium star which subsequently undergoes a second Case B mass transfer (so called Case BB evolution). The ultimate product of such an evolution is either a CO-white dwarf with a mass up to the Chandrasekhar limit or a neutron star. The formation of CB's via Case BB evolution requires the binary to undergo at least one, most probably two separate phases of spiraling-in in a common envelope.

## Introduction

A cataclysmic binary (CB) is a short-period system consisting of a white dwarf primary and a low mass secondary which is filling its critical Roche-volume and is presumably a main-sequence star (Warner, 1976). According to the conventional theories of binary evolution, such a system is the descendant of an originally much wider binary which has experienced a highly nonconservative evolution of low mass Case B or Case C (Ritter, 1975, 1976; Webbink, 1975; Paczyński, 1976). The respective primary remnants are a low mass helium white dwarf with a mass of less than 0.45 $M_\odot$ and a CO white dwarf with a mass above 0.55 $M_\odot$. The mass of the white dwarf formed in a Case C evolution is determined mainly by the core evolution of the primary and in this aspect the formation of white dwarfs via Case C and in single stars are similar; the mass spectra of the white dwarfs produced in both cases therefore should also be similar. The observations of single white dwarfs show that most of the white dwarfs have a mass of

*Z. Kopal and J. Rahe (eds.), Binary and Multiple Stars as Tracers of Stellar Evolution, 447–451.*
*Copyright © 1982 by D. Reidel Publishing Company.*

(0.6 ± 0.1) $M_\odot$ and that massive white dwarfs with masses above ∼1 $M_\odot$ are extremely rare (Koester, Schulz, Weidemann, 1979). However, massive white dwarfs seem to be formed much more frequently in CB's than in single stars (see e.g. Robinson, 1976; Warner, 1976). Since Case C evolution probably cannot account for all of the observed massive white dwarfs in CB's, these white dwarfs are likely to be formed along other evolutionary paths. The purpose of this paper is to discuss such an alternative evolution which can lead to the formation of high mass white dwarfs in CB's.

## THE FORMATION OF MASSIVE WHITE DWARFS VIA CONSERVATIVE MASS TRANSFER

In a recent paper Delgado and Thomas (1981) - hereafter referred to as DT - have shown how massive white dwarfs can be formed in binaries. The evolution involves first a conservative mass transfer of high mass Case B which yields a helium star as an intermediate product. The formation of high mass white dwarfs is then a consequence of the fact that a helium star in the mass range 0.85 $M_\odot$ < M < 4 $M_\odot$ becomes a red giant during its shell helium burning phase (Paczyński, 1971) and therefore can reduce its mass below the Chandrasekhar limit in a second mass transfer. This type of mass transfer has been referred to as Case BB.

## THE FORMATION OF MASSIVE WHITE DWARFS VIA NONCONSERVATIVE CASE BB MASS TRANSFER

Since CB's are believed to be formed in a nonconservative evolution (Ritter, 1975, 1976; Webbink, 1975; Paczyński, 1976), we investigate the consequences of a Case BB mass transfer in binaries which, after the onset of the first mass transfer, went through a common envelope evolution (Meyer and Meyer-Hofmeister, 1979).

A) the Formation of Helium Stars via Nonconservative Case B Mass Transfer

A common envelope is likely to be formed whenever the primary, at the onset of the mass transfer, has a deep outer convective envelope (Webbink 1979). At present, however, the loss of the common envelope at the end of the "spiraling in" is not understood by theory. Therefore the orbital parameters of the emerging post common envelope binary (hereafter PCEB) cannot be determined theoretically. Nevertheless, due to the short duration of the common envelope phase, the mass of the helium star formed via the common envelope is not significantly different from what is obtained in a conservative evolution. Therefore, with respect to the mass of the helium star formed, we can make use of the results obtained by conventional conservative computations. The other properties of the PCEB are assumed to be qualitatively similar to those of the observed objects which are believed to be PCEB's (Law and Ritter, 1981). Accordingly, a PCEB is characterised by the following properties:

   1) The primary is a helium star of roughly the same mass as predicted by a conservative evolution.
   2) The secondary's mass is much lower than that in the conservative

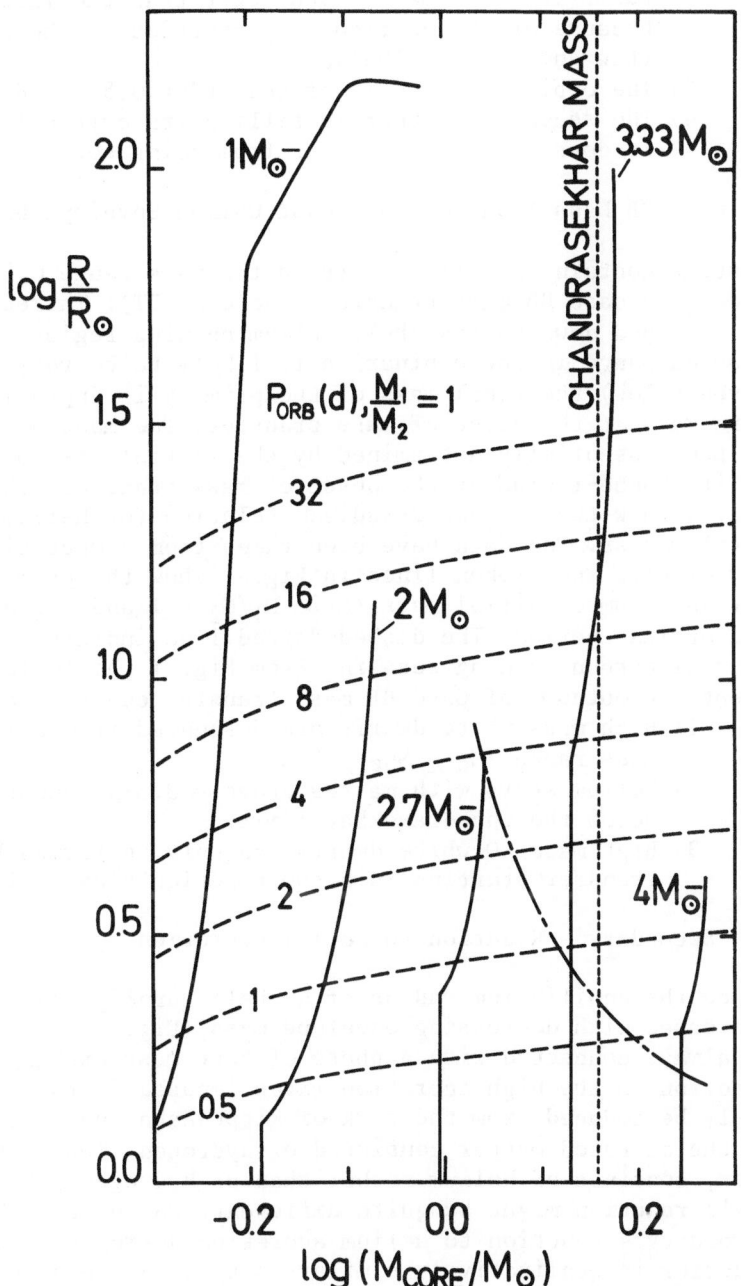

Fig. 1: Core-mass-radius-relation of shell burning helium stars
of different total mass (full lines). The broken lines
show the Roche-radius for an assumed mass ratio $M_1/M_2=1$
for various orbital periods. The dashed-dotted line
indicates the radius at which central carbon burning
sets in.

case. It is probably even lower than the mass of the helium star
because of the particular properties of the common envelope phase
(Law and Ritter, 1981).
3) The orbital period is of the order 0.5 - 1 day.
4) The secondary either is filling its critical Roche-volume or if
   it does not it is not far from doing so.

B) Case BB Mass Transfer in a Post Common Envelope Binary

Systems containing helium stars in the mass range $0.85M_\odot \lesssim M_{He} \lesssim 4M_\odot$
undergo a case BB mass transfer (see e.g. DT). Thereby the helium star
is stripped down to its shell helium burning region. Since the phase of
mass exchange in these binaries is likely to be very short (for reasons
given below) the final mass of the primary is approximately equal to its
core mass at the onset of mass transfer. The mass of the primary remnant
is then essentially determined by the initial mass of the helium star and
by its Roche-radius at the onset of mass transfer. The full lines in
Fig. 1 show the core-mass-radius relation for helium stars of various
total masses. The data have been taken from computations by DT and by
Law (1981). The broken lines in Fig. 1 show the primary's Roche-radius
for an assumed initial mass ratio $M_1/M_2 = 1$ and for different values of
the orbital period. The dashed-dotted line indicates the radius at which
central carbon burning sets in. From Fig. 1 the following conclusions
about the outcome of Case BB mass transfer can be drawn:
1) High mass white dwarfs are descended from helium stars in the
   mass range $2M_\odot \lesssim M_{He} \lesssim 3M_\odot$.
2) Helium stars with masses above $\sim 3.5M_\odot$ cannot reduce their mass
   below the Chandrasekhar limit.
3) High mass CO-white dwarfs can only be formed by a Case BB mass
   transfer starting in a short period binary, i.e. $P \lesssim 2^d$.

The Secondary's Reaction to Helium Accretion

Since the equilibrium radius of a shell burning helium star strongly
increases with decreasing envelope mass (Fig. 1), Case BB mass transfer
is always connected with a phase of fast mass exchange. The secondary's
reaction to the high accretion rates imposed by the primary could immediately be deduced from the work of Kippenhahn and Meyer-Hofmeister (1977),
if the accreted matter consisted of hydrogen. The exchanged matter, however, consists of helium rather than of hydrogen. Therefore, the secondary's reaction might be quite different. We suggest, however, that the
secondary's reaction to helium accretion during the early phases of mass
transfer is qualitatively similar to that in the case of hydrogen accretion. (Law and Ritter, 1981). In the cases considered by DT the secondary's
reaction is then negligible. This is because, as a consequence of the
conservative evolution, the secondary has a high mass and a thermal time
scale which is much shorter than the accretion time scale. However, in
the PCEB's considered here, the secondary has a low mass. Its thermal
timescale is much longer than the accretion timescale for the mass exchange
rate of $\sim 2 \cdot 10^{-4}$ $M_\odot/yr$ found by DT. Under these circumstances the secondary's radius increases sufficiently to bring the binary very quickly into

a second common envelope situation. The binary emerging from the second common-envelope phase will have an even shorter orbital period, a secondary of even lower mass and a primary which will become either a high-mass white dwarf or a neutron star. The ejected second common envelope will look to a distant observer like a helium rich planetary nebula.

# REFERENCES

Delgado, A.J., Thomas, H.-C.: 1981, Astron. Astrophys. 96, 142.
Kippenhahn, R., Meyer-Hofmeister, E.: 1977, Astron. Astrophys. 54,539.
Koester, D., Schulz, H., Weidemann, V.: 1979, Astron. Astrophys. 76, 262.
Law, W.-Y.: 1981, preprint, MPI/PAE Astro 277.
Law, W.-Y., Ritter, H.: 1981, in preparation.
Meyer, F., Meyer-Hofmeister, E.: 1979, Astron. Astrophys. 78, 167.
Paczyński, B.: 1971, Acta Astron. 21,1.
Paczyński, B.: 1976, in "Structure and Evolution of Close Binary Systems", IAU Symp. No. 73, ed. P. Eggleton, S. Mitton, and J. Whelan (Dordrecht: D. Reidel), p. 75.
Ritter, H.: 1975, Mitt. Astr. Ges. 36,93.
Ritter, H.: 1976, Monthly Notices Roy. Astron. Soc. 175, 279.
Robinson, E.L.: 1976, Astrophys. J. 203, 485.
Warner, B.: 1976, in "Structure and Evolution of Close Binary Systems", IAU Symp. No. 73, ed. P. Eggleton, S. Mitton, and J. Whelan (Dordrecht: D. Reidel), p.85.
Webbink, R.F.: 1975, Ph.D. Thesis, University of Cambridge.
Webbink, R.F.: 1979, in "White Dwarfs and Variable Degenerate Stars", IAU Coll. No. 53, ed. H. van Horn and V. Weidemann, (University of Rochester, Rochester, New York), p. 426.

## REFERENCES

# UV SPECTROSCOPY OF THE NOVALIKE VARIABLE TT ARIETIS

W. Wargau, H. Drechsel, J. Rahe
Remeis-Sternwarte Bamberg, Astronomisches Institut
Universität Erlangen-Nürnberg / F.R.G.
G. Klare, B. Wolf
Landessternwarte Heidelberg / F.R.G.
J. Krautter
European Southern Observatory
8046 Garching bei München / F.R.G.
N. Vogt
Institut für Astronomie und Astrophysik der Universität
München, Universitäts-Sternwarte / F.R.G.

TT Ari was detected by Strohmeier et al. (1957) and is classified as a novalike variable. It was hitherto unclear whether TT Ari is a special type of dwarf nova (Warner, 1976) or an old nova (Cowley et al., 1975). Our group obtained a total of four IUE spectra between 1979 and 1981 in the short and long wavelength region. The first spectrum was taken in July 1979, when the system had a visual brightness of 11.3 magnitudes. The following two IUE observations in November 1980 revealed TT Ari in the lowest optical state (V = $14^m.3$) observed so far. The fourth spectrum was obtained during the rise to maximum in January 1981, when the system had an apparent magnitude of V = $11^m.8$. From this behavior, Krautter et al. (1981) concluded that TT Ari is a dwarf nova with extremely extended standstills as they are typical for Z Cam stars.

During standstill, the spectra are dominated by broad absorption lines (Duerbeck et al., 1980), which, in quiescence, are present in emission. The spectrum taken in January 1981 is showing that absorption is also present during the rise to maximum, although weaker than at standstill; some lines can still be observed in faint emission (e.g. C IV at $\lambda$ 1550 Å). The standard deviation of the absolute fluxes of the C IV line from its mean value amounts to about 40 % during the different activity phases (determined from the three SWP spectra), while the continuum flux undergoes variations by a factor of 10 to 15.

The absolute flux distribution of the UV continuum can be interpreted as arising from two black body radiators with very different temperatures: the accretion disc with about 20000 K, and a hot component with 150000 - 200000 K. The latter, might be the boundary layer between accretion disc and white dwarf or, as an alternative, the hot

*Z. Kopal and J. Rahe (eds.), Binary and Multiple Stars as Tracers of Stellar Evolution, 453–454.*
*Copyright © 1982 by D. Reidel Publishing Company.*

spot. The cool secondary contributes in the UV less than 0.1 % to the total radiation and can therefore be neglected. The disc temperature is constant in all activity stages; the temperature of the hot component shows only marginal variations within the above mentioned range. However, both components undergo significant variations of their radiating areas: the disc radius from 0.035 $R_\odot$ (quiescence) to 0.15 $R_\odot$ (standstill), the hot component from 0.005 $R_\odot$ to 0.015 $R_\odot$, respectively.

REFERENCES

Cowley, A.P., Crampton, D., Hutchings, J.B., Marlborough, J.M.: 1975, Astrophys. J. 195, 413
Duerbeck, H.W., Klare, G., Krautter, J., Wolf, B., Seitter, W.C., Wargau, W.: 1980, Proc. "Second European IUE Conference", Tübingen, 26-28 March 1980, ESA SP-157, 91
Krautter, J., Klare, G., Wolf, B., Wargau, W., Drechsel, H., Rahe, J., Vogt, N.: 1981, Astron. Astrophys. 98, 27
Strohmeier, W., Kippenhahn, R., Geyer, E.: 1957, Kl.Veröff.Sternwarte Bamberg Nr. 18
Warner, B.: 1976, in "Structure and Evolution of Close Binary Systems", IAU-Symp. No. 73, eds. P. Eggleton et al. (Dordrecht: Reidel Publ.), 85

This work was supported by DFG grant (Ra 136/8).

# THE SUBDWARF ECLIPSING BINARY LB3459 (AA Dor)

R.W.Hilditch
University Observatory, St Andrews, Scotland.
Graham Hill
Dominion Astrophysical Observatory, Victoria, Canada.
D.Kilkenny
South African Astronomical Observatory,Cape Town,South Africa

A review was presented of the currently available observational data on this evolved system, namely, uvby photometry (Kilkenny et al. 1978; paper II) radial velocities (Kilkenny et al. 1981; paper III) and a non-LTE atmosphere analysis of the primary component (Kudritski et al. 1981). Kudritski et al. conclude that $T(pr) = 40000 \pm 2500^{\circ}K$ and $\log g = 5.3 \pm 0.2$ and from the spectroscopic data and light curve analysis (papers II, III) derive masses and radii for the primary and secondary components of $M(pr) \simeq 0.25\,M_{\odot}$, $M(sec) \simeq 0.04\,M_{\odot}$, $R(pr) \simeq 0.16R_{\odot}$, $R(sec) \simeq 0.09R_{\odot}$ respectively. Thus the primary component is a normal sdO star whilst the secondary component is a most enigmatic object, perhaps a normal composition degenerate dwarf but of too low a mass to be on the main sequence, perhaps an evolved degenerate object but then the radius is too large. The published evolutionary models for this system (Paczynski 1980; Conti et al 1981) do not adequately describe its current status.

An estimate of the secondary component's temperature is clearly an important datum but with a contribution in V light of 0.01 of the total light of the system it is difficult to determine. Indeed the uvby light curves do not contain any significant information about the intrinsic luminosity of the secondary, the only contribution from the secondary being that of reflection of the primary component's light. Accordingly, in an attempt to obtain further information about the secondary component, David Kilkenny has obtained VRI observations on two nights in 1980/81. The complete light curves have been solved simultaneously using the LIGHT program (Hill 1979) to yield a value of $T(sec) = 8500 \pm 1500^{\circ}K$ provided that, in order to explain the reflection effect, the bolometric albedo is greater than unity and is of the order of 2-3. The source of this additional energy is unknown (perhaps X-ray heating?). Certainly, it cannot be due to any extra reflection of primary component radiation from an ionised hydrogen cloud surrounding the system since, not only would the required electron density be so great that absorption effects would dominate, but also the amount of reflected light increases with increasing wavelength and hence cannot be due to

*Z. Kopal and J. Rahe (eds.), Binary and Multiple Stars as Tracers of Stellar Evolution, 455–456.*

Thomson scattering.    Additionally, the existence of the reflection
effect demonstrates that the secondary component is rotating
approximately synchronously and therefore cannot be rotationally
flattened(which would thereby suggest an apparently larger radius
as derived from primary eclipse).

The VRI observations obtained on these separate nights display
some differences which seem to indicate that the secondary component
is intrinsically variable.    Further VRI and perhaps JHK observations
are planned for the next observing season.    A full version of this
paper, including the observational data, is being submitted to
Mon. Not. R. Astr. Soc.

References
Conti, P.S., Dearborn, D. & Massey, P. 1981. Mon.Not.R.Astr.Soc.195,165
Hill, G. 1979, Publ.Dom.Astrophys. Obs. 15, 297.
Kilkenny, D., Penfold, J.E. & Hilditch, R.W., 1978.
                                           Mon.Not.R.Astr.Soc.187,1.
Kilkenny, D., Hill, P.W. & Penfold, J.E., 1981.
                                           Mon.Not.R.Astr.Soc.194,429
Kudritski, R.P., Simon, K.P., Lynas-Gray, A.E., Kilkenny, D. &
Hill, P.W. submitted to Astr.Astrophys.
Paczynski, B. 1980 Acta Astr. 30, 113.

# VARIABILITY OF SOFT X-RAY EMISSION OF EX HYDRAE OBSERVED WITH EINSTEIN OBSERVATORY

A. Kruszewski[1,2], R. Mewe[3], J. Heise[3], T. Chlebowski[4],
W. van Dijk[3], R. Bakker[5]

[1] Warsaw University Observatory, Warsaw, Poland
[2] European Southern Observatory, Garching, FRG
[3] Space Research Laboratory, Utrecht, The Netherlands
[4] Harvard-Smithsonian Center for Astrophysics, Cambridge, Mass., USA
[5] Astronomical Institute, University of Amsterdam, Amsterdam, The Netherlands

The cataclysmic variable star EX Hydrae has been observed with the High Resolution Imager (HRI) and the Imaging Proportional Counter (IPC) onboard the Einstein Observatory. The X-ray position is coincident within 3 arcsec of the optical position as measured on Schmidt survey plates. During a 15 1/2 hour observation with IPC we have searched for a modulation of the X-ray flux. Strong evidence for a 67 min period (one of two known optical periods) has been found in the energy range 0.1-3.5 keV with the IPC. The time dependence of modulations is used to discuss a model and evolutionary status of this close binary system.

## 1. INTRODUCTION

The cataclysmic variable EX Hydrae is known to be a spectroscopic and eclipsing binary (Kraft and Krzeminski, 1962). Its short period of 98.26 min places this object into the ultra short-period subgroup of cataclysmic variables (Patterson, 1979c; Vogt, 1980).

The discovery of an additional periodic variation in the optical brightness of EX Hydrae with a period of 67 minutes which has remained stable for over ten years (Vogt et al., 1980) has emphasised its unique character.

Various suggestions have been offered concerning the origin of 67 min variations (Vogt et al., 1980; Papaloizou and Pringle, 1980; Sherrington et al., 1980; Breysacher and Vogt, 1980; Cowley et al., 1981; Warner and McGraw, 1981). They include rotation of the white dwarf, disc instability and periodic mass transfer.

EX Hydrae is known to emit during optical quiescence both soft X-rays (0.7-2 keV) (Cordova and Riegler, 1979) and hard X-rays (2-10keV)

457

*Z. Kopal and J. Rahe (eds.), Binary and Multiple Stars as Tracers of Stellar Evolution, 457–466.*

(Watson et al., 1978 and references cited herein) at a flux level at Earth of $10^{-10}$ erg/cm$^2$/s in each energy interval. The total X-ray flux is of the same order as the optical flux. In fact, EX Hydrae seems to be one of the brightest X-ray sources among all dwarf novae in quiescence. The early X-ray observations did not show any clear evidence for flux variations with either 98 min orbital or additional 67 min period.

In this paper we report the observational results obtained with the high-resolution imager (HRI) and the imaging proportional counter (IPC) onboard the Einstein Observatory.

## 2. OBSERVATIONS

EX Hya was at first observed for approximately 3.9 hrs (net 1.04 hrs) centered around 10 42 UT on Jan 14 with the HRI onboard Einstein. A detailed description of the instrumentation characteristics can be found in Giacconi et al. (1979).

The high spatial resolution of the HRI permits an accurate position determination within a few arcsec (the 1σ error radius is typically 4."5 (Grindley 1980)). The measured HRI position of EX Hya is: α(1950) = 12 h 49 m 42.4 s and δ(1950) = −28° 58' 41."4, which is within 3" from the optical position (12 h 49 m 42.41 s ; −28° 58' 38."7).

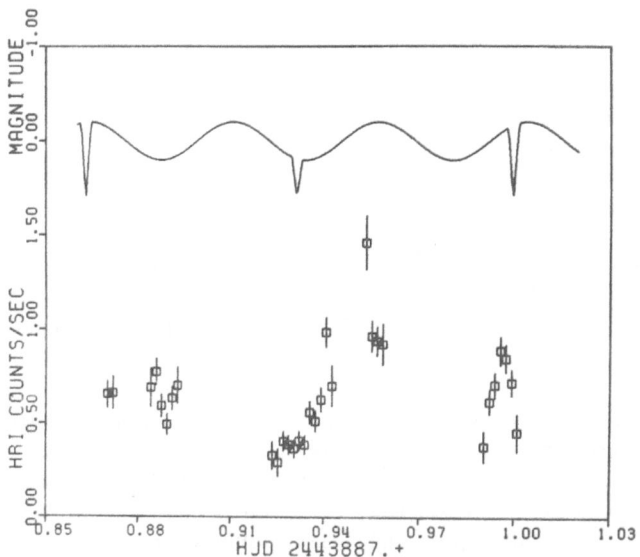

Figure 1. X-ray light curve of EX Hya as measured with HRI onboard Einstein Observatory on January 14, 1979. On top is drawn a highly schematic optical light curve, showing the phase of the periodic variations and the times of eclipses, which repeat every 98 min due to the orbital motion.

The optical position was measured on two Palomar Sky Survey prints and on two ESO Quick Blue Survey glass copies. Its accuracy is about 0."5, and excellent agreement between Palomar and ESO values indicate that annual proper motion of EX Hya is smaller than 0."05/year. The agreement between the optical and X-ray positions confirms very well the previous identification of the X-ray source with its optical candidate (Warner, 1972; Watson et al., 1978; Schwartz et al., 1978).

In the lower half of Figure 1 we present the observed HRI counting rate as a function of time. The counts are binned in 150 s intervals in order to smooth the Poisson noise. For comparison, the upper half represents a schematic optical light curve with eclipse minima, the phase of which was calculated with the epochs and periods from Vogt et al. (1979. Large X-ray variations are visible but the HRI observations

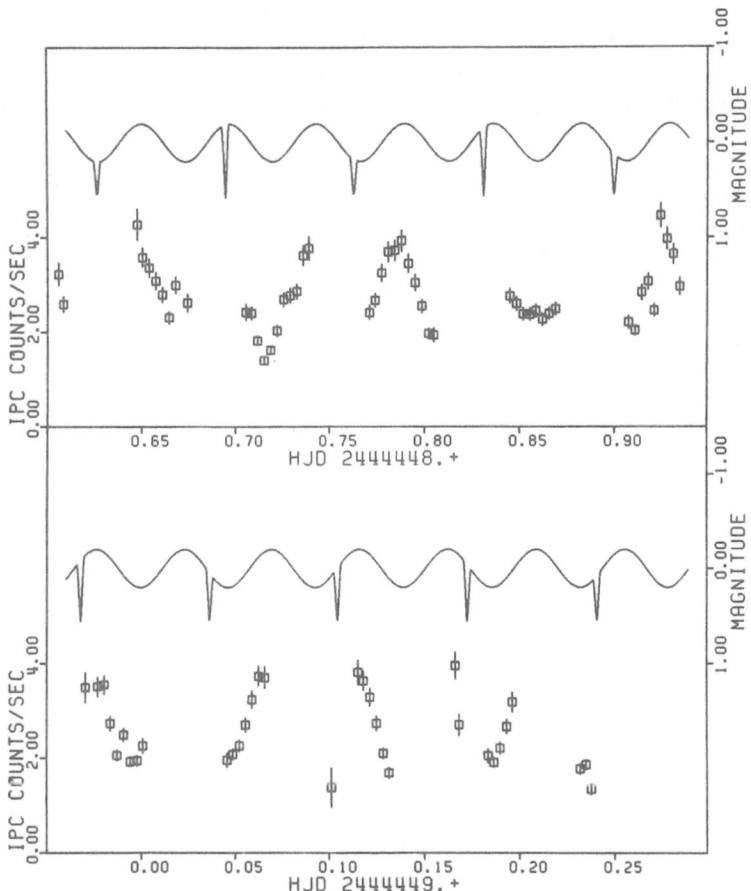

Figure 2. X-ray light curve (0.1-5 keV) in 300 s bin integrations as measured with IPC onboard Einstein Observatory on July 28, 1980. A highly schematic optical light curve is also shown for comparison.

cover a too short time interval for drawing any firm conclusions about
relations between the optical and X-ray variations. A new, over 15 hrs
long set of Einstein observations was obtained on July 28, 1980. This
time the IPC was used as a detector. Figure 2 shows these observations
binned in 300 s intervals compared with a schematic optical light curve.
The positive correlation between the X-ray and optical modulations is
well visible. Fig. 2 leaves no doubts that the 67 min variations are
also present in the soft x-ray spectral region.

Unfortunately in the phases around the times of the photometric
eclipses, there are gaps due to the South Atlantic Anomaly, so hardly
anything can be concluded about the presence of eclipses in the X-ray
flux.

Figure 3 presents a power spectrum of the IPC observations calculated with the help of the Deeming (1975) method. The power spectrum
of the original uncorrected observations binned in intervals of 100 s
is presented in the middle part of Figure 3. In the upper part a spectral window, shifted to have its highest peak at the frequency of the

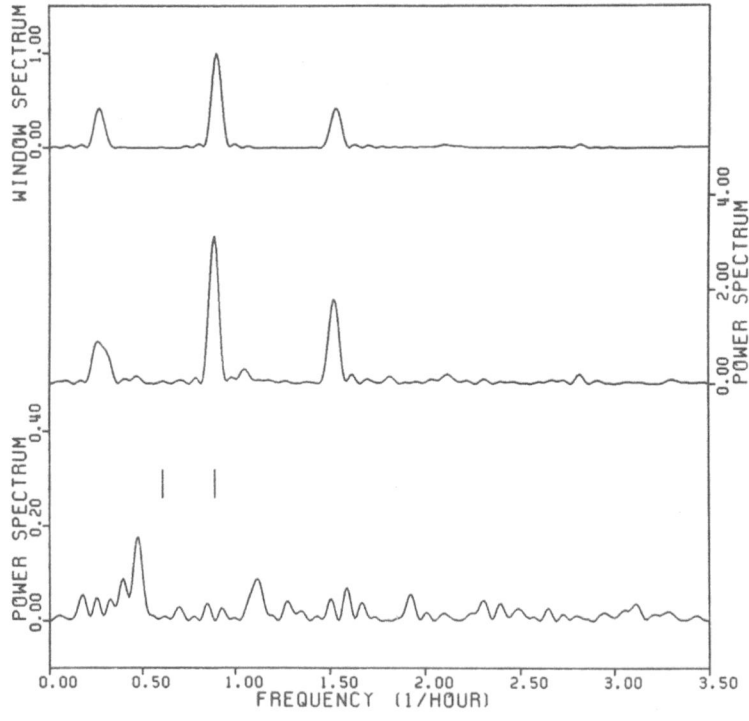

Figure 3. Power density spectrum of the X-ray light curve of Figure 2
          (middle) accompanied by the observational data window spectrum
          (top), and by power density spectrum calculated after the
          67 min periodic component has been removed.

67 min variations, is presented. Two sidelobes in the power spectrum correspondes closely to similar sidelobes in the spectral window. The differences in their heights and shapes are most likely due to the superimposed noise. But because of the closenes of the sideral satellite period (94 min), which is reflected in the spectral window, to the orbital period of the binary system (98 min), this difference may also indicate the existence of sidelobe variations in the system. A weak trace of a first harmonic of the fundamental 67 min variations is also present. No signs of higher harmonics and of variations with the orbital frequency can be seen.

The absence of the orbital frequency in the power spectrum speaks nothing about the absence or presence of X-ray eclipses because corresponding phases have not been propely observed, but it indicates instead, that there are no variations between eclipses with the orbital period.

The lower part of the Fig. 3 presents a power spectrum of the IPC signal calculated after the 67 min fundamental and its first harmonic have been removed. The vertical scale has been 10 times enlarged. We see that both sidelobes have been almost completely removed, and that the power spectrum is now dominated by noise. It indicates therefore, that the sidelobes seen in the middle part of this Figure are due to the beating with the satellite siderial period, and that there is no additional periodicity in the object.


3.  DISCUSSION

EX Hya is not the only object among cataclysmic binaries which shows an additional coherent periodic variations. It has been known since a long time (Walker, 1956) that an eclipsing post-nova DQ Her shows coherent short-term oscillations with a period of 71 s. Recently three more objects with similar periods have been discovered. They are WZ Sge with a period of 28 s (Robinson et al., 1978), V533 Her with a period of 63 s (Patterson, 1979a) and AE Aqr with a period of 33 s (Patterson, 1979b). These periods are around 3 orders of magnitude shorter than the corresponding orbital periods. On the other hand there exists a class of objects, called AM Her type objects or white dwarf magnetic binaries or polars, where brightness variations are caused by a white dwarf rotating synchronously with the orbital motion (Chiappetti et al, 1980; Stockman et al., 1981). EX Hya was the first object found to be situated between these two extrema with its 98 min orbital period and additional 67 min variations (Vogt et al., 1980). Two more were added recently. The system H2252-035 shows a 215 min orbital period and 14 min optical brightness oscillations (Patterson and Price, 1981). And finally another cataclysmic binary 2A0526-328 is found to have a photometric period only little shorter than the orbital period (Motch, 1981; Hutchings et al., 1981). We are listing here only objects with stable periods and not those, like SU UMa type objects or nova V1500 Cyg, where variations are transient and periods highly variable.

Out of 12 objects with stable optical variations, not caused by orbital motion, 10 are observed as X-ray sources (Chiappetti et al., 1981; Patterson et al., 1980; Schwartz et al., 1979; Patterson and Price, 1981; Patterson, 1981), and 6 out of these (AM Her, AN UMa, 2a0311-227, VV Pup, AR Aqr, EX Hya) show X-ray flux variations with periods equal to the optical periods. In addition, the source H2252-035 shows X-ray flux variations with a frequency that is larger than the optical variations frequency by exactly a single orbital frequency (White and Marshall, 1980; Patterson and Garcia, 1980), thus proving that the variable optical flux is caused by X-rays coming from a rotating compact component and reprocessed in an atmosphere of a secondary star. For most of these objects there are good reasons to believe that such periodic variations are caused by rotation of the white dwarf. It is natural then to consider such an explanation also for EX Hya. We shall look now into the background informations about EX Hya in order to infer what these informations tell us about the rotating white dwarf hypothesis.

Broad band photometric data is summarised by Vogt, Krzeminski and Sterken (1980). The observed eclipses have variable depth ranging from 0.3 mag to 0.8 mag and they have also variable width and shape. There is only a small hump observed to appear just before the eclipse (Mumford, 1967) and therefore the "hot spot" can not be very conspicious.

The narrow eclipses and small humps which repeat with the orbital cycle do not dominate the optical light variations of EX Hya. The more conspicious is the 67 min periodic variability together with superimposed fast flickering. Vogt, Krzeminski and Sterken (1980) saw 67 min variations in almost every observational run in a time interval of 14 years. The amplitude is variable, even from cycle to cycle, ranging from 0.05 mag to 0.9 mag, and shows a tendency to increase secularly from 0.2 mag in 1962 to 0.4 mag in 1976. This last statement is contradicted by Quinley et al. (1980) who have not seen 67 min variations in more recent observational material. There are several multicolor observing runs obtained, but up to now very little informations are published on the color dependence of the 67 min variations. In particular, Sherrington et al. (1979) publish in their Fig. 5 the V and K measurements folded with the 67 min period. One can see from that Figure that the amplitude with the K filter ($2.2\mu$) is about twice smaller than with the V filter ($0.55\mu$).

Vogt et al. (1980) say that the 67 min period is stable, what means that there are no difficulties with keeping cycle count and that a linear ephemeris is sufficient to predict nearly all times of maxima with an accuracy better than 10 minutes. The observations reported in this paper give clearly significant deviations from the Vogt et al. (1980) ephemeris, therefore it is interesting to take a closer look at the secular behaviour of the 67 min period. At first it was tried if there is any sign of period variability in Table 4 of Vogt et al. (1980) which contains times of maxima of 67 min variations. The parabolic fit has resulted in the following ephemeris:

$$HJD(maximum) = 2437699.8896 + 0.046546508 \times N - 6.1 \times 10^{-13} \times N^2$$
$$\phantom{HJD(maximum) = 2437699.88} +/- 6 \qquad\quad +/- 36 \quad +/- 3.5$$

One can see that there is only a marginal evidence for a decrease of the 67 min period.

Shortly before conducting the IPC observations, the star was observed for a total of 2.9 hrs on July 10, 11, and 12 with the Dutch 90 cm optical telescope equipped with the Walraven photometer at the site of the European Southern Observatory in La Silla, Chile. The visual magnitude of EX Hya varied between 13.4 and 13.9 mag (i.e., the star was in optical quiescence) at the time of these observations. Three maxima of the 67 min period were determined.

We can add new optical timings in order to repeat the parabolic fit on an extended time base. One more time of maximum has been derived from the Fig. 5 of Sherrington et al. (1980). It is HJD=2443986.496 . Because of the way the data was published there is an ambiguity of a few full cycles in the above value. Figure 4 gives all these times of maxima pictured as crosses together with two timings of X-ray observations shown as squares. The parabolic fit with an extended time base, but using only optical data, gives the following result:

$$HJD(maximum) = 2437699.8894 + 0.046546549 \times N - 9.4 \times 10^{-13} \times N^2$$
$$\phantom{HJD(maximum) = 2437699.88} +/- 6 \qquad\quad +/- 25 \quad +/- 2.1$$

Now the conclusion about the decrease of the 67 min period is on the 4.5σ significance level and therefore it can be considered as a good

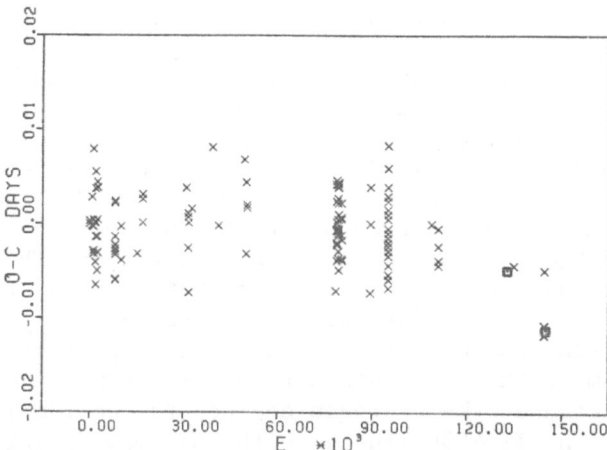

Figure 4. O-C timings of the maxima of the 67 min optical brightness periodic variations (crosses). The last four data points are added to those of Vogt et al. (1980). Also two X-ray timings of the flux maxima are shown (squares).

working hypothesis. The quantity P/P is equal $-3.17 \times 10^{-7}$ $yr^{-1}$ and the time scale for the decrease of the 67 min period is $3.1 \times 10^{6}$ years.

It is interesting to verify if the rotating white dwarf hypothesis is consistent with the derived spin-up rate. Rappaport and Joss (1977) have derived a relevant formula in case of disc accretion on a rotating compact object

$$\dot{P}/P = -3 \times 10^{-5} \times f\ f\ (P/1s) \times (L \times 10^{37} erg/s)^{(6/7)}\ yr^{-1}$$

where P is the period of rotation, L is the luminosity of the compact object due to the accretion, and f is a numerical factor whose value was estimated by Rappaport and Joss as 0.003 in the case of white dwarfs. Accepting P=4022 s and $\dot{P}/P=-3.17 \times 10^{-7} yr^{-1}$ we may solve this equation for L what gives $L=3 \times 10^{33}$ erg/s what in turn is not unreasonable for an object like EX Hya. Therefore we can say that the derived rate of the decrease of period is consistent with the rotating white dwarf hypothesis.

The ultraviolet and optical spectrophotometry (Bath et al., 1980) gives for EX Hya a spectrum which is consistent with that predicted by a steady accretion disc model. The extension of the measured spectrum into the infrared (Sherrington et al., 1980) results in the conclusion that the accretion disc is quite extensive and in particular it extends down nearly to the white dwarf surface. Sherrington et al. even conclude from the derived disc dimensions that the central white dwarf must be of radius smaller than $5 \times 10^{8}$ cm what in turn implies a large, close to the Chandrasekhar limit, mass (1.4 $M_{\odot}$). This conclusion can be relaxed somehow by taking into account possible contributions to the spectrum from both stellar components, but it seems likely that the white dwarf mass is high and that there is not much free space left around it.

Spectroscopic observations (Kraft, 1962; Breysacher and Vogt, 1980; Cowley et al., 1980) show the presence of wide emission lines which have a double structure and an "S-wave" component. The measurements of the emission line wings resulted in deriving the parameters of the or-bital motion of the primary (Breysacher and Vogt, 1980; Cowley et al., 1980). The resulting masses range from 0.7 to 1.5 $M_{\odot}$ for the primary and from 0.16 to 0.19 $M_{\odot}$ for the red secondary component. The obser-vations of Breysacher and Vogt (1980), made in 1976, show a strong dependence of the emission lines intensities on the phase of the 67 min variations. There is some phase-shift present with respect to the used ephemeris. The maximum of the line intensity occurs at phase 0.9 of the 67 min variations in case of the hydrogen and neutral helium lines. The less accurate data for He II 4686 line give the maximum of intensity at phase 0.72 . The observations of Cowley et al. (1980), made in 1980, do not show a clear dependence of emission line intensities on the 67 min phase. Certainly their observations of Hβ do not fit to a cosine curve with a maximum at phase 0.9, but the fit is much better if the phase of maximum is shifted to 0.7. And this can be treated as an independent confirmation of the 67 min period decreasing.

An important observational constraint is the measured total width
of the emission lines which gives the rotation speed of the innermost
part of the accretion disc to be 3500 km/s (Cowley et al., 1980).
Using the Shipman (1977) tabulation of the mass/radius relation for
white dwarfs, and assuming high orbital inclination, it is possible to
derive a lower limit 0.7 M for the white dwarf mass. In this limiting
case the disc extends down to the white dwarf surface. For higher
masses there may be some free space between disc and the white dwarf.

The above reasoning are leading us to a tentative conclusion about
a rotating white dwarf with a nonuniform surface brightness being a
source of optical and X-ray 67 min variations of EX Hya. The greatest
difficulty that we have encountered so far is little space that is left
to form an accretion column between disc and star surface. There is
another difficulty originating from not detecting any linear or circular
polarization in the optical spectral range (Krzeminski et al., 1981;
Knoechel and Vogt, cited by Breysacher and Vogt, 1980). By analogy with
AM Her type objects, one can expect that the variable optical component
in EX Hya is polarized. This prediction is not so firm, however. Detailed models of radiation processes in magnetic binaries (Masters et al.,
1977; King and Lasota, 1979) give a picture of polarized cyclotron
radiation competing with thermal bremsstrahlung radiation, with relative
roles depending on the accretion rate and the stellar magnetic field
strength. For high accretion rates and week magnetic fields the brem-
strahlung dominate and no large polarization is expected.


4.  EVOLUTIONARY ASPECTS

A decrease in the 67 min period may have an important evolutionary
significance. Of course one cannot be sure that the derived time scale
of 3 million years really corresponds to a secular evolution of the
period. But a rough agreement of this time scale with the expected one
for the case of an accreting white dwarf entitle us to consider its
consequences for the evolution of its parent binary system.

Extrapolating such a shortening of period back over about 1 million
years we obtain equality of both periods. If the white dwarf magnetic
field was sufficiently strong at that time, then the white dwarf rotation could be magnetically coupled to the orbital motion. We get therefore a tentative picture of EX Hya being for long time a bright AM Her
type object, whose magnetic field got dissipated some one million years
ago. The white dwarf lost at that time its hold on a companion star and
started to spin up. The derived rate of spin up is not large enough in
the present balance of the angular momentum in the system, but ultimately, after many millions of years, an appreciable fraction of the
system angular momentum can be stored in the white dwarf rotational
motion

REFERENCES

Bath, G.T., Pringle, J.E., Whelan, J.A.J.: 1980, Monthly Notices Roy.
    Astron. Soc. 191, 185.
Breysacher, J., Vogt, N.: 1980, Astron. Astrophys. 87, 349.
Chiappetti, L., Tanzi, E.G., Treves, A.: 1980, Space Sci. Rev. 27, 3.
Cordova, F.A., Riegler, G.R.: 1979, Monthly Notices Roy. Astron. Soc.
    188, 103.
Deeming, T.J.: 1975, Astrophys. Space Sc. 36, 137.
Cowley, A.P., Hutchings, J.B., Crampton, D.: 1981, Astrophys. J. 246,489.
Giacconi, R. et al: Astrophys. J. 230,540.
Grindlay, J.E.: 1980, private communication.
Hutchings, J.B., Crampton, D., Cowley, A.P., Thorstensen, J.R., Charles,
    P.A: 1981, preprint.
King, A.R, Lasota, J.P.: 1979, Monthly Notices Roy. Astron. Soc. 188,653.
Kraft, R.P., Krzeminski, W.: 1962, in Annual Report of the Director of
    Mount Wilson and Palomar Observatories (1961-1962), p. 20.
Krzeminski, W., Priedhorsky, W., Tapia, S.: 1981, private communication.
Masters, A.R., Pringle, J.E., Fabian, A.C., Rees, M.J.: 1976, Monthly
    Notices Roy. Astron. Soc. 178, 501.
Motch, C.: 1981, ESO preprint no. 135.
Papaloizou, J., Pringle, J.E.: 1980, Monthly Notices Roy. Astron. Soc.
    190 (Short Comm.), 13P.
Patterson, J.: 1979a, Astrophys. J. Letters 233, L13.
Patterson, J.: 1979b, Astrophys. J. 234, 978.
Patterson, J.: 1979c, Astron. J. 84, 804.
Patterson, J.: 1981, Santa Cruz Summer Workshop on Cataclysmic Binarias.
Patterson, J., Branch, D., Chincarini, G., Robinson, E.L.: 1980, Astro-
    phys. J. Letters 240, L133.
Patterson, J., Garcia, N.: 1980, IAU Circ. No. 3514.
Patterson, J., Price, C.M.: 1981, Astrophys. J. Letters, 243, L83.
Quinley, R., Africano, J.L., Rogers, W.: 1980, Bull.Am.Astr.Soc. 12,848.
Rappaport, S., Joss, P.C.: 1977, Nature 266, 683.
Robinson, E.L., Nather, R.E., Patterson, J.: 1978, Astrophys. J. 219,168.
Schwartz, D.A., Bradt, H., Biel, U., Doxey, R.E., Fabbiano, G., Griffiths,
    R.E., Johnston, M.D., Margon, B.: 1978, Astron. J. 84, 1560.
Sherrington, M.R., Lawson, P.A., King, A.R., Jameson, R.F.: 1980,
    Monthly Notices Roy. Astron. Soc. 191, 185.
Shipman, H.L.: 1977, Astrophys. J. 213, 138.
Stockman, H., Liebert, J., Tapia, S., Green, R., Williams, R., Ferguson,
    D., Szkody, P.: 1981, IAU Circ., No. 3616.
Vogt, N.: 1980, Astron. Astrophys. 88, 66.
Vogt, N., Krzeminski, W., Sterken, C.: 1980, Astron. Astrophys. 85,106.
Walker, M.F.: 1956, Astrophys. J. 123, 68.
Warner, B.: 1972, Monthly Notices Roy Astron. Soc. 158, 425.
Warner, B., McGraw, J.T.: 1981, Monthly Notices Roy. Astron. Soc. 196
    (Short Comm.), 59P.
Watson, M.G., Sherrington, M.R., Jameson, R.F.: 1978, Monthly Notices
    Roy. Astron. Soc. 184 (Short Comm.), 79P.
White, N.E., Marshall, F.E.: 1980, IAU Circ. No. 3514.

ON THE POSSIBLE SHORT-PERIOD IRREGULAR LIGHT
FLUCTUATIONS OF V1357 CYG =CYG X-1

M.I. Kumsiashvili
Abastumani Astrophysical Observatory,     Georgian S.S.R.
U.S.S.R.
Z. Kraicheva
Section of Astronomy, Bulgarian Academy of Sciences

ABSTRACT
     Photoelectric observations in U, B, V, were carried out to study
the possible rapid (during a night) fluctuations of the close binary
system V1357 Cyg, including the X-ray source Cyg X-1.  Assuming the
existence of irregular fast light fluctuations in the X-ray source
Cyg X-1, the conclusion can be drawn that during these observations
in 1979 the amplitude of the variations did not exceed 0.04 mag.

     There is at present no general consensus about possible short-
period (i.e., within a few hours) irregular light fluctuations of
V1357 Cyg in the optical region.  (Cherepashchuk et al., 1974) conclude that the light undergoes irregular variations from night to
night only and that it does not exhibit any rapid fluctuations in
excess of $0^{m}008$ within time intervals from 15 min to 5 hours.  On the
other hand, Kardopolov et al.,(1978) imply that the question is still
open because they observed light variations with an amplitude of
$0^{m}04$ near maxima in three cases.  It is, however, not impossible that
these fluctuations reflect at least partly observational errors.

     A recent paper by Natali et al. (1978) discusses anomalous light
curves of Cyg X-1 which were obtained in April-May 1975 during its
high state.  From an analysis of their optical observations, the
authors conclude that with the increase of the radiation in the X-ray
band, the system's optical component (i.e., HD 226868) shows large
fluctuations, mainly in V, with an amplitude ranging from $0^{m}06$ to
$0^{m}10$ within an interval of 20 to 40 minutes.  These fluctuations
disappear with the return of the X-ray radiation to the lower state.
Thus, the observed fluctuations are likely to be related to the high
state of the X-ray source.

     Three colour photoelectric photometry of Cyg X-1 have been carried
out at the Abastumani Astrophysical Observatory in order to investigate
in detail its photometric properties.  The observations were made from
1975 to 1979.  An extensive paper dealing with the reduction of the
data and a discussion of the results is in progress.  We shall here

467

*Z. Kopal and J. Rahe (eds.), Binary and Multiple Stars as Tracers of Stellar Evolution, 467–472.*
*Copyright © 1982 by D. Reidel Publishing Company.*

only report on one aspect of our observations.

Within the framework of a collaborative program on massive close
binary systems, we have undertaken to study possible, rapid irregular
fluctuations of $V_{1357}$ Cyg, i.e., fluctuations occurring during one
night. Accordingly, photoelectric observations were performed with the
AZT-14 48 cm telescope at Abastumani during some nights in 1979.
The observations were carried out in our UBV system which is very close
to the standard one. A pulse-counting photoelectric photometer was
used and the stars "a" and "c" from Liutij (1972) served as comparison
stars. For stars of this magnitude, our usual r.m.s. errors are of
the order of $0^{m}005$ to $0^{m}007$ for one observation.

An examination of the observations permits us to draw some preliminary conclusions. In all cases, the differences between the two
comparison stars $(m_a - m_c)$ as well as between the variable $V_{1357}$ Cyg
and star "a" $(m_v - m_a)$ show larger fluctuations than the differences
between the variable and star "c" $(m_v - m_c)$. This clearly indicates
a possible variability of the comparison star "a". It should be noted
that this star has been used by various investigators of Cyg X-1
but that little is known about its possible variability. Walker and
Quintanilla (1978) mention a variability of star "a" with a period of
tens or hundreds of days and also that it is double. Furthermore, it
exhibits periodic micro-variability. According to their data, the
amplitudes of the periodic variations are $0^{m}0037$ and $0^{m}0032$, corresponding to periods of 1.3608 and 0.8055 days, respectively.

Our observations indicate that the amplitude of star "a" may
be more than quoted above. To illustrate this, we shall discuss
briefly the observations for each night separately.

Figure 1 shows the differences $m_v - m_c$, $m_v - m_a$ and $m_a - m_c$ in the
V-band during the night of June 24-25, 1979; the abscissa values here
and in the following diagrams are given in U.T. The magnitude differences were corrected for extinction in the Earth's atmosphere and
we shall only consider nights with stable transparency. is the
phase of the orbital motion, computed from

$$\text{Min} = \text{JD } 2441166^{d}22 + 5^{d}60125 \text{ E}$$

The peak-to-peak scatter in $(m_a - m_c)$ is about $0^{m}03$ and is
greater than that of $(m_v - m_c)$. This is also clearly seen in the
B-band during the same night (Figure 2). Moreover, the $(m_a - m_c)$ and
$(m_v - m_a)$ differences vary in opposite sense, indicating an increase
in the light of star "a". In B, the variation of $(m_a - m_c)$ is about
$0^{m}05$ and that of $(m_v - m_c)$ is about $0^{m}04$.

A similar state of affairs was also observed on June 29-30, 1979.
The observations in V are shown in Figure 3. The peak-to-peak amplitude $(m_a - m_c)$ is of the order of $0^{m}05$ and that of $(m_v - m_c)$ is $0^{m}04$.
Thus, during this night, the Cyg X-1 variation differs somewhat from
that of June 24-25, 1979.

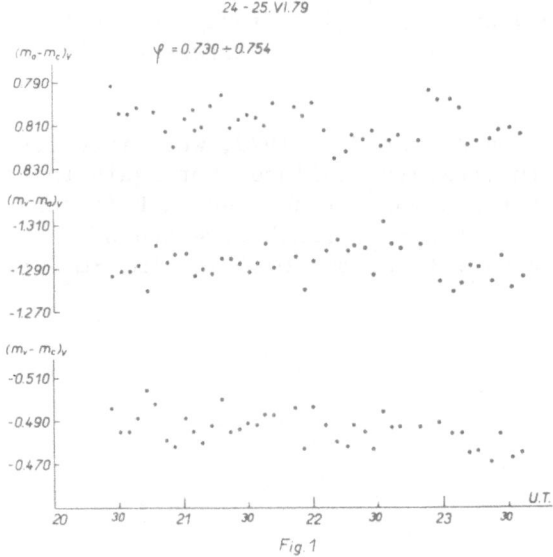

Fig. 1

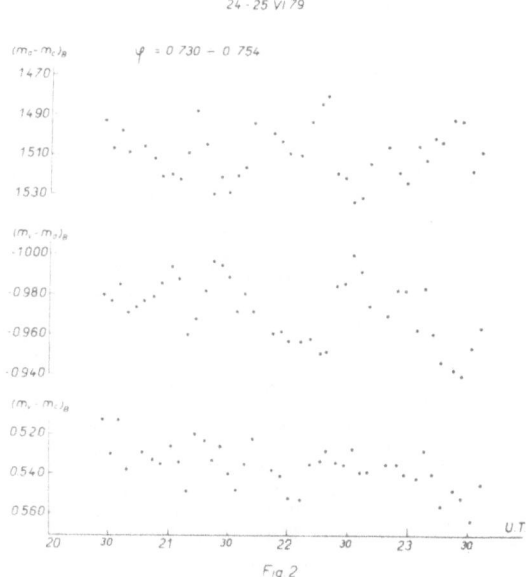

Fig 2

The duration of observations on July 19-20, 1979, is shorter (Figure 4 and 5). This night there were more marked fluctuations between the comparison stars $(m_a - m_c)$ in V, amounting to $0^m\!.05$; the trends of the $(m_a - m_c)$ and $(m_v - m_a)$ differences are the same. In the B-band, the variation of all three differences is the same and does not exceed $0^m\!.035$.

The observations on August 2-3, 1979, were interrupted at midnight because of an instrumental failure, but again the fluctuations in $(m_a - m_c)$ are greater and more pronounced in B (Figures 6 and 7). In V the fluctuations of $(m_v - m_c)$ reach $0^m\!.04$ and of $(m_a - m_c)$ about $0^m\!.045$, while in B the variations are $0^m\!.05$ for $(m_v - m_a)$ and even $0^m\!.07$ for $(m_a - m_c)$.

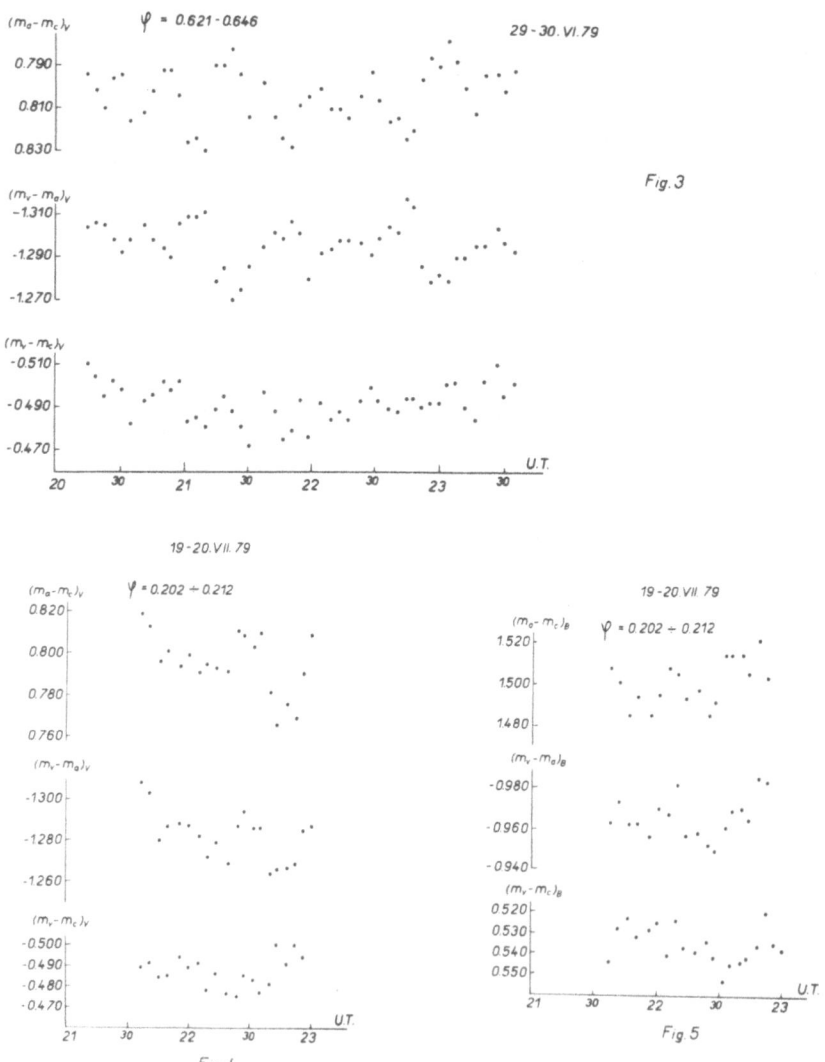

Fig. 3

Fig. 4

Fig. 5

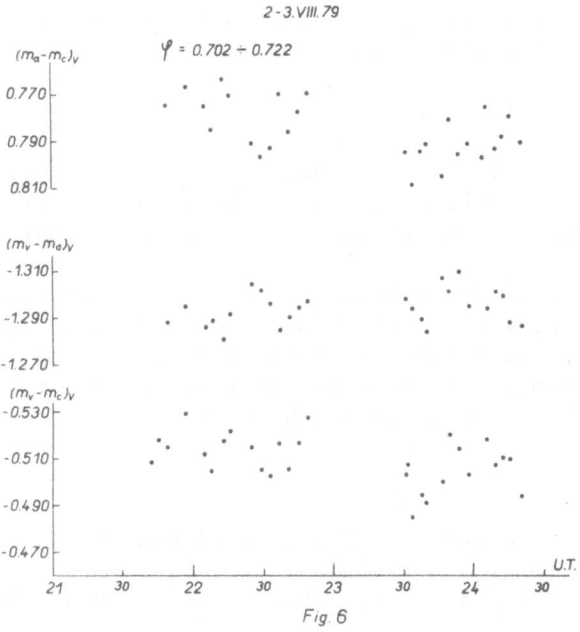

Fig. 6

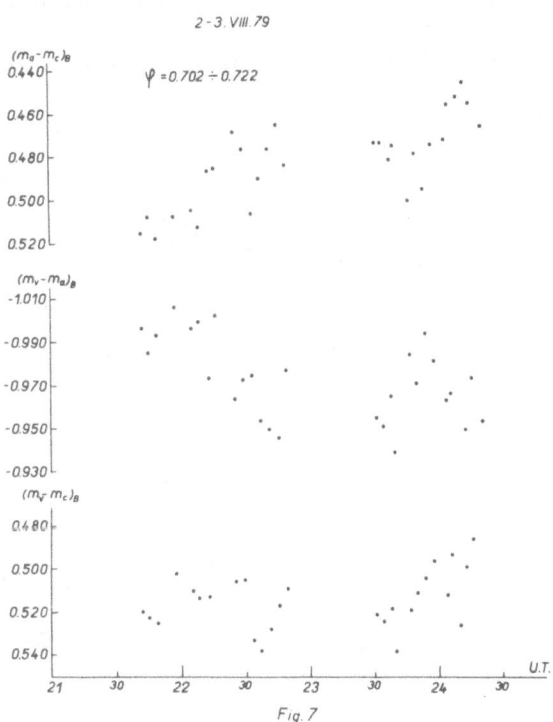

Fig. 7

In conclusion, we find that the amplitude of possible, irregular fast light fluctuations of Cyg X-1 does not exceed 0ṃ04 in V and B, at least at the time of our observations.

It would be desirable:

1)  To carry out further, accurate photoelectric photometry of star "a" in order to study the nature of its variability.  Similarly, a spectroscopic study would be useful;

2)  To perform accurate, high-speed photoelectric photometry of Cyg X-1, continuously during one or more nights, for a further investigation of possibly present fast light fluctuations.  It would be particularly desirable to do so when the star is in its high state.

## REFERENCES

Cherepashchuk, A.M., Kovalenko, V.M., Kovalenko, O.N., Mironov, A.I.:
    Peremennie Zviozdi, Vol. 19, No. 4, 3+5, 1974.
Kardopolov, V.K., Filipiev, G.K.:  Peremennie Zviozdi, 20, No. 6, 501,
    1978.
Liutij, V.M.:  Peremennie Zviozdi, 18, No. 5, 417, 1972.
Natali, G., Fabrianesi, R., Messi, R.:  Astron. and Astrophys., 62,
    No. 1/2, L1-L3, 1978.
Walker, E.N., A. Rolland Quintanilla:  Mon. Not. R. Astr. Soc., 182,
    No. 2, 315, 1978.

# PU VULPECULAE (OBJECT HONDA-KUWANO 1979) - A POSSIBLE
# SHORT-PERIOD RELATIVE OF THE SYMBIOTIC STARS

D. Chochol
Astronomical Institute, Slovak Academy of Sciences,
Tatranská Lomnica/Czechoslovakia
J. Grygar
Institute of Physics, Czechoslovak Academy of Sciences,
Rez/Czechoslovakia

## 1. PHOTOMETRIC HISTORY

PU Vul was discovered by Honda (1979) and Kuwano (1979) due to
its conspicuous increase of brightness in the years 1977-79. The
analysis of the archive data by Liller and Liller (1979) revealed
that the star was about $16.0^m$ - $16.5^m$ since the year 1898 until
October 1977, with occasional brightening to $15^m$ in 1926 and 1955.
Since November 1977 it started to flare up by $5^m$ until April 1979
when it reached a maximum of $9.0^m$. After the flat maximum lasting
to June 1979, the brightness of the object started to diminish
(Nakagiri and Yamashita, 1980). It reached $14.0^m$ in the autumn of
1980, but then it flared up again, and was again of $9.0^m$ in the
summer of 1981 (Belyakina et al., 1981). The latter authors have
found that the B-V index diminished in the 1980 minimum.

UBV photometry was conducted in Brno and Skalnaté Pleso observatories during the 1979 maximum (Chochol et al., 1981) and the authors
searched for possible periodicity in the V data, compiled from their
own observations as well as from data in the I.A.U. Circulars. The
best period found is 76.4 days, but periods up to 80 days are also
possible. Over this period, the object changes its V magnitude within
about $0.^m2$ - $0.^m25$.

## 2. SPECTROSCOPIC OBSERVATIONS

The spectroscopic evolution of the object was described by Honda
et al. (1979), Mochnacki (1979), Bensammar et al. (1980) and Hric
et al. (1980). Until September 1978, the star displyed an M-type
spectrum. In the April 1979 maximum, the spectrum was classified as
A4 from the appearance of the absorption lines, while the infrared
continuum was still well represented by a black-body radiation with
T = 3200 K, equal to spectral class M0 III. Only $H\alpha$ and $H\beta$ lines
were seen in the emission, $H\alpha$ exhibiting P Cyg profiles implying a
velocity of mass ejection close to 50 km s$^{-1}$. The absorption spectrum
changed to F5 in September 1979 and was again of a mid-M type during

473

*Z. Kopal and J. Rahe (eds.), Binary and Multiple Stars as Tracers of Stellar Evolution, 473–474.*
*Copyright © 1982 by D. Reidel Publishing Company.*

the photometric minimum in the second half of 1980.  IUE spectrum
obtained in the August 1981 flare-up, corresponded to the spectral
type A9 (Cassatella et al., 1981).  No emissions were seen in the
UV part of the spectrum.

3.  A TENTATIVE MODEL

    Several authors (Bensammar et al., 1980; Belyakina et al., 1982;
Honda et al., 1979) proposed that PU Vul is actually an interacting
binary.  The M star is a giant or supergiant shedding mass on the low
luminosity companion.  The light curve originally resembled slow
novae, but the spectral evolution is certainly different; the absence
of typical emission lines is particularly striking.

    It seems probable that both outbursts, in the years 1979 and 1981,
were the consequences of enhanced accretion of matter onto the low-
luminuous component.  For this, the Roche lobe overflow of the giant
or supergiant component is apparently necessary.  The 76-80 days
period found by Chochol et al. (1981) could be well explained as an
orbital period.  If this is the case, the system PU Vul may be regarded
as a short-period relative to the "normal" symbiotic stars.

    Most probably we may expect a prolonged stage of an enhanced activity
of this "down-scaled" symbiotic object.  Thus, it is to be expected, that
while at this Colloquium the star was a subject of a short note, on the
next Bamberg meeting it may be in the focus of attention.

    Our thanks are due to the I.A.U. and to Prof. J. Rahe for his
support, which enabled us to attend the meeting, and also to Prof.
M.J. Plavec for valuable advice.

REFERENCES

Belyakina, T.S., Chuvaev, K.K., Gershberg, R.E., Petrov, P.P.: 1980,
    IAU Circ. No. 3494.
Belyakina, T.S., Gershberg, R.E., Efimov, Y.S., Krasnobabtsev, V.I.,
    Pavlenko, E.P., Petrov, P.P., Chuvaev, K.K., Shenavrin, V.I.:
    1982, IAU Colloq. No. 70, p. 221.
Bensammar, S., Friedjung, M., Assus, P.: 1980, Astr. Astrophys. 83, 261.
Cassatella, A., Ponz, D., Friedjung, M., Viotti, R.:  1981, private
    commun.
Chochol, D., Hric, L., Papousek, J.:  1981, private commun.
Honda, M., 1979, IAU Circ. No. 3348.
Honda, M., Ishida, K., Noguchi, T., Norimoto, Y., Nakagiri, M.,Soyano, T.,
    Yamashita, Y.: 1979, Tokyo Astr. Bull., IInd. ser., No. 262, 2983.
Hric, L., Chochol, D., Grygar, J.:  1980, Inform. Bull. Var. Stars
    No. 1835.
Kuwano, Y., 1979, IAU Circ. No. 3344.
Liller, M.H., Liller, W.:  1979, Astron. J. 84, 1357.
Mochnacki, S.:  1979, IAU Circ. No. 3350.
Nakagiri, M., Yamashita, Y.:  1980, Tokyo Astron. Bull., IInd ser.,
    No. 263, 2993.

# DID SU UMa UNDERGO A CLASSICAL NOVA OUTBURST?

M.F. BODE[1], A. EVANS[1] and A. BRUCH[2]

(1)  Dept. of Physics, University of Keele, U.K.
(2)  Astronomical Institute, Westfälische Wilhelms-Universität, Münster, F.R.G.

ABSTRACT

Recent observations of an apparent soft X-ray halo around the dwarf nova SU UMa have led to speculation that this may well be evidence of the object having undergone a <u>classical</u> nova-like outburst within historical times (Cordova and Mason 1980). By combining the relationship between quiescent X-ray luminosity and speed class for classical novae and the observed X-ray luminosity of SU UMa we derive a distance dependent apparent magnitude at outburst for the object. Distance estimates for SU UMa and absolute magnitude ranges for classical novae then determine the apparent magnitude of an outburst more exactly. From the angular size of the halo and the absolute magnitude - ejection velocity relationship for classical novae we derive the approximate date of any outburst. Comparison with historical records does not reveal any promising candidates. An alternative interpretation of the halo in terms of scattering of soft X-rays from dwarf nova outbursts by interstellar grains is suggested.

## 1.  INTRODUCTION

The dwarf nova SU UMa was observed by Cordova and Mason (1980) to possess a soft X-ray halo, the properties of which are suggested as being consistent with the interaction with the interstellar medium of material ejected in a <u>classical</u> nova outburst several centuries ago. If this is the case then it is potentially an important clue to the evolution of cataclysmic binaries (see Vogt, 1982). However, the detection of the halo was marginal and we therefore set out in this paper to determine how bright a classical nova eruption at the distance of SU UMa might have appeared; we also estimate its duration of visibility and the most likely outburst date, and compare this with the historical record.

## 2. DISTANCE

As a first step we require a good estimate of the distance of SU UMa.

*Z. Kopal and J. Rahe (eds.), Binary and Multiple Stars as Tracers of Stellar Evolution, 475–481.*
*Copyright © 1982 by D. Reidel Publishing Company.*

Bruch (1981) has used two independent methods to derive this quantity and the first of these, and potentially the most accurate, relies on the findings of Vogt (1981 ) that dwarf novae at outburst maximum have essentially the same visual luminosity. The absolute magnitude measured by a distant observer then depends in a well known way only on the inclination of the system which in this case is ∿ 21°. Thus with $m_V$ = 14.51 ± 0.23 (taking the average of a variety of observations) and neglecting interstellar extinction, which will be small for an object at the distance and Galactic latitude of SU UMa, we find 213 ≲ d (pc) ≲ 421 and $M_V$ ≃ 7.2 ± 0.8 at minimum. As <u>classical</u> novae at outburst have -5 ≲ $M_V$ < -9 (Bath, 1978) then if SU UMa underwent such an outburst it would have had $m_V$ ≃ 3 at worst. Stephenson (1981) has suggested that objects of this magnitude and brighter are highly likely to have been noted in Eastern records. Lundmark (1921) does indeed cite such records of possible novae of this magnitude and brighter. We now proceed to define the probable apparent magnitude of any outburst more exactly and to determine its observable duration and date

## 3. BASIC RELATIONSHIPS

Recently Becker and Marshall (1981) have discussed a relationship between the X-ray (0.15 keV - 4.5 keV) luminosity of classical novae at <u>quiescence</u> $L_X$ and their optical luminosity at <u>outburst</u> $L_0$ (more specifically their speed class $\overset{\circ}{m}_V$), the common factor apparently being the accretion rate onto the white dwarf. These authors show that evidence for such a relationship is provided by X-ray studies of old classical novae by Cordova et al (1981). Although the nova DQ Her does not appear to follow the general run of the $L_X$ - $\overset{\circ}{m}_V$ relationship, Becker and Marshall (1981) note that its high inclination may suppress any X-ray emission. However, we note here that the soft X-ray luminosity that is required (cf. Ferland and Truran, 1981) to maintain the excitation of the nebula surrounding DQ Her is significantly greater than the upper limit  obtained by Cordova et al (1981). While the X-ray luminosity of HR Del ($\overset{\circ}{m}_V$ ≃ 0.008 mag $d^{-1}$) adjusted from the data of Hutchings (1980) to the same waveband follows the general trend of the Becker-Marshall relationship well, with $L_X$ ∿ 3 x $10^{31}$ erg $s^{-1}$. The data presented by Becker and Marshall (1981)are suggestive of a $L_X$ - $\overset{\circ}{m}_V$ relationship and the additional data presented above seem to strengthen their argument. The Becker-Marshall relationship plays a central role in the estimation of the date of, and apparent magnitude at, any classical nova outburst SU UMa may have undergone.

In order to proceed, we write:-

$$\log L_X = \alpha_X \log \overset{\circ}{m}_V + \beta_X \qquad (1)$$

$$\log L_0 = \alpha_0 \log \overset{\circ}{m}_V + \beta_0 \qquad (2)$$

where equation (1) represents the Becker-Marshall relationship and equation (2) is the usual relationship between optical luminosity at maximum and speed class for classical novae (cf. McLaughlin, 1960). We take $\beta_X$ = 32.2, as suggested by a linear least squares fit to the data on novae detected at X-ray wavelengths as illustrated in figure 2 of Becker and Marshall together with that on HR Del from Hutchings (1980), and $\alpha_O$ and $\beta_O$ from the standard relationship in McLaughlin (1960). The X-ray flux of SU UMa at quiescence is 1.3 x $10^{-11}$ erg cm$^{-2}$ s$^{-1}$ (Cordova and Mason, 1980). We can now use equations (1) and (2) to compute the value of $m_V$ at maximum of classical nova outburst. Similarly we can compute the date of outburst by means of the relationship between ejection velocity V and speed class (cf. McLaughlin, 1960):

$$\log V = \alpha_V \log \dot{m}_V + \beta_V \qquad (3)$$

In the equation (3) we take for $\alpha_V$ and $\beta_V$ the values appropriate for the principal spectrum (McLaughlin, 1960), as this generally contains the bulk of the ejecta (e.g. Gallagher, 1977).

## 4. RESULTS

Although the data presented by Becker and Marshall (1981) suggest a value $\alpha_X \sim 0.4$, more X-ray data on slow novae at quiescence are required to determine the exact form of the relationship, and here we compute $m_V$ and V at maximum for various values of distance d and $\alpha_X$. The relationship between $m_V$ at classical nova maximum and distance is shown in Figure 1 for several values of $\alpha_X \gtrsim 0$. For $\alpha_X = 0$, the distance (d = 316 pc) is uniquely determined by the X-ray properties of SU UMa at quiescence; while for $\alpha_X = \infty$, the distance is determined by the optical properties alone, $m_V \propto 5 \log d$. The effect of any uncertainty in $\beta_X$ is indicated in Figure 1 by the short horizontal arrow : an uncertainty of 0.2 in $\beta_X$ results in an uncertainty of $\sim$ 25% in the distance and a corresponding shift of all lines by the amount indicated. Also indicated on Figure 1 is the region of the $m_V$ - d diagram most probably occupied by SU UMa. The limits delineated by the parallelogram are determined by (a) the distance estimates for SU UMa and (b) the range of visual absolute magnitude of classical novae at outburst (see above).

Figure 1 suggests that values $\alpha_X \gtrsim 0.5$ are ruled out. The least squares fit suggests $\alpha_X \simeq 0.4$ and we find from Figure 1 that, at maximum, SU UMa had -2.1 $\gtrsim m_V \gtrsim$ -2.3 at classical nova outburst; the corresponding distance is 213 <d(pc) <224. Also, we note that this distance range implies a visual decay rate of 0.24-0.31 mag d$^{-1}$ typical of fast novae (Payne-Gaposchkin, 1957). Such novae normally have rather smooth light curves, and SU UMa would have faded to naked eye invisibility in $\sim$ 1-2 months.

The relationship between ejection velocity at classical nova outburst and distance is shown in Figure 2, in which similar considerations apply to the values $\alpha_X = 0, \infty$. The range of V values for classical novae lies in the range 200 km s$^{-1}$ for slow novae to 2000 km s$^{-1}$ for fast (McLaughlin, 1960). These limits on V, together with the limits on d from Figure 1, define the parallelogram in Figure 2 in which SU UMa is most likely to lie. Again Figure 2 suggests that large values of $\alpha_X$ are ruled out, although the upper limit in this case is $\alpha_X \lesssim 2$. Taking

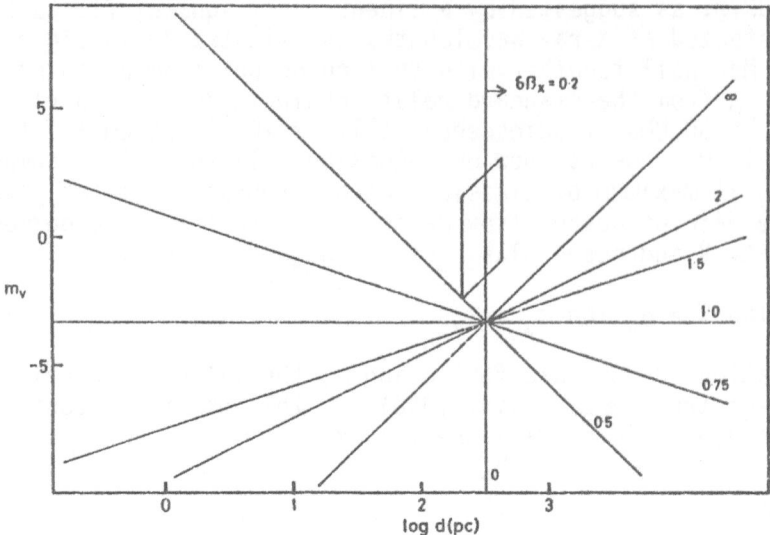

Fig. 1   Relationship between apparent visual magnitude at classical nova maximum and distance for SU UMa; $\alpha_x$ values as indicated.   See text for details.

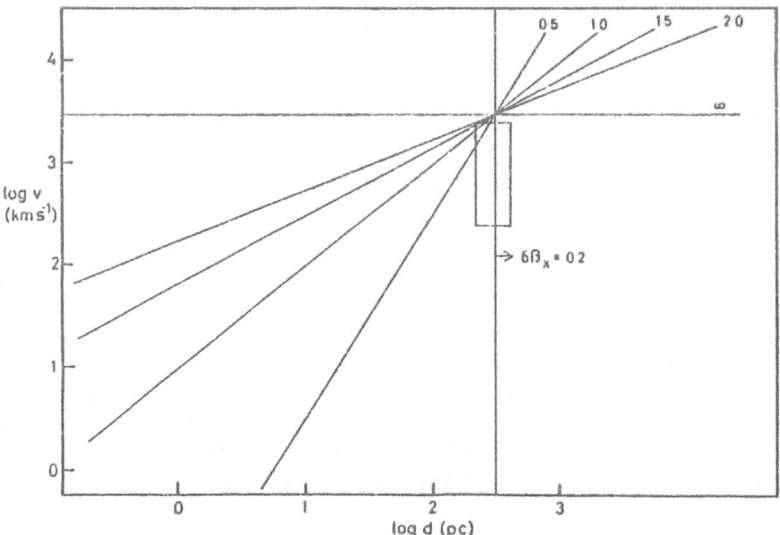

Fig. 2   Relationship between ejection velocity at classical nova maximum and distance for SU UMa; $\alpha_x$ values as indicated.   See text for details.

as above, the specific value $\alpha_X \simeq 0.4$ suggested by the Becker-Marshall (1981) relationship, we find the limits $1100 \lesssim V(\text{km s}^{-1}) \lesssim 2000$ and $213 \lesssim d \text{ (pc)} \lesssim 280$. The limits on V are consistent with the above classification of SU UMa as a fast nova.

From the point of view of confirming a historical classical nova outburst of SU UMa, the quantities of interest are the apparent magnitude at, and date of, eruption. The determination of the latter requires the angular radius of the X-ray halo around SU UMa, given by Cordova and Mason (1980) as $\phi = 14$ arcmin. Obviously the uncertainties in our numerical results (date and $m_v$) will arise mainly from the intrinsic scatter in the relationships (1)-(3); while the uncertainty in the date will be further aggrevated by our assumption of uniform outflow since outburst.

The dependence of $m_v$ at classical nova maximum on date of outburst is shown in Figure 3; we note that the uncertainty induced by the uncertainty in $\alpha_X$ is eliminated as this relationship is fortuitously independent of $\alpha_X$. However, Figure 3 shows one unfortunate trend : as the classical nova outburst goes further into the past the apparent magnitude at maximum increases. Even so, an outburst at 1 A.D. would have had $m_v \simeq 0.4$ and SU UMa would have ranked fifth in brightness in the northern sky.

## 5. COMPARISON WITH THE HISTORICAL RECORD

The above limits on $m_v$ and V together with Figure 3, suggest an outburst in the period 1360-1410 (it is of interest to note that Cordova and Mason 1980, arrived at an expansion age $\sim 500$ y). Unfortunately none of the objects in the list of Lundmark (1921) for this period meets the positional requirement. A recent discussion by Imaeda and Kiang (1980) does however suggest the occurrence of several "guest stars" in the relevant period. Alternatively Stephenson (1981) has expressed the view that the only known candidate may be an object classified by Lundmark as being very probably a classical nova, which was observed in A.D. 369. This had naked-eye duration of six months and $m_v \simeq -3$. According to Figure 3 SU UMa should have had $m_v \simeq -0.1$ and duration $\sim$ six years in A.D. 369. Also, although the declination of the A.D. 369 nova is in satisfactory agreement with that of SU UMa the right ascension is more uncertain and is likely to differ from that of SU UMa by $\sim 8^h$ (cf. Lundmark, 1921).

If the dwarf nova SU UMa did indeed undergo a classical nova outburst sometime in the past then the evidence for a generic relationship between these two types of cataclysmic binary would be greatly strengthened. However the apparent lack of historical records of a long duration ($\sim$ two month ) bright ($m_v \simeq -2.2$) outburst in the most likely period can only add to doubts about the origin of the Cordova-Mason (1980) halo.

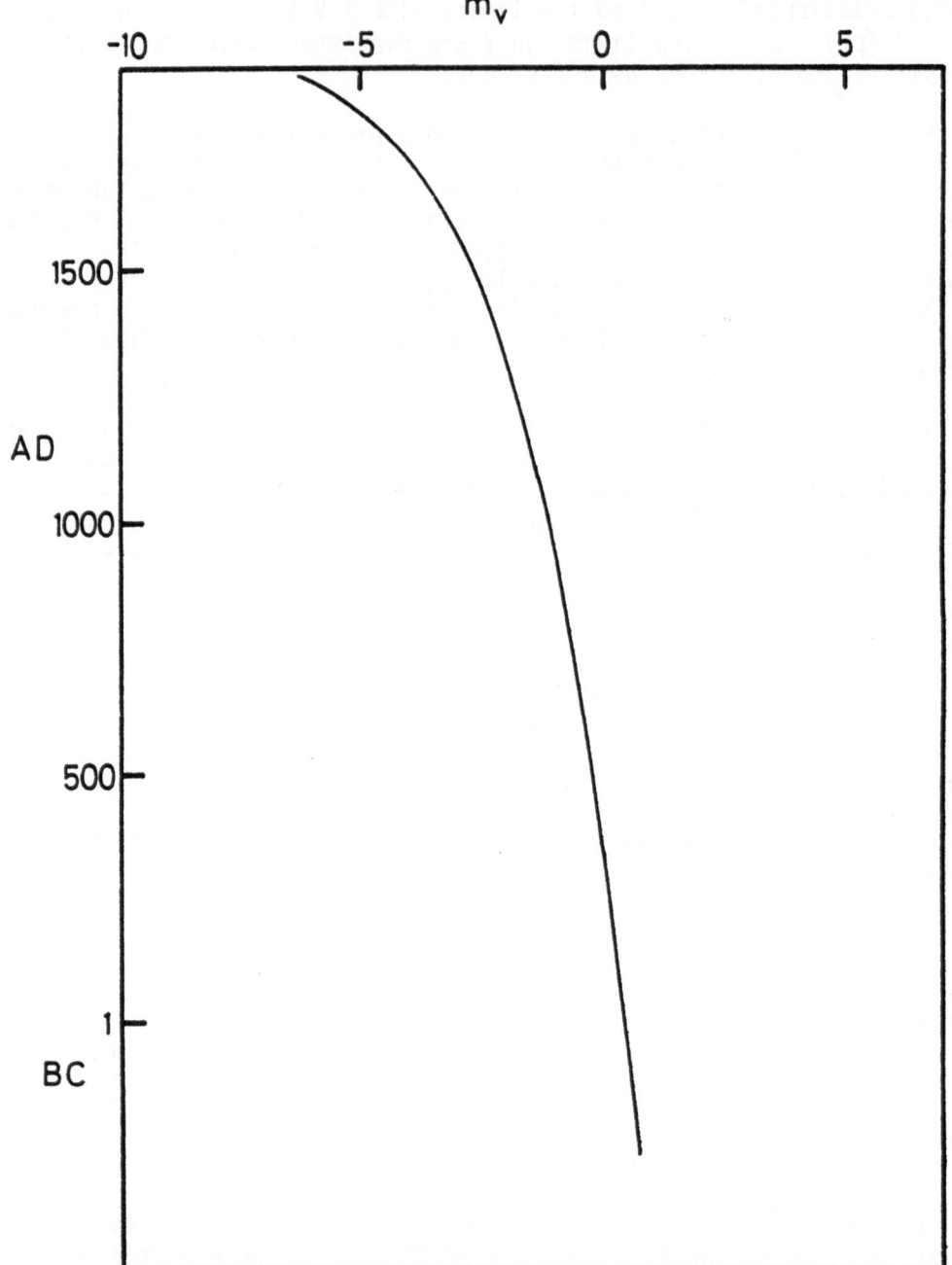

Fig. 3 Relationship between date of outburst and apparent visual magnitude at classical nova maximum for SU UMa. See text for details.

## 6. A POSSIBLE ALTERNATIVE EXPLANATION OF THE HALO

Assuming that, despite its marginal detection the halo was not due to statistical fluctuations (which would not in any case be expected to account for its symmetric disposition about SU UMa) and that it was not a purely instrumental effect (as it did not occur for similar objects at similar exposures (Cordova and Mason, 1981))one other explanation can perhaps be suggested. A localised cloud of small ($\leq$0.05 μm) inter-stellar grains lying between us and SU UMa, with resulting forward scattering of its soft X-ray flux may produce a halo about the object (Martin, 1978; Rolf, 1980). Perhaps the relatively high flux and ring-like nature of the halo are due to such grains scattering the X-ray flux of an outburst or, more plausibly, a superoutburst of the system. The halo would then be expected to expand and dissipate very rapidly. Further soft X-ray observations of SU UMa and related objects after outburst would be extremely useful to explore this possibility.

## Acknowledgments

Drs. R.H. Becker, F. Cordova, T. Kiang, D. Rolf, R. Stephenson, N. Vogt, have provided valuable comments and information on this work. MFB is supported by the SERC.

## References

Becker, R.H. and Marshall, F.E., 1981, Astrophys. J. Letters, 244, L93.
Bruch, A., 1981, preprint
Cordova, F.A. and Mason, K.O., 1980, Nature, 287, 25.
Cordova, F.A., Mason, K.O. and Nelson, J.E., 1981, Astrophys. J., 245, 609.
Ferland, G.J. and Truran, J.W., 1981, Astrophys. J., 244, 1022
Gallagher, J.S., 1977, Astron. J., 82, 209.
Hutchings, J.B., 1980, Publ. Astron. Soc. Pacific, 92, 458.
Imaeda, K. and Kiang, T., 1980, Journ. Hist. Astron., 11, 77.
Lundmark, K., 1921, Publ. Astron. Soc. Pacific, 33, 225.
Martin, P.G., 1978, Cosmic Dust, 67 (Oxford).
McLaughlin, D.B., 1960, in Stars and Stellar Systems, 6, 585.
Payne-Gaposchkin, C., 1957, The Galactic Novae (Dover).
Rolf, D., 1980, Ph.D. Thesis, University of Leicester
Stephenson, F.R., 1981, private communication.
Vogt, N., 1981, Habilitationsschrift, Bochum.
Vogt, N., 1982, 'Remarks on the Evolutionary Status of Cataclysmic Variables', this volume, 415.

# DOES THE CATACLYSMIC BINARY Z CHA CONTAIN A BLACK DWARF SECONDARY?

John Faulkner
Lick Observatory, Board of Studies in Astronomy and Astro-
physics, University of California, Santa Cruz
Hans Ritter
Max-Planck-Institut für Physik und Astrophysik, Garching

ABSTRACT

It is shown that the assumption of a black dwarf secondary in Z Cha
leads to a number of contradictions with well established theoretical
and observational facts. In particular the model predicts the radial
velocity $K_1$ to be much smaller than is observed, the radius of the
accretion disk to be smaller than is physically possible and a white
dwarf mass which is inconsistent with the white dwarf's radius derived
from eclipse analysis. Using the same arguments as for Z Cha it is also
possible to exclude a black dwarf secondary in the similar systems
OY Car and HT Cas.

## 1. INTRODUCTION

Z Cha is an eclipsing cataclysmic binary (hereafter CB) with an orbital
period of about 107 min. According to the standard model of CB's the
binary consists of a white dwarf primary and a low mass secondary which
fills its critical Roche-volume (Robinson, 1976; Warner, 1976). The
secondary spills mass through the inner Lagrangian point $L_1$ and thereby
gives rise to the formation of an accretion disk around the primary. At
the point where the matter, coming from $L_1$, impacts the disk a shock
front is formed which is usually referred to as the hot spot. In Z Cha
the hot spot and the center of the accretion disk, i.e. the white dwarf
undergo phase shifted total eclipses (Warner, 1974, Bailey, 1979).

There is considerable observational evidence, mainly from investigations of CB's which have longer orbital periods, that the secondaries
in CB's are main sequence stars, or at least very nearly so. (For a discussion of this point see Ritter, 1980d). On the other hand recent
theoretical studies on the consequences of gravitational radiation in
short period CB's predict that gravitational radiation forces the secondary to mass loss which eventually transforms it into a black dwarf
(Pacyński and Sienkiewicz, 1981; Joss, Rappaport and Webbink, 1981).
In course of its evolution from a CB containing a main sequence secondary

483

*Z. Kopal and J. Rahe (eds.), Binary and Multiple Stars as Tracers of Stellar Evolution, 483–488.*
*Copyright © 1982 by D. Reidel Publishing Company.*

to a CB containing a black dwarf, the binary's orbital period first decreases, then goes through a minimum of about 80 min and finally increases as the secondary becomes a degenerate star. In the context of
these findings the recent observation that the·orbital period of Z Cha
is currently increasing (Cook and Warner, 1981) has led to the speculation that Z Cha's secondary might be a black dwarf. In this paper we investigate whether the assumption of a black dwarf secondary is consistent
with other well established observational facts about Z Cha.

## 2. PREDICTIONS AND CONFRONTATION WITH OBSERVATIONS

The computations carried out below have been made under the following
assumptions:
1) The binary's orbit is circular.
2) The secondary fills its critical Roche-volume.
3) The secondary is a cold degenerate dwarf obeying the mass-radius-
   relation given by Zapolski and Salpeter (1969).
4) For the eclipse analysis, the eclipsing and the eclipsed objects have
   a circular shape and a uniform surface brightness. For details see
   Ritter and Schröder (1979).

The orbital period of Z Cha is $P = 0\overset{d}{.}07449927$ (Warner, 1974; Bailey,
1979; Cook and Warner, 1981). For a given chemical composition, the
assumptions 1), 2) and 3) together with the binary's orbital period P
and Kepler's third law determine the secondary's mass $M_2$ and radius $R_2$
uniquely. Numerical values of $M_2$ and $R_2$ for two values of the hydrogen
mass fraction X (= 0.7 and = 0.75 resp.) are listed in Table 1. As can
be seen, a possible black dwarf secondary of Z Cha is an object of extremely low mass.

| X | $M_2/M_\odot$ | $R_2/R_\odot$ |
|------|--------|--------|
| 0.70 | 0.0149 | 0.0841 |
| 0.75 | 0.0162 | 0.0871 |

Table 1. Mass $M_2$ and radius $R_2$ of a possible
black dwarf secondary of Z Cha for two
different chemical compositions X.

The half width of the white dwarf's eclipse, $\Delta t_{1/2}$, (see Fig. 1) yields
a unique relation between the orbital inclination i and the binary's
mass ratio $q = M_1/M_2$:

$$i = \arccos\left\{\left(\frac{C}{(1+q)^{1/3}}\right)^2 - \left(\frac{\pi\Delta t_{1/2}}{P}\right)^2\right\}^{1/2} \tag{1}$$

For deriving Eq. (1) we have made use of an approximation for the
secondary's Roche-radius (Pacyński, 1971)

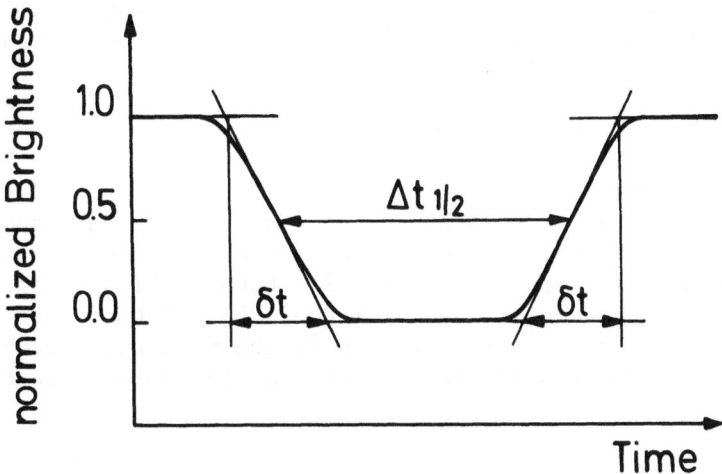

Figure 1.  Idealised light curve of a total eclipse.
The definitions of the eclipse's half width $\Delta t_{1/2}$
and of the ingress/egress time $\delta t$ are shown.

$$\frac{R_{2,\text{Roche}}}{A} = C(1+q)^{-1/3} \quad , \quad C = \frac{2}{3^{4/3}} \quad , \quad q \gtrsim 2 \quad , \qquad (2)$$

where A denotes the binary's orbital separation. Taking $\Delta t_{1/2}$ = 344 s
(Bailey, 1979) we find an upper limit of the mass ratio $q \lesssim 20$ and an
upper limit for the white dwarf's mass $M_1 \lesssim 0.32 \ M_\odot$.

The primary's radial velocity amplitude

$$K_1 = \frac{2\pi R_2}{PC(1+1)^{2/3}} \sin i \qquad (3)$$

depends on q and (via $R_2$) on the secondary's chemical composition. For
any reasonable value of $q \lesssim 20$, however, the predicted value of $K_1$ is
always much smaller than the observed $(87\pm14)$ km/s (Vogt, 1981).

The fractional radius of the accretion disk (assumed to be in
Keplerian roation)

$$\frac{R_{\text{disk}}}{A} = q(1+q)\left(\frac{K_1}{v \cdot \sin i}\right)^2 \qquad (4)$$

turns out to be smaller than the corresponding (minimum) radius of a
viscosity free disk (Flannery, 1975; Lubow and Shu, 1976) for any value
of $q \lesssim 20$ and even for the lowest published value of $v \cdot \sin i \approx 600$ km/s
(Vogt, 1981). The values derived from Eq. (4) are also in contradiction
with a determination of the disk's radius which is based only on a light
curve analysis (Ritter, 1980c).

The duration $\delta t$ of the white dwarf's ingress into and egress from total eclipse (see Fig. 1) provides information about the white dwarf's radius $R_1$. Following Ritter and Schröder (1979) (and neglecting effects of limb darkening) we find

$$R_1 = R_2 \frac{4\pi}{P^2} \Delta t_{1/2} \, \delta t \, \frac{(1+q)^{2/3}}{c^2} \qquad (5)$$

Eq. (5) together with the relation $M_1 = q \, M_2$ yields a mass-radius relation for the white dwarf which can be compared with the corresponding theoretical mass-radius-relation. This is shown in Fig. 2. The value of $\delta t = (44\pm10)s$ has been determined from published light curves (Bailey, 1979). As can be seen from Fig. 2, even in the most extreme case, namely $q \approx 20$, the white dwarf's radius is still smaller than it ought to be for its mass.

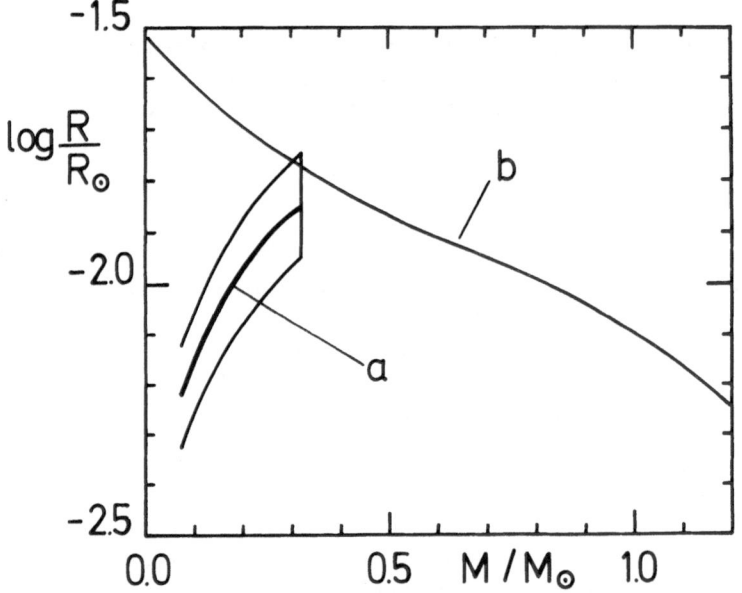

Figure 2. a) Mass-radius relation of the white dwarf derived from eclipse analysis (Eq. 5).
b) Theoretical mass-radius-relation of white dwarfs.

The fact that the white dwarf undergoes an observable eclipse indicates that the orbital inclination is at most 87°. For higher inclinations the white dwarf is permanently eclipsed by the disk's outer rim. This is due to the finite thickness of the disk (see e.g. Meyer and Meyer-Hofmeister, 1981). Therefore the upper limits for the mass ratio and the white dwarf's mass are reduced to $q \lesssim 17$ and $M_1 \lesssim 0.27 \, M_\odot$ respectively. On the other hand the observed width of the base of the emission lines indicates a rotational velocity at the white dwarf's surface of at least 2300 km/s (Vogt, 1981). This, however, corresponds to a lower limit of the white dwarf's mass of $M_1 \gtrsim 0.42\ M_\odot$.

If gravitational radiation drives the mass exchange, the predicted mass exchange rate is of the order $10^{-13}\ M_\odot$/yr. This has to be compared with the mass exchange rate predicted for the corresponding system with a main sequence secondary which is $\sim 4\cdot10^{-11}\ M_\odot$/yr. In fact a mass exchange rate of this order was determined from observations by Ritter (1980b). Z Cha's luminosity is almost entirely due to accretion. Therefore, if Z Cha has a black dwarf secondary, the low mass transfer rate implies that its distance is at most a tenth of the current estimate of about 100 pc. However, at a distance of about 10 pc, Z Cha would be one of the nearest stars. It should have a large parallax and probably a high proper motion. Thus even astrometrical observations could contribute to deciding whether Z Cha's secondary is a degenerate object or not.

Finally, we mention that although the assumption of a black dwarf secondary accounts for the sign of the period change, the predicted time scale of the period change of $10^{18}$ yrs is at least three orders of magnitudes longer than is actually observed (Cook and Warner, 1981).

## 3. CONCLUSIONS

We have shown that the assumption of a black dwarf secondary in Z Cha leads to contradictions with a number of well established observational and theoretical facts. Therefore, Z Cha's secondary is most likely not a black dwarf but rather a low mass main sequence star. The sign and the time scale of the observed period change (Cook and Warner, 1981) still remains to be explained. Applying the same arguments to the CB's OY Car (Ritter, 1980d; Schoembs and Vogt, 1981; Bailey and Ward, 1981) and HT Cas (Patterson, 1981; Young and Schneider, 1981), we can probably also exclude a black dwarf secondary in these systems.

Acknowledgements
This paper resulted from the stimulating discussions held at the Summer Workshop in Astronomy and Astrophysics 1981 on "Cataclysmic Variables and Related Objects" in Santa Cruz.

References

Bailey, J.: 1979, Monthly Notices Roy. Astron. Soc. 187, pp. 645-653.
Bailey, J., Ward, M.: 1981, Monthly Notices Roy. Astron. Soc. 194, pp. 17P-23P.
Cook, M.C., Warner, B.: 1981, Monthly Notices Roy. Astron. Soc. 196, pp. 55P-57P.

Flannery, B.P.: 1975, Monthly Notices Roy. Astron. Soc. 170, pp. 325-331.
Joss, P.C., Rappaport, S., Webbink, R.F.: 1981, preprint.
Lubow, S.H., Shu, F.H.: 1976, Astrophys. J. 198, pp. 383-405.
Meyer, F., Meyer-Hofmeister, E.: 1981, preprint MPI-PAE/Astro 271.
Paczyński, B.: 1971, Ann. Rev. Astron. Astrophys. 9, pp. 183-208.
Paczyński, B., Sienkiewicz, R.: 1981, Astrophys. J. Letters 248,
    pp. L27-L30.
Patterson, J.: 1981, Astrophys. J. Suppl. 45, pp. 517-539.
Ritter, H., Schröder, R.: 1979, Astron. Astrophys. 76, pp. 168-175.
Ritter, H.: 1980a, Astron. Astrophys. 85, pp. 362-364.
Ritter, H.: 1980b, Astron. Astrophys. 86, pp. 204-211.
Ritter, H.: 1980c, Astron. Astrophys. 91, pp. 161-164.
Ritter, H.: 1980d, ESO Messenger 21, pp. 16-18.
Robinson, E.L.: 1976, Ann. Rev. Astron. Astrophys. 14, pp. 119-142.
Vogt, N.: 1981, ESO preprint No. 138.
Vogt, N., Schoembs, R., Krzeminski, W., Pedersen, H.: 1981, Astron.
    Astrophys. 94, pp. L29-L32.
Warner, B.: 1974, Monthly Notices Roy. Astron. Soc. 168, pp. 235-247.
Warner, B.: 1976, IAU Symp. No. 73, pp. 85-140.
Young, P., Schneider, D.P., Shectman, S.A.: 1981, Astrophys, J. 245,
    pp. 1035-1042.
Zapolski, H.S., Salpeter, E.E.: 1969, Astrophys. J. 158, pp. 809-813.

# CONCLUDING REMARKS

Zdeněk Kopal
Department of Astronomy
University of Manchester

In conclusion of our conference, the last duty which I am being called upon to perform is to summarize what we have heard in the last few days, to outline the present state of the subjects under discussion, and to single out the most important ones deserving further attention in the future.  To do so is obviously not easy; for so much has been brought up in the past few days that any kind of more detailed summary would exceed the time available for these remarks.  Nevertheless, in what follows I shall attempt to do so by subjects discussed in the past sessions - in the hope that this overview may stimulate our thinking in the future.

Before doing so I should, however, like to comment on a topic not specifically discussed in any one single session, but which over-shadowed them all; namely, an unprecendented impact of new observational data forthcoming in recent years from the telescopes operating in space.  While the study of wide binaries (at least its astrometric aspects) may have to await the consumation of the European project Hipparcos to experience such an impact, at least half of all new data presented at this colloquium on close binary systems have been based on observations obtained by use of the International Ultraviolet  Explorer. And when, in the latter part of the present decade, the existing facilities will be augmented by the U.S. Large Orbiting Telescope, the influx of new and unique data will become simply overwhelming.

Are we sufficiently well prepared to cope with their output; or are we at least taking the necessary steps to put us in this position?  Even at the present time (as we have witnessed, in particular, in the 3rd session of our colloquium), their examination has been more qualitative than quantitative; for to exhaust all information implied in these new data would claim more time - and greater scientific manpower - than is available so far.  Is this increased manpower "in the pipelines" to be ready when they are needed?  It is a sad comment on the contemporary situation in our field that - if anything - the opposit is the case; and that while we still continue to spend tens or hundreds of millions of dollars (or marks) on the design, construction and launch of space

*Z. Kopal and J. Rahe (eds.), Binary and Multiple Stars as Tracers of Stellar Evolution, 489–498.*
*Copyright © 1982 by D. Reidel Publishing Company.*

vehicles for which astronomical requirements are used as a pretect of
the injustification, no adequate steps are being planned to train
scientific manpower needed to cope with their output - with the disturbing prospect that a large part of the new data being obtained in
space may gather but dust on the Earth; or receive only a cursory
attention.  If so, posterity would no doubt condemn us severely for this
lack of foresight in planning for the future; and will wonder why we
were so much better at hardware engineering than human education - is
there still time to bring them both to closer harmony?

But to return from future prospects to the present, and from
space back to the ground:  what did we learn at this colloquium about
the major problems, some of which were mentioned in my introductory
remarks?  The main one of these - concerning the "evolutionary paradox"
met so frequently among certain types of binary systems - is still very
much with us; and I dare say in an even more disturbing manner.  The
principal evidence aggravating the situation has been provided (and
keeps forthcoming) from wide binary systems, the properties of which
were under discussion in Session II.

From all data we possess today (and for their fuller survey cf.
Agayev, Guseinov and Novruzova, 1982) it transpires that the evolutionary paradox - which stares us in the face in visual binary systems
like Sirius or Procyon - is replicated in too many other to regard
these latter stars as exceptions to the rule.  The above-mentioned
catalogue of white dwarfs by Agayev et al discloses that, among almost
500 such objects, 62 (or 13%), including Sirius and Procyon, are
components of wide binary systems; and, in view of the small absolute
brightness of such objects (i.e., with observational selection
hampering their discovery), the actual percentage of binary systems with
white dwarf components attending Main Sequence stars is probably much
higher.  Such pairs are, therefore, by no means unusual or exceptional
phenomena; and to account for their existence has become an important
astrophysical problem.

In considering the implications of this problem which I raised
already many years ago (cf. Kopal, 1959; pp. 542-543), let us adhere
to the view that white-dwarf (or, generally, more evolved) components
of such systems had once to be more massive of the two, and lost
excess mass at subsequent stages of their evolution.  If any of these
stars were initially more massive than 2-3 $\odot$ - and many of these must
have been if they belonged to Population I or disc-type population -
they must have got rid of excess mass to suppress the remainder below
the Chandrasekhar limit as a <u>condition sine qua non</u> for them to be able
to become degenerate.

Some investigators (e.g., Lauterborn, 1970) considered a possibility that the excess mass may have "overflown" on to their mates; but
conditions necessary for this to happen are so extremely specialized
as to make them scarcely of astrophysical interest.  In order to
demonstrate this, it is sufficient to recall that the velocity of

escape from the gravitational field of a star is as a rule 10-100
times higher than those necessary for the zero-velocity surfaces surrounding the binary system (regarded as a rotating gravitational dipole)
to remain closed. For instance, in the case of Sirius, the mass of
the A0-component is equal to 2.3 $\odot$; and its radius, 1.76 $\odot$; while those
of its attendant white dwarf are 0.98 $\odot$ and 0.022, respectively. The
relative orbit of the two stars is markedly eccentric (e = 0.59); and
its semi-axis A = 7.62 astronomical units. As a result, at the mean
distance of both components, the fractional radius of the A0 star is
only 0.0011; and its angular diameter, as seen from the secondary
component, 7.5 arc minutes. The velocity of escape from the gravitational field of Sirius A is in excess of 700 km/sec; and although it may
have been less for the present Sirius B at the peak of its post-Main
Sequence expansion, it is by almost two orders of magnitude higher than
that necessary to make the surface of zero velocity closed around the
systems as a whole. As a result, no dynamical necessity exists for
ejected matter to wander within the system until captured by the
companion star. It can escape the system altogether,and this is what
it probably did in this (and other similar) case.

But Sirius - while a typical example of the "evolutionary paradox"
- much more conspicuous than Algol or U Sagittae - is still not the
worst objection to the conventional interpretation of this paradox.
At least in Sirius (or Procyon) the present white-dwarf component proves
to be the less massive of the two. But - still within our neighbourhood in space - in the visual systems of o$^2$ Eridani ($m_{1,2}$ = 0.42 $\odot$ +
0.20 $\odot$) or Stein 2051 ($m_{1,2}$ = 0.48 $\odot$ + 0.22 $\odot$) the white-dwarf
components are more massive of the two; and if the entire mass of the
present secondaries were transferred on to their white-dwarf primaries,
their total masses would still be too small for reaching the white-
dwarf stage during the entire age of our Galaxy! Or consider the quadruple system of Giclas 107-69 and 70, consisting of two red and two
white dwarfs of total mass not exceeding that of our Sun.

Moreover, it is not only in wide binary systems that we encounter
white dwarfs among their components; but in close (spectroscopic)
binaries as well. Consider, for instance, the close pair V 471 Tauri
(discovered in 1970 by Nelson and Young), with an orbital period of
only 0.5212 days, which (on account of a chance direction from which
we see it) happens to be also an eclipsing variable. It consists of
two components of combined mass equal to approximately 1.4 $\odot$ (cf.
Yound and Lanning, 1975), one of which is a K0 V dwarf on (or still
contracting towards) the Main Sequence, and the other a white dwarf;
separated from each other by less than three solar radii. Their masses
are nearly equal (cf. Young and Lanning, op. cit), but the dimensions
are not; and an ingress (or egress) of the eclipse of the white dwarf
by its Main Sequence mate lasts only about 40 seconds!

The particular significance of this system rests on the fact that
not only are its component stars of manifestly the same age, but that
also the absolute value of this age is known. As (by its proper motion)

V 471 Tauri is a member of the Hyades cluster, the system - in common
with all other stars of this cluster - cannot be much older than 600
million years (i.e., the time which elapsed since the commencement of
the Paleozoic era on the Earth).  During this time our Sun - a much
older star - managed to burn only a few per cent of its internal hydrogen supply; yet V 471 Tauri - a system whose combined mass amounts at
present to 1.4 ⊙ - evolved far enough for one of its components to
become a white dwarf!  Obviously the mass of its progenitor must once
have been much larger than it is today - but its excess was not merely
transferred on to its mate (whose present mass is too small even today
to have served for such a receptacle), but must have escaped the system
altogether -- as it almost certainly did in the case of Sirius and
Procyon.

It should be emphasized that the real cause of mass ejection in
more advanced evolutionary stages of the stars is intrinsic to the star
itself; and its onset is to be sought in its own internal structure;
it does occur even if the star is single (stellar winds!); and whenever
direct observational evidence is on hand to disclose the actual velocity of escape (such as furnished by spectroscopic observations of
giants, Wolf-Rayet stars, or of the Novae), it proves to range from
several hundred to a few thousand kms per second.  Matter escaping with
such speeds would pay but scant attention to the presence of any
companion in binary systems - especially in wide binaries where
companion stars become but little more than passive onlookers of dramatic phenomena which produce the mass loss.

But even in close binaries, in which post-Main Sequence expansion
of the components can bring stellar surfaces to actual contact with
their static Roche limits (which can never happen in wide binaries
of the Sirius-Procyon type!), such a phenomenon can at best facilitate
mass escape (by reducing gravity in the neighbourhood of the inner
Lagrangian point), or render the loss non-isotropic; but cannot by itself cause it.  In close binaries of the Algol -U Sge (or U Cep-type)
some of the escaping matter may be detained by hydrodynamical reasons
to linger in the system for some time, and give rise to spectroscopic
phenomena observed in many such systems.  But that the bulk of it
- and it may be from one-half to nine-tenths of the original mass of the
star - is ejected from the gravitational field of the systems seems to
be its likeliest fate; at least it offers the simplest hypothesis which
can be brought in harmony with all observed facts; and none are known
to be contrary to it.

More detailed aspects of this situation have already been discussed
by the present speaker on pp. 415-430 or 472 of his Dynamics of Close
Binary Systems (Kopal, 1978), and need not be repeated in this place
- beyond stressing again that a postulate of more complicated processes - such as low-velocity mutual transfer of significant fraction of
stellar masses between the components - is not only physically unlikely, but unnecessary; for the observed facts can be just as well re-
conciled with a physically simpler (and observationally better-founded)

process of high-velocity mass escape by stellar winds.  To insist
- in the fact of such a situation - on low-velocity mass transfers
(or exchange) between the components provides, to my mind, a good
example of "Procrustean Science", in which by chopping off, (or turning
our backs to) various phenomena, or accumulating superfluous hypo-
theses, we not only strive to fit the observed picture on to the Bed
of Procrustes of our preconceived opinions, but also offend the spirit
of "Occam's razor" requiring that "entia non sunt multiplicanda praeter
necessitatem".

   However, even if we are willing to dispense with unncessary hypo-
teses which may obstruct our way towards fuller understanding of the
observed facts provided by binary systems, this still does not mean
that we are out of the woods.  That stars, at certain stages of their
post-Main Sequence evolution, are likely to divest themselves of 50% -
90% of their original mass appears to be attested by evidence provided
by binary stars almost without doubt; and that the bulk of this loss
occurs at high speed is very probable; but the specific source of
energy necessary to bring it about is still obscure; though the demands
on it are considerable.

   To demonstrate this on a specific example, consider the well-known
semi-detached binary system of Algol, whose principal (Main-Sequence)
component of spectral type B8 possesses a mass close to 3.8 $\Theta$, while that
of its evolved component of spectrum gK0 IV is only 0.82 $\Theta$ (Tomkin and
Lambert, 1978).  If (to account for its present evolutionary stage)
Algol B originally possessed a mass close to (say) 5 $\Theta$ - i.e., was more
massive than the present Algol A - a removal of some 80% of its original
mass from its present size of 3.4 solar radii to infinity would have
called for an expenditure of energy of the order of

$$\frac{3}{2} G \frac{(4.2 \ m_{\Theta})^2}{3.4 \ R_{\Theta}} = 3.5 \times 10^{49} \ \text{ergs}$$

($G = 6.67 \times 10^{-8}$ cm$^3$/g sec$^2$ representing the gravitation constant), equal
to the present nuclear energy output of $2.4 \times 10^{34}$ ergs/sec of that
star for some 50 million years - a tall requirement, but not incon-
ceivably so; and our main task (as yet unfulfilled) remains to
identify its mechanism of release.

   But this is not the only task challenging the students of stellar
evolution; for observations continue to dangle before us a series of
other facts which we cannot yet explain; and these concern mainly the
distribution of double stars in time.  As is well known, binary
systems (of all separations) constitute at least 70-80% of stellar
population in the neighborhood of the Sun; and from their various
characteristics we infer that most of the latter belong to Population I
stars, younger than $10^9$ years (i.e., about one-tenth of the age of our
Galaxy).  But we mentioned already that, on dynamical grounds, such
binary pairs - not only close (i.e., spectroscopic or photometric), but

also wide (i.e., visual or astrometric) - are virtually undissolvable
for time intervals of the order of $10^{10}$ years. And this, in turn, is
bound to give rise to the question: where are those older than $10^9$
years - where are the binaries of the disk-type population contemporary
with our Sun?

And the inquiry becomes all the more perplexing to us when we
turn to the old stars of Population II: where are the binaries
(photometric, to be sure; for no other could be discovered) in the
globular star clusters? Eclipsing variables of W UMa-type (about which
more will be said later on) could be as easily discovered in globular
clusters as short-period cepheids (especially of Bailey's c-type).
Sawyer's 1975 Catalogue of 1421 Variable Stars in Globular Star Clusters
lists only 3 eclipsing variables in 3 different clusters (NGC 3201,
5139 and 6338) which, however, are probably all foreground stars.
Attempts made (e.g., by Batten, 1973) to explain away this disparity
by evolutionary effects, which could render ageing binaries more immune
to observational detection, are unconvincing. No; most probably the
disparity is real, but its cause remains so far obscure.

And the same is true of the conspicuous disparity in the frequency
of occurrence of close binaries among absolutely brightest stars of
young Population I in our Milky Way system and in the neighbouring spiral
galaxies - such as the Andromeda and Triangulum nebulae; or (to a lesser
extent) in the Magellanic clouds. What is the cause of this behaviour?
Is it a different type of interstellar substrate, or of interstellar
magnetic fields? We do not yet know; and as long as this is the case,
we cannot but acknowledge the fact that, in following the tracks of
nuclear evolution of the binary stars of constant mass, somewhere in
the latter parts of the post-Main Sequence stage we have probably lost
the way.

And if this is true of evolved components of binary systems at the
time of their principal mass loss, it is equally true to say that as
regards the second and, in many respects, even more enigmatic group
of predominantly dwarf objects much discussed in Session III of our
Colloquium and usually classified as close binaries of W Ursae Maioris
type - stars exhibiting well-nigh continuous variation of light, suggestive of the fact that these systems - if binary - consist of components which are in virtual contact (or even surrounded by a common
envelope). Although the W UMa-stars have been the subject of more
communications presented in Session III than any other group of close
binaries, no two investigators agree about models which could account
for all aspects of the observational evidence - a fact from which we
can only conclude that we are still some distance from a fuller
understanding of their real nature.

There are several reasons for this situation which deserve special
attention. First, the extraordinary abundance of the W UMa-type stars
in space. Already more than thirty years ago Shapley (1948) pointed

out that these are 20-30 times as numerous in the sky as all other
types of eclipsing variables lumped together - an astonishingly high
frequency, confirmed subsequently by Kraft (1965) or Eggen (1967)
who concluded that one out of 1000 - 2000 stars of the same spectral
class is a variable star of W UMa-type, corresponding to about two
such binaries per million cubic parsecs. One of them - i Boo, at a
distance of 12.6 parsecs - belongs, in fact, among the nearest stars;
with VW Cep only 18.9 parsecs away, being the second nearest W UMa-
type star to us in space.

Secondly, the frequently encountered instabilities of their light
- and velocity changes, variations of periods, etc., strongly suggest
that W UMa-stars constitute secularly unstable configurations evolving
on the Kelvin time-scale, with lifetimes of the order of $10^7$ years.
If so, however, the total number of stars in the Galaxy which may have
passed through the W UMa-stage at one time or another may be $10^2$ - $10^3$
times higher than the number of those we see now in the act - possibly
as high as one such star per $10^5$ cubic parsecs (which corresponds to
about one star in $10^4$ in our neighborhood).

Third, known stars of the W UMa-type are found to cluster (albeit
rather loosely) around the Main Sequence, and are, therefore, presumably
hydrogen-burning objects. They are, moreover, found anywhere along
the Main Sequence - from B-type stars (such as EM Cep or V 701 Sco)
to those of late K dwarfs; the majority belonging to spectral classes
F and G of luminosity Class V. These facts, perhaps, do not deserve
undue emphasis; for the fact that only few W UMa-type objects are known
to be of early spectral classes may be due to the rapidity of their
evolution; and the paucity of K or M stars among them may again be due
to observational selection (i.e., low intrinsic luminosity of such
objects).

Fourth, quite a number of W UMa-type systems prove to be components
of wide binaries - such as i Boo (=ADS 9494B), AK Her (=ADS 10498A); or
constitute common proper-motion pairs (such as VW Cep with HD 199476, or
W UMa itself with BD +55°1351). Of greater importance is, however,
the fact that many occur also in star clusters of known age. None was,
to be sure, found in any globular cluster so far; but many galactic
(open) clusters are known to contain them in considerable numbers.

To give some examples, the southern cluster IC 2994 which provides
the celestial home for BH and LW Cen (Eggen, 1967) is so young that
its stars of spectral class later than B3 are still contracting to the
Main Sequence (cf. Thackeray, 1964); so that the variables just quoted
cannot be older than $10^7$ years. TX Cnc in the Praesepe cluster cannot,
on the other hand, be younger than $6-8 \times 10^8$ years (which is the age of
that cluster); while variables like EP to ES Cep (cf. Efremov et al.,
1964; Hoffmeister, 1964; or Kurochkin, 1965) in an old galactic cluster
NGC 188 must be at least $5 \times 10^9$ years old (Eggen and Sandage, 1969;
or Demarque, 1979). The W UMa stars in our neighbourhood - judging
from their kinematic characteristics (cf. Schatzman and Rigal, 1954;

Rigal, 1955; Artiukhina, 1964; or Popov, 1964) - appear again to
belong to the disc-type population of the Galaxy, of age comparable
with that of our Sun.

All these facts taken together rule out certain avenues of approach to the interpretation of observed phenomena exhibited by
W UMa-type stars, and weaken others. They virtually eliminate
(cf., e.g., Van't Veer, 1980) a possibility that such stars constitute
contact configurations, in which both components of a detached close
binary expanded towards their Roche limits as a result of incipient
hydrogen shortage. For quite apart from the fact that these stars continue to cluster around the Main Sequence (and are, in particular, no
subgiants) some such binaries which can be dated by the cluster to
which they belong, could not have reached the state of hydrogen
exhaustion since their birth, on account of their small mass. Consider,
for instance, the star TX Cnc which - as a member of the Praesepe
cluster - cannot be older than some 600 - 800 million years. The
combined mass of this star (deduced from spectroscopic observations)
is close to 1.9 0, which (if divided between the two components) is
not large enough to compel them to emark on post-Main Sequence expansion
since the time when our Sun - a star of comparable mass - was at the
commencement of the Paleozoic era; with long future still ahead of it
on the Main Sequence. It has been pointed out at this colloquium by
Dr. Van't Veer that not all stars of any given cluster need to be of
the same age; but surely none can be older than the cluster itself!

But quite apart from problems arising in this connection, the
observed facts pointed out earlier give rise to the following question
whose importance overshadows all others - and one which becomes the
Skylla and Charybdis for all theories on the evolutionary significance
of variable stars of the W UMa-type. With so many such stars filling
the sky (especially if the variable phase represents only a transient
stage of their evolution) where are the progenitors of these objects,
or descendants of those which may already have passed through their
variable stage in the past? Those which we observe today cannot, in
particular, have descended from any other know type of variable stars
- for any such hypothetical parents would be by orders of magnitude
too few for their offspring! In fact, the only way to avoid this
embarrassing predicament - and which would seem to be able to provide
ample reservoir for ancestry as well as of the descendants - would be
to put forward a tentative hypothesis that the present W UMa-type
variables are really single stars, which can temporarily stimulate
close binaries in all manifestations which this may entail; and
eventually return to the stage at which they will shine again with
constant light, and thus cease to attract attention.

Let us develop what we mean in a few more words. Suppose, for
the sake of argument, that the W UMa-stage of variability of the
respective star is preceded by a contraction (rather than expansion
to the Roche limit) on the Kelvin time-scale - fast enough for the
angular momentum of axial rotation to be conserved in its course.

This would, in turn, be bound to increase the angular velocity of
axial rotation - possibly beyond the stage at which the initially
spheroidal configuration will acquire three-axial form.   Theory
of stellar rotation can neither prove, nor deny, such a possibility
so far.   But once such a configuration has attained the form of a
pear-shaped figure, it would become a variable star; and, moreover,
photometric as well as spectroscopic observations at a distance could
scarcely distinguish phenomena exhibited by a rotating dumb-bell
figure from those produced by a contact pair.

While the surface manifestations of these alternative models
could, we repeat, be very much the same, dynamically their difference
would be profound.   For whereas a binary star (close or wide) represents
a dynamical system formed by an irreversible process - and the components of which possess two independent centres of gravity - a rotating
dumb-bell figure possesses only one centre of gravity; and could,
therefore, revert to a less extreme form by despinning.   If, moreover,
such a configuration can be de-spun below the limit at which it will
return to spheroidal form, its light would cease to be variable; and
the object would lose its identity as a W UMa-type star.   The requisite
de-spinning could, in turn, be brought about if (as is likely) the
dumb-bell configuration did not rotate as a rigid body.   For if so,
a gradual dissipation of kinetic energy of axial roation into heat
through viscous friction could lessen the spin and help the configuration to revert to spheroidal star shining once more with constant
light; with only a diminished store of potential energy to draw
upon in the future.

The foregoing "scenario" remains, of course, still wholly hypothetical - though not any less likely than many others which have
been put forward in recent years to explain the characteristics of
W UMa-type stars - and may deserve further consideration.   However,
its more detailed elaboration must be left to a more courageous
individual, with more years ahead of him than may be vouchsafed to
your present speaker.   Observations alone are, alas, not likely to
provide a more direct answer to our inquiry in the foreseeable future.
For consider again the variable star i Bootis, which is the nearest
W UMa-type star to us in space.   At a distance indicated by its
trigonometric parallax of $0\overset{''}{.}079 \pm 0.005$, its apparent angular diameter
should be no greater than $0\overset{''}{.}001$ - i.e., still at least an order of
magnitude below the limit at which we could begin to discern its shape
by speckle interferometry or any other method of direct observation;
and this situation may not change at least for many years to come.

And at this not too optimistic note, the time has come for me
to stop and bidd all future investigators of these problems God speed.
If, in my opening remarks, I ventured to quote some words of Friedrich
Schiller from his Ode an die Freude, now I am almost tempted to take
issue with his optimism, reflected in his words (echoed from the
same source), "Brüder, überm Sternenzelt muss ein lieber Vater wohnen".
Sometimes, in wrestling with our problems concerning double stars

(and not these alone) we may ask ourselves, in the midst of our per-
plexities, why did the good Father have to make things so difficult
for us to read His work with fuller understanding?  Maybe, He has
done so only to test our mettle; and if so, we should not fail to
meet the challenge and persevere in our efforts to unravel the
celestial wonders (at least in so far as they concern the topic of our
present colloquium) until all basic problems exercising us at present
will be solved - and the way open to other problems, of which we may
as yet have no inkling, and which will test the mettle of our
descendants.  And, in the meantime,..."Froh, wie seine Sonnen fliegen,
durch des Himmels prächt'gen Plan; Wandelt, Brüder, eure Bahn, freudig
wie ein Held zum Siegen".

## REFERENCES

Agayev, A.G., Guseinov, O.H. and Novruzova, H.I.:  1982, Astrophys.
    Space Sci., $81$, 5.
Artiukhina, N.M.:  1964, Per. Zvjozdy, $15$, 127.
Batten, A.H.:  1963, in Binary and Multiple Star Systems, Pergamon
    Press, London.
Demarque, P.:  1979, in IAU Sympos., No. 85; p. 281.
Efremov, Y.N., Kholopov, P.N., Kukarkin, B.V. and Sharov, A.S.:
    1964, Inf. Bull. Var. Stars, No. 76.
Eggen, O.J.:  1967, Mem. Roy. Astr. Soc., $70$, 111.
Eggen, O.J. and Sandage, A.:  1969, Astrophys. J., $158$, 669.
Hoffmeister, C.:  1964, Inf. Bull. Var. Stars, No. 67.
Kopal, Z.:  1959, Close Binary Systems, Chapman-Hall and John Wiley
    London and New York.
Kopal, Z.:  1978, Dynamics of Close Binary Systems, D. Reidel Publ. Co.
    Dordrecht and Boston.
Kraft, R.P.:  1965, Astrophys. J., $142$, 681, 1588.
Kurochkin, N.E.:  1965, Inf. Bull. Var. Stars, No. 79.
Lauterborn, D.:  1970, in Proc. IAU Colloq. No. 6, Copenhagen;
    pp. 190-192.
Nelson, B. and Young, A.:  1970, Publ. Astr. Soc. Pacific, $82$, 699.
Popov, M.V.:  1964, Per Zvjozdy, $15$, 115.
Rigal, J.L.:  1955, C.R. Acad. Paris, $240$, 50.
Sawyer, H.H.:  1975, David Dunlap Obs. Publ., $3$, No. 6
Schatzman, E. and Rigal, J.L.:  1954, C.R. Acad. Paris, $238$, 2392.
Shapley, H.:  1948, in Harvard Centennial Symposia (Harv. Obs. Mono.
    No. 7), pp. 249-260.
Thackeray, A.D.:  1964, in "The Galaxy and Magellanic Clouds", IAU-
    URSI Sympos., No. $20$, pp. 18-22.
Tomkin, J. and Lambert, D.L.:  1978, Astrophys. J., $222$, L119.
Van't Veer, F.:  1980, Acta Astron., $30$, 381.
Young, A. and Lanning, H.H.:  1975, Publ. Astron. Soc. Pacific, $87$,
    461.

# INDEX

AA Doradus 455f

AB Andromedae 338

Absolute magnitudes of evolved stars 123

Accretion 191, 450

Accretion disk 193, 219ff, 261, 413, 453, 483

Algol evolution 239ff

Algol-type stars 159, 183ff, 187ff, 493

AM Herculis-type stars 399

Am stars 33

Angular momentum 115

Angular momentum loss 289, 413

Apsidal motion 37

Atmospheric eclipse 153, 163

AW Ursae Majoris 343

Barium stars 171

Beta Lyrae 159, 261ff

Binary systems
    among nearby stars 145ff
    astrometric companions
    degenerate components
    evolution 447
    formation 116
    frequency 115, 493
    in nearby galaxies 145ff
    perturbations 92
    proper motion 88
    X-ray sources 373

Black dwarf secondary 483ff

Black holes 73

44i Bootis 337

BV Draconis 338

Cataclysmic binaries 399, 403, 415, 447ff, 475, 483ff
    hydrogen-rich 413f

Cepheid binaries 23ff

Close binary systems 159ff, 219ff, 317, 321, 351ff, 457, 467, 489ff
    statistical models 199ff

Contact binary systems 279ff, 289ff, 351ff
    evolutionary effects 275ff

CO-white dwarfs 450

CW Eridan 217f

Cygnus X-1 467ff

Degenerate stars 403

Delta Scuti-type stars 27, 33ff

Detached binary systems 289

Disk instability 19ff

Double-contact binary 262

DQ Herculis-type stars 399

Dwarf novae 191, 219, 415, 475

Early-type contact system 205

Eclipsing binary systems 37, 47ff, 202, 455f

Epsilon Aurigae 165

Evolutionary paradox 490

Evolved stars 123

EX Hydrae 457ff

Expanding circumbinary envelope 205ff

Gas stream 321

GR Tauri 345ff

HD 165590 67ff

Helium stars 450

Helium white dwarf 447

Honda-Kuwano 1979 object 473f

Hot spot 191

Hubble-Sandage variable 19ff

Interacting binary systems 159, 205, 474

IR sources 105

Luminosity determinations 123

Magnetic fields 115

Main-sequence stars: internal structure 37ff

Mass flow 317ff, 321

Massive white dwarfs 448

Mass loss 187ff, 205, 473, 490

Mass transfer 67, 159, 261, 413, 448

Metal abundances 409

Mg II emission feature 309

Minimum times 47ff

Molecular clouds 105ff

Multiple stars 27ff, 47ff, 61, 105ff, 373

   among binary systems 94

   among nearby stars 145

   in nearby galaxies 145

   trapezium-type 109ff

Neutron stars 73

NGC 188 279ff

Novae 415

   outburst 475ff

Nova-like variable 453f

OB stars 105

P-Cygni profiles 473

Period changes 47ff, 389

Period decrease 205

Photographic astrometry 84

Planetary nebula: central star 403ff

Polarimetry 399ff

Polarization 231

Post-main sequence stage evolution 3ff

Praesepe 239

Precessional period 389

Pulsars 417ff

Pulsating stars 27ff

PU Vulpeculae 473f

Quadruple systems 51

R Arae 321ff

Relativistic stellar models 73ff

Rotational velocity 33

RS Canum Venaticorum-like binaries 327

Run-away stars 417ff

RY Tauri 231ff

S Cancri 239ff

Scorpius X-1: optical burst 445

Semi-detached binary systems 162, 239

Sirius 491

Soft X-ray emission 457ff

Space-time continuum 75

Spectroscopic binary systems 34, 67, 129, 200, 217
    eccentricity distribution 133
    orbits 119

SS 433 373ff, 389ff

Star clusters 61, 109, 239, 279

Star formation 61, 109ff, 115

Star spots 305ff

Stellar associations 61ff, 109

Stellar companion of the Sun 101

Stellar evolution 109

Stellar wind 210

Subdwarfs 455f

Supernovae 417ff

SU Ursae Majoris 475ff

SV Centauri 205ff

Symbiotic stars 159, 168, 473